PROCESSES OF VEGETATION CHANGE

TITLES OF RELATED INTEREST

PROCESSES OF VEGETATION CHANGE

Colin J. Burrows

Department of Plant and Microbial Sciences,
University of Canterbury,
Christchurch, New Zealand

London
UNWIN HYMAN
Boston   Sydney   Wellington

Published by the Academic Division of
Unwin Hyman Ltd
15/17 Broadwick Street, London W1V 1FP, UK

Unwin Hyman Inc.,
955 Massachusetts Avenue, Cambridge, Mass. 02139, USA

Allen & Unwin (Australia) Ltd,
8 Napier Street, North Sydney, NSW 2060, Australia

Allen & Unwin (New Zealand) Ltd in association with the
Port Nicholson Press Ltd,
Compusales Building, 75 Ghuznee Street, Wellington 1, New Zealand

First published in 1990

British Library Cataloguing in Publication Data

Burrows, Colin J.
    Processes of vegetation change.
1. Plants. Ecology
I. Title
581.3′8

ISBN 0–04–580012–X
ISBN 0–04–580013–8

Library of Congress Cataloging-in-Publication Data

Burrows, C. J. (Colin James), 1931–
    Processes of vegetation change / by Colin J. Burrows.
        p. cm.
Bibliography: p.
Includes index.
ISBN (invalid) 0–04–580012–X (alk. paper). – ISBN (invalid)
0–04–580013–8 (pbk.: alk. paper)
1. Vegetation dynamics.   I. Title.
QK910.B87 1989
581.5–dc20   89–14672
                                                              CIP

Typeset in 11 on 13 point Times by Computape (Pickering) Limited,
North Yorkshire and printed in Great Britain by
Butler & Tanner Ltd, Frome and London

Dedicated to the memory of:

R. Hult
E. Warming
H. C. Cowles
F. E. Clements
G. E. Nicholls
W. S. Cooper
F. Shreve
L. G. Ramensky
H. A. Gleason
A. G. Tansley
J. E. Weaver
J. T. Curtis
R. H. Whittaker

Contributors to the development of the theory of vegetation change.

' ... no pleasure is comparable to standing upon the vantage ground of Truth ... and to see the errors and wanderings and mists and tempests in the vale below.'

Francis Bacon, *Essays: Of Truth*, 1625

Preface

This book is about ideas on the nature and causes of temporal change in the species composition of vegetation. In particular it examines the diverse processes of interaction of plants with their environment, and with one another, through which the species composition of vegetation becomes established. The first chapter considers the general nature of vegetation and the ways in which vegetation change is perceived by ecologists. Chapters 2 and 3 provide essential background about the relationships between plants and their abiotic and biotic environment. Anyone who is familiar with the fundamentals of plant ecology may prefer to pass over Chapters 2 and 3 which, of necessity, cover their subject matter very briefly.

Sequences of development of vegetation on new volcanic rocks, sand dunes and glacial deposits, respectively, are outlined in Chapters 4, 5 and 6. Chapter 7 is about the patterns of vegetation change which occur in severe habitats around the world, and Chapter 8 discusses wetlands. Chapter 9 discusses the diverse responses of temperate forests to a variety of disturbing influences, and Chapter 10 deals with change in the species-rich forests of the Tropics. Chapter 11 treats, in detail, the empirical and inferential data on the biological processes occurring during vegetation change sequences. Chapter 12 considers the plant community phenomena which are implicated in the development of theory about vegetation change. The final chapter, Chapter 13, draws the diverse themes together into a unified theoretical structure by which the vegetation change phenomena may be understood.

Although the ecological literature is well endowed with material on aspects of vegetation change, there is still scope for coverage of the fundamental causal processes, and for a revision of the conceptual framework for the subject. Older schemes for describing the nature of vegetation change are unsatisfactory because they do not encompass certain vegetation change phenomena observed in nature. Some earlier schemes are erroneous or philosophically suspect. To be adequate a theory for vegetation change must overcome these problems and offer opportunity for future progress.

The text tends to be descriptive; this is deliberate, because ecologists need to have available a wide database before attempting to understand the universal truths of their subject. Examples are drawn from varied vegetation around the world, although, regrettably, there is little information from the literature in languages other than English. Mathematical treatment of the phenomena described is absent or minimal, partly because of my personal inclination in that direction, and partly because the value of attempting to apply great precision, or attempting a simplified description of ecological complexity by mathematical means, is doubtful at the present stage of understanding of the subject. In fact, the diversity, variability and apparent random influence and response in ecosystems suggests that an appropriate general systems approach to the subject may be through chaos theory; that being so, they are not readily amenable to straightforward mathematical analysis.

Acknowledgements

I am grateful to Robert McIntosh, Donald Walker, Eric Godley, Peter Wathern, Colin Webb, John Spence and Nicholas Brokaw for critical readings of chapters of the manuscript. Many ecologists around the world have contributed reference and other material, for which I am most grateful. In particular Sturla Fridriksson, Donald Lawrence, John Flenley, Robert Whittaker, Robert Peet, John Matthews, C. H. Thompson, Nicholas Brokaw, Miron Heinselman, Robert McIntosh, Peter Raven, John Dransfield, W. T. Swank, G. R. Stephens and Jerry Baskin very generously helped with published or unpublished material. I am especially indebted to ecologists who contributed photographs: Dwight Billings, Robert Peet, John Vankat, Donald Lawrence, Nicholas Brokaw, Howard Crum, Michael Barbour, Robert Whittaker, John Flenley, Peter Wardle, Paul Glaser and Hörð ur Kristinsson. I thank Richard West, Peter Grubb and other friends at the Botany School, Cambridge University, for support during my library search there. I also am grateful to the University of Canterbury and Lincoln College librarians for their willing help and to the University administration for financial assistance through research grants and in various other ways. I thank the various publishers who gave permission for material to be reproduced.

I thank my wife Vivienne, our children Adam and Julia and my family, friends and colleagues warmly for their patience, forbearance, encouragement and material help during the long gestation of the book. Finally, I thank the excellent and long-suffering typists Nancy Goh, Janet Warburton, Julie Patterson and Jane Chambers, who prepared the script, and the skilled artists Tim Galloway, Sabrina Malcolm and Patty Altmann, who prepared the plant drawings.

We are grateful to the following individuals and organizations who have kindly given permission for the reproduction of copyright material (figure numbers in parentheses):

L. Olsvig-Whittaker, reproduced from Whittaker, *Science* **163**, 150–60, Copyright 1969 A.A.A.S. (1.1); H. Wilson (part: 1.7, 7.24); Kluwer Academic Publishers (1.8, 1.9, 1.11, 8.16, 11.15, 12.6, 12.7, Tables 10.10 (part), 13.3); Sidgwick & Jackson (1.10); Prentice-Hall, Worth Publishers (Table 1.1); W. Larcher and Springer-Verlag (2.3, 2.13, 2.15, Table 2.7); Figures 2.8, 2.13, 2.16 from *Productivity of Forest Ecosystems*, © Unesco 1971; A. H. Fitter (2.10); Institut Royal des Sciences Naturelles de Belgique (2.12); Department of Scientific and Industrial Research (NZ) (2.17); Springer-Verlag (2.20, 2.21, 5.11–13, 8.12, 9.6, 9.7, 11.30, Tables 2.5, 5.4, 7.11); H. C. Fritts (2.22); Masson S.A., Paris (2.23); Figure 2.24 reproduced by permission from Holdridge et al, *Forest Environments in Tropical Life Zones*, © 1971 Pergamon Press PLC; Ecological Society of America (Tables 2.4, 6.4, 6.7, 9.5); G. E. Likens (Table 2.5, Figs 9.6, 9.7); Carnegie Institution of Washington (3.2, 3.3); P. Dansereau, Université de Montreal and John Wiley & Sons (3.5); John Wiley &

Sons (3.7, Table 3.4); Ecological Society of America (3.17, 7.10, 7.13, 7.17, 7.26, 8.8, 11.11, 12.4, 12.5, Table 9.4); S. Fridriksson and Butterworths (4.2); University of Hull, Department of Geography (4.5, 4.6, 4.10, Table 4.3); Figure 4.14 reproduced from Atkinson, *Pacific Science* **24**, 1970 by permission of the University of Hawaii Press; Table 4.2 reproduced from *Arctic and Alpine Research* **19**, 1987 by permission of the Regents of the University of Colorado; University of Chicago Press (5.1, 5.2, 5.4, 5.6–10, 12.1); CSIRO (5.11–13, Table 5.4); T. W. Walker and Elsevier (5.14); Blackwell Scientific Publications (2.11, 7.16, 7.31, 7.34, 10.7, 11.2, 11.3, 11.5 (part), 11.6, 11.9, 11.21, 12.16, Tables 5.2, 5.3, 10.6–8, 11.15); F. Ugolini and U.S. Department of Agriculture (6.8–11, Tables 6.4, 6.5); F. Ugolini (6.14); H. J. Birks and Academic Press (6.15, Table 6.6); *New Zealand Journal of Botany* (6.16, 6.17); Elsevier (7.1, 10.5); Figures 7.18 and 7.19 reproduced from Matthews, *Boreas* **7**, 1978 by permission of Norwegian University Press and J. Matthews; the Editor, *Arctic* (7.20); Elsevier and W. G. Beeftink (7.23); M. O. Collins Ltd (7.25); Blackwell Scientific Publications (Australia) (11.34, Table 7.8); P. H. Zedler (Table 7.11); P. D. Moore (Table 8.1); the Editor, *Tellus*, E. Gorham, part reproduced from Gorham, *Geological Society of America Bulletin* **72**, 795–846 (Table 8.2); New Phytologist Trust (8.3, 8.4); J. Godwin and The Royal Society (8.5); A. Danielsen (8.6, 8.7); Koninklijke Nederlandse Naturhistorische Vereniging (8.17b); Figure 9.5 reproduced from Forcier, *Science* **189**, 808–9, Copyright 1975 AAAS; F. Bourliere and Elsevier (Table 10.3) the Editor *Selbyana* (Fig 10.6d, e, f, j); A. H. Gentry and Plenum Publishing (10.10); F. A. Bazzaz, reproduced with permission from the Annual Review of *Ecology and Systematics,* **10**, © 1979 Annual Reviews Inc (11.1, Table 11.10 (part)); the Editor, *Journal of Forestry* (Table 11.9); Chapman & Hall (Table 11.10 (part)); R. K. Peet and Springer-Verlag (11.15; 11.33 (part), 11.36); Figures 11.19 (part) and 11.20 reproduced from H. S. Horn, *The Adaptive Geometry of Trees,* © 1971 Princeton University Press; H. H. Shugart and Springer-Verlag (11.25, 12.14, 12.15); Figure 11.26 by permission, © University of Wisconsin Press; L. Olsvig-Whittaker, reproduced from Olsvig-Whittaker, *Science* **147**, 250–60, Copyright 1965 AAAS (11.27); K. D. Woods (11.30); South Illinois University (Tables 11.12, 11.16); E. P. Odum, reproduced from Odum, *Science* **164**, 262–70, Copyright 1969 AAAS (Table 12.1); H. H. Shugart, reproduced from *Bioscience* **37**, 119–27, Copyright 1987 the American Institute of Biological Sciences (Table 12.4); J. Jenik and Netherlands Centre for Agricultural Publishing and Documentation (12.10); MIT Press (12.12, 12.13).

Contents

xiv

List of tables

1
The nature of vegetation and kinds of vegetation change

INTRODUCTION

Before tackling the description and explanation of the processes of vegetation change, we need to set the stage by discussing the ways in which ecologists perceive the general nature of vegetation and vegetation change. The **flora** of a region (or of the world), is the complete range of plant species found there. The **vegetation** consists of assemblages of plant species forming a green mantle, an almost continuous and conspicuous plant cover over the land surface, except in dry or cold deserts such as the Sahara or the Antarctic continent. On land most mature plants are firmly fixed in position. The vegetation of aquatic habitats contains some conspicuous fixed plants, but many of the plant species are microscopic and floating.

Although very little of the world's vegetation is unmodified by human activities, enough remains in places where natural processes predominate to enable its fundamental nature to be examined. Such natural or semi-natural vegetation (unmanaged or only lightly managed by people) is considered in this book, with a main emphasis on terrestrial sites in the temperate zone.

The study of vegetation is part of the wider subject of **plant ecology**. Ecology in its widest sense is the scientific discipline concerned with the relationships of all kinds of organisms with their environment. The word ecology is derived from the Greek words *oikos* = a house and *logos* = study (i.e. 'a study of organisms in their homes'). Vegetation ecology is sometimes called **plant synecology** (i.e. the ecology of groups of plants living together). **Plant autecology** is the study of the environmental relationships of individual plant species.

KINDS OF ORGANISMS COMPRISING THE VEGETATION

Figure 1.1 shows an outline of the five kingdoms of living organisms, the procaryotic **Monera**, and, emerging from the undifferentiated eucaryotic, unicellular **Protista**, the three groups **Fungi**, **Plantae** and **Animalia**. Table 1.1 is a classification of the moneran, fungal and plant kingdoms, and shows the approximate numbers of species in each group. Some botanists treat the algae as a separate kingdom; here they are regarded as plants.

The world's natural or semi-natural vegetation is composed of about 400 000 plant species. The most numerous and prominent of these are the vascular plants (angiosperms, gymnosperms and ferns, and a few other small groups). Most emphasis is

1

The kingdoms of organisms

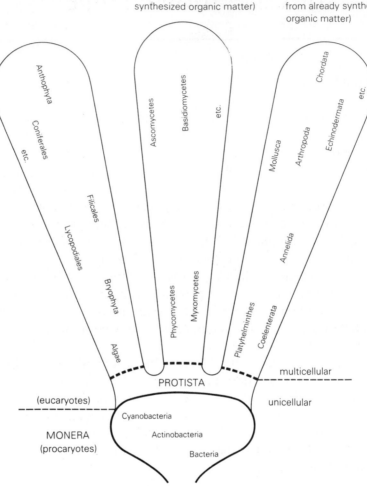

PLANTAE	FUNGI	ANIMALIA
Specialize on autotrophic nutrition (photosynthesis) and also mineral absorption	Specialize on heterotrophic nutrition (extracellular enzymatic solution and absorption from already synthesized organic matter)	Specialize on heterotrophic nutrition (ingestion and intracellular enzymatic solution and absorption from already synthesized organic matter)

Figure 1.1 A five-kingdom subdivision and general presumed phylogeny of organisms. The subdivision is based, first, on the absence or presence of cell nuclei, then on the unicellular versus the multicellular condition, and finally on the modes of nutrition of the organisms. Some biologists classify the Algae as a separate kingdom (after Whittaker 1969a).

placed here on the angiosperms, the main group of seed plants. Vegetative form ranges from giant trees with woody trunks 100 m tall or more, to herbs, lacking woody tissues, the smallest of which are 1 cm high or less. Other angiosperms are much-branched woody shrubs, or climbing vines with long, flexible, woody or non-woody stems. Some trees may live for 1000 years or more; some tiny herbs may live for only a few months, but produce enough seeds in that time to perpetuate themselves (Table 1.2). Many of the herbaceous forms and some of the woody ones customarily reproduce, also, by vegetative means.

2

Table 1.1 A classification of the moneran, fungal and plant kingdoms (modified from Bold 1977, Raven *et al.* 1985).

Kingdom	Main group		Approximate numbers of species	Some particular characteristics
Monera (procaryotic)[*]	Schizonta	(bacteria)	2 500	Most are heterotropic; the cyanobacteria and a few bacteria are photo-autotrophic; a few bacteria are chemo-autotrophic; many are microscopic.
	Cyanobacteria	(blue-green algae)	2 000	
	Actinobacteria	(star moulds)	1 500(?)	
Fungi (eucaryotic)[†]	Myxomycetes	(slime moulds)	80 000	
	Phycomycetes	(algal fungi)		
	Zygomycetes	(bread moulds, etc.)		
	Ascomycetes	(cup fungi)		
	Basidiomycetes	(club fungi)		
	etc.			
	Lichenes	(lichens)	17 000	associations of fungi and green algae or cyanobacteria
	Algae			All but a few, which are total saprophytes or partial or total parasites, are photo-autotrophic.
	Rhodophyta	(red)	4 000	
	Pyrrophyta[‡]	(dinoflagellates, etc.)	1 100	
	Chrysophyta[‡]	(diatoms, etc.)	10 000	
	Phaeophyta	(brown)	1 500	
	Euglenophyta[‡]	(euglenoids)	800	
	Charophyta	(charoids)	300	
	Chlorophyta	(green)	7 000	
	etc.			
	Bryophytes			
	Hepaticae	(liverworts)	6 000	
	Anthocerotae	(hornworts)	100	
	Musci	(mosses)	14 000	
	Pteridophytes			
	Psilophytales	(*Psilotum, Tmesipteris*)	30	
	Equisitales	(horsetails)	20	
	Lycopodiales	(lycopods)	1 000	
	Selaginellales			
	Isoetales	(quillworts)	100	
	Phylloglossales	(*Phylloglossum*)	1	
	Filicales	(ferns)	9 500	
	Gymnosperms			
	Cycadales	(cycads)	100	
	Welwitschiales	(*Welwitschia*)	1	
	Gnetales	(*Gnetum*)	70	
	Ephedrales	(*Ephedra*)		
	Ginkgoales	(*Ginkgo*)	1	
	Coniferales	(conifers, etc.)	550	
	Angiosperms			
	Anthophyta (flowering plants)		235 000	
	(Dicotyledons)		(170 000)	
	(Monocotyledons)		(65 000)	

Plantae (eucaryotic) — Embryophytes — vascular 'higher' plants — Spermatophytes (seed plants)

Total 400 000 (approx.)

[*] Procaryotic: cells lack nuclei.
[†] Eucaryotic: cells possess nuclei.
[‡] Protista: single-celled eucaryotes.
[§] Include some protista as well as multi-celled forms.

Table 1.2 The lifespan of some very short-lived and some very long-lived seed plant species.

		Lifespan (months)	Habitat	Locality
Euphrasia frigida A*	Arctic eyebright	4–5	open Arctic–alpine	northern Europe
Gentiana tenella A	slender gentian	4–5	open Arctic–alpine	northern Europe
Koenigia islandica A	Iceland purslane	3–5	open Arctic–alpine	northern Europe
Cerastium atrovirens A	dark mouse-ear	4–6	dunes, gravel	Europe
Erophila verna A	whitlow herb	4–6	dunes, gravel	Europe
Saxifraga tridactylites A	rue-leaved saxifrage	4–6	dunes, gravel	Europe
Amsinckia intermedia A		4–6	desert	Arizona
Erodium texanum A	Texas crane bill	4–6	desert	Arizona
Kallstroemia grandiflora A		4	desert	Arizona
Lesquerella gordonii A		4–5.5	desert	Arizona
		(yr)		
Agathis australis G	kauri	> 1000	dense lowland forest	northern New Zealand
Pinus longaeva G (= *P. aristata*)	bristlecone pine	4500	open hillside forest	White Mountains, California
Sequoiadendron giganteum G	big tree	> 3000	dense hill forest	inland California
Sequoia sempervirens G	redwood	> 2000	dense hill forest	coastal California
Taxus baccata G	yew	about 1000	open forest	Britain, Europe, North Africa, West Asia
Dracaena draco AM	dragon tree	> 2000	open hillside forest	Canary Islands

* A, *dicotyledonous angiosperm*; G, *gymnosperm*; AM, *monocotyledonous angiosperm*.

Nearly all gymnosperms, a heterogeneous group of seed plants, are trees or shrubs. Among these are the tallest of all trees and the oldest known living individuals. Ferns are mostly herbaceous, although there are some tree-like species with tall stems and a few vines. They and the lycopods, and other small vascular plant groups, reproduce by means of spores, through complex life-cycles.

Other green plants which normally are present in the vegetation are the non-vascular liverworts, mosses, algae and a few other small groups. All of these are usually less conspicuous than the vascular plants, except in some habitats that are too extreme for the latter – for example, in the sea and in some freshwater habitats (algae) or in some mires – or in mountain habitats (mosses). Important, though often cryptic, associates of plants in the vegetation are the bacteria, cyanobacteria (blue-green

Figure 1.2 The habit of growth of the main kinds of plants. (a) A lichen, *Hypogymnia* (fungus with an algal symbiont). (b) A green alga, *Chlamydomonas* (microscopic, aquatic). (c) An alga (diatom), *Pinnularia* (microscopic, aquatic). (d) A red alga, *Pterocladia* (marine). (e) A brown alga, *Fucus* (marine). (f) A liverwort, *Scapania*. (g) A moss, *Polytrichum*. (h) A lycopod, *Lycopodium*. (i) An equisitalean, *Equisetum*. (j) A fern, *Woodsia*. (k) A cycad gymnosperm, *Zamia*. (l) A gymnosperm, *Ginkgo*. (m) A conifer gymnosperm, *Tsuga* (hemlock). (n) An angiosperm tree, *Castanea* (sweet chestnut)*. (o) An angiosperm shrub, *Corylus* (hazel), in winter*. (p) An angiosperm woody vine, *Lonicera* (honeysuckle)*. (q) A monocotyledonous angiosperm herb, *Asphodelus*. (r) An angiosperm herb, *Saxifraga*. (s) An annual angiosperm herb, *Koenigia*. (t) An annual angiosperm herbaceous vine, *Vicia*. (u) A leafless angiosperm 'herb', *Opuntia* (cactus, with swollen stems)*. (v) A cushion-forming angiosperm 'herb', *Raoulia*. *Perennial dicotyledonous species. Dimensions are plant heights above ground level (or lengths for the aquatic species).

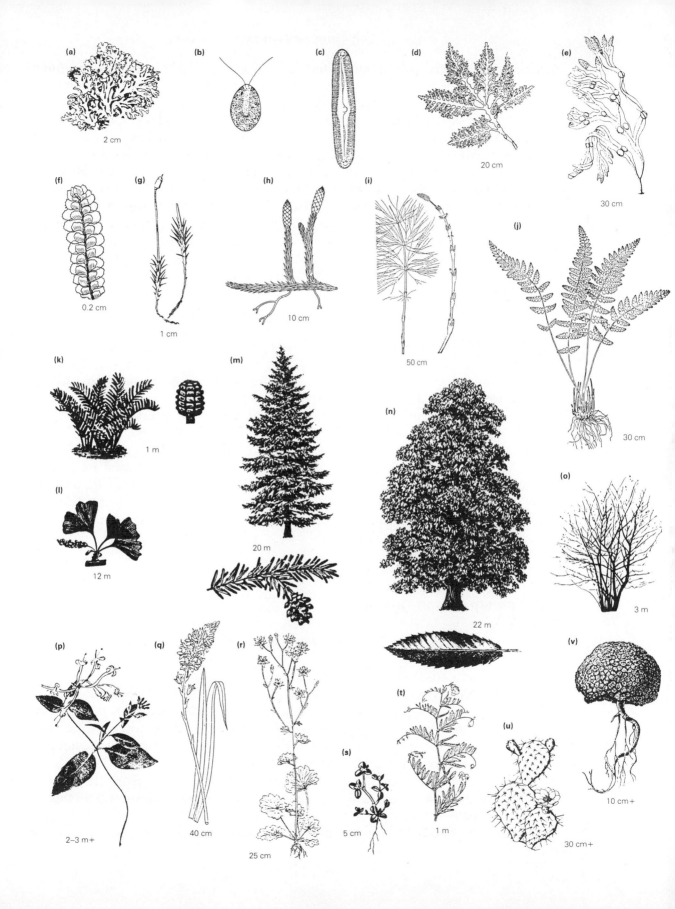

(a) 2 cm

(b)

(c)

(d) 20 cm

(e) 30 cm

(f) 0.2 cm

(g) 1 cm

(h) 10 cm

(i) 50 cm

(j) 30 cm

(k) 1 m

(l) 12 m

(m) 20 m

(n) 22 m

(o) 3 m

(p) 2–3 m+

(q) 40 cm

(r) 25 cm

(s) 5 cm

(t) 1 m

(u) 30 cm+

(v) 10 cm+

algae), actinobacteria (star moulds) and fungi. The bacteria, star moulds and fungi function in the decay of dead plant and animal material, or in various loose or intimate relationships with plants, where their action may be beneficial, neutral or harmful. Lichens – thalloid organisms which combine fungal and algal, or fungal and cyanobacterial, components – are not plants. However, they behave like plants in some respects and are considered, along with plants, as components of the vegetation. Lichens are most prominent in extreme habitats like rock outcrops, mountain tops and in parts of the Antarctic and Arctic regions. Although hereafter the main reference will be to the larger vascular, green plants – the seed- or 'higher' plants and ferns (Table 1.1) – it is implicit that other plant groups and monera and fungi, including microscopic forms, are present in the vegetation. Animals are arbitrarily excluded from consideration as part of the vegetation, although they may have important influences on it, to be explained later. Table 1.3 and Figure 1.2 indicate the main kinds of life-form of plants.

PLANT POPULATIONS

Notwithstanding the emphasis here on vegetation, the ultimate units to be considered are individual plants. Each species comprises local **populations** of adults (able to reproduce), juveniles (young seedling or adolescent plants, not yet reproducing, but potentially capable of replacing adults when they die, or of forming colonies in new sites) and propagules (seeds or spores and other means, including vegetative offsets, by which the population can be maintained, increased and, under favourable conditions, dispersed). Propagules such as seeds and spores are also the means by which a species can outlast adverse periods during which the adult and juvenile part of a population may have become locally extinct. It is in the populations of plants that the fundamental processes of vegetation change take place (cf. Silvertown 1982).

PROPERTIES OF VEGETATION

Among the easily recognizable variable properties of vegetation are the following.

1 *Species composition*. Composition ranges from simple to very complex, depending on the habitat conditions and the relative richness of the local flora.
2 *Structure*. Some vegetation is structurally very simple, but often a degree of structural complexity and organization is evident. The structural patterns arise from the different stature and growth forms of the constituent plant species and their spatial disposition relative to one another.
3 *Physiognomy*. The general appearance of vegetation results from the relative abundance of species possessing distinctive stature, form, colour and texture of shoot systems and foliage.
4 *Spatial patterns*. The species composition of vegetation varies in space because the

Table 1.3 The main kinds of life-forms of adult plants (excluding algae).

1 *herbs*: *annual* plants (adults live for one year or less), or *perennial* plants (adults live for at least two years; some are *monocarpic* and die after flowering, others are *polycarpic* and live for from several to many years, flowering annually, or at less frequent intervals) with very little or no development of woody tissue; among the herbs may be included the flowerless, spore-producing bryophytes and most pteridophytes, as well as grasses and grass-like monocotyledonous angiosperms (graminoids), leafy dicotyledonous angiosperms (forbs) and some angiosperms with more-specialized shoot and leaf form; types of herbs may be differentiated further according to their above-ground stature; the general habit of their shoot systems; leaf sizes and form; the periodicity of their growth behaviour; the positions of resting buds or shoot apices with respect to the ground surface; the form of underground organs (shoots, roots), etc. (see Raunkaier 1934, Dansereau 1957)

2 *shrubs*: perennial plants (a few are monocarpic, most are polycarpic) with more or less stiff, woody aerial stem systems, ranging from a few centimetres to 6 m tall; they include a few pteridophytes and some monocotyledonous angiosperms with fibrous, hard, 'woody' stems; types of shrubs may be differentiated, as for the herbs

3 *trees*: perennial plants (very few are monocarpic, most are polycarpic) with more or less stiff, erect woody stem systems, taller than 6 m; they include a few pteridophytes (tree ferns) and monocotyledonous angiosperms (palms, joshua trees and bamboos) with fibrous, hard, 'woody' stems; various types of trees may be differentiated according to stature, stem form, leaf form, periodicity, etc.

4 *vines*: annual or perennial plants, herbaceous or woody, with rather weak, elongated stems which climb on the stems or branch systems of other plants by means of specialized clinging roots, or modified leaves or other organs (tendrils, suckers or hooks) or by the twining activity of the stem; they include a few pteridophytes

component species respond differently to sets of habitat conditions, which are, themselves, spatially variable.

5 *Temporal patterns*. The species composition of vegetation varies with time. The explanation of this is the main purpose of this book.

More about each of these vegetation attributes is outlined later, in this and subsequent chapters.

VEGETATION CLASSIFICATION AND TERMINOLOGY

A simple physiognomic classification of vegetation subdivides it according to the form and stature of the most conspicuous plants (the **physiognomic dominants**). Most of the reference in this book is to the characteristic dominants of various kinds of vegetation. **Forest** (treeland, silva) is vegetation dominated by woody plants 6 m tall or more. **Scrub** (shrubland, frutica) is dominated by woody plants less than 6 m tall. **Herb vegetation** (herbland, herba) is dominated by non-woody plants, including the often prominent grasses (Fig. 1.3). Barrens are areas where the substrate is more prominent than the plant cover (Fig. 1.4).

Mixtures of plants of widely different growth form are usual; a forest will normally contain shrubs and herbs as well as trees. Scrub may contain vines and herbs, as well as shrubs. Viewed from above, as in an aerial photograph, vegetation may appear structurally uniform, consisting of plants of one physiognomic class, or it may be more complex. In plan view other kinds of spatial patterns are, for example, mosaics of patches of trees, shrubs and herbs, or shrubs dotted in a predominant matrix of herbs (Fig. 1.5). **Closed vegetation** forms a continuous cover over the ground surface;

Figure 1.3 Mosaics of forest (mainly lodgepole pine, *Pinus contorta* and some quaking aspen *Populus tremuloides*); scrub (sagebrush, *Artemisia* sp.); and grassland with forbs, Rocky Mountain National Park, Colorado, USA. Early spring scene (photo C. J. Burrows).

open vegetation is discontinuous, with patches of bare ground (or plants of much smaller stature) between the more-prominent plants.

Seen in elevational view, structural spatial patterns (layers or strata) may be differentiated, more or less distinctly, in the more-complex kinds of vegetation. Four or more strata may be evident in forest. Herbaceous vegetation and scrub are usually simpler (Fig. 1.6). The **canopy** or overstorey, of the vegetation is the uppermost more or less continuous layer with foliage. The **subcanopy**, or understorey, consists of those layers of vegetation beneath the canopy. Some forests have a complex understorey structure, with one or more subcanopy tree layers, a shrub layer and a field layer (herbs, including ferns, bryophytes, algae and lichens). Epiphytes (angiosperms, ferns, bryophytes and lichens) may clothe the branches and stems of the trees. Climbing vines may creep on the tree stems, or hang to the ground from the canopy branches.

More-complex classifications, or representations, of vegetation are possible, using criteria other than physiognomy alone (cf. Shimwell 1971, Whittaker 1962, 1967). Most differentiate kinds of vegetation according to their floristic composition.

Patches of vegetation (**stands**) of similar species composition are found in different places. Similar stands are often classed into abstract groups and called **communities**.

Figure 1.4 Barrenland. (a) A river floodplain, Rangitata River, Canterbury, New Zealand. (b) Scree and cliffs, Maruia River, Nelson, New Zealand (photos C. J. Burrows).

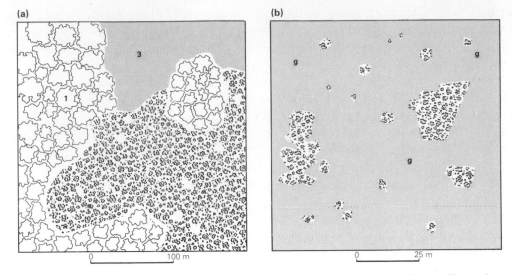

Figure 1.5 (a) Plan of an area of vegetation consisting of a mosaic of forest, scrub and grassland. 1, Forest; 2, scrub; 3, grassland. (b) Plan of an area of vegetation consisting of patches of scrub and individual shrubs in a grassland matrix (g).

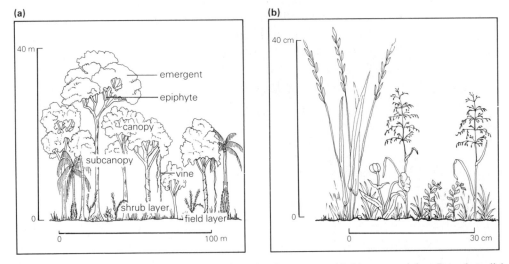

Figure 1.6 (a) Profile view of temperate rainforest where the plants are stratified into several tiers. Bryophyte, lichen and angiosperm epiphytes and ground species are present. (b) Profile view of grassland with some forbs, stratified into a few tiers.

THE VEGETATION CONTINUUM

The plant community concept that is followed here maintains that communities are arbitrary subdivisions of a continuously varying pattern of species composition (**continuum**) (Gleason 1926, Whittaker 1956, Curtis 1959). Repeated species combinations found in similar stands arise from the responses of the plants present in a region to similar sets of habitat conditions. The species composition of different stands is never identical, however, because of the vast array of permutations and combinations of environmental conditions, to which each species responds individualistically. Interactions between species, chance factors of distribution and temporal differences are further causes of vegetation diversity.

It is recognized that, at times, it may be useful to class similar stands into communities. Some of the species in a community may have close (possibly mutual) interdependences; thus, there may be a degree of community integration.

In spite of the emphasis here on continuity of species composition, spatial discontinuities may be evident in the vegetation. They can arise in at least three ways.

1 As a result of sharp boundaries in the substrate; e.g. different rock types, or wet or dry conditions.
2 Through disturbances such as fires which destroy some vegetation. As it regenerates, the disturbed vegetation appears distinct from the undisturbed.
3 Because of mutual exclusion either side of a boundary (**ecotone**) by dominant species of different physiognomy (e.g. forest–grassland). Regeneration of each dominant occurs only on its own side of the boundary.

None of these discontinuities violates the concept of continuous variation in the species composition of vegetation. The third pattern comes closest to doing so, but usually subordinate species overlap the physiognomic boundary.

Ecosystems are communities of plants, animals, fungi and monera, with different degrees and kinds of interdependence among the component species. The abiotic climatic, atmospheric and soil components of the habitat (including light, gases, water and nutrients) are usually included within the concept of the ecosystem (Odum 1971).

WHY STUDY VEGETATION CHANGE?

Since the time of F. E. Clements' (1916) publication on **plant succession** (sequential changes in species composition on a site), the notion of regular patterns of vegetation change has been one of the most important conceptual frameworks in vegetation ecology. It has been used as an organizational scheme and as a predictive tool. Clements, whose work is described at greater length in Chapter 12, stressed the 'progressive' nature of plant succession and its predictability. He emphasized the inevitability of the return of vegetation to stability after disturbance, and developed an elaborate terminology to explain his views. Much of his terminology was never generally accepted and many of his ideas have been abandoned by most vegetation ecologists in recent times. However, the basic concept of sequential development of vegetation on bare surfaces (first a colonizing phase, followed by immature ('seral') phases and culminating in a mature and stable ('climax') phase) is firmly embedded in the literature of vegetation ecology and in the minds of many plant ecologists (see Ch. 12 for further discussion of this topic).

It is natural that concepts of regular and predictable patterns of vegetation change would be of great interest to plant ecologists for, through them, the past history of a site might be reconstructed and its future predicted. In a forest composed of potentially long-lived trees the timespan encompassed in its development might be 1000 years or more. To ecologists with a lifespan of but three score years and ten, or less, such possibilities of examining the distant past and the future (as well as the present) are exciting. There are also important implications in these concepts for the management of vegetation, either for human use or for conservation purposes.

Other important biological matters, inherent in the successional standpoint, relate to ecosystem development itself. Some ecologists and soil scientists are interested in the enormously complex interactions of all biota in living systems, and their inter-relationships with the inorganic components by way of energy exchange and nutrient fluxes. The development of ecosystems on new sites, such as those provided by the eruption of volcanic debris, or the abandonment of ground by a retreating glacier, is of special concern to soil scientists and to ecologists who wish to understand how substrates support plant life.

Other biologists are interested in vegetation development from the viewpoint of the evolution and functioning of plant species in communities. The origin and mainte-nance of species diversity in developing plant communities, the closely allied concept of the niche and concepts of ecosystem stability form part of the subject matter of evolutionary population as well as ecology.

Vegetation change is thus an important, but not well-understood, part of the discipline of plant ecology in which ecologists must continue to take a lively interest. As is discussed later in this chapter, there are many more facets to it than those arising out of the plant succession concepts alone.

OBSERVING VEGETATION CHANGE

Although temporal change is ubiquitous in vegetation, it is not always easily observed. An obvious means of doing so is to examine, over a period, newly formed land surfaces, or places where vegetation disturbance has occurred. Suitable sites of the latter kind are areas where soil has been turned over or vegetation burnt, heavily grazed, or damaged by logging or windthrow. Changes in the kinds and proportions of plant species on the sites will be apparent. A case in point would be a garden in a temperate region such as New Zealand, which usually had been carefully dug and weeded, but was then, for some reason, abandoned. The first colonists are usually annual herbaceous weeds such as annual poa grass (*Poa annua*), shepherd's purse (*Capsella bursa-pastoris*), spurge (*Euphorbia peplus*), chickweed (*Stellaria media*), fathen (*Chenopodium album*) and speedwell (*Veronica persica*). In one growing season they may cover most of the bare ground. In the next growing season perennial herbs such as dock (*Rumex crispus*), dandelion (*Taraxacum officinale*), ribwort plantain (*Plantago lanceolata*), white clover (*Trifolium repens*) and creeping thistle (*Cirsium arvense*) become well-established. In the third and subsequent years a close cover of tall, or dense sward-forming perennial herbs, such as the grasses brown top (*Agrostis capillaris*), cocksfoot (*Dactylis glomerata*) and couch grass (*Agropyron repens*), and the leafy angiosperm herbs (forbs) red clover (*Trifolium pratense*), and yarrow (*Achillea millefolium*) will have choked out most of the annuals and reduced the cover of the earlier perennials. The dormant seeds of the species which have been ousted from the site will probably be present on the soil surface or just under it. If seed sources are available, further change may include invasion by shrubs such as broom (*Cytisus scoparius*), gorse (*Ulex europaeus*) or the low tree elderberry (*Sambucus nigra*) and, subsequently, taller trees such as birch (*Betula pendula*) (Fig. 1.7). All of these species originated from Britain or western Europe, so similar sequences could be expected in abandoned gardens there, also.

A similar course of events, with different constituent species, occurs in abandoned gardens or fields around the world. In wet tropical regions, because of the continuously suitable growing conditions, change proceeds more rapidly. Within a year of abandonment of a garden in a forested area woody plants form tall scrub or low forest. Within three years tall transitional forest develops, and in subsequent years the site returns to something approximating the mature forest (Richards 1964, Flenley 1979). Why do such changes occur? Are there any common or universal patterns? What are the underlying ecological factors and processes which control the patterns of change? These are the questions to which answers are sought in later chapters. It is necessary to go to the detailed field evidence (and any supporting experimental work) to do this.

Other evidence for vegetation change may be seen when you walk through an old forest (Fig. 1.8). A 'freeze' of the time sequence is often present. This results from the long life-history of the tree species present and the easily discernible differences in the various phases of growth from seedling to adult. Furthermore, something of the history of a site may be obtained from the ages of individuals, recorded by their annual growth layers. There will be signs of the decline and death of old trees, some of which will have fallen over. In the gaps so formed occur groups of young plants, often species different from the canopy dominants. They will be of various ages and, correspondingly, spaced at a range of densities. Some of the species present may be colonists, analogous to the colonist annual weeds in the abandoned garden. These have requirements for high light intensity and can withstand relatively little root competition. They are usually eliminated as stands become denser. Other species will include those able to withstand the shade and root competition present as the forest canopy closes. Even so, crowding of the young plants, as they struggle for existence, will thin their ranks. Within forest stands there may be large patches created by disturbances such as windthrow or felling. These will exhibit similar developmental patterns, with the different phases containing increasingly less-dense populations of trees with increasingly older and larger individuals, demonstrating a return to the stature and composition of the mature forest.

In shrubby or herbaceous vegetation, such as grassland, bog or the herb cover on the forest floor it is less easy to observe, at any one visit, signs of change consequent on death and replacement of the old plants, but changes can be discovered by repeated observation (Figs 1.9–11) (cf. A. S. Watt 1947, Austin 1980, Persson 1980, Beeftink 1980, Falinski 1986).

(a)

(i)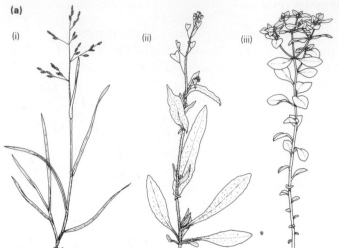

(ii)

(iii)

(iv)

(v)

(b)

(vi)

(vii)

(viii)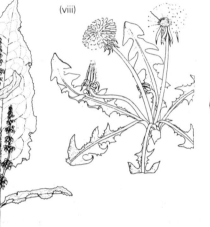

(ix)

(x)

(c)

(xi)

(xii)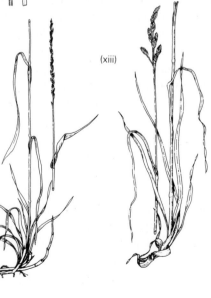

(xiii)

(xiv)

(xv)

(xvi)   (xvii)   (xviii)   (xix)

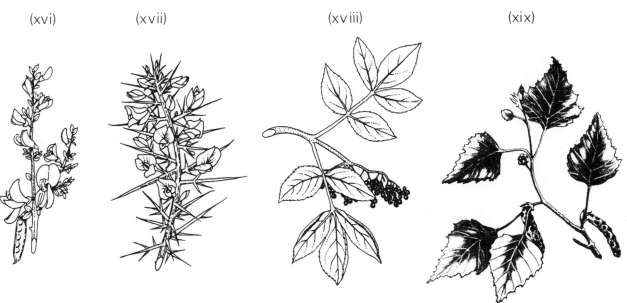

Figure 1.7 The weedy plants of a neglected garden in New Zealand. (a) First year colonists: (i) Annual poa grass, *Poa annua*; (ii) Shepherd's purse, *Capsella bursa-pastoris*; (iii) Spurge, *Euphorbia peplus*; (iv) Chickweed, *Stellaria media*; (v) Fathen, *Chenopodium album*: (vi) Speedwell, *Veronica persica*. (b) Second year colonists: (vii) Dock, *Rumex crispus*; (viii) Dandelion, *Taraxacum officinale*; (ix) White clover, *Trifolium repens*; (x) Creeping thistle, *Cirsium arvense*. (c) Third and subsequent year colonists: (xi) Browntop, *Agrostis capillaris*; (xii) Couch grass, *Agropyron repens*; (xiii) Cocksfoot, *Dactylis glomerata*; (xiv) Red clover, *Trifolium pratense*; (xv) Yarrow, *Achillea millefolium*; (xvi) Broom, *Cytisus scoparius*; (xvii) Gorse, *Ulex europaeus*; (xviii) Elderberry, *Sambucus nigra*; (xix) Silver birch, *Betula pendula*.

The most satisfactory way of demonstrating change in vegetation is, therefore, by the use of permanent study plots, containing marked or mapped individuals, which are evaluated at intervals over a period of years. The relative potential longevity of the species present and the rate of change will determine how frequent such observations need to be. The causes of change may or may not be readily apparent, so that prolonged study of the causal ecology of vegetation change is needed. This will involve observations of plant life-histories and experimentation on the requirements of the species concerned for seed dispersal, seed germination, and growth of the seedlings and young plants, alone and in competition. It will also involve observation of the patterns of vegetative spread. It may require some manipulation of the field environmental conditions to try to simulate the suspected causes of change.

In this approach to recording vegetation change there are obvious limitations, due to a lack of continuous observation, and the relatively short active life of any one observer. Few accurate, detailed, long-term records, in the form of numerical data, have been made on any site. Before about 1880 AD few data of any sort are available. Some older, qualitative information exists, which includes descriptions, maps and photographs. Clearly, standard methods for taking records are desirable so that the results of different workers may be compared.

An extension of this kind of empirical study of vegetation change is to use existing data sets for population changes, observed for relatively short periods, to carry out

Figure 1.8 Regeneration phases after treefall in linden–oak–spruce (*Tilia–Quercus–Picea*) forest, Bialowieza National Park, Poland. (1) Mature forest with some emergent *Picea abies* (spruce), an open canopy of *Tilia cordata* (linden), *Quercus robur* (oak) and *Picea* and subcanopy *Carpinus betulus* (hornbeam). (2) Pole stand on a site disturbed many years previously. Mature *Betula pendula* (birch) and *Alnus glutinosa* (alder) being replaced by submature *Picea*, *Tilia* and *Quercus*. (3) Shrub and sapling stand on a recently disturbed site. Shrubs include *Salix* sp. (willow) and *Populus tremula* (aspen). Young *Betula*, *Alnus*, *Picea* and *Tilia* are also present. After Falinski (1986).

computer simulations of possible future developments (G. R. Stephens & Waggoner 1970, Emanuel *et al.* 1978). In this way predictions of future trends are possible.

Another approach to observing vegetation change is the study of the fossils present in continuous sedimentary sequences. Both microfossils (pollen and spores) and macrofossils (fragments of leaves, seeds, wood, etc.) are useful. The only material of direct relevance to modern vegetation is usually that for the past 10 000 years or less. In practice only the changes in the local vegetation of lakes, mires or other sedimentary locations may be examined reasonably fully, because there are numerous vagaries in the preservation of fossils of more-widespread terrestrial vegetation. Fossil pollen at one site may be derived from a range of different vegetation types spread over a wide area. Discontinuities of sediment deposition, difficulties of dating and a fossil complement representing only part of the flora and vegetation which occupied the site, make difficult the interpretation even of the lake and mire vegetation sequences (D. Walker 1970, Burrows 1974). It is usually also difficult to separate out the effects of **allogenic** (caused by factors extrinsic to plants) and **autogenic** (caused by the plants themselves) processes. Nevertheless, under ideal conditions a good picture of change in the mire vegetation may be obtained. A broad

16

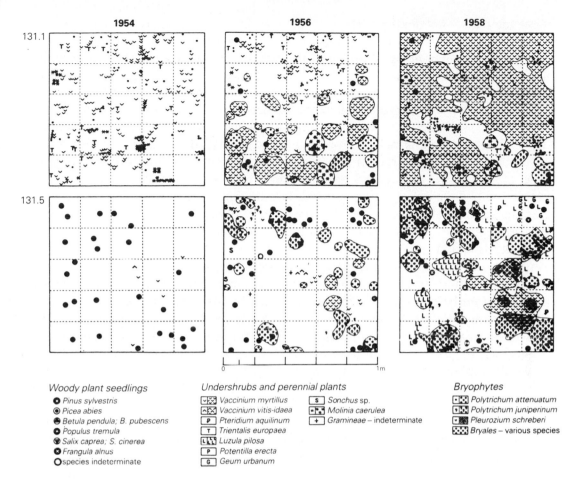

Figure 1.9 Development of ground cover in two permanent quadrats in pine forest after a fire, Bialowieza National Park, Poland (after Falinski 1986).

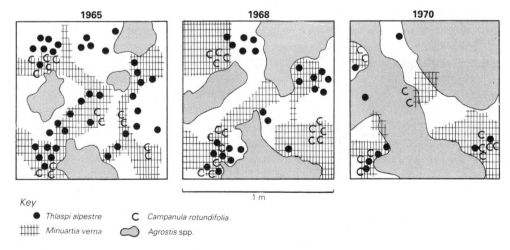

Figure 1.10 Changes in the cover and distribution of plant species in a permanent quadrat on a lead mine spoil heap, in Derbyshire, England (after Shimwell 1971).

view may also be obtained (from pollen spectra) of changes in the terrestrial vegetation of a region. These matters are considered further in Chapter 8.

A further, commonly used method for observing vegetation change is to examine, in a region with uniform climate, a sequence of sites of different age which have the same landform and substratum conditions, i.e. a **chronosequence**. The sites are assumed to have undergone the same sequence of vegetation changes. Despite criticism of the use of this device (e.g. by Miles 1979), it is a valid means of determining the long-term developmental trend, provided that great care is taken to ensure the initial uniformity of geomorphic and substratum conditions, to date the time of formation of surfaces and to apply tests to verify the likely course of vegetation change on such sites. Indeed, it is the only means available at present of gaining an understanding of former processes and vegetation composition in truly terrestrial sites in sequences whose duration has been more than about 200 years. A difficulty is that the sites where vegetation development has been proceeding for long periods will almost certainly have been subject to allogenic disturbances sufficient to interrupt and possibly redirect the sequence in some way; the details of this will often be obscure. The method is tested by, first, making repeated observations, at intervals, of the sequence on young sites of known history, as described above, to establish the developmental trends. Future trends are indicated by the occurrence of young populations of plant species which, when well-grown, will probably replace the pre-existing adult species population. Secondly, by ageing woody plants on the younger surfaces, rates of change may be established. Thirdly, from evidence of the plant populations which formerly grew on the site (by macrofossil or pollen analysis) an impression of the vegetation sequence can be gained. Circumstantial evidence for the validity of the method is that, in places where good dating control is possible, many land surfaces of similar age have

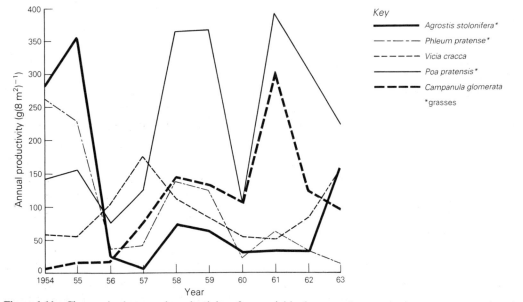

Figure 1.11 Changes in the annual productivity of perennial herbaceous plant species in permanent quadrats in a wet meadow, Oka River, USSR (data from Rabotnov 1966).

Table 1.4 Methods for the study of vegetation change.

Method	Approximate maximum period of detailed data so far available	Amount of data available
(a) direct observation of marked populations	80 years	very few
(b) manipulation of marked populations	20–30 years	very few
(c) use of photographs and less-detailed observations on sites of known history	100–150 years	many
(d) predictions based on species composition of replacement stages (usually confined to forest)	50–150 years	few
(e) observation of vegetation on land surfaces of different age but the same landform and substrate type and the same sequential history	5000+ years	few–many but very few carefully verified and dated
(f) pollen and macrofossil analysis of sediments, peat (useful data confined to wetland vegetation)	10 000 years	many, but few well-dated

comparable vegetation and soil development. Table 1.4 lists the main methods for observing vegetation changes.

Spatial sequences have sometimes been taken as a model for temporal sequences. For example, the arrangement of species in the successive bands formed by encroachment of peat-forming vegetation at the edge of a lake are supposed to emulate their occurrence in a time sequence. This assumption is not justified unless there is independent, well-dated evidence for it. This corroboration must be sought by the careful study of the sediments underlying the spatial vegetation sequence.

STYLES OF VEGETATION CHANGE

Regular changes in plants, correlated with the seasons, include the sequence of annual or other periodic variations which are part of the normal cycle of growth. It is usual for each of these **phenological** changes (e.g. leaf emergence, growth flushes, flowering, fruiting, leaf-shedding and die-down of plants in unfavourable seasons) to occur simultaneously in a plant population. Thus, they influence the seasonal aspect of the physiognomy of the vegetation but, except in the case of annuals, biennials or other monocarpic species (which flower once, then die) do not usually result in changes in plant populations. Hereafter, unless otherwise indicated, reference is made to the populations of perennial plant species.

Within certain limits, changes in the number of individuals in populations of any plant species on areas of ground are constantly occurring, in the form of birth and death phenomena, but, unless some or all of the adults are severely affected by some concurrent disturbance or severe stress, the overall composition of the species populations does not necessarily change as a result of this. The replacement of old individuals by younger members of the same species is known as **regeneration** (implying also the normal growth of individuals through the sequence youth → maturity → old age).

Vegetation changes which occur as plant populations respond to changes in habitat conditions are the main subject of this book. The initial event(s) which set up the potential for change to occur and the subsequent plant population response(s) involve distinct, though possibly interlinked processes. Commonly recognized styles of change are the following.

Fluctuations

Year-to-year or longer-period environmental variations (climatic or otherwise) cause variations in seed production, seedling establishment and survival, or in gross productivity or vegetative proliferation by mature plants. The numbers of flowers and seeds, and the amounts of root, leaf and stem tissue, or organs of vegetative spread produced by each species, can vary considerably from year to year. Each species will be affected to a different degree. These fluctuations in seedling survival and adult growth and productivity may not necessarily result in changes in the numbers of adults. However, the environmental variations may be sufficiently severe, over a number of years, to limit or kill mature individuals of some species, thus causing local extinction of adult plants and giving other species an advantage at their expense. In following years the pendulum may swing back in favour of the original species. The recovery of populations in these circumstances is often aided by the occurrence of dormant **seed banks**. Such changes are called **fluctuations** (Rabotnov 1974). A return to exactly similar species composition, or species proportions, on the same patch of ground is unlikely, because of the random nature of both the environmental shifts and of plant factors like seed survival, seed dispersal or vegetative proliferation. Over long periods (relative to the lifespan of the component species) the species composition of most stands of vegetation probably varies in this way as a result of environmental oscillations. However, long-lived adult perennial plants may be able to outlast environmental shifts which severely affect their juveniles, or shorter-lived, herbaceous plants.

Cycles

Recurrent sequences of species populations on a site are known as **cycles** or **cyclic successions** (Barbour *et al.* 1980). They may consist of a simple alternation of one or a few species with bare ground, or alternation of two species or groups of species, or more-complex sequences with several more or less distinct phases, each marked by dominance of one or more species. Some cycles can be related to the interaction of plant growth and reproductive patterns, or autogenic effects, or both, with strong environmental forcing factors. In other instances plant ageing factors seem to be involved. The concurrent changes in environment create conditions where prevailing plant populations are competitively disadvantaged and the successor populations are advantaged, but eventually the original prevailing populations are favoured once more.

Successions

If a *directional* sequence of populations of different plant species occurs on a site (i.e. the end-point differs from the starting point), this pattern of change is known as **plant succession**. Often quite long sequences are apparent, with more or less distinct phases, characterized by the dominance of different species. Change in the plant populations may be wrought by the establishment of new plants from seedlings or by vegetative proliferation. In the past sequential changes like these have been distinguished in several ways. An often-used distinction is that between **primary succession**, occurring on new sites initially devoid of all vegetation and lacking soil, and **secondary succession**, occurring when established vegetation is severely disturbed but the soil and at least some of the original plant populations are not completely destroyed (Clements 1916). The death of plants in established vegetation, as a result of stress or disturbance, occurs at various scales of magnitude. Gaps of different size may be created by the death of single plants, small groups of plants or vast areas of vegetation. It is often considered that secondary succession culminates in a return to vegetation with species composition, structure, etc., resembling the original, established vegetation. This is not necessarily always the case, because habitat changes may prevent the original established species from re-inhabiting the site.

Where some disturbing or stress factor causes the death of prevailing populations of plants and they are replaced permanently by other species (gradually, one-by-one, or more or less all together over a short period), the sequences of change have been called **allogenic succession** (Tansley 1929). Some likely causes of permanent changes of this sort are shifts in climatic conditions, or the introduction of grazing animals and diseases to which the original plants are not accustomed. Where the causes of change are alterations of the habitat conditions by the vegetation itself (e.g. gradual soil fertility modification, or changes in the moisture relations created by litter accumulation), the process has been called **autogenic succession** (Tansley 1929). In this instance, as the prevailing species die of old age, they are replaced by other, more-competitive species.

In some relatively species-poor communities, after disturbance, a pattern sometimes known as **direct succession** often occurs. One or more species are both the initial colonists of the disturbed sites and the main constituents of the subsequent mature vegetation. In this instance there can be no sequential replacements – in effect the pattern is the same as regeneration.

Two other possible forms of vegetation change may be envisaged. There may be the development of a changed species composition of vegetation in a region as a result of immigration of species not previously present. Examples of this are the invasion of mixed forest and its replacement, by beech (*Nothofagus*) forest in New Zealand (Burrows & Greenland 1979), or the invasion of shrublands in South Africa by aggressive, introduced Australian species of the genus *Hakea* (Specht 1981). A second possibility is that new species may evolve and pre-empt some of the sites that were formerly occupied by others. An example is the relatively new allopolyploid species *Spartina anglica*, a grass of salt marshes which has taken over much of the habitat of other salt marsh species in Britain because of its vigorous growth and rapid vegetative spread (Salisbury 1961).

The view may be taken that the commonly recognized styles of vegetation change are arbitrarily separated portions of a complex continuum of vegetation change. Fluctuations, some cycles and the patterns of change known as succession have points in common, for example.

The term 'succession' has been used in many different ways, and sometimes in ways which do not accord with the original definition, earlier this century (cf. Cowles 1911, Clements 1916), so that nowadays it is no more meaningful than the non-committal term **vegetation change**. Vegetation change will be the term used hereafter in this book to refer to any dynamic pattern, where dominant populations of one or more species are being replaced on the same area of ground by new populations of the same or different species. To avoid confusion with other usages, or commitment here to any particular concept of the nature of vegetation change patterns, non-committal terms will be used. **Direct replacement** refers to situations where a species population is replaced by members of the same species. **Sequential replacement** (or **change**) refers to situations where a species population is replaced by members of different species. **Cyclic replacement** (or **change**) refers to a cycle of changes beginning and ending with the same species. It may involve only two species, or a step-by-step series of several to many species. **Fluctuating replacement** (or **change**) means a series of rather unpredictable shifts in species composition.

A very large number of variables is associated with the phenomena of vegetation change. They can be described, simply, as the numerous properties of plants, interacting with habitat properties, which may include exceptional disturbing influences, as well as the 'average' conditions for any site. The description of the processes resulting from these interactions, operating through time, will form a common thread in the chapters of this book.

2
Plants and their abiotic environment

One of the main aims of plant ecology is to try to understand the causes for patterns of plant distribution. Plants that persist in any site must be in equilibrium with their environment (i.e. with all external forces which influence their establishment, survival, function, growth, form, abundance and replacement) (Fig. 2.1, Table 2.1). Inherent features possessed by plants are the structures, physiological phenomena, and growth and reproductive processes which enable them to respond to the environmental forces and so maintain their integrity through all stages of their lives (Fig. 2.2).

Ecologists must therefore study the variable environmental phenomena which affect plants in various ways (light, temperature, water, atmospheric gases, wind, substrate conditions, nutrient supply, and beneficial and harmful effects of animals,

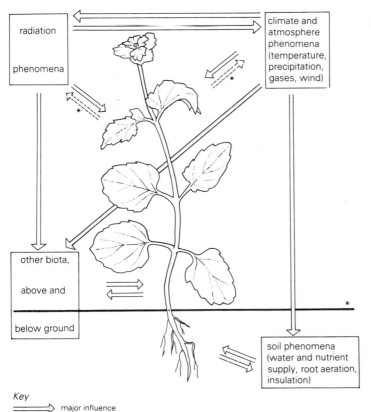

Figure 2.1 The main interrelationships of a plant with its environment.

radiation

phenomena

climate and atmosphere phenomena (temperature, precipitation, gases, wind)

other biota,

above and

below ground

soil phenomena (water and nutrient supply, root aeration, insulation)

Key

⟹ major influence

┈┈⟹ relatively minor influence

*there are profound influences of the vegetation cover on climate and light beneath canopies of leaves

23

Table 2.1 The main physical, chemical and exogenous biotic variable factors affecting plants.

Element	Specific property that varies	Plant processes and activities affected	Important interactions
1. Solar radiation			
A. light	intensity (varies with latitude, time of day, time of year, opacity of atmosphere, cloudiness, shading by plants*)	germination of seeds; leaf expansion; photosynthesis, growth, form, reproduction	
	duration, periodicity (vary with time of day, time of year)	leaf and flower initiation; flowering; fruit-ripening; hardening of tissues to tolerate cold; dormancy of buds/bud growth; leaf abscission; stomatal movements	with temperature, spectral quality, intensity
	direction (varies with time of day, shading by plants*)	stem directional growth; tropic and nastic movements	spectral quality
	spectral quality (varies according to opacity of atmosphere, degree of shading by plants and type of leafy canopy*)	seed dormancy; germination and seedling growth; stem extension; flower induction; leaf expansion; stem growth movements	
	extremes (vary with topographic exposure, shading by plants*)	survival in the face of excess light or insufficient light	
B. temperature	intensity, duration (vary spatially with latitude, altitude, slope, aspect, shading by plants*, temporally with time of day, time of year and irregularly due to weather and climate variation)	photosynthesis, respiration and growth; maturation; form; germination, dormancy/growth initiation, leaf emergence	with light, water relations
	periodicity (varies with time of day, time of year)	germination, shoot growth; hardening of tissues to tolerate cold, heat; initiation of bud, seed dormancy/growth, leaf abscission	with light
	extremes (vary with general geographic, topographic position, altitude, protection by plants*)	survival in the face of excessively high or low temperatures	

	Factor	Effect
2. Atmosphere		
A. precipitation (and condensation)	form, frequency, duration (vary with temperature, pressure and other atmospheric conditions)	
	(i) rain, dew, fog	supply to soil (or direct to plants) of water used for life processes, photosynthesis; within-plant transport of nutrients and other solutes; nutrient inputs; growth, form, reproduction
	(ii) snow	insulation of plants in winter, shortening of growing season in spring
	extremes	survival in the face of excess or insufficient supply of water, or too prolonged snow cover
B. gases	density and concentration of CO_2, O_2 in air, soil spaces (vary with altitude, pressure, temperature, absolute amounts present)	gas exchange for photosynthesis, respiration; growth; form, reproduction
	density and concentration of water vapour in air, soil spaces (vary with pressure, absolute amounts present, temperature)	transpiration; water and nutrient uptake and transport; maintenance of water balance; growth and form
	wind (varies with pressure difference, topographic exposure)	transpiration; growth and form; seed dispersal
3. Soil	soil morphology (texture, structure) and soil colloids (clays, humus) (vary with parent material, climate, site history, especially leaching, weathering, vegetation*)	root water and gas relations; nutrient uptake
	soil nutrients (vary with parent material, colloidal content, leaching rates, weathering rates, organic content, rates of nutrient cycling through biological activity, inputs from the atmosphere, inputs from symbioses)	plant nutrient supply; many plant physiological processes (photosynthesis, osmoregulation, translocation) growth and development, form, reproduction

temperature, wind

Table 2.1 (*cont.*)

	pH and other chemical conditions including toxic substances (H^+, Al^{3+}, S^{2-}, etc.)	affect availability of nutrients; survival
	extremes	survival in the face of excess or insufficiency of particular nutrients
4. Abiotic disturbing factors	wind, salt spray, flooding, wave action, fire, ice glaze, snow, freeze and thaw, landslide, glaciation, vulcanism, etc. (vary with geographical position, climatic conditions, etc.)	
	variables concerned with damage to plants, site disturbance through erosion of substrate or deposition of new substrate; some disturbances are regular, frequent or incessant, others are infrequent and irregular; occasional exceptionally extreme and destructive disturbances may occur	maintenance or survival in the face of the various kinds of disturbance
5. Exogenous biota		
A. animals	population densities (vary with general ecological conditions which may include time of year, plant development, and population inertia)	negative influences – removal of tissues by grazing, affecting other plant processes adversely; trampling and other plant damage, ground disturbances of various kinds affecting site stability; positive influences – soil humus formation, nutrient cycling from living and dead plant tissues providing fresh supplies of nutrients; pollination; seed dispersal, etc.
B. fungi, monera	population densities (vary as for animals)	negative influences – pathogenic attacks, competition for nutrients; positive influences – soil humus formation, nutrient cycling from dead plant tissues providing fresh supplies of nutrients; nitrogen fixation; mycorrhizas

* It is impossible to consider some abiotic influences (especially light and nutrients) without also taking plant influences into account.

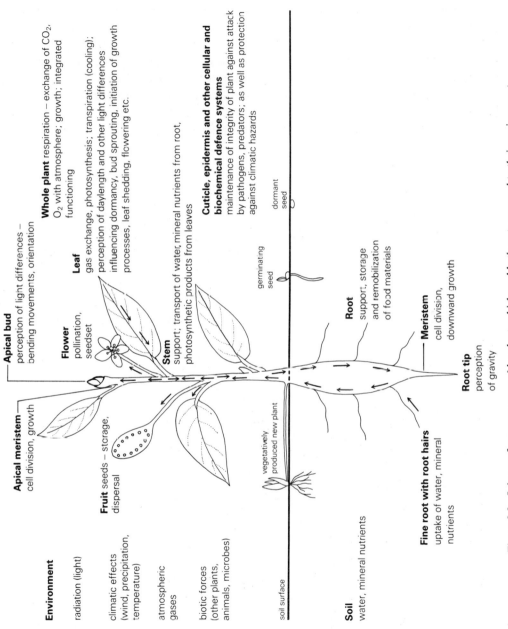

Environment

radiation (light)

climatic effects
(wind, precipitation,
temperature)

atmospheric
gases

biotic forces
(other plants,
animals, microbes)

soil surface

Soil
water, mineral nutrients

Apical meristem —
cell division, growth

Apical bud
perception of light differences —
bending movements, orientation

Fruit seeds – storage,
dispersal

Flower
pollination,
seedset

Whole plant respiration – exchange of CO_2,
O_2 with atmosphere; growth; integrated
functioning

Leaf
gas exchange, photosynthesis; transpiration (cooling);
perception of daylength and other light differences
influencing dormancy, bud sprouting, initiation of growth
processes, leaf shedding, flowering etc.

Stem
support; transport of water, mineral nutrients from root,
photosynthetic products from leaves

**Cuticle, epidermis and other cellular and
biochemical defence systems**
maintenance of integrity of plant against attack
by pathogens, predators; as well as protection
against climatic hazards

dormant
seed

germinating
seed

vegetatively
produced new plant

Root
support, storage
and remobilization
of food materials

Meristem
cell division,
downward growth

Root tip
perception
of gravity

Fine root with root hairs
uptake of water, mineral
nutrients

Figure 2.2 Inherent features possessed by plants which enable them to respond to their environment.

fungi and other biota) as well as the many kinds of plant phenomena which react to the external stimuli. Although their origins and functioning are independent of plants, the impacts on plants of factors such as solar radiation, wind, moisture or nutrients are strongly modified by the presence of a vegetation cover (Fig. 2.3).

This chapter emphasizes the coupling of plants with their abiotic environment, the modulating influences of the vegetation on the exogenous factors and feedback processes which arise from plant activities. Chapter 3 is concerned more with plant population phenomena, ecological amplitude and niche proportions of plants, and the interactions of plants with their own kind and with other biota.

Since, throughout the world, the expression of the many environmental factors varies widely and there is a very large number of different kinds of plants and plant responses, a full study of plant–environment interactions would be an immense task.

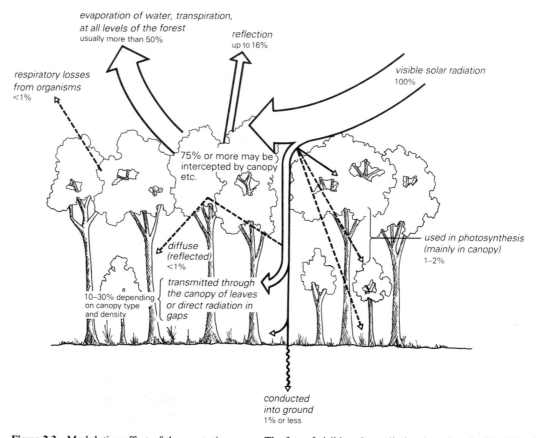

Figure 2.3 Modulating effect of the vegetation cover: The fate of visible solar radiation (wavelengths 0.4–0.7 μm) received by a temperate forest on a clear day in summer. Most of the radiation absorbed at and beneath the canopy is used in the evaporation of water or in transpiration. About 6–16 per cent of visible radiation (mainly green (0.5 μm), some orange and red (0.6–0.75 μm)) as well as about 70 per cent of the infrared (0.75–1 μm) is reflected. At ground level the distribution of radiation is spatially very uneven. Large canopy gaps may be well lit for part of the day. Beneath small gaps sunflecks may reach particular spots on the ground briefly (changing continuously during windy conditions). Other places under the densest part of the canopy will receive only weak transmitted and diffuse (reflected) light, possibly reduced to 2 per cent of full sunlight or less, and with spectral composition very low in the portion (0.4 and 0.7 μm) useful for photosynthesis, somewhat higher in the green (0.55 μm) and high in far-red (0.75 μm). (from Larcher 1980 and various other sources).

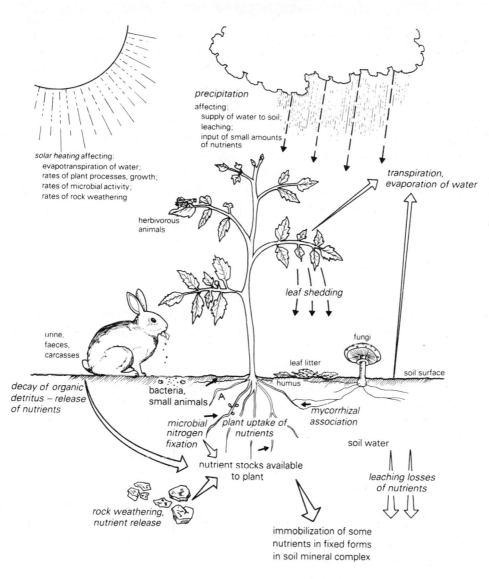

Figure 2.4 Examples of interactions of the main environmental factors influencing plants.

It would encompass almost all of the subject of plant ecology. Only a few brief approximations can be presented here. This is done by considering a specific example and, in later chapters, by emphasizing the responses of different plants to other sets of habitat conditions.

THE ENVIRONMENTAL COMPLEX

Physical, chemical and exogenous biological factors affecting plants (Billings 1974a, C. A. Black 1968, Slatyer 1967, Scott 1974, Nix 1978, Chapin 1980, Etherington 1982, Lange *et al.* 1981, Grace 1983, Grace *et al.* 1981, Mooney 1975, Larcher 1980, Precht *et al.* 1973) are shown in Table 2.1. Figure 2.4 emphasizes potential factor interactions. To simplify the consideration of these many phenomena they are usually each treated separately, but the complex interrelationships of the various components must always be borne in mind.

Some of the items from the list of variables in Table 2.1 are qualities of the milieu within which a plant functions. They determine whether and when physiological processes occur, and at what rates (e.g. temperature, light intensity and daylength). There is often a strong seasonal effect. In extreme form some of them determine the survival of plants. Solar radiation is the energy source which makes most plant processes possible, through photosynthesis. Other items, also variable in space and time, take the form of raw materials needed by the plants to perform physiological processes and as cellular structural materials (water, mineral nutrients and gases). Again, extremes of their occurrence may affect survival. Other factors are those which disturb or destroy plants or disturb the substrate in which they are rooted. Biotic factors (see Ch. 3) are many and diverse, and interact in complex ways with the abiotic components of ecosystems.

Plants operate in the **micro-environment**, defined as the zone within the limits of plant shoot and root systems. It is implicit that the impact on plants of external forces will be modified by the presence of plant structures and through plant activities, including the emission of materials into their surroundings.

A distinction is often made between the influences on plants which are independent of the plants themselves (exogenous, extrinsic or allogenic factors such as precipitation, wind, soil texture and animal predators) and those which arise from the presence and activities of plants (endogenous, intrinsic or autogenic factors). The latter include modifications of light and moisture conditions by vegetation canopies or modifications of soil properties by litter accumulation and acid plant exudates. The allogenic–autogenic distinction, though arbitrary, is worth retaining in the context of analysis of the processes of vegetation change because it is precisely these kinds of differences which must be investigated. However, in real situations, allogenic and autogenic effects are likely to be tightly interwoven and it may be difficult or impossible to decide which is cause and which effect. Co-evolutionary relationships between plants and animals are a case in point (Gilbert & Raven 1975).

PLANT VARIABLES

Differences between plants in a forest community, with respect to form, stature, seasonal activity and reproduction, are indicated in Table 2.2 and Figure 2.5. Contrasts in plant form and life-history in the main plant groups are evident from Table 1.3 and Figure 1.2, and differences in plant attributes are emphasized further in Chapter 3. Esau (1965), Bold (1967) and McLean & Ivimey-Cook (1973) give descriptions of the enormous range of structural and functional diversity in the plant world. Deciduous forest in temperate western Europe is taken as a particular model here. Similar conditions would apply in temperate deciduous forest elsewhere.

The specific example, chosen because its ecology is relatively well-known, is an oak–beech forest at Virelles, Belgium (Duvigneaud 1971a, b, Duvigneaud &

Denaeyer-de Smet 1970, 1971, Denaeyer-de Smet 1971, Duvigneaud *et al.* 1971, Duvigneaud & Kestemont 1977, Froment *et al.* 1971). The young forest stands are dominated by oak (*Quercus robur*) and beech (*Fagus sylvatica*) trees, 85–90 years old and up to 23 m high. The lesser trees are young hornbeam (*Carpinus betulus*) and beech and a patchy distribution of adult ash (*Fraxinus excelsior*), linden (*Tilia platyphyllos*), bird cherry (*Prunus avium*) and field maple (*Acer campestre*). The big trees have widely spreading branches forming a nearly continuous canopy with a dense summer leaf layer which creates deep shade. A sparse scattering of 1–6 m high shrubs in the forest includes dogwood (*Cornus sanguinea*), hazel (*Corylus avellana*), hawthorn (*Crataegus oxyacanthoides*), spindleberry (*Euonymus europaeus*) and buckthorn (*Rhamnus catharticus*). All are deciduous. *Hedera helix* (ivy), an evergreen climber, covers the ground in places and creeps up the stems of older trees by means of specialized roots. Sometimes it envelops the whole bole of its host tree. Herbaceous plants which form the forest field layer (all perennial) are species of *Anemone, Scilla, Narcissus, Primula, Cardamine, Ranunculus, Mercurialis, Potentilla, Viola, Phyteuma, Epipactis, Vicia, Campanula, Lamiastrum, Arum, Orchis, Polygonatum, Carex* and *Melica*.

The data outlined in Table 2.2 show that there is some overlap in size, form, phenology and reproductive behaviour of some of the species of the tree, shrub and herb groups of plants, but a considerable degree of difference is also evident within each group. There are also interspecific differences in productivity (Table 2.3) and other properties, which are considered in the next section and in Chapter 3.

PRODUCTIVITY

The addition of material (dry matter, biomass) to plants is achieved by the photosynthetic process. The gross annual primary production (GPP) for a species is the total amount of material fixed by photosynthesis each year, per individual or (more usually) per unit area of land surface occupied. The units of measurement are either dry matter in grams or energy in gram-calories. Because the basic unit is small, values are usually expressed in kilocalories. On average 1 g of dry matter = about 4.2 kcal of fixed energy, but different plant tissues have different calorific value (Table 2.4). The net annual primary production (NPP) is the GPP minus losses through plant respiration. It is the amount of annual production left at the end of the growing season plus the amounts removed by consumer organisms. The NPP differs for different species (30–70 per cent of GPP).

In a forest it is difficult to measure the total NPP, because some of the material produced each year is laid down in stems and roots. It is easier to measure leaf, flower and fruit production, but very difficult to measure the amount of material consumed by herbivores. The plant biomass of the Belgian oak–beech forest was measured by taking samples for the drying and weighing of:

1 Whole trees and shrubs. These were reduced to components (buds, flowers, fruit, leaves, twigs, small branches, large branches, trunks (wood, bark), small roots, large roots, etc.).

Table 2.2 Stature, leaf, flower and fruit phenology and reproductive features of plants of an oak–beech forest, Virelles, Belgium (from Froment *et al.* 1971, Ellenberg 1974, and various other sources).

Species	Plant height (m)	Gain leaves	Flowers	Fruit ripe	Shed leaves	Periodic mass flowering of trees	Pollination*	Seed or fruit dispersal agency†	Seed dormancy/ germination time	Shade tolerance‡ (seedling)	Vegetative proliferation
trees											
Acer campestre field maple	12–15	May	May–June	Sept.–Oct.	Oct.	sporadic	w	w	over winter or two winters/spring 1 or 2 years after shedding	(M)	stump sprouts, root suckers
Carpinus betulus hornbeam	12–15	Apr.–May	Apr.–May	late Sept.	Oct.	flower annually	w	w	over winter or two winters/spring 1 or 2 years after shedding	(S)	stump sprouts
Fagus sylvatica beech	15–22	late Apr.	May	Oct.	Oct.	2–5 yearly intervals	w	gravity, squirrels	over winter/spring	(S)	some stump sprouts
Fraxinus excelsior ash	15–18	mid-May	Apr.–May	Sept.– following spring	mid-Oct.	approx. 2-yearly intervals	w	w	most 18 months, some up to 6 years/spring	(S)	stump sprouts, root suckers
Prunus avium bird cherry	12–15	May	Apr.–May	July	Oct.	flower annually	i	b	over winter or several years if buried/spring	(S)	stump sprouts, root suckers
Quercus robur oak	15–20	May	May	Sept.–Oct.	Oct.–mid-Nov.	3–4 yearly intervals or more	w	gravity, squirrels, pigeons, jays	over winter, (as sprouted epicotyl)/ spring	(L)	some stump sprouts
Tilia platyphyllos linden	15–18	Apr.	June	Sept.–Oct.	Oct.	flower annually	i	w	over winter, or for several years/spring	(S)	stump sprouts
shrubs											
Cornus sanguinea dogwood	1–3	Mar.–Apr.	May–July	Sept.–Oct.	Oct.–Nov.		i	b	over winter/spring	L	stump sprouts
Corylus avellana hazel	2–6	Apr.–May	Feb.–Mar.	Aug.–Sept.	Nov.		w	squirrels, jays	over winter/spring	M	stump sprouts, coppice
Crataegus oxyacanthoides hawthorn	2–6	Apr.	Apr.–May	Sept.–Oct.	Oct.		i	b	18 months/ early spring	M	stump sprouts
Euonymus europaeus spindleberry	2	Mar.	May–June	Sept.–Oct.	Oct.–Nov.		i	b	over winter/spring	M	stump sprouts, root suckers
Rhamnus catharticus buckthorn	2–6	Mar.–Apr.	May–June	Sept.–Oct.	Sept.		i	b	some germinate after shedding, others over winter/spring	L	stump sprouts

						Die-down§						
Hedera helix	climber ivy	0.3–0.5 on ground >10 on trees	evergreen	Sept.–Nov.	winter or May–June of following year	—		i	b	over winter/spring	S	vigorous layering on ground, trees
herbs												
Anemone nemorosa	wood anemone	0.05–0.2	Feb.–Mar.	Mar.–May	May	June–July	G	i	ants	overwinter/spring	X	short rhizomes
Arum maculatum	cuckoo pint	0.3	Mar.	May	July–Aug.	June–July	G	i	b	over winter/spring	S	tuber
Campanula trachelium	bell flower	0.5–0.8	Apr.–May, summergreen	June–Aug.	Sept.–Oct.	Oct.	H	i	n	n.d.	S	
Cardamine pratensis	lady's smock	0.15–0.5	evergreen	Apr.–June	May–July	—	H	i	projected	over winter/spring	S	short rootstock, tuber
Carex digitata	fingered sedge	0.05–0.15	evergreen	Apr.–May	June–July	—	H	w	ants	n.d.	S	short rhizomes
Lamium (Lamiastrum) galeobdolon	yellow archangel	0.2–0.4	evergreen	May–June	Aug.–Oct.	—		i	ants	over winter/spring	S	somewhat tuberous rootstock, long stolons
Melica uniflora	wood melick	0.3–0.6	Mar. summergreen	May–June	July–Sept.	Oct.	G,H	w	ants	n.d.	S	rhizomes
Mercurialis perennis	dog's mercury	0.15–0.3	Jan.–Feb. summergreen	Feb.–May	June–Sept.	Oct.–Nov.	G,H	w	ants	over winter/spring	D	rhizomes
Narcissus pseudonarcissus	daffodil	0.1–0.35	Mar.	Mar.–Apr.	early June	June–July	G	i	n	Nov., Dec. of same year	L	bulb
Polygonatum officinale (= *multiflorum*)	Solomon's seal	0.3–0.6	Apr.–May	May–June	July–Oct.	Sept.	G	i, selfed	b	n.d.	D	short rhizomes
Potentilla sterilis	white cinquefoil	0.05–0.1	evergreen	Apr.–May	June–Sept.	—	H	i, (later selfed)	n	n.d.	M	stolons
Primula veris	cowslip	0.05–0.25	evergreen	Apr.–May	Aug.–Sept.	—	H	i	n	over winter/spring	L	
Ranunculus auricomus	goldilocks	0.1–0.3	Mar.–Apr., summergreen	Apr.–May	June–Sept.	Oct.–Nov.	H	i, but apomictic	n	n.d.	M	
Scilla bifolia	two-leave squill	0.05–0.15	Apr.	Mar.–Apr.	May	May–June	G	i	ants	n.d.	M	bulb
Viola reichenbachiana	dog violet	0.05–0.1	evergreen	Apr.–May	June–Sept.	—	H	i, (later selfed)	projected	n.d.	S	stolons

* w, wind; i, insect.

† w, wind; b, birds; n, no apparent specialization.

‡ The shade tolerance levels are based on the system of Ellenberg (1974), which applies to the growing season.
D, tolerate deep shade (Ellenberg 1 & 2); S, tolerate shade (Ellenberg, 3 & 4); M, tolerate semi-shade (Ellenberg, 5 & 6); L, tolerate very light shade (Ellenberg, 7 & 8); F, require full light (Ellenberg 9); X, wide range.

§ By die-down is meant complete death of aerial shoot, at least to ground level. Most evergreen herbs undergo some dieback to aerial shoots in winter. G, geophyte (buds overwinter below ground level); H, hemicryptophyte (buds overwinter at ground level); C, herbaceous chamaephyte (buds overwinter above ground level).

n.d., No data available.

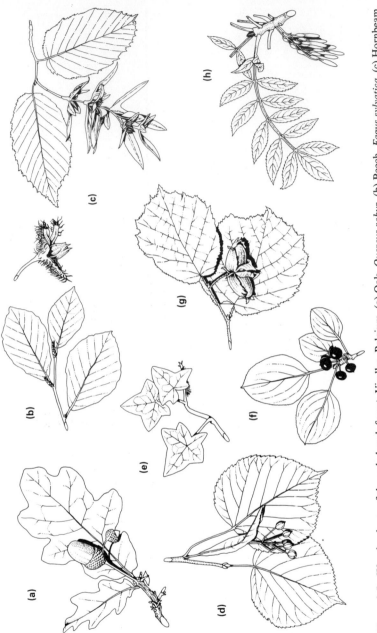

Figure 2.5 Woody plants of the oak–beech forest, Virelles, Belgium. (a) Oak, *Quercus robur*. (b) Beech, *Fagus sylvatica*. (c) Hornbeam, *Carpinus betulus*. (d) Linden, *Tilia platyphyllos*. (e) Ivy, *Hedera helix*. (f) Buckthorn, *Rhamnus catharticus*. (g) Hazel, *Corylus avellana*. (h) Ash, *Fraxinus excelsior*.

Table 2.3 Standing crop biomass and productivity of plants in an oak–beech forest, Virelles, Belgium, (after Duvigneaud et al. 1971, Duvigneaud & Denaeyer-De Smet 1970).

Median age of large trees (years)	No. of trees per hectare		Tree height (m)	Mean basal area of trees (m² ha⁻¹)	Total trunk volume (m³)	Leaf area index (m m⁻² of ground area)
	Dominants	Subdominants				
35–75	260	1226	13–20	21.2	129	6.77

Aerial biomass (kg ha⁻¹)

	Trees	Shrubs	Green leaves	Herbs
Wood and bark	trunks 73 282 branches 33 978 twigs < 1 cm dia. 4947	2513		
Green leaves			3458	
Herbs				2189
subtotals	112 207	2513	3458	2189
			total	120 367

Underground biomass (kg ha⁻¹)

Trees and shrubs	34 600	
Herbs	668	
	total	35 268
	total living plant biomass	155 635

Annual litter production (kg ha⁻¹ . year⁻¹)

Leaves	Inflorescences, fruits, scales, etc.	Dead branches	Dead roots	Dead wood on ground from previous years
3165	672	720	690	1950

Aerial productivity (kg ha⁻¹ . yr⁻¹)

Table 2.3 *(cont.)*

Aerial productivity (kg ha^{-1} . year^{-1})

	Wood and bark		Dead leaves (end of season)	Herbs	Inflorescences, fruits, scales, etc.
Trees		Shrubs			
trunks 2977					
branches 2098					
twigs < 1 cm dia. 1045		216	3165	658	672
subtotals	6120	216	3165	658	672
					total 10 831

Underground productivity (kg ha^{-1} . year^{-1})

Trees and shrubs	Herbs
2000	332
	total 2332

Calorific value (kcal m^{-2} . year^{-1})

total annual living plant production 13 163   6252

total annual living plant and litter production 14 573   6980

Standing crop biomass and annual productivity of aerial parts of some of the individual plant species

	Quercus robur S/C	Quercus robur A/P	Fagus sylvatica S/C	Fagus sylvatica A/P	Carpinus betulus S/C	Carpinus betulus A/P	Acer campestre S/C	Acer campestre A/P	Carpinus betulus (sapl.) S/C	(sapl.) A/P	Crataegus oxyacanthoides S/C	A/P	Cornus sanguinea S/C	A/P	Corylus avellana S/C	A/P	Hedera helix S/C	A/P	Mercurialis perennis S/C	A/P	Other S/C	A/P
			Adult trees						**Shrubs and Carpinus saplings**								**Vine**		**Herb.**			
leaf area index (m² m⁻²)	1.39		2.15		3.05		0.18															
no. of trees ha⁻¹	195		87		1135		69															
total basal area (m² ha⁻¹)	8.03		4.80		7.65		0.73															
standing annual crop production (kg ha⁻¹)																						
green leaves	1030	1030	1040	1040	1300	1300	88	88														
twigs	1395	378	1252	275	2300	392	n.d.	n.d														
branches (wood & bark)	12 435	754	9454	539	9069	755	n.d.	n.d.														
trunks (wood & bark)	28 061	651	18 841	923	26 380	1403	n.d.	n.d.														
total aerial	41 891	1783	29 547	1787	37 749	2550	3020	n.d.	2075	108	165	4.05	45	12.5	285	34.8	1455	296	627	157	271	91
leaf litter		856		1155		1000		88														

(The leaf area index values are bracketed with a total of 6.77.)

n.d., not determined.

Table 2.4 Calorific values for plant tissues (after Golley 1961).

Leaves	Stems, branches	Roots	Seeds	Litter
4.229	4.267	4.720	5.065	4.298

Average energy values, based on determinations from 57 species (kcal g^{-1} dry weight)

2   Whole plants of ground-cover herbs.

Nutrient analysis was also done on these samples. The total biomass of the forest was estimated (by extrapolation from the samples) to be 156 tonnes ha^{-1}.

The NPP was established by measuring the annual crop of leaves, flowers, fruit, etc., and one year's increment of wood and bark from the biomass samples. The estimated annual net productivity of the forest is 14.43 tonnes ha^{-1}, but this does not take into account the amounts eaten by consumers. Values for some individual species and the vegetation in general are given in Table 2.3. Production contrasts with other kinds of vegetation are shown in Table 2.5.

THE ROLE OF PHYSICAL AND CHEMICAL VARIABLES

The proper functioning of plant physiological and growth processes (such as the establishment and breaking of the dormancy of buds and seeds, seed germination, photosynthesis, respiration, nutrient and water uptake, translocation of materials, storage and remobilization of materials within the plant, cell division, morphogenesis, extension growth, movements of organs, flowering, fertilization, seed development and leaf abscission) requires that the processes be integrated with one another and with the appropriate physical and chemical conditions (Fig. 2.6). The ecophysiological processes can only be indicated here briefly. Texts such as Larcher (1980) and Fitter & Hay (1981) give more detail.

Radiation

Solar radiation determines the light and temperature conditions for plants (Fig. 2.3). Figure 2.7 shows the part of the electromagnetic spectrum that is relevant to plants, Table 2.6 the units for measurement of radiant energy and Figure 2.8 the radiation regimes for forest in summer and winter (see also Fig. 2.3). The three components influencing plants are direct, diffuse (reflected) and transmitted radiant energy.

In temperate latitudes the average solar radiation received at the ground surface at noon in open, level sites (and at the outer canopy of leaves) during the growing season is about $1.3–1.5\,g\,cal\,cm^{-2}\,min^{-1}$ ($= 900–1200\,W\,m^{-2}$). The amounts vary from day to day, according to cloudiness.

Table 2.5 Standing crop biomass and productivity of major terrestrial (and freshwater) vegetation types (after Whittaker & Likens 1975).

Vegetation type	Approximate extent (10⁶ km²)	Dry matter, standing crop biomass			Mean leaf area index (m² m⁻²)	Total leaf surface area (10⁶ km²)	Dry matter, annual net productivity		
		Total (10⁹ tonne)	Mean (kg m⁻²)	Range (kg m⁻²)			Total (10⁹ tonnes year⁻¹)	Mean (g m⁻² year⁻¹)	Range (g m⁻² year⁻¹)
tropical rainforest	17.0	765	45	6–80	8	136	37.4	2200	1000–3500
tropical seasonal forest	7.5	260	35	6–60	5	38	12.0	1600	1000–2500
subtropical and temperate evergreen forest	5.0	175	35	6–200	12	60	6.5	1300	600–2500
temperate deciduous forest	7.0	210	30	6–60	5	35	8.4	1200	600–2500
boreal forest	12.0	240	20	6–40	12	144	9.6	800	400–2000
woodland and scrub	8.5	50	6	2–20	4	34	6.0	700	250–1200
savannah	15.0	60	4	0.2–15	4	60	13.5	900	200–2000
temperate grassland	9.0	14	1.6	0.2–5	3.6	32	5.4	600	200–1500
tundra and alpine	8.0	5	0.6	0.1–3	2	16	1.1	140	10–400
desert and semi-desert scrub	18.0	13	0.7	0.1–4	1	18	1.6	90	10–250
extreme hot and cold desert	24.0	0.05	0.02	0–0.2	0.05	1.2	0.07	3	0–10
swamp and marsh	2.0	30	15	3–50	7	14	6.0	3000	800–6000
lake and stream	2.0	0.05	0.02	0–0.1	n.d.	n.d.	0.8	400	100–1500
cultivated land	14.0	14	1	0.4–12	4	56	9.1	650	100–4000
total	149.0	1837	12.2	–	4.3	644	117.5	782	–

n.d., not determined.

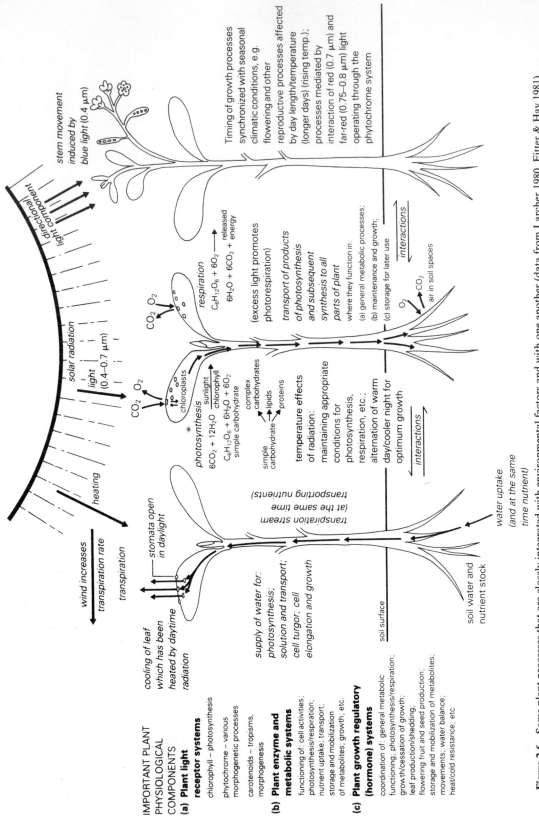

Figure 2.6 Some plant processes that are closely integrated with environmental factors and with one another (data from Larcher 1980, Fitter & Hay 1981).

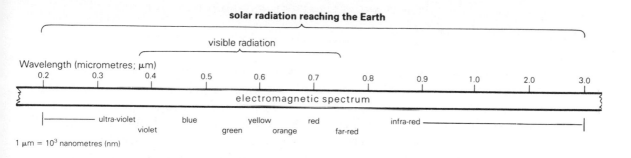

Figure 2.7 The portion of the electromagnetic spectrum which is relevant to plants.

Table 2.6 Units for the measurement of radiant energy.

I.   *radiometric units* (for measurement of the purely physical phenomena of radiant energy)

   dyne   1 dyne (dyn) is the force which gives a mass of 1 gram an acceleration of $1\,cm\,s^{-2}$ ($1\,dyn = 10^{-5}$ newton; 1 newton (N) = $1\,kg\,m^{-1}\,s^{-2}$)

   erg $1\,erg = 1\,dyn\,cm^{-1}$, i.e. the work done when a force of 1 dyn moves the point of application through 1 cm

   joule   1 joule (J) = $10^7\,erg$; 1 kilojoule (kJ) = $10^3\,J$

   watt   1 watt (W) = $1\,J\,s^{-1}$ of power (work done over a period); 1 kilowatt (kW) = $10^3\,W$

   gram-calorie (usually abbreviated to calorie)   1 calorie (cal) is the energy required to raise 1 gram of water through 1°C; $1\,cal \approx 4.18\,J$; 1 kilocalorie (kcal) = $10^3\,cal \approx 4.18 \times 10^3\,J = 1.163\,W\,h^{-1}$

   equivalents $1\,J = 1\,N\,m = 1\,kg\,m^{-2}\,s^{-2} = 1\,W\,s^{-1} \approx 0.239\,cal = 10^7\,erg$

       $1\,W\,h^{-1} = 3.6\,kW\,s^{-1} = 3.6\,kJ - 0.86\,kcal$

   watt m^{-2} (unit of energy flux density) $1\,W\,m^{-2} = 1\,J\,m^{-2}\,s^{-1} \approx 1.43 \times 10^3\,cal\,cm^{-2}\,min^{-1}$

   micrometre (unit of measurement of wavelength, λ)   1 micrometre (μm) = 10^3 nanometres (nm); 1 mm = $10^3\,\mu m$

   Einstein (a unit of energy which varies with the wavelength)   1 einstein (E) = $1.7 \times 10^5\,J$ at $\lambda = 700\,nm$, and $3 \times 10^5\,J$ at $\lambda = 400\,nm$

   PAR (photosynthetically active radiation)   measured in $W\,m^2$ in the 0.4–0.7 μm (400–700 nm) waveband

II.   *photometric units* (for measurement of psychophysical phenomena of visible radiant energy)

   lumen (lm)   a unit of illumination

   lux   1 lux (lx) = $1\,lm\,m^{-2} \approx 5\,erg\,cm^{-2}\,s^{-1}$; 1 kilolux (klx) = 4–$10\,W\,m^{-2}$, depending on the wavelength

Light

Light is a complex factor affecting many plant physiological processes (Whatley & Whatley 1980). The quantity (intensity) of light and the length of time for which a particular intensity is maintained, as well as the wavelength, are important for photosynthesis. In Belgium the growing season, when most plants are in leaf and photosynthesizing, is about 155 days long, from May to October. Only that part of the spectrum of wavelength 380–710 nm (about 0.4–0.7 μm), amounting to 500–1000 $W\,m^{-2}$ energy input on sunny days, and 50–200 $W\,m^{-2}$ on overcast days, is useful for photosynthesis. Other wavebands are used by plants for various other purposes (Fig. 2.9) (Larcher 1980, Fitter & Hay 1981).

On sunny days much of the canopy of leaves of forest trees is exposed to far more light than is needed for photosynthesis. In fact, the brightly lit leaves of the canopy are likely to be oversupplied with light. Photosynthesis in most temperate zone plants is performed by the relatively inefficient C3 pathway (Larcher 1980). On bright, warm days photorespiration will use up about 20 per cent or more of the photosynthetic product (Fig. 2.10). On cloudy days the lower illumination is usually quite sufficient for maximum photosynthetic rates.

The leaf area index of the forest is 6.8 (i.e. each unit area of ground is covered by the projection of 6.8 times that amount of leaf area). Light intensity is attenuated beneath the canopy by the interposition of stems, branches and leaves. Some direct radiation is absorbed by leaves (heating them); some is reflected out into space; a smaller but significant amount, is reflected from leaves or the atmosphere to positions beneath the canopy (the diffuse component) and a small amount, with much-altered spectral qualities, is transmitted through leaves (Figs 2.3 & 11).

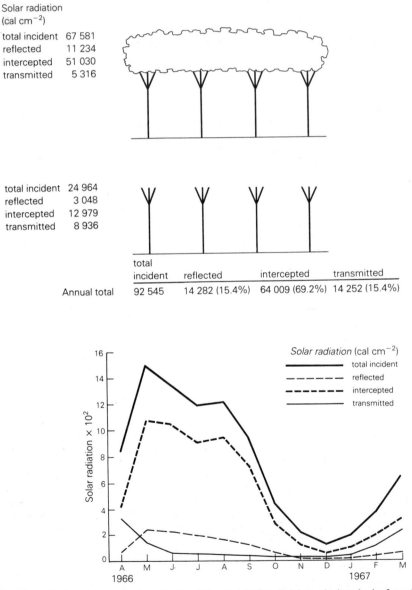

Figure 2.8 The radiation regimes for the oak–beech forest, Virelles, Belgium, during the leafy and leafless periods (after Galoux 1971).

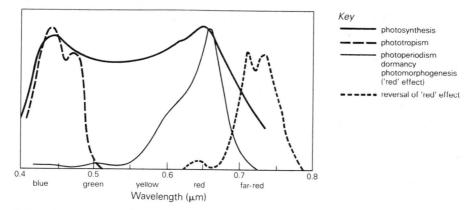

Figure 2.9 Action spectra for plant physiological processes (after Leopold 1964).

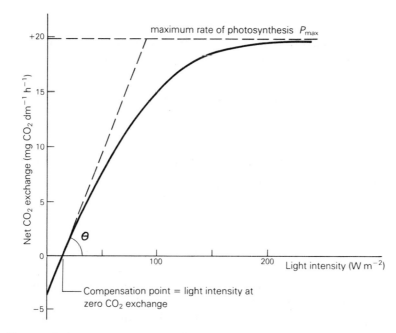

Figure 2.10 Photosynthetic response curve for a C3 plant (representative of nearly all temperate and most subtropical and tropical species), showing the compensation point (the light intensity at which assimilation of CO_2 by plants is balanced by respiratory losses). Levelling off of the curve with increasing light intensity occurs because the photosynthetic mechanism becomes light-saturated. In high light intensities and temperatures photorespiration occurs in C3 plants, causing loss of assimilated carbon. The problem may be overcome to some extent by an increase in CO_2 concentration in the atmosphere. Some subtropical and tropical grasses, and some other plant families employ a different (C4) photosynthetic pathway which is not limited by high light intensities and high temperatures. It is, also, a system that is more economical of water use than is the C3 system (modified from Fitter & Hay 1981).

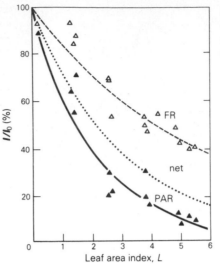

Figure 2.11 Attenuation of light quantity and spectral modification beneath canopies of leaves with increase in leaf area index (leaf area per unit area of ground surface). (....) Net radiation; (▲) PAR, photosynthetically-active radiation; (△) FR, far-red radiation. I/I_o (%) is the irradiance in each waveband, expressed as a percentage of that incident at the top of the canopy (after Szeicz 1974).

Depending on their position and the season, leaves or plants beneath the canopy usually experience much lower light levels than does the canopy. Light intensities beneath the canopy, even in forests where no treefall gaps or larger open areas are present, are, in fact, spatially and temporally very uneven on sunny days because of: (a) the presence of different amounts of stems and branches and different thicknesses of leaf layers; (b) the occurrence of small gaps which allow sunflecks to penetrate to the ground; and (c) the daily progression of the Sun across the sky, and shifts in the foliage, due to wind and periodic shading by the passage of clouds. On cloudy days subcanopy illumination (by diffuse light only) is more even, although intensities are lower than on brightly lit days. The most densely shaded sites – for example, beneath clumps of beech trees – may have light reduced to fractions of 1 per cent of full daytime levels. Few or no vascular plants inhabit such sites. Some bryophytes can cope with extremely low light intensities.

If the light intensity is inadequate for leaves to reach compensation point (Fig. 2.10), then respiration uses up carbohydrate faster than it can be synthesized. Shaded vascular plants will die under these circumstances. Shaded individual leaves must draw on materials that are synthesized by other leaves, or they too will die. Plants which live beneath deciduous forest canopies have several options for coping with the shading problem. They can:

1   Carry out their activities before the canopy trees gain their leaves (e.g. *Anemone*, *Narcissus* and *Scilla*, which produce shoots and flower in early spring, receiving a growth boost from food stored in underground organs) then die down in early summer.
2   Occupy relatively well-lit microhabitats such as canopy gaps where trees have died, or places where sunflecks habitually occur, e.g. species such as bracken, *Pteridium aquilinum* or *Rhamnus*.
3   Perch on boughs or boles of living or dead trees. In the Belgian forest the only epiphytes are mosses and liverworts.

44

4   Scramble or creep up other plants to reach well-lit sites (e.g. *Hedera*).
5   Operate efficiently at lower light intensities.

Some plant species can adjust their leaf photosynthetic capacity up to a point, according to the degree of shading. However, many species have relatively narrow tolerance for light intensity on the gradient from fully lit to deeply shaded. Obligate 'sun' and 'shade' species are at the extremes. Shade tolerance is achieved by having one or more of: (a) leaves with relatively large surface area, thin cuticles and high concentration of chlorophyll and accessory pigments; (b) thin leaves with single layers of palisade cells; (c) relatively low compensation point and light saturation levels; and (d) a specialized physiological mechanism, operating through the phytochrome system, which allows them to grow efficiently in the shade (Grime 1966, Grime & Jarvis 1975, Vogel 1968, H. Smith 1981). Red and blue wavebands are strongly absorbed by green leaves, so that light transmitted through the canopy is often five to ten times higher in the far-red wavelengths than is unfiltered sunlight. When far-red light is high in proportion to red light, the extension growth of 'sun' plants is poor. Extension growth is reduced much less, proportionately, in 'shade' plants under these conditions.

'Shade' plants can be damaged when subjected to bright sunlight, although often high temperature effects and inability of the leaves to control water loss are implicated. 'Shade' plants also adjust their physiological processes very quickly to take advantage of sunflecks.

By contrast with 'shade' plants, species that are tolerant of well-lit conditions have more-efficient water use and generally more-active metabolism. Their leaves are often smaller and thicker, with several layers of palisade cells. They often have leaf surface features (white hairs or lustre) that facilitate the reflection of direct radiation. The leaves are not damaged by high temperatures. Compensation point levels and light saturation rates are higher. Water loss is controlled by stomatal closure and thick leaf cuticles.

Leaves and the branchlets that support them are arranged so as to present leaves beneath the canopy optimally to the solar light source. The growing shoots respond by bending towards the strongly lit side through increased cell elongation (**phototropism**). Leaves and flowers of some plant species also adjust their position daily in relation to the Sun's position by short-term (**nastic**) movements, operated by turgor changes in petioles, or stems.

The germination of seeds of many species requires exposure to light. The light quality is also important, as seeds of many species germinate only when red light predominates over far-red. In deciduous forest their germination is restricted to the period before the leaves unfurl or after they are shed. Otherwise they germinate only in well-lit microsites. The seeds of other species which germinate in the shade either do not require light for this process or are indifferent to the low red:far-red ratio.

The photoperiod is the means by which some diurnal, as well as longer-term, developmental and seasonal plant processes are regulated and synchronized (often in interaction with contrasts in temperature). A simple photoperiodic effect is the control of daytime stomatal opening and closure at night. The interaction of many morphogenetic processes requires brief exposure to light of specific spectral range. Passage into and out of dormancy (buds and whole plants), fruit ripening and abscission of leaves in autumn, in the temperate deciduous forest plants, are moderated by changing daylength and temperature. Initiation of flowering in many temperate-zone plant species requires the long days of summer (Larcher 1980).

Temperature

The range of temperature effects on plants is nearly as complex as that for light. Temperature often interacts with light to influence physiologic processes, as was noted earlier. Some temperature properties which are important for plants are listed in Table 2.7.

Leaving aside diurnal and seasonal temperature differences for the moment, general temperature conditions for any locality are influenced by the condition of the prevailing air mass, as well as by solar radiation, which heats leaves, stems, the soil and the air adjacent to plants. The temperature conditions of any site are also affected by slope, aspect, altitude, windiness, soil colour, and vegetation structure and colour. Figure 2.12 shows the atmospheric temperature regime at the Belgian oak–beech forest site, and an open meadow site in winter and summer (Schnock 1967a, b).

In spring, rising temperature (and increasing daylength) initiate the breaking of bud dormancy and the unfolding of leaves of deciduous species. Warm days also permit seeds of many species to germinate. Different plant species require particular temperature levels for seed germination and the operation of the photosynthetic and respiratory processes. Some general values are summarized in Table 2.7, which also includes information on resistance to temperature extremes.

Temperatures in the summer vary according to the clarity of the atmosphere and cloudiness. On bright, sunny days in high summer, heating by direct solar radiation at the top of the canopy is sufficient to kill leaves which are not being rapidly cooled by transpiration. Most of the solar radiation absorbed by vegetation is dissipated by evaporation and transpiration. Wind assists in the cooling of leaves by removing heated air, saturated with water vapour, from close proximity to leaf surfaces, so that the transpiration rate increases. As a result of shading and heat dissipation, temperatures within the forest in summer are cooler than those at the upper surface of the leaf canopy. The canopy has the effect, also, of maintaining a relatively even temperature within the forest, as night temperatures there remain high compared with those above the canopy.

The mean air temperature of the Belgian oak–beech forest, during the growing season, is about 14.6°C and the mean monthly temperature range is from 1.5°C (January) to 16.5°C (July). All of the plant species present operate within this range, but those restricted to shaded sites prefer the cooler end of the scale. The root systems of all the plants experience relatively cool, even temperatures. Seed germination can take place within much the same range, although the maximum temperatures tolerated by germinating seeds tend to be lower.

To complete all their processes and to prepare for winter and the following

Table 2.7 Temperature conditions that are important for plants (mainly from Larcher 1980, Precht et al. 1973, J. M. Baskin & G. C. Baskin 1988, unpublished data).

	Temperature for net photosynthesis at light saturation (°C)			Temperature for germination of seeds (°C)*			Resistance of mature plants to extreme temperatures (values are 50% injury after exposure to low temperature for 2h or more or to high temperature for 0.5h)	
	Minimum	Optimum	Maximum	Minimum	Optimum	Maximum	Cold injury (winter)	Heat injury (summer)
temperate deciduous trees and shrubs	− 3 to − 1	15–25	40–45	< 10†	20–30	40–45	− 25 to − 40	about 50
herbaceous 'sun' plants	− 2 to 0	20–30	40–50	(a) 'low temperature' species 0	2–20(25)	25–30	− 10 to − 20 (− 30)	48–52
				(b) 'high temperature' species 20–25	30–40	45		
herbaceous 'shade' plants	− 2 to 0	10–20	about 40	0	5–20(25)	30	− 10 to − 20 (− 30)	40–45

* Fluctuating temperature is important for germination, especially of annual and biennial herbs.
† After pre-chilling treatment.

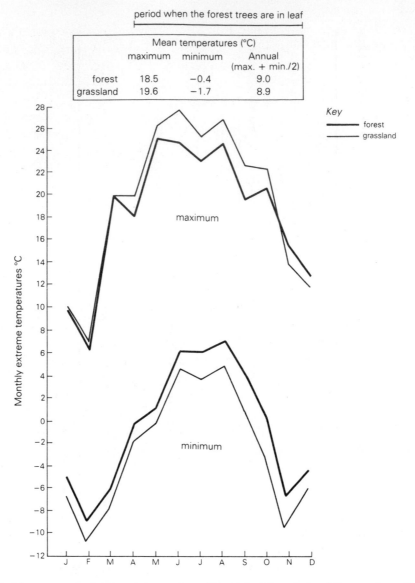

period when the forest trees are in leaf

Mean temperatures (°C)			
	maximum	minimum	Annual (max. + min./2)
forest	18.5	−0.4	9.0
grassland	19.6	−1.7	8.9

Key
—— forest
—— grassland

maximum

minimum

Figure 2.12 Maximum and minimum air temperatures (°C, at 2.0 m) for oak–beech forest and an adjacent grassland site, Virelles, Belgium, during 1965 (from Schnock 1965).

growing season, the plants of temperate regions must experience a definite period above a threshold temperature (Larcher 1980). The initiation of flower primordia, and of new buds is often connected with the thermoperiod.

Some tree species (e.g. *Fagus sylvatica*) seed very heavily (and synchronously) at intervals of a few years, with poorer seed years between them. There is evidence that the heavy seeding (mast year) phenomenon is connected with particularly warm summer conditions in the preceding summer (Silvertown 1982b). Mast seeding is also believed to be an evolutionary response connected with the satiation of seed predators (Silvertown 1980) but the causes may be more complex. Seed production in most species varies with the climatic conditions.

Alternation or variation in temperature is necessary for some plant functions. Temperature alternation between day and night promotes germination and shoot growth. Many temperate species require a diurnal thermal cycle of 5–10°C range for optimal growth (Larcher 1980).

The autumn decline in temperature, interacting with declining daylength, is linked with the preparation by plants for leaf-shedding, induction of bud-dormancy, and hardening of tissues against the cold and winter desiccation. Leaf abscission by shrubs and trees and the die-down of perennial herbs to ground level is a protective measure. Freezing of unhardened plants causes ice to form within cells, dehydrating and disrupting the cytoplasm. The loss of foliage and above-ground shoots enables plants to avoid the severe winter period. Hardening is induced by a gradual lowering of minimum temperatures to below 5°C. Hardening is accompanied by lowered cell water content and higher osmotic pressure as the vascular and cytoplasmic solute concentration increases. These changes protect plants against ordinary frost. Several other mechanisms seem to be involved with hardiness for extreme low temperatures. These include changes in the cell membrane lipids and in the cytoplasm, which permit supercooling of the tissues (Lyons 1973, George *et al.* 1977).

In winter the forest is inactive. Snow lies on the ground at times, and frosts are heavy. Only ivy and some herbs remain green. They and the leafless trees and shrubs and herbs (most with resting buds at or beneath the soil surface) (and many seeds) lie dormant.

Most, if not all, of the plants require a period of cold treatment before bud dormancy and seed dormancy can be broken and before they will flower. This prevents premature sprouting or flowering if mild temperatures persist for a prolonged period in autumn. Acorns of *Quercus robur* begin to sprout in autumn, but cannot complete germination unless they experience a period of chilling.

Atmospheric gases

No specific data are available on CO_2 and O_2 concentrations in the Belgian oak–beech forest or its soils. In some dense vegetation, at times, it is possible that CO_2 supplies to leaves are limited, in small, confined spaces. Soil CO_2 concentrations may be sufficiently high and O_2 concentrations sufficiently low at times to inhibit root growth or seed germination. However, the Belgian oak–beech forest soils are porous and probably well-aerated.

It may be assumed that usually there are adequate amounts of the gases in the atmosphere to support maximum rates of photosynthesis and respiration. The most usual bottleneck for photosynthesis is the slow rate of diffusion of the gas into leaves and through the leaf tissues to chloroplasts, rather than low supplies of the resource (Fitter & Hay 1981).

Water

The Belgian oak–beech forest is on a plateau with no supply of ground water from aquifers. Precipitation for the year, on average, is about 97 cm, about 49 cm falling during the growing season. Of the precipitation, 17 cm is intercepted by plants, and most of this evaporates; 80 cm reaches the soil. The stony substrate is free-draining, and 45 cm drains away. The upper soil layers are rich in humus; small amounts of water are held in the humus or as a film over fine soil particles and within pore spaces. About 33 cm of the annual precipitation is returned from the soil to the atmosphere by evaporation and transpiration. Less than 1 per cent is retained in the biomass (Schnock 1971) (Fig. 2.13).

The soil can be maintained at field capacity (when water potential is about −0.15 bar and water easily available to plant roots) for only short periods after rain (Figs 2.14 & 15). This means that, unless rainfall is well spread, at times in the summer moisture supply to plant roots will be sub-optimal. Evapotranspiration rates will also be highest then. This will in turn cause reduced plant growth. It is unlikely that there

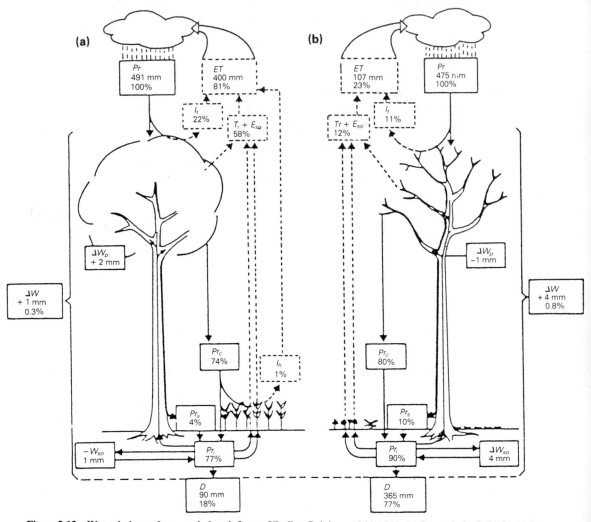

Figure 2.13 Water balance for an oak–beech forest, Virelles, Belgium, 1964–1968. (a) Forest in leaf; (b) forest bare, in winter. Pr, Total incident precipitation; Pr_c, canopy throughfall; Pr_s, stemflow; Pr_i, infiltration (water soaking into the soil); D, drainage water; ET, evapotranspiration; Tr, transpiration of the stand; E_{so}, evaporation from the soil; I_t, interception by the tree stratum; I_h, interception by the herbaceous stratum; ΔW, total water content of the stand; ΔW_p, water content of the phytomass; ΔW_{so}, water content of the soil (simplified from Schnock 1971, Larcher, 1980).

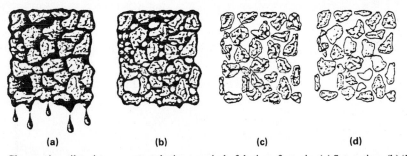

(a)   (b)   (c)   (d)

Figure 2.14 Changes in soil moisture content during a period of drying after rain. (a) Saturation. (b) 'Field capacity' when capillaries are well charged but air-spaces are present. (c) 'Permanent wilting point' the only water present is a thin film, tightly bound to soil particles. (d) A very dry soil with only hygroscopically bound water present. After Buckman & Brady (1969).

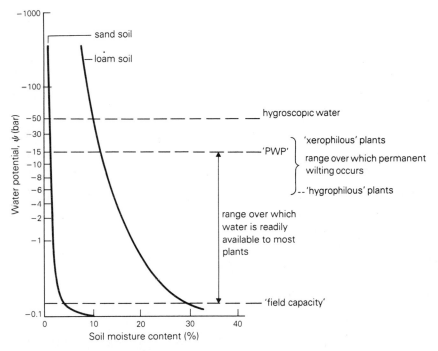

Figure 2.15 Differences between a loam (25% clay, 25% silt, 50% sand) and a sand soil with respect to soil water availability over a range from 'field capacity' (when soil interstices are well charged with water) through to the stage where all capillary water is gone and hygroscopic water only is present. Water retention depends on pore sizes. The loam soil, with many small pores (field capacity -0.15 bar) is better able to retain water than is the sand soil, as shown by the area under the curve. 'PWP', permanent wilting percentage' (at -15 bar) is only a generalized, arbitrary value; plants wilt according to their specific adaptations for resisting drought (after Larcher 1980).

will be oversupply of water such as to cause soil aeration problems for roots, except during very brief periods of heavy rain in winter, or during the thaw of snow.

Germinating seeds are very vulnerable to water stress (Harper 1977), but the seeds of most species germinate in the litter or uppermost humus layers of the soil, which are moist for most of the time. Also, many seeds germinate in spring and autumn, when moisture is abundant.

Table 2.8 Mineral content of important rock types* (Birkeland & Larson 1978)

Mineral elemental composition	Quartz SiO_2	Feldspars		Micas		Pyroxenes	Amphiboles	Olivine $(Fe,Mg)_2SiO_4$	Chlorite complex group of hydrous silicates containing Mg,Al and other metal ions
		Orthoclase $KAlSi_3O_8$	Plagioclase $NaAlSi_3O_8 \cdot CaAl_2Si_2O_8$	Muscovite $KAl_3Si_3O_{10}(OH)_2$	Biotite $K(Mg,Fe)_3 \cdot ASi_3O_{10}(OH)_2$	Hornblende, etc. $Ca_2\text{-}Na(Mg,Fe,Ti)_3(Al,Fe,Ti)_3Si_6O_{22}(O,OH)_2$	Augite, etc. $Ca(Mg,Fe,Al)(SiAl)_2O_6$		
i.									
igneous (originally molten)									
plutonic									
A. (formed deep in the Earth)									
granite (acidic)	+	+	+(Na)	+	+	○			
diorite (intermediate)			+(Ca)		+	+	+		
gabbro (basic)			+(Ca)			+	○	+	
peridotite (ultrabasic)							+	+	
B. *volcanic* (extruded at the Earth's surface)									
rhyolite (acidic)	+	+	○	+	○	○			
andesite (intermediate)			+(Na,Ca)		+	+	+		
basalt (basic)			+(Ca)			+		+	
volcanic ash	various, but often rhyolitic, with abundant finely-divided, glassy quartz								
breccia, tuff	various, depending on the volcanic source								
II. *metamorphic* (altered through heat and pressure)									
gneiss	various, according to the original source rocks; usually much quartz, feldspar, amphibole, biotite								
schist	various, according to the original source rocks; micas common and minerals such as quartz, pyroxenes, chlorite often present								
serpentine (ultrabasic)									

III. *sedimentary*
   (deposited
   clastic,
   precipitated or
   organic
   material)

A. *consolidated*
   conglomerate   various, according to the original source rocks
   sandstone   various, according to the original source rocks; usually abundant quartz
   mudstone, shale various, according to the original source rocks; particles very finely divided
   limestone   composed mainly of the calcareous exoskeletons of marine animals. Calcite, $CaCO_3$, is the dominant mineral
   coal   composed of compressed remains of plants

B. *unconsolidated*
   colluvium
   alluvium
   glacial till   } various, according to the original source rocks
   dune sand
   loess
   lake mud   remains of aquatic plants and animals
   peat   composed of the remains of plants

---

* Only the most abundant minerals are listed (+, very abundant, ○, less abundant). Various other mineral species may be present in particular rock types in small and variable amounts.

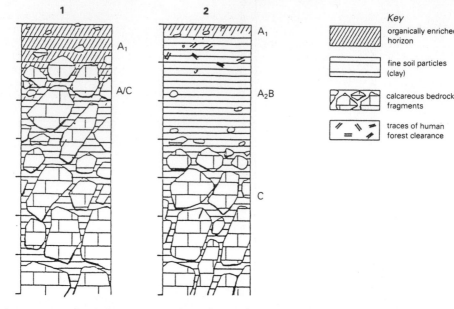

Figure 2.16 Soil profiles for the oak–beech forest site, Virelles, Belgium. (1) Shallow, immature profile. (2) Deeper, more mature profile (after Froment *et al.* 1971).

The soil

The soil of the oak–beech forest site is relatively immature, but fertile (Fig. 2.16). It is calcareous and stony, with weak horizon development.

We must digress here, to discuss the general nature of soil. Soil is the outermost 'skin' of the Earth, modified by biological activity. The soil matrix is composed of mixtures of more or less finely divided rock particles (**clasts**) ranging in size from boulders and cobbles to pebbles, sand, silt and clay. Interspersed with this matrix, particularly near the surface, is the organic component, living and dead.

Local rocks, with different origins and altered in different ways, are the parent materials of soils in different regions. They will have inherently different chemical composition, residing in their minerals (Table 2.8). At the Belgian oak–beech forest site the parent material is the bedrock – a rather soft limestone – which is weathering *in situ*.

In soils the minerals in rock fragments are more or less strongly chemically altered (weathered). Some minerals weather faster than others do. In any soil the degree of weathering of each mineral is a function of both the climatic conditions and time since the rock became exposed. Plant-derived acids and carbonic acid hasten weathering. Warm, moist climates and old, undisturbed sites tend to have the most strongly weathered minerals; cold, dry climates and youthful or disturbed sites tend to have relatively unweathered minerals. Substrate disturbance (which may include the overturning of trees, which lift rock particles caught up among their roots) causes fresh, relatively unweathered rock material to be brought within range of biological activity and strong weathering conditions.

Weathering of minerals releases nutrient ions and, ultimately, gives rise to clay

Table 2.9 Clay minerals. Primary minerals weather in soils, by oxidation, hydration, etc., to form secondary *clay minerals*; some are amorphous, but many are complex, platy aluminium silicates. The individual clasts are of very small (< 0.002 mm) *colloidal* size and clays, thus, have a relatively large surface area. Among their important effects are the maintenance of soil structure, affecting water-holding properties and the ability to *adsorb* plant nutrient cations (and some anions), either on their surface or, in some crystalline clays, internally. The concentrations of adsorbed ions are in equilibrium with the soil solution and the ions are passed into solution in response to concentration gradients, often as a result of uptake by plant roots. Organic colloids, formed by weathering of dead plant and animal detritus to amorphous *humus*, also have important ion adsorbing properties.

	Cation exchange capacity (milli-equivalents $(100 \text{ g})^{-1}$)
amorphous	
goethite (hydrous oxide FeO(OH))	
limonite (sesquioxide Fe_2O_3)	small
gibbsite (hydroxide $Al(OH))_3$	
crystalline	
1 : 1 (1 silica, 1 alumina sheet joined by shared O-atoms)	
kaolinite	3–15
halloysite	
2 : 1 (2 silica, 1 central alumina sheet, with Fe, Mg)	
non-expanding	
illite	10–40
chlorite	
expanding	
montmorillinite	80–150
vermiculite	100–150
organic colloids (not clays)	100–150 (400)

minerals (Table 2.9). These are small **colloidal** particles, with very important roles in binding together coarser particles (affecting soil texture and the water-retaining and drainage properties of soils) (Fig. 2.17) and in maintenance and supply of nutrient ions to plants.

The leaves and other plant and animal debris which fall from the trees, plant roots which penetrate the soil and the animals, microscopic plants, fungi and other microbes which inhabit them contribute dead remains and exudates which eventually are broken down by animals, fungi and bacteria to form the dark amorphous material **humus**. Humus is also colloidal, and is extremely important as part of the ion-exchange complex (as also are the clay minerals). It is also a reservoir of plant nutrients. In well-drained soils, such as that of the oak–beech forest, there is an equilibrium between the addition of fresh supplies of detritus and its eventual microbial reduction and oxidation to its constituent components (CO_2, O_2, water and mineral ions), so that the amount of humus remains constant. In waterlogged soils anaerobic conditions often permit the accumulation of organic matter (see Ch. 8).

Some of the organic residues and exudates (especially organic acids) participate in the mineral weathering process, so that weathering proceeds relatively rapidly under a vegetation cover, if the climate is mild and humid. The litter from different plant species differs in this respect. For example, litter from *Quercus* is less acid, breaks down more quickly and is more fertile than the litter of *Fagus* (Heath *et al.* 1966). This creates local site differences in many temperate forest soils, according to the distribution of individual trees. The extremes of relative fertility and litter accumulation (see profiles 1 and 3, respectively, in Fig. 2.18) are known as **mull** and **mor** soils. The soil of the Virelles oak–beech forest is a mull. Mor soils occur in many gymnosperm forests, whereas intermediate **moder** soils occur in other forest types (Etherington 1982).

As a soil develops *in situ* it becomes stratified into distinct **horizons**, conventionally called A, B and C (Fig. 2.18). This results partly from the impact of the biota. Their effects are experienced most intensely nearest the soil surface, so organic matter accumulation and microbial and root activity are greatest there. There is a gradation from most-weathered to least-weathered mineral material with increasing depth. In part the stratification arises from the influence of leaching (or washing of dissolved and finely divided particulate matter down through the mineral soil as water percolates downwards after precipitation). The A horizon layers within and just beneath the humus-rich layers near the surface are usually most strongly leached. The B horizon is a place of accumulation of clay particles, organic particles and dissolved mineral ions which have been washed down through the profile.

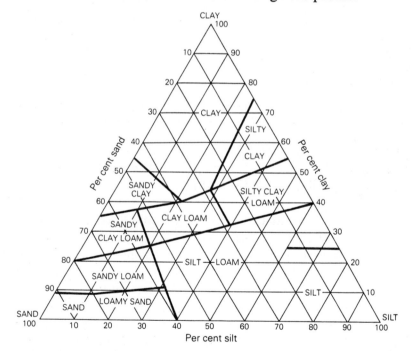

Figure 2.17 Classification of soil texture according to the proportions of the clast sizes sand, silt and clay. The relative amounts of each influence soil structure in important ways. *Loam* soils, (approx. 50% sand, 25% silt, 25% clay) have the clasts aggregated in such a way that the soils are porous and relatively well-aerated and free-draining, but with sufficient fine interstices to maintain water accessible to plant roots and enough clay to impart favourable ion-exchange properties. *Sandy* soils are very free-draining. Clay soils are impervious to water and not well-aerated because the colloidal clays are tightly packed and swell when wet, so that interstices are poorly developed (after N. H. Taylor & Pohlen 1962).

56

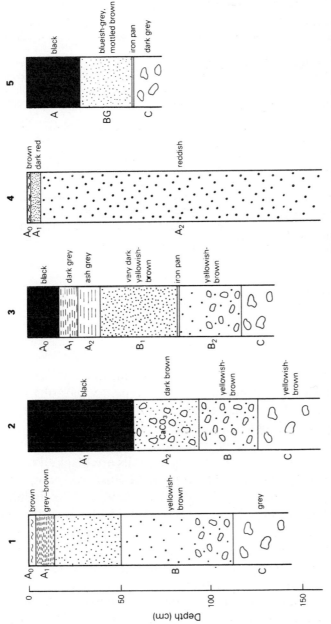
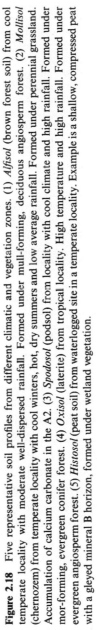

Figure 2.18 Five representative soil profiles from different climatic and vegetation zones. (1) *Alfisol* (brown forest soil) from cool temperate locality with moderate well-dispersed rainfall. Formed under mull-forming, deciduous angiosperm forest. (2) *Mollisol* (chernozem) from temperate locality with cool winters, hot, dry summers and low average rainfall. Formed under perennial grassland. Accumulation of calcium carbonate in the A2. (3) *Spodosol* (podsol) from locality with cool climate and high rainfall. Formed under mor-forming, evergreen conifer forest. (4) *Oxisol* (laterite) from tropical locality. High temperature and high rainfall. Formed under evergreen angiosperm forest. (5) *Histosol* (peat soil) from waterlogged site in a temperate locality. Example is a shallow, compressed peat with a gleyed mineral B horizon, formed under wetland vegetation.

Table 2.10 Main soil groups according to the US Department of Agriculture classification* and some approximate equivalents in other classifications (cf. Buol *et al.* 1973).

| Soil order | General characteristics related to soil genesis | Approximate equivalents | | |
		older US	western European	USSR
inceptisols	young soils on new surfaces			lithosol, regosol
entisols	recent, or undeveloped soils	Arctic brown	ranker	lithosol, regosol
vertisols	mixed or inverted soils			
aridisols	desert and sub-arid soils (and some associated saline and alkaline soils)	grey desert, arid brown		sierozem (solonchak, solonetz)
mollisols	subhumid calcimorphic grassland soils	chestnut, chernozem		chernozem (rendzina)
spodosols	leached, acid, forest soils in cool to cold moist climates	podsol	podsol	podsol
alfisols	less leached, less-acid forest and meadow soils in cool, moist climates	grey-brown podsolic, prairie	grey forest, brown forest, brown podsolic	
ultisols	strongly weathered, iron sesquioxide-rich soils in warm, moist to seasonally dry climates	red-brown podsolic, red-yellow podsolic, lateritic	terra rossa, red loam, brown loam	red-yellow podsolic, krasnozem
oxisols	very strongly weathered, iron sesquioxide-rich soils in very warm, moist to seasonally dry climates	laterite		latosol
histosols	organic soils in waterlogged sites	peat		

* Soils characterized by special controlling features in any soil region (e.g. waterlogged gley soils, saline soils and serpentine rock soils) are associated with the regional soil order. At the level of *suborder* and *group* there is an elaborate and precisely defined terminology.

Soil types with characteristic profiles occur in different climatic and vegetation zones of the world, but local conditions such as rock type, topography or drainage profoundly influence the soils of particular sites. Soil profiles in any locality therefore vary in their morphology according to many specific site conditions.

The colours of the horizons are useful indicators of the conditions which have prevailed as the soils developed. Dominantly organic soils are black, dark brown or reddish brown. Strongly leached soil horizons are pale because iron compounds have been removed, leaving quartz and colourless organic colloids. The B horizons of mature, relatively well-drained soils in temperate, humid climates are yellowish, yellow-brown, orange or brown, due to the accumulated, weathered iron compounds. Tropical soils often are red in colour, because free iron oxides are very abundant. Aluminium oxides, fine quartz and kaolinite impart white coloration to some tropical soils. Waterlogged (gleyed) mineral soils have pale, grey, bluish or greenish colours, from the abundant reduced iron compounds (Etherington 1983). Regionally developed soil classifications are compared in Table 2.10 with a modern American classification (Soil Survey Staff 1975, Bunting 1967), which attempts to accommodate all of the soils of the world. The bases of the classification are the internal properties affecting soil genesis or derived from it.

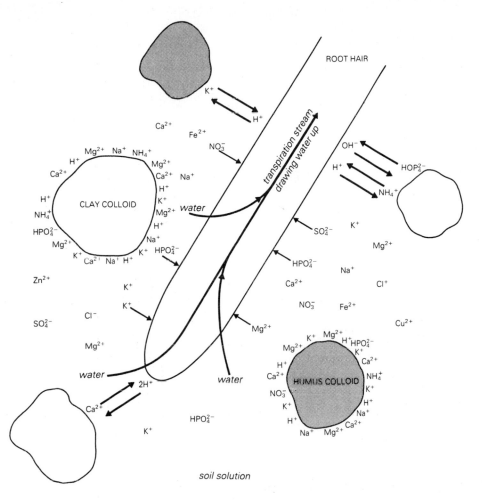

Figure 2.19 The environment of plant root hairs with respect to water and nutrient supply. When soils are moist the soil solution consists of a weak concentration of plant nutrient cations and anions. Much higher concentrations of cations and some anions are adsorbed on the clay and organic colloids. They are in equilibrium with the soil solution concentrations, so that if the soil solution nutrients are depleted (e.g. by removal by plant uptake) adsorbed ions go into solution to restore the equilibrium. Ions reach root hair surfaces by: (1) direct contact of root hair with colloids (2) mass flow in water which moves towards roots, impelled by the gradient resulting from the transpiration stream, (3) diffusion along concentration gradients, (4) ion exchange (cations for H^+ ions, some anions for OH^- ions, produced by plant processes), and (5) adsorption. The complex ion-uptake processes are not considered here.

Soils, nutrients and plants

The most important role of the soil, for plants, is to supply water, mineral nutrients and oxygen to roots. The nutrients are maintained as a 'pool' in the soil minerals and raw organic matter. They are released by weathering of the former and mineralization of the latter (see Ch. 3), and are held in ionic form in the colloidal clay and humus exchange complex (Fig. 2.19). The nutrient ions are released from the exchange complex to the soil solution (maintained in interstices between soil particles, and as a film over their surfaces), and are taken up by plants with the water which is drawn into roots through root hairs.

The main nutrients required by plants are listed in Table 2.11. A discussion of the

Table 2.11 The main nutrients required by plants (after Larcher 1980).

Nutrient element		Original sources	Forms used by plants		Quantity of element in plants (g kg⁻¹ dry matter)		Mean quantity of element in soil (g kg⁻¹ dry matter)
					Mean	Range	
macronutrients (needed in relatively large amounts)							
carbon	C	atmosphere	carbon dioxide gas	CO_2	approx. 450	–	not applicable
hydrogen	H	water	water	H_2O	approx. 60	–	very variable
oxygen	O	atmosphere	oxygen gas	O_2	approx. 450	–	variable
		water	water	H_2O			
nitrogen	N	atmosphere	nitrate (anion)	NO_3^-	20	10–50	1
			ammonium (cation)	NH_4^+			
potassium	K	rocks	potassium (cation)	K^+	10	5–50	14
calcium	Ca	rocks	calcium (cation)	Ca^{2+}	10	5–50	14
magnesium	Mg	rocks	magnesium (cation)	Mg^{2+}	2	1–10	5
phosphorus	P	rocks	phosphate (anion)	$H_2PO_4^-$	2	1–8	0.7
			phosphate (anion)	HPO_4^{2-}			
sulphur	S	rocks	sulphate (anion)	SO_4^-	1	0.5–8	0.7
micronutrients (needed in small amounts)							
iron	Fe	rocks	iron (cation)	Fe^{2+}	0.1	0.05–1	38
manganese	Mn	rocks	manganese (cation)	Mn^{2+}	0.05	0.02–0.3	0.9
zinc	Zn	rocks	zinc (cation)	Zn^{2+}	0.02	0.01–0.1	0.05
copper	Cu	rocks	copper (cation)	Cu^{2+}	0.006	0.002–0.02	0.02
molybdenum	Mo	rocks	molybdate (anion)	MoO_4^{2-}	0.0002	0.0001–0.001	0.002
boron	B	rocks	borate (anion)	HBO_3^{2-}	0.02	0.005–0.1	0.01
chlorine	C	rocks	chloride (anion)	Cl^-	0.01	0.2–10	0.1

Table 2.12 Quantities of nutrients in vegetation and soil in an oak–beech forest, Virelles, Belgium (from Duvigneaud & Denaeyer-De Smet 1970).

		Soil	Leaves	Trees, Wood & bark	Shrubs, Shoots	Trees & Shrubs, Roots	Field layer		All plants, Flowers, fruits, scales	Total Plant
							Shoots	Roots, etc.		
		$(t\ ha^{-1})$	$(kg\ ha^{-1})$	$(kg\ ha^{-1})$	$(kg\ ha^{-1})$	$(kg\ ha^{-1})$	$(kg\ ha^{-1})$	$(kg\ ha^{-1})$	$(kg\ ha^{-1})$	$(kg\ ha^{-1})$
Biomass $156\ t\ ha^{-1}$	K	26.8	36	164	5.3	92	41	4.8	3.3	342
Productivity $14.4\ t\ ha^{-1}$	Ca	133	54	776	35	372	35	8.6	2.5	1248
	Mg	6.5	4.6	72	2.0	19	3.9	1.5	0.4	102
	N	4.5	73	295	14	121	32	6.4	5.6	533
	S	—	6.7	38	2.3	29	5.5	1.1	0.6	81
	P	0.9	4.7	23	1.2	10	3.4	1.7	0.5	44

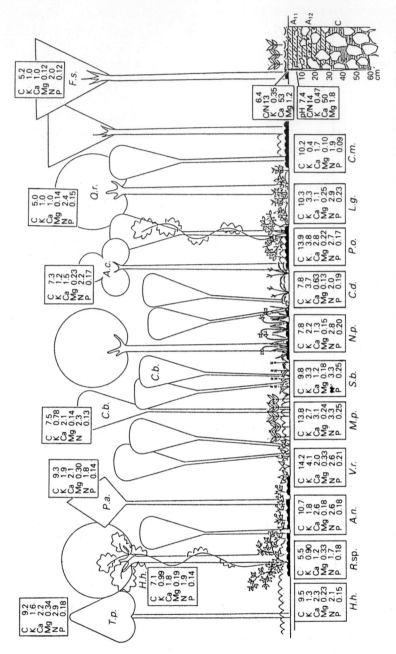

Figure 2.20 Nutrient content of foliage of plants in oak–beech forest, Virelles, Belgium. Foliar ash (C) and nutrient content are given as a percentage of plant dry weight. Values for exchangeable cations in the soil are milli-equivalents per 100 g of soil. pH, acidity; C/N, carbon : nitrogen ratio. *F.s.*, *Fagus sylvatica*; *Q.r.*, *Quercus robur*; *C.b.*, *Carpinus betulus*; *A.c.*, *Acer campestre*; *T.p.*, *Tilia platyphyllos*; *P.a.*, *Prunus avium*; *H.h.*, *Hedera helix*; *R.* sp., *Rubus* sp.; *A.n.* *Anemone nemorosa*; *N.p.*, *Narcissus pseudonarcissus*; *V.r.*, *Viola reichenbachiana*; *S.b.*, *Scilla bifolia*; *M.p.*, *Mercurialis perennis*; *P.c.*, *Polygonatum officionale*; *L.g.*, *Lamium galeobdolon*; *C.d.*, *Carex digitata*; *C.m.*, *Ctenidium molluscum* (moss).

physiology of water and nutrient uptake is beyond the scope of this discussion. It is covered fully by Larcher (1980). Numerous factors, including soil moisture, temperature and microbiological activity, affect the availability of nutrients to plants. Different plant species have distinct requirements for nutrients. Some soil nutrient properties and the amounts of macro-nutrients obtained by analysing the foliage of some of the plant species of the Belgian oak–beech forest are given in Table 2.12 and Figure 2.20. The differences between species, consistent from site to site, are suggestive of their specific nutrient requirements. Careful experimental testing of the specificity of nutrient requirements has been undertaken for only a limited number of plant species (Larcher 1980).

Heterogeneity of habitat conditions with respect to the soil nutrient supply to plants can be maintained literally by substantial site-to-site differences in nutrient concentrations or availability (Tilman 1982). Different species require different amounts of nutrients, but even if the nutrient requirements of species in a stand were very similar (or identical), demands on the nutrient supplies can be spread by the spatial separation of root systems, both vertically and horizontally, and by differences in the timing of growth activities.

Nutrient cycling

The plants of a relatively stable community take up and release about the same amount of each mineral nutrient each year. Quantities of nutrients are sequestered for a time in plant tissues, and are released when individual organs, or whole plants, die and decay. Table 2.12 lists the nutrients maintained in the total biomass for some of the species in the Belgian oak–beech forest. Figure 2.21 summarizes the annual mineral budget.

Interacting factors

Although not all of the abiotic variable factors listed in Table 2.1 have been discussed in relation to the Belgian oak–beech forest (the data are incomplete), enough has been covered to show the enormous complexity of plant–environment relationships. The complexity arises not only because of the large number of separate exogenous factors which influence plants in varying intensities (and the variety of plant responses), but also because of the various interactions of the exogenous factors. For example, summer drought causes growth disruptions of temperate deciduous forest plants in several ways. Water is needed for photosynthesis (as a raw material), for cooling, as a solute for soil nutrients and as a solute for other materials being translocated to all parts of plants. It is also needed to maintain cell turgor, and hence for cell expansion and plant growth. Extreme drought will cause wilting, scorching by overheating and the death of leaves, or whole plants. Even moderate drought will slow or stop the general growth of the plants through limitation of processes such as translocation of photosynthetic products or uptake of nutrients. The intricate relationships between two climatic variables – plant variables and the formation of narrow growth rings in trees – is shown in Figure 2.22.

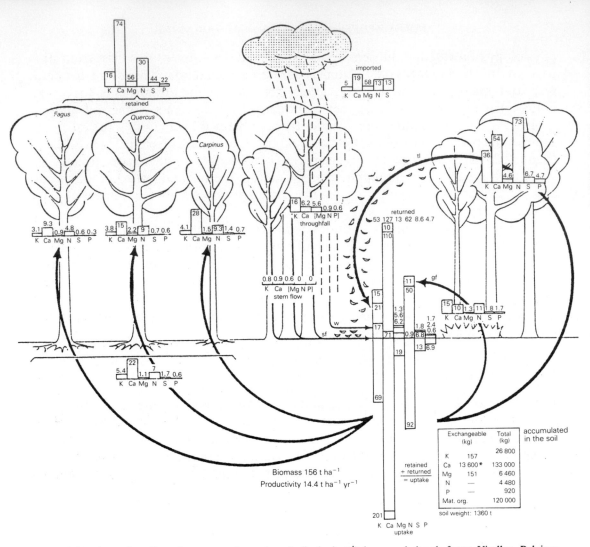

Figure 2.21 Annual cycling of macronutrient minerals (in kg ha^{-1}) in an oak–beech forest Virelles, Belgium. Retained: in the annual wood and bark increment of roots and aerial parts of each species. Returned: by tree litter (tl), ground flora (gf), washing and leaching of the canopy (w) and stem flow (sf). Imported: by incident rainfall (not included). Macronutrients contained in the crown leaves when fully grown (July) are shown on the right-hand side of the figure; these amounts are higher (except for Ca) than those returned by leaf litter. Exchangeable and total nutrient content in the soil are expressed in relation to air-dry soil weights of particles <2 mm (after Duvigneaud & Denaeyer-De Smet 1970).

Biotic relationships within a forest are also complex, as seen in Chapter 3. Figure 2.23 and Table 2.13 record biota other than flowering plants occurring in the Belgian oak–beech forest, and important impacts that some of them have on the plants. The effect of grazing by insects may be cited as an example of interaction between the biotic and abiotic components of a forest system. Grazing of the leaf canopy may diminish the productivity of the trees. However, the thinning out of leaves can permit sufficient light penetration to stimulate the germination of seedlings and the growth of juveniles. It can also speed up the cycling of nutrients from plant sources, as the large amounts of insect faeces are readily reduced by microbes (Mattson & Addy 1975).

MASTER FACTORS

Now we may consider some environmental relationships of vegetation in general. As a first approximation to the explanation of vegetation distribution on the Earth's surface, the system of Holdridge (1967) may be employed (Fig. 2.24). Three master variables – temperature, precipitation and potential evapotranspiration – define the habitat conditions. Locally other variables such as soil nutrients, fire, grazing, etc., will be important determinants of vegetation patterns. The control of plant distribution and abundance must be ascribed to the interaction of many environmental factors. Often none can be especially singled out, but sometimes a single *master factor* has overriding effects. Examples are the lack of water in arid regions, low temperature in alpine and polar regions and hard substrate on rock outcrops (see Ch. 7).

FACTOR GRADIENTS

Many of the essential factors which affect plant growth and reproduction may be thought of as existing in the form of continuous gradients; for example, of temperature, water availability or the concentration of particular nutrients in the soil. Each plant species will have a distinct optimum and range of tolerance for each factor, manifested in its productivity or some other growth attribute. The relationship of any

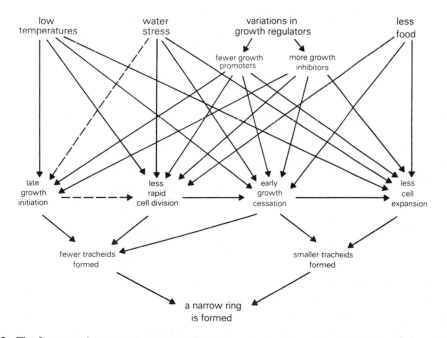

Figure 2.22 The factors and connected events which can cause reduced growth of cambium and wood cells in a conifer tree, eventually resulting in the formation of a narrow annual growth ring. The arrows indicate cause and effect, as well as the interactions among processes and variables. Relaxation of stress from any or all of the factors will permit greater cambial activity and formation of wider growth rings. (after Fritts 1976).

Figure 2.23 The main groups of animals of Western European oak–beech forest: see Table 3.13 for a fuller list of herbivorous animals (modified from Duvigneaud 1967).

		Feeding habits			Feeding habits
I Vertebrates					
MAMMALS[a]			BIRDS		
1 fox	*Vulpes vulpes*	C	14 buzzard	*Buteo buteo*	C
2 stoat	*Mustela erminea*	C	sparrow hawk	*Accipiter nisus*	C
polecat	*Putorius putorius*	C	15 tawny owl	*Strix aluco*	C
3 feral cat	*Felis domesticus*	C	16 pheasant	*Phasianus colchicus*	H(O)
4 wild pig	*Sus scrofa*	O	17 wood pigeon	*Columba palumbus*	H(O)
5 red deer	*Cervus elephas*	H	18 green woodpecker	*Picus viridus*	C
fallow deer	*Dama dama*	H	cuckoo	*Cuculus canorus*	C
roe deer	*Capreolus capreolus*	H	jay	*Garrulus glandiarius*	O
6 badger	*Meles meles*	O	magpie	*Pica pica*	O
7 hedgehog	*Erinaceus europaeus*	C	blackbird	*Turdus merula*	O
8 pipistrelle bat	*Pipistrellus* spp.	C	songthrush	*T. ericetorum*	O
noctule bat	*Nyctalus noctula*	C	19 chaffinch	*Fringilla coelebs*	O
9 squirrel	*Sciurus vulgaris*	H	bullfinch	*Pyrrhula pyrrhula*	O
10 hare	*Lepus capensis (europaeus)*	H	robin	*Erithacus rubecula*	C
11 mole	*Talpa europaea*	C	great tit	*Parus major*	O
12 shrews	*Sorex* spp.	C	wood warbler	*Phylloscopus sibiliatrix*	C
13 voles, mice	*Cleithrionomys* spp. *Arvicola* spp.	H(O)	dunnock	*Prunella modularis*	C
	Sylvaemus spp.	H(O)	etc		
REPTILES					
20 adder	*Vipera berus*	C			
AMPHIBIA[b]					
toad	*Bufo bufo*	C			
newt	*Triturus* spp.	C			
II Invertebrates[c]					
MOLLUSCA	snails, slugs	H	INSECTA		
			Hymenoptera	bees, wasps, ants	H,C,X,Y
ARACHNIDA	spiders, mites △	C,H,D,X,Y	Aphaniptera	fleas	X
	harvestmen	C			
INSECTA			MYRIAPODA	millipedes, centipedes	C,D,H
Collembola	springtails △	H,D			
Orthoptera	cockroaches, locusts	H	CRUSTACEA	slaters, hoppers	D,H
Psocoptera	book lice	D			
Anopleura	sucking lice	C,X	ANNELIDA	earthworms	D
Thysanoptera	thrips	H,D			
Hemiptera	cicadas, plant bugs, aphids, mealy bugs	H,Y	ROTIFERA	rotifers°	D
Lepidoptera	moths and butterflies	H	NEMATODA	roundworms°△	D,C,H,X,Y
Diptera	flies	H,C,D,X,Y			
Coleoptera	beetles	H,C,D	PLATYHEL-MINTHES	flatworms	C,D,X
			PROTOZOA	amoeba etc.°	D,X

[a] The larger mammals are present only in extensive areas of forest, or in remote localities.

[b] Only in localities where there is standing water.

[c] Only the main groups are listed. Very large numbers of species and of individuals from certain groups are present. They are not shown on the diagram but, broadly, spiders and certain insect groups are arboreal, while most of the rest occur on the ground or in the soil. In particular, many species of insects are present, filling many different roles. Often their larvae feed in very different ways from the adults.

△, submicroscopic; o, microscopic; C, mainly carnivorous; O, omnivorous; H, herbivorous; D, detritivorous; X, parasites on animals; Y, parasites on plants.

Table 2.13 Non-flowering plants, fungi and monera of western European oak–beech forest

I. **Plants**

A. Filicinae (ferns)
Dryopteris filix-mas
Pteridium aquilinum
and others

B. Musci (mosses)
Ctenidium molluscum
Eurhynchium striatum
Fissidens sp.
Hypnum cupressiforme E*
Orthotrichum spp. E
Thamnium alopecurum
Thuidium tamariscinum
and others

C. Hepaticae (liverworts)
Frullania spp. E.
Lepidozia sp. E
Lophocolea sp. E

D. Algae (chlorophyta)
Hormidium sp. E
Pleurococcus sp. E
and various soil-inhabiting species
including desmids

II. **Fungi**

A. Lichens
Buellia sp. E
Cladonia spp.
Hypogymnia sp.
Lecanora spp. E
Lecidea spp. E
Opegrapha sp. E
Parmelia spp. E
Pertusaria spp. E
Physcia spp. E
Usnea spp. E
and others

B. Basidiomycetes (gill fungi and others)
Agaricales (most are saprophytes, many form
mycorrhizas, some are parasites on plants)
Amanita spp.
Armillaria sp.
Boletus spp.
Collybia spp.
Coprinus spp.
Gloeoporus spp.
Hypholoma spp.
Inocybe spp.
Lactarius spp.
Mycena spp.
Pleurotus spp.
Polyporus spp.
Russula spp.
Tricholoma spp.
and others

Other Basidiomycete groups
Clavulina sp.
Dacrymyces sp.
Exidia sp.
Geastrum sp.
Hydnum sp.
Lycoperdon spp.
Ramaria spp.
Stereum spp.
Tremella sp.
and others including parasitic smut fungi

C. Ascomycetes (cup fungi and others)
(saprophytes; some form mycorrhizas,
some are parasites on plants)
Chlorociboria sp.
Coryne sp.
Helvella sp.
Nectria sp.
Otidia sp.
Peziza sp.
Tuber sp.
Ustulina sp.
Various yeast species, mildews and others

D. *Phycomycetes* (algal fungi)
(saprophytes, some form mycorrhizas,
some are parasites on plants)
Pythium
Glomus and many others

E. Fungi imperfecti
Includes the parasitic rust fungi and many mould
fungi, some of which are parasitic on plants

F. Myxomycetes (slime moulds)
Lycogala sp.
Stemonitis sp. and others

III. Monera

A. Cyanobacteria (Blue-green algae)
Nostoc and others in soil

B. Actinobacteria (star moulds)
Various soil species

C. Schizonta (bacteria)
Many species in a wide variety of habitats

* E, epiphyte on trees.

plant species (or individual plant) to a particular gradient may be expressed, in terms of relative *favourability* and *unfavourability*, by means of a normal distribution curve (Fig. 2.25).

The favourable range of intensity of each factor is that in which the species performs best. Along each gradient, proceeding in both directions from the optimum, the factor becomes decreasingly favourable until lethal limits are reached. In the marginal zones the unfavourable conditions will cause physiological **stress** and, thus, depressed growth and reproduction. The species will be less able to cope with competition from better-adapted neighbours or with attack by pests and diseases.

In reality the relationships of plants with environment are never as simple as this, mainly because of the interactions between the different controlling variables. In any case it becomes difficult to depict the tolerance ranges of species in relation to the large numbers of gradients which exist. Means of doing so are discussed in Chapters 11 and 12.

STRESS

Those who live in the temperate zone tend to think of plant–environment relationships in terms of relatively well-watered conditions and alternating mild summers, when plants grow best, and cold winters, when plants are dormant. Stressful periods arise from drought, unseasonable cold or excessive heat, as well as from influences such as nutrient deficiency, or the presence of toxic soil conditions or waterlogging. If we are considering plants in other places, we need to have different perspectives according to whatever climatic conditions prevail.

Plants can accommodate to a degree of periodic stress normal for the locality, by acclimation (hardiness in response to cold, heat- or water-deficiency, for example, acquired through gradual application of the stress-causing factor). Plants in some tropical situations could be killed by a single extreme stressful episode (e.g. frost, or drought in a rainforest) which would not harm plants of temperate or drier locations where such events occur more frequently. Highly specialized plants in some other situations cope with environments which would be far too stressful for most temperate zone plants (e.g. the xerophytes of deserts, the high Arctic cold-tolerant herbs, halophytes of saline soils and submerged aquatic plants on the beds of lakes). Stress is a relative matter. This is also so on a finer scale than has so far been considered. Within the temperate zone, or a desert, there are degrees of ability of the different plant species present to cope with the relative stresses which the environment places on the vegetation as a whole.

DISTURBANCE

As well as being limited by factors that directly influence their physiology, plants are affected by agencies that cause them physical harm by damage to or removal of tissues or materials, or both. Such influences may operate only occasionally, or continually.

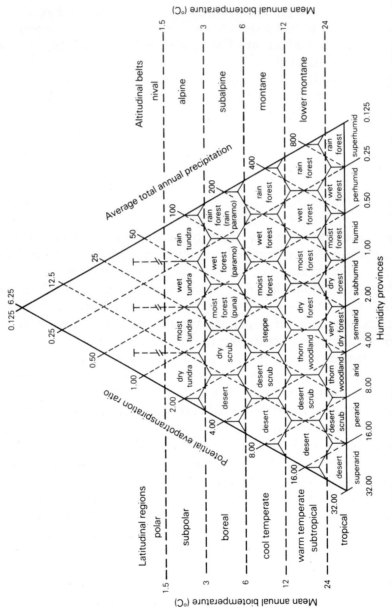

Figure 2.24 The Holdridge system for classifying vegetation types of the world in relation to three climatic variables. Mean annual biotemperature is calculated from monthly mean temperatures after converting means below freezing to 0°C. The potential evapotranspiration ratio is the potential evapotranspiration divided by the precipitation; the ratio increases from humid to arid regions (simplified from Holdridge *et al.* 1971, 1967).

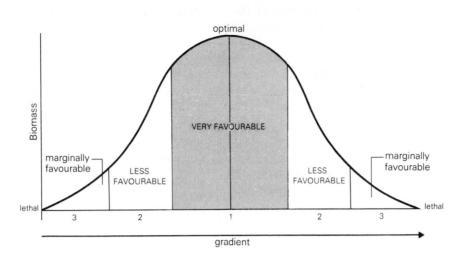

Figure 2.25 Relationship of a plant species to an environmental gradient (such as air temperature, soil moisture or soil nutrient content) as reflected by the biomass of the plant populations.

The disruption will have indirect physiological effects but, provided that damage is relatively minor, plants can recover by the regrowth of tissues. Other harmful effects affecting plant substrates are the erosion of soil and consequent exposure of roots, or the burial of plants by debris. All such influences may be considered to be **disturbances**. Stress and disturbance which generally affect plants in negative ways, may be grouped together under the heading of **environmental pressures** (Table 2.14). The intensity of disturbance may also be expressed as a gradient for each disturbing factor, ranging from minimal to lethal. An apparent difference between disturbing factors and stress caused by lack of, or overabundance of, essential factors which affect plant physiology and growth, is that disturbance factors may be totally absent without any harmful effects. However, some toxic stress-causing substances, if present, have unfavourable or lethal effects through their direct effects on physiology. If they are absent, then there are no such effects. Also, in some instances, as is seen in Chapter 7, disturbing factors are virtually essential for the survival of some plants. This applies to many grasses and grazers, and to many other plants and fire, for example. It also applies to species which are tolerant of disturbance, through features of their life-history or growth responses, but intolerant of crowding.

Some of the biotic effects grouped among disturbances in Table 2.14 resemble stresses in their manifestation. For example, the feeding by aphids on the phloem stream of a plant, or a fungal pathogen attack causing the death of leaves or decay of wood will have considerable immediate physiological impact. The natural endogenous processes of plants such as senescence of their flowers, fruit, leaves or some branches, at particular phases of their growth cycles, also resemble the effects of stress and damage. However, these changes are integrated with the general plant physiology and are not harmful.

Severe stress and disturbance affect vegetation on widely variable scales of space and time. They may affect only one plant or the populations of one species, or patches of various sizes or vast areas. They may occur only once, occasionally, frequently or continuously. These differences, interacting with plant properties, determine the types of vegetation responses. Very strong, continuous stress or very frequent disturbance of vegetation leads to the development of the 'stress-specialized' or 'disturbance-specialized' flora and vegetation types, which are described in Chapters 7–9 and, to some degree, in several other chapters.

Table 2.14 Environmental pressures which influence plants in negative ways.

---

1. *stress* (extremes of factors of which all except toxic substances and possibly wind are essential for plant growth, and which directly affect physiological processes)

    cold
    heat
    shortage of water
    excess water
    wind effects on temperature, water loss
    shortage of nutrients
    excess nutrients
    limitation of access to light (e.g. buried seed)
    excess light
    presence of toxic substances

    the intensity of any of these effects may be *unfavourable* to plants (causing considerable limitations to performance), or *lethal* (causing death)

2. *disturbance* (damage to plants, erosion or burial resulting in removal or death of plant parts or material)

    fire
    breakage by wind, ice, etc.
    abrasion by wind or water-carried sand, ice, etc.
    effects of grazers, fungal pathogens
    erosion of soil from roots
    partial or complete burial

    the effects may be *unfavourable* (partial removal, burial etc., causing impaired performance) or *lethal* (causing destruction and/or removal of whole plants); some plant species seem to be benefited in various ways by fire and grazing; human activities such as logging, ploughing or the generation of toxic wastes may also be disturbing or stressful

---

PLANTS IN THEIR ENVIRONMENT

Any plant in a natural community is usually the descendant of a long line of individuals which have lived in broadly similar conditions in the same region for long periods. Plant populations are genetically variable, and this variability is expressed anew in each generation. Natural selection winnows out the individuals (and genotypes) which cannot survive under the prevailing conditions. The plants which survive are therefore those that are adapted to the environment in which they find themselves. However, they maintain a pool of genetic variation which permits each new generation to adjust to shifts in environmental conditions. Each plant is also phenotypically plastic to some degree, and can adjust to the variable conditions which prevail as it grows. By acclimation it can also adjust to a degree of environmental variation throughout the year. A considerable degree of latitude is also present in the physiological processes of plants, allowing them to adjust to day-to-day environmental variations.

CHANGES IN ENVIRONMENTS

The environment with respect to prevailing plant populations has so far been discussed as if it is virtually constant (although undergoing regular seasonal cycles of change and occasional, periodic extremes such as drought or unseasonal frost which may harm or even kill some plants). Environments for any particular locality do not necessarily remain constant; established plant populations may be pushed out of equilibrium if climatic or other conditions shift too far. Adverse effects for the prevailing dominant species in a locality might be caused, for example, by a systematic decline in temperature or precipitation, the introduction of a new pest or fungous disease, or the addition of more nutrients by the deposition of alluvium. Changes like this may shift conditions on particular factor gradients, from favourable to unfavourable for the resident population and from unfavourable to favourable for other species, which then replace the prevailing dominants.

The vegetation itself can cause habitat modifications which, particularly at the level of establishing seedlings, will cause shifts in the species composition of the plant populations on a site. Particular instances are the alteration of light and moisture conditions as a plant cover develops and creates shade, and the modification of soil properties with time, caused by the accumulation of organic matter on and in the soil, and subsequent changes in the soil physical conditions and chemistry.

Plants and vegetation are products of their environment. The opposite is true, also, for the plant cover on the Earth's surface to some degree modifies environmental conditions. The consequences of these matters for change in vegetation are considered at greater length in subsequent chapters.

3

Plants and their biotic environment

Plants, monerans, fungi and animals live together and interact in many different ways. This chapter describes how plants are influenced by their neighbours, with emphasis on plant–plant interactions and some plant–animal, plant–moneran and plant–fungal relationships which are of very great significance for plant life. It is necessary to examine certain aspects of the nature of plants in greater detail than hitherto, because they have important effects on plant responses to neighbour influences at the levels of individuals, populations, stands and large ecosystems.

Ecologically significant plant variables are listed in Table 3.1 (cf. Table 2.1). The items listed under the subheadings in each of Tables 2.1 and 3.1 are not all mutually exclusive. Some that have already been briefly noted are considered at greater length here.

REGENERATION AND PLANT POPULATIONS

Reproductive modes differ between algae, bryophytes, pteridophytes, gymnosperms and angiosperms (see Table 1.1). When considering their regeneration, allowance must be made for these differences (reproduction by means of spores, alternation of generations in the 'lower' plants, contrasting with the seed habit in angiosperms and gymnosperms). In this chapter reference is specifically to seed plants, and in particular flowering plants, unless otherwise indicated.

The increase of plant populations and replacement of old plants which die (regeneration) is accomplished by seeds and by the production of vegetative off-shoots. Many plants reproduce only by means of seeds, but many others also *proliferate* by one or more vegetative means (Harper 1978, J. White 1979) (Table 3.2). Proliferation implies both reproduction and spread.

Annual seed production is normal for many plant species, but some plants seed at longer intervals. Flowering is determined by the physiological state of the plant and environmental features such as daylength and temperature. Pollination is accomplished by vectors such as wind or insects or some other animals (Faegri & van der Pijl 1971). From the fertilized ovules develop seeds, the numbers that are available to germinate varying with the species and with environmental effects (climate, pollination success, nutrition and predators) (Salisbury 1942, Fenner 1985). The seeds of most short-lived and colonizing species are produced in large numbers and are small. The seeds of longer-lived species of crowded communities are often larger and fewer in number. Some species produce very few or no seeds; they reproduce mainly by vegetative means.

Seeds are dispersed to varying distances from their parents by agencies such as

Table 3.1 Ecologically significant plant variables.[*]

A. *plant form and function*
   woodiness – non-woodiness
   dimensions of root and shoot systems
   general habit of root and shoot systems, architecture, leaf orientation
   leaf dimensions and properties with respect to light interception, gas exchange, water loss, energy fixation
   root properties with respect to water and nutrient uptake
   support and transport organs
   storage organs

B. *periodicity and vegetative growth*
   seasonal development and growth of vegetative organ systems
   evergreen-ness – leaf shedding (or die-down in unfavourable season)
   resting organs
   inherent growth rate
   amount and rate of vegetative productivity
   storage of materials
   vegetative spread

C. *reproduction and population properties*
   age at first flowering
   periodicity of flowering, fruiting and seed production
   pollination
   seed productivity
   seed size and storage of materials in seed
   seed dispersal
   longevity and dormancy of seed
   microsite requirements for germination and successful seedling establishment
   development and maturation of juveniles
   modes and rates of vegetative proliferation
   longevity of adults

D. *ecophysiological amplitude*
   tolerance, under the prevailing range of conditions, for the radiative, climatic, soil and other habitat variables;
   this results from inherent physiological efficiency in processes such as photosynthesis, respiration, nutrient and
   water uptake, translocation of metabolic products, production of tissues, storage, control of water loss, etc.;
   acclimation; relative resistance to environmental hazards – growth responses, life-history responses; production
   of resistant or deterrent organs, tissues, materials (alexins, toxic compounds); strength and durability of tissues

E. *competitive power*
   ability to compete in crowded conditions; this results from inherent physiological efficiency in processes, as
   above, when in contest with other species for resources; it also depends on features of life-history, reproduction,
   growth and form

[*] Items under the separate headings are not necessarily mutually exclusive.

gravity, wind, water and animals. The specializations of fruit and seeds for these purposes are numerous (Ridley 1930, Van der Pijl 1969). Small, light seeds are likely to travel relatively long distances; wind is an effective dispersal agent. However, larger, bird-dispersed, mammal-dispersed or stream-dispersed seeds may also be carried for considerable distances. Some seeds are well-suited for flotation and withstand immersion for long periods in sea water, so can be widely dispersed by ocean currents. Exceptional events sometimes cause very long distance dispersal (Ridley 1930, J. M. B. Smith 1978). However, experiments have shown that, on land, most seeds, especially the heavier ones, fall not very far from their parents (Silvertown 1982, Fenner 1985).

Table 3.2 Modes of vegetative proliferation in vascular plants.[*]

1. production of simple offset branches, as in many tufted grasses, bulbous plants, shrubs and some trees (e.g. willows)
2. new branch and root development on toppled plants
3. sprouts from stem bases, with different degrees of vigour and survival as new plants
4. layering – roots develop on prostrate stems – some herbs and vines or creepers extend continuously in this way[†]
5. tubers or root bulbils that extend away from parent
6. sprouts from rhizomes – more or less long underground stems[†]
7. stolons – specialized overground organs of extension, with tips that take root and develop new plants[†]
8. sprouts from roots (suckers) extending away from parent[†]
9. re-establishment of whole plants (on slopes or in water) after displacement[†to‡]
10. growth of detached parts (leaves, branches or specialized plantlets); some small, whole aquatic plants, e.g. *Lemna*, customarily reproduce vegetatively by budding and are dispersed long distances by flotation or on aquatic birds[‡]
11. formation of specialized aerial buds or bulbils which become detached and serve as seed-like propagules; turions of Lemnaceae are one such case[‡]
12. seed development without any sexual process having occurred (apomixis)[‡]

[*] Many perennial vascular plants have some means of vegetatively increasing themselves, although often the new parts of such clones establish close to the parent and the parental connection may not be broken for a long time. Vegetative growth often results from damage to the root or stem of woody plants.

[†] Potential for extension of new ramets moderately long distances (0.5–5 m or more) from parent by these modes.

[‡] Potential for extension of new ramets long distances (> 5 m to many km) by these modes.

Some seeds fall on unsuitable sites, are eaten by predators such as rodents or insects, or are killed by fungi. The seeds of some perennial species retain their viability for only brief periods, so they must germinate immediately. The seeds of many other species, especially annuals and biennials (but also many perennials), are viable (and **dormant**) for longer periods. Some are dormant only through the winter following their formation. Others remain so for many decades, especially if buried deeply in the soil. This may apply even when adults of a species have become locally extinct. In some species dormant seeds are retained for a time on the parent plant, in dry and often tough-walled fruit.

From the dormant **seed bank** new plant populations can arise when circumstances are favourable (Fig. 3.1). The seed dormancy phenomena are very complex. Table 3.3 summarizes the main causes of delayed germination and the ways by which dormancy is broken in natural circumstances. Table 2.3 lists the known dormancy and other reproductive properties for plant species in the Belgian oak–beech forest.

The conditions required for successful germination of seed and establishment of seedlings are often very specific (Harper 1977, R. E. Cook 1980). Light, moisture and other microsite conditions must be just right. When germination occurs the tiny new plants may be killed by the inability of shoots to reach well-lit positions, by climatic hazards such as drought, by predators such as snails and insect larvae, or by fungal pathogens. Seedlings from small seeds, with only a small supply of stored food, generally require open, well-lit, uncrowded sites for successful establishment. In sites which at ground level are crowded and shaded, seedlings from larger seeds are more likely to be successful. The greater store of food enables them to extend roots and shoots into positions where habitat requirements will be satisfied. It may be almost

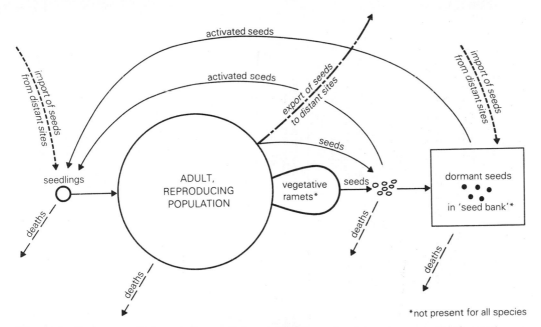

Figure 3.1 Components of plant species population regeneration and maintenance on an area of ground.

impossible for seedlings to become established in very crowded sites. The establishment phase is, thus, normally, the most critical in the life of a plant; there is usually an enormous loss among populations of newly germinated seedlings.

As the surviving young plants begin to increase in stature, they experience strong competition from neighbours of their own kind and from other species. Competition is considered more fully later in this chapter. During growth towards adulthood the adolescent plants of any cohort (derived from one season's seeds) are thinned by competition and predation by grazing animals. Adult densities (numbers per unit area) are often thousands of times less than those of seedlings (cf. Silvertown 1982).

The regeneration patterns of species that reproduce by means of seeds are affected by: the periodicity of seeding (annually, or at longer intervals); the numbers of viable seeds produced; the numbers of seedlings which become established (and the conditions needed for this); the time taken to grow from seedlings to reproductive age; mortality rates through these stages; and the longevity of the adults. Relative difference in lifespan influence many other aspects of plant ecology very profoundly, as is seen later.

It is important to understand why plants die. Annuals, biennials and perennial monocarpic species (which die after flowering) exhibit inherent senescence. Other perennials are of two sorts. Some seem to be determinate, having characteristic lifespans (although there are few well-documented examples of inherent senescence). In practice most of these succumb from the combined effects of disease, attack by grazers, stress and damage caused by the weather. Other plants which customarily increase by vegetative proliferation seem to be virtually immortal, although they may shift their location and replace their tissues. Of course, seeds are the means by which most species ensure their potential immortality.

Table 3.3 Phenomena associated with delayed seed germination

	Cause of delay*	Delay overcome by
dry seed	insufficient water available; water-impenetrable testa	wetting puncture or erosion of testa to facilitate entry of water
dry† or imbibed seed	immaturity of embryo	elapse of time
imbibed seed	temperature; gas, light conditions unsuitable (or gas- or light-impenetrable testa; or mechanically restricting testa)	appropriate temperature, gas, light conditions (or puncture of testa to facilitate gas exchange, access to light, or expansion of embryo)
dry or imbibed seed	blocking of metabolic pathways determined by inhibitory factors in pericarp or other maternal tissues; or in testa, endosperm, perisperm or some part of the embryo; the blocking may be complete under the full range of prevailing conditions, or may apply to all but a narrow range of conditions	appropriate treatment to remove blocks (e.g. leaching, chilling, heating, light of the right quality or quantity, puncture or removal of testa); in nature elapse of particular periods of time, together with specific amounts of leaching, or passage through seasonal climatic shocks, are needed to achieve this; timing is related to seasonal climatic, daylength or other natural cycles or events

* Several of the causes may operate simultaneously, in the same seed. In batches of seed from individual plants or in species populations there may be polymorphisms for different degrees of biochemical blocking, or the biochemical blocks may be developed over a range of degrees. Embryos differ in degrees of immaturity, depending on species. Seeds of some species may come out of a dormant state and, if the conditions for their germination are not met, go back into the dormant state.

† It is implicit that seeds must be imbibed before germination can occur.

Population recruitment may be more or less continuous (from each year's seeding) or at intervals which depend on periodic seeding (e.g. the mast year phenomenon; Silvertown 1980b) or special requirements for establishment. The death of adults may be necessary to provide sites for establishment of new juvenile populations (Fig. 3.1).

Among the many species which customarily reproduce by vegetative means, some have specialized on the vegetative mode and produce seeds rarely or not at all. Vegetative offsets from shoot or root may ultimately become independent of the parent. These offsets (ramets, forming clones) are usually thought to be genetically identical with the parent, whereas seedlings are not (unless the parent is obligately self-fertilized or apomictic) (Table 3.3). Recent work shows that, in fact, somatic mutation can give rise to genetic variation in clones (D. E. Gill 1986, Sutherland & Watkinson 1985). Vegetative offsets of some species are detachable units, shed as soon as they are produced and dispersed like seeds. They include various kinds of bulbil (whole miniature plants), shed leaves or branch- or root-fragments, which easily develop new roots. In other species vegetative offsets remain attached to the parent but may spread several to many metres in a growing season (cf. Turkington & Harper 1979). Eventually the parental links are broken by death of the connections. Some herbaceous species migrate continuously by means of stolons or rhizomes and extend a leading edge forward while dying at a trailing edge behind.

Many woody and herbaceous species can survive severe damage (e.g. from grazing or fire) by the activation of dormant buds on stems, stem bases or roots. Although some woody plants spread widely by means of root sprouts, sprouts of many species remain close to the original parent. Allowing for some somatic mutation, plants of very similar genotype may thus persist on the same piece of ground for long periods, in some cases thousands of years (Vasek 1980a). This applies especially to species with specialized woody underground or semi-underground bases which can survive total destruction of the above-ground parts.

The discipline of demography studies numerical changes in populations over time (Hickman 1979, Silvertown 1982, J. White 1985). This can be done relatively easily with annual or other short-lived, discrete plants, but it is more difficult with long-lived plants. It is very difficult in many kinds of perennial vegetation such as grassland, where it is usually impossible to count the discrete, vegetatively proliferating individuals. If plants are modular in form (i.e. if the branch systems are approximately equivalent in size) counts of the modules may be made (cf. Harper 1978, J. White 1979). However, some of the concepts of demography may not be easily applied to 'super-plants', in which connections are retained between offsets and parents. A similar problem may be met in some forests, in which adjacent individual trees of a species (or in some instances different species) can form root grafts and exchange materials, including nutritive substances (Graham & Bormann 1966, Bormann 1966, F. W. Woods 1970).

ECOPHYSIOLOGICAL AMPLITUDE

Different plant species respond unequally to the same sets of environmental conditions. The range of tolerance of a species for the set of conditions which it experiences is known as its **ecophysiological amplitude**. The differing ecophysiological amplitude of different species is manifest to some degree in differences in their form, growth, timing of activities and relative ecological behaviour. Such differences can be examined on natural environment gradients. However, the relative environmental requirements and tolerances of different species, including their physiological differences, can best be tested when plants are grown under controlled conditions.

There is not space here to describe all of the components of plant ecophysiology which determine plant behaviour. They are summarized well by Etherington (1982), Larcher (1980) and Fitter & Hay (1981). Instead, important aspects of the relative ecology of plant species are considered in relation to interactions with neighbours.

A complicating factor in the matter of ecophysiological amplitude is genetic variation within species. Species are defined as being composed of individuals which potentially can all interbreed and produce fertile offspring. This definition will suffice for most purposes, although it applies to outbreeding species only. Exceptions are plants which do not rely much, if at all, on outcrossing (habitual inbreeders, usually self-fertilizing) or on the sexual reproductive process (apomicts, which produce seed without undergoing any sexual process, and plants which always, or nearly always, proliferate vegetatively; cf. Stebbins 1950).

Each outbreeding species is composed of populations of individuals growing adjacent to one another. The genotypes (the intrinsic genetic complement controlling their essential character) of all individuals of a population are usually very similar. Thus, if individuals of a population are grown in a uniform environment, their phenotypes (their outward observable form and physiological behaviour) are relatively uniform, although some variations of stature, shape, colour, etc., will be evident due to minor genetic differences, as they are in people. In some species there is considerable variation between populations (cf. Harper 1977). This variation is due to the normal genetic variability inherent in outbreeding species. Individuals from genetically uniform clones, grown in different environments, may have very different phenotypes. Those grown in cold, arid or nutrient-poor conditions will be very small and stunted, and will usually produce limited numbers of flowers and seeds. Others grown in warm, well-watered, fertile conditions will be large and lush, and flower and seed profusely (Fig 3.2). The productivity, or any other attribute of a plant, thus depends partly on its inherent nature and partly on the moulding influence of environment. Some species are phenotypically (and genotypically) more flexible than others.

Any region will have its own distinctive flora, usually well-adapted to the local conditions, through a long process of evolution there. Any species which occupies a range of varied habitats will, in fact, have locally adapted races, each genetically somewhat distinct (Clausen 1962, Clausen *et al.* 1948, McMillan 1959, 1969, Antonovics *et al.* 1971) (Fig. 3.3). The local races of wide-ranging, genetically flexible species differ both physiologically and morphologically. For example, high-altitude races of

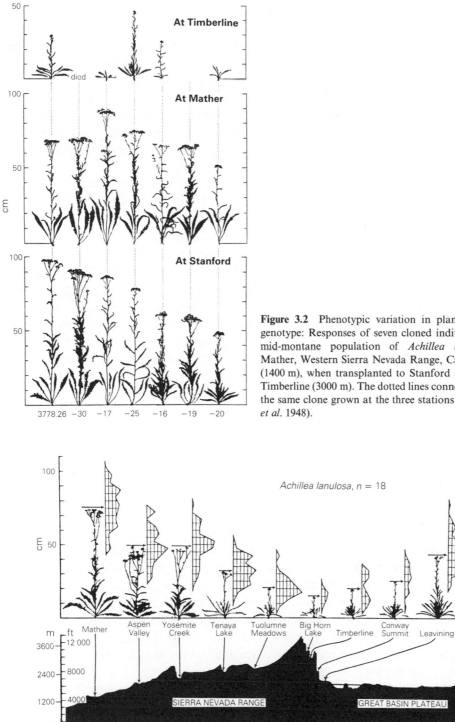

Figure 3.2 Phenotypic variation in plants of uniform genotype: Responses of seven cloned individuals from a mid-montane population of *Achillea lanulosa* from Mather, Western Sierra Nevada Range, California, USA (1400 m), when transplanted to Stanford (sea level) and Timberline (3000 m). The dotted lines connect members of the same clone grown at the three stations (after Clausen *et al.* 1948).

Figure 3.3 Locally adapted races of *Achillea lanulosa* from sites at various altitudes in the Sierra Nevada Range, California, USA, grown in uniform conditions in a garden in Stanford. The frequency diagrams indicate variation in height within the populations and the plant silhouettes represent the means (after Clausen *et al.* 1948).

a species in a temperate climate will be reduced in size, tolerant of low temperatures and intolerant of high temperatures. Low-altitude races of the same species will be larger, intolerant of cold and tolerant of heat. Wide-ranging species will have races adjusted to the local concentrations of soil nutrients, or to the local variations in water supply or other limiting factors. Some species have both tree forms, in mild environments, and shrub forms, in severe environments. Clearly, the wide ecophysiological amplitude of such species consists of numbers of distinct, but overlapping, narrower amplitude ranges of the component local populations. In wide-ranging species genetic variation in some plant properties may be continous along environmental gradients.

A further complicating factor is that individual plants may consist of mosaics of genetically different parts (D. E. Gill 1986, Sutherland & Watkinson 1986). In perennial species with efficient means of vegetative spread, this may be a very effective way for plants to adjust to environmental variation.

The broad ecophysiological amplitude of many species allows them to cope with a wide range of environmental conditions. This flexibility, in the face of variable environment, is extended by acclimation, the adjustment of the phenotype of a plant to cope with more-severe stress than it usually experiences, after gradual exposure to those conditions for a time. Acclimation may apply differently to different parts of a plant, e.g. the acquisition of distasteful or toxic properties by leaves only near an area which has been grazed by a herbivore (Rhoades 1979). Often it operates seasonally, e.g. the development of distasteful properties by leaves just before the main irruption of grazing herbivores (Feeny 1970).

DIFFERENCES BETWEEN PLANT SPECIES

Although groups of plant species can be found which are similar in many respects, species differ in general form and stature, seasonal activity, growth rates, reproductive patterns, aspects of their physiology and various other attributes. A very wide range of permutations and combinations of characteristics is possible and, as far as is known, each species is uniquely different from every other species in at least some respects. This means that each species must respond uniquely to the ecological situation in which it occurs, compared with other species exposed to the same set of conditions. In fact, as was seen earlier, genotypic variation within species and even within individuals will cause some ecological differences between populations of the same species.

It would be difficult to summarize, succinctly, all of the variable properties for even the species in the flora of one region. Attempts to describe plant diversity in very generalized ways include the relatively simple ecomorphologic-functional systems of Raunkaier (1934) and Dansereau (1957) (Figs 3.4 & 5), the application to plants by Whittaker (1975b) of the two-dimensional r–K 'strategy' approach of McArthur & Wilson (1967) (Fig. 3.6), and the more-elaborate three-dimensional method of Grime (1979) (Fig. 3.7 and Table 3.4). Wells (1976) also developed a special-purpose 'climax-index' method for use in vegetation change studies in eastern American forests (see Ch. 11). None of these approaches is very satisfactory for expressing the whole gamut of ecologically meaningful plant diversity. Examination of plant responses on environmental gradients (e.g. Parrish & Bazzaz 1982a, b) or the genotype–environment analysis method of Garbutt & Zangerl (1983) may be more useful, but data collection is difficult. Grubb (1985) has recently provided a critique of methods for expressing the ecological diversity of plants. Before examining these matters further in terms of the concept of the ecological niche, we need to know how neighbours, especially plants, affect the expression of the ecophysiological amplitude of any plant.

PLANT–NEIGHBOUR RELATIONSHIPS

The other plants, microbes and animals which surround any plant are a very important part of its environment (Zadoks 1978, Crawley 1986). Various direct or indirect relationships are possible. Table 3.5 summarizes these, with some examples. All of the relationships in Table 3.5 except the first, involve some impact on the growth, reproduction, general success or survival of at least one of the two interacting species. The possibilities are: 0 (no appreciable interaction); + (a beneficial effect); and − (an inhibitory effect). The scheme is a modification of one developed by Burkholder (1952) and adapted by Odum (1971). The subscript for an obligate relationship

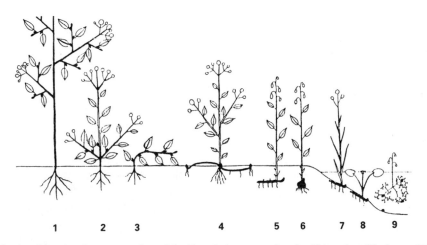

Figure 3.4 An abbreviated representation of the Raunkaier system for classifying plant life-forms. Persistent axes and buds are emphasized in the diagrams. (1) *Phanerophytes* (perennating buds or shoot apices borne on aerial shoots). Subdivisible according to stature, evergreenness or deciduousness, bud protection, etc. (2, 3) *Chamaephytes* (perennating buds or shoot apices borne very close to the ground). Subdivisible according to perennial nature or not of aerial shoot systems and their attitude (erect, prostrate, tightly grouped into cushions, etc). (4) *Hemicryptophytes* (perennating buds at ground level and aerial parts dying back at the onset of unfavourable conditions). Includes plants with long prostrate axes, rosettes, mat plants, etc. (5–9) *Cryptophytes* (perennating buds below ground level or submerged in water). Includes species with rhizomes, tubers, bulbs, corms and other underground stem systems, as well as aquatic species rooted below water level, with emergent, floating or submerged shoot systems. Not shown are *Therophytes* (annual species which complete their life-history within the favourable season of the year and survive the unfavourable season as seeds) (after Raunkaier 1934).

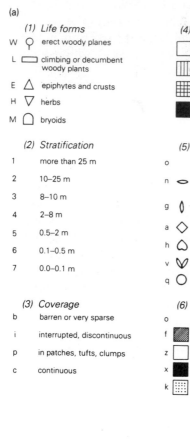

(a)

(1) Life forms

W ⚲ erect woody planes

L ⬜ climbing or decumbent woody plants

E △ epiphytes and crusts

H ▽ herbs

M ⌂ bryoids

(2) Stratification

1   more than 25 m

2   10–25 m

3   8–10 m

4   2–8 m

5   0.5–2 m

6   0.1–0.5 m

7   0.0–0.1 m

(3) Coverage

b   barren or very sparse

i   interrupted, discontinuous

p   in patches, tufts, clumps

c   continuous

(4) Function

⬜ deciduous or ephemeral

⊟ semideciduous

▦ evergreen

■ evergreen-succulent or evergreen-leafless

(5) Leaf shape and size

o   leafless

n ⬭ needle, spine, scale or subulate

g ◊ graminoid

a ◇ medium or small

h ⌂ broad

v ⋁ compound

q ◯ thalloid

(6) Leaf texture

o   leafless

f ▨ filmy

z □ membranous

x ■ sclerophyll

k ▦ succulent or fungoid

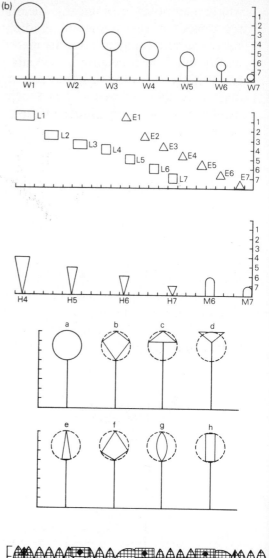

(b)

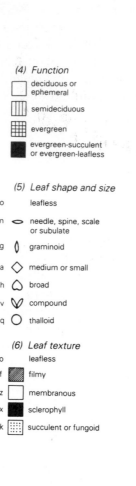

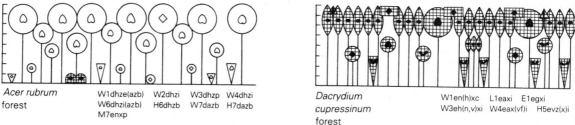

Acer rubrum forest   W1dhze(azb)   W2dhzi   W3dhzp   W4dhzi
W6dhzi(azb)   H6dhzb   W7dazb   H7dazb
M7enxp

Dacrydium cupressinum forest   W1en(h)xc   L1eaxi   E1egxi
W3eh(n,v)xi   W4eax(vf)i   H5evz(x)i

Figure 3.5 (a) The Dansereau system for classifying plant life forms by six basic sets of criteria and depicting them by standard symbols. (b) The symbols combining *life-form* (1 in (a)) and *stratification* (stature) (2 in (a)). Crown outlines are shown below for tall woody species. (c) Application of the Dansereau system to two forests: (A) an *Acer rubrum* (red maple) forest in eastern North America; (B) a *Dacrydium cupressinum* (rimu)–*Weinmannia racemosa* (kamahi) forest in Westland, New Zealand. After Dansereau (1958).

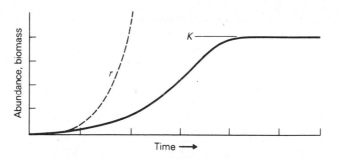

Figure 3.6 The exponential and logistic growth curves. The broken curve shows exponential growth of a population with a rate of increase r in an environment that does not limit its growth (rate of increase expressed by the formula $dN/dt = N_0 r$, where N_0 is the initial population density). The sigmoid (logistic) full curve is that for a population where conditions (resource availability or other constraints) cause stabilization at the carrying capacity K (rate of increase expressed by the formula $dN/dt = rN(1 - N/K)$, where N is the population density). All organisms are ultimately constrained by conditions but those, including plants, which at first show tendencies to follow the r course (rapid population growth, production of large numbers of light, well-dispersed seeds, often small adult plant size and short life) are said to be *r-selected* – specializing on reproduction. Those which follow the K course (slow population growth, production of fewer, heavy, seeds, dispersed only short distances, often large adult plant size and long life) are said to be *K-selected* – specializing on maintenance of the mature plant and persistence. There is, of course, an array of intermediate conditions on the *r-K continuum* (modified from Whittaker 1975).

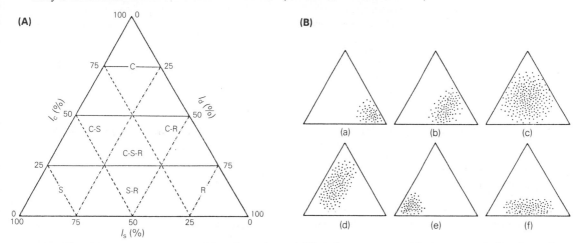

Figure 3.7 (a) Grime's model describing the various equilibria between competition, stress, and disturbance in vegetation and the location of primary and secondary 'strategies'. I_c, relative importance of competition (——); I_s, relative importance of stress (– – –); I_d, relative importance of disturbance (–·–). A key to the symbols C, S, R for the strategies is included in Table 3.4. (b) Diagrams describing the range of 'strategies' encompassed by (a) annual herbs, (b) biennial herbs, (c) perennial herbs and ferns, (d) trees and shrubs, (e) lichens and (f) bryophytes. Intergrades between the primary 'strategies' are: (1) *'competitive ruderals'* (C–R) – adapted to circumstances in which there is a low impact of stress and competition is restricted to a moderate intensity by disturbance; (2) *'stress-tolerant ruderals'* (S–R) – adapted to lightly-disturbed, unproductive habitats; (3) *'stress-tolerant competitors'* (C–S) – adapted to relatively undisturbed conditions experiencing moderate intensities of stress; (4) *'C–S–R strategists'* – adapted to habitats in which the level of competition is restricted by moderate intensities of both stress and disturbance. After Grime (1977).

Table 3.4 Some characteristics of 'competitive', 'stress-tolerant' and 'ruderal' plants (adapted from Grime 1977)

	'Competitive' (C)	'Stress-tolerant' (S)	'Ruderal' (R)
(i) *morphology*			
1. life-forms	herbs, shrubs and trees	lichens, herbs, shrubs and trees	herbs
2. morphology of shoot	high dense canopy leaves; extensive lateral spread above and below ground	extremely wide range of growth forms	small stature, limited lateral spread
3. leaf form	robust, often mesomorphic	often small or leathery, or needle-like	various, often mesomorphic
(ii) *life-history*			
4. longevity of established phase	long or relatively short	long–very long	very short
5. longevity of leaves and roots	relatively short	long	short
6. leaf phenology	well-defined peaks of leaf production coinciding with period(s) of maximum potential productivity	evergreens, with various patterns of leaf production	short phase of leaf production in period of high potential productivity
7. phenology of flowering	flowers produced after (or, more rarely, before) periods of maximum potential productivity	no general relationship between time of flowering and season	flowers produced early in the life-history
8. frequency of flowering	established plants usually flower each year	intermittent flowering over a long life-history	high frequency of flowering
9. proportion of annual production devoted to seeds	small	small	large
10. perennation	dormant buds and seeds	stress-tolerant leaves and roots	dormant seeds
11. regenerative 'strategies'*	V, S, W, B_s	V, B_Y	S, W, B_s
(iii) *physiology*			
12. maximum potential relative growth rate	rapid	slow	rapid
13. response to stress	rapid morphogenetic responses (root–shoot ratio, leaf area, root surface area) maximizing vegetative growth	morphogenetic responses slow and small in magnitude	rapid curtailment of vegetative growth, diversion of resources into flowering
14. photosynthesis and uptake of mineral nutrients	strongly seasonal, coinciding with long, continuous period of vegetative growth	opportunistic, often uncoupled from vegetative growth	opportunistic, coinciding with vegetative growth

Table 3.4 *(cont.)*

15.	acclimation of photosynthesis, mineral nutrition and tissue hardiness to seasonal change in temperature, light and moisture supply	weakly developed	strongly developed	weakly developed
16.	storage of photosynthate, mineral nutrients	most photosynthate and mineral nutrients are rapidly incorporated into vegetative structure, but a proportion is stored and forms the capital for expansion of growth in the following growing season	storage systems in leaves, stems and/or roots	confined to seeds
(iv)	*miscellaneous*			
17.	litter	copious, often persistent	sparse, sometimes persistent	sparse not usually persistent
18.	palatability to un-specialized herbivores	various	low	various, often high

* Key to regenerative 'strategies': V, vegetative expansion; S, seasonal regeneration in vegetation gaps; W, numerous small wind-dispersed seeds or spores; B_s, persistent seed bank; B_Y, persistent seedling bank.

indicates that it is an absolute necessity for at least one of the participating organisms. All of the other relationships are facultative.

The table is a simplistic representation of the gamut of possible connections. In many instances several kinds of effect will occur at the same time; the intensities of effects will vary considerably, according to circumstances (for example, the influence of one species on another on different occasions may range from $-,-$ to $0,-$ or $+,-$); and there may be complex influences of third, or fourth, parties. For example, the grazing of plant species A by an animal may fortuitously benefit plant species B.

Competition

In the thin layer of the Earth's skin and atmosphere that is occupied by plants, inherently low amounts of some resources may occur. **Resource use** (r.u.) competition occurs when plants pre-empt some or all of the supplies of particular substances or factors, thereby interfering with the growth (or other functions) of other plants.

Vascular plants and their plant neighbours have similar environmental requirements; their root and shoot systems are often in close proximity, if not in contact. They cannot move away to avoid this, except by seed dispersal or, in the case of some species, by growth of vegetative offshoots (which is usually relatively slow). This means that plant neighbours must often be in contest for resources such as water, light and nutrients.

Organisms in other trophic levels (fungi, bacteria and animals), which obtain their

Table 3.5 Kinds of relationship between pairs of neighbouring species

Term for interaction	Species		Influence	Example(s) and occurrence
	A	B		
neutralism	0	0	neither affects the other	clear-cut examples are difficult to find; certain small, undemanding plants and animals probably live with larger plants in this way
amensalism	0	−	one species is unaffected, the other suffers	large plants suppress smaller plants (or inhibit some animals) by shading or other kinds of habitat modification; common
resource use competition	−	−	mutually negative	plant species in contest for nutrients, water and, in some circumstances, light, gases; common
direct interference competition	+	−	one species benefits, the other suffers	one species suppresses another by exudation of toxic substances or leaf deposition; epiphyte load on leaves causes less-efficient photosynthesis; fostering of insect or fungus by one species to the detriment of another; shading by one species causes unfavourable temperature conditions for another; common
herbivory (or predation)	+	−	as above	animals feeding on foliage, wood, roots, seeds; common
parasitism*	+	−	as above	mistletoe or fungus affecting host plant; plant gall-forming animals; common
commen-salism*	+	0	one species benefits, the other is unaffected	certain obligate epiphytes and animals sheltered by plants; common
benefaction	+	0	as above	facultative epiphytes; seedlings requiring shade cast by a canopy of leaves; plants protected from browsing by another, spiny plant; some types of seed dispersal by animals; common
mutualism*	+	+	mutually beneficial effects	lichen and legume symbioses, mycorrhizas; specific pollination relationships; protective ant colonies in certain plants; fruit-eating animals which disperse seeds; common
mutual benefaction	+	+	as above	non-specific pollination, fruit-eating and seed dispersal relationships; stimulation of plant growth by herbivory, dung and urine deposition; nesting by insectivorous birds; common

* Obligate and often very specific relationship for at least one of the pair.

energy in other ways, are generally thought not to compete with vascular plants. This is not always strictly true. For example, animals in aquatic habitats may compete with submerged plants for oxygen; epiphyllous lichens may compete with leaves for light, and bacteria may compete with roots for nitrogen.

Usually the list of resources includes nutrients, water, gases and light (as an energy source for photosynthesis). Other items might be considered as resources in this context (e.g. heat, which can be modified in the same way as light, and light of particular quality, quantity or duration, as a trigger for physiologic processes like movements, flowering and germination). Living space may be a resource, but it seems likely that when space is limited the main things being contested by plants are resources such as light or nutrients. Pollinators and seed-dispersing animals (if their numbers are low) may also be considered to be resources for which plants must compete. The distinctive nature of each of the main resources and the modes by which plants compete for them are outlined in Table 3.6.

The processes of competition may be very complex; often the specific details are difficult to unravel. It is usually assumed that interference effects between two species in contest are mutual, but this is not necessarily so; some species are inherently strong competitors (dominants) ousting others which could occupy the site in their absence.

As individuals of the same species are genetically very similar, they are strong competitors, and this determines their spacing and productivity in stands of vegetation. Different species have evolved so as to avoid direct competition as much as possible, but the struggle for existence is a strong selective pressure favouring some species over others under particular sets of prevailing conditions.

Species can gain advantage over one another as a result of differences in seed production, in shoot and root growth rates, in stature, in modes of displaying leaves and by being efficient at the physiological processes of nutrient and water uptake and gas exchange. Competition begins with seedlings. They are very vulnerable to environmental pressures. The relative success of individuals from a group of plants of the same (or different) species, beginning life from seeds at the same time, depends on the amount of energy contained in the stored material in the seeds and how quickly it is mobilized during germination. The individuals that germinate first and grow rapidly will pre-empt sites. Larger-seeded individuals will be able to sustain growth for longer than those with smaller seeds. Through weak root growth the seedlings from relatively small seeds will be unable to exploit the soil for water and nutrients as well as those with vigorous root growth do. In turn, the shoots of the weaker species will be poorly developed. Because they cannot grow and expand their leaf area sufficiently quickly, they will be likely to succumb in the low light beneath the canopy of leaves. Competition continues during all phases of a plant's life, subsequently, but its effects on population numbers are probably most severe at the seedling stage (cf. Grubb 1977).

Mechanisms for interplant competition for resources are outlined in Table 3.6 and are examined further in Chapter 11. Some other implications of resource use competition are considered by Tilman (1982).

Another form of competition, **direct interference**, is invoked by some ecologists to explain interspecific relationships. Direct interference (d.i.) competition occurs when plants add materials to, or alter, the environment in some way that limits other plants (but which does not involve direct contest for resources). Examples are the production of toxic compounds or deposition of deep litter which suppresses seedlings.

There is not much difference between some of the phenomena usually classed as either r.u. or d.i. competition. Litter deposition is a case in point. Both the development of canopies of leaves and the development of litter layers can cause limitation of

Table 3.6 Resource-use competition between plants in terrestrial habitats

Resource*	Resource availability in natural habitats, allogenic limitations	Plant part or process involved with competition	General competition modes; plant-caused (autogenic) limitations
light	instantly available in daytime in much greater quantities than required for photosynthesis, but no residual pool available at night; *limitation*, in already shaded situations, by cloudiness, physical barriers and low angle of solar beam	leaf canopy and green shoots which engage in photosynthesis – any plant parts which obscure these	germinating seedlings – large size of seed food store; older plants – rapid growth and versatility in displaying a leaf and green shoot canopy so as to intercept light most efficiently: efficient photosynthetic – respiratory, growth and other processes ensure competitive advantage; relative shade tolerance; limitation of other plants by pre-empting of space through shading by canopy, stems or litter affecting both quantity and quality of light received
mineral nutrients	Ca, Mg, K, P, S, etc.; for seedlings – present in seed; otherwise present as a vast reservoir of nutrients locked up in the minerals of unweathered rocks; a large temporary reservoir also occurs in undecayed organic material; released by weathering and decay to form a pool of 'available nutrients' in the clay and humic colloidal exchange complex of the soil and in the soil solution; some also derives from recycling within plants; except P, small inputs in rainfall and by import of organic material; losses by soil erosion, runoff, burning and export of organic material N present as a gaseous reservoir in virtually unlimited amounts in atmosphere, but this is unavailable until fixed by micro-organisms; a large pool present in undecayed organic material; labile and thus not retained readily in soluble forms; ammonium a more stable form than nitrate; *limitation*: undersupply where rocks are inherently infertile, or in old systems where leaching and various forms of export have reduced fertility; oversupply to toxic proportions can occur where certain animals congregate, in arid areas or places where abundance of other minerals makes supply of required nutrients difficult; limitations also during drought or in waterlogged conditions	roots and to a limited extent leaves, which can absorb nutrients from direct precipitation and drip	germinating seedlings – large size of seed food store; older plants – rapid growth and versatility in extending roots and root hair surfaces so as to absorb nutrients efficiently; presence of mycorrhizal associations (especially P) and associations with N-fixing organisms (usually in roots, sometimes leaves, stems); efficient absorptive, growth and internal recycling processes ensure competitive advantage; modifications of root environment affecting exchange complex; relative tolerance of limited nutrient supply; limitation of other plants by pre-emption of soil volumes by roots

water	present as a recurrent supply in precipitation (rainfall, fog, dew and snow); an often steadier supply at depth, in ground water, and irregularly during flooding; limitation through regular or periodic failure of precipitation, ground water; flooding and persistent waterlogging cause other problems noted above and below	roots and to a limited extent, leaves which absorb moisture	germinating seedlings – large size of seed food store; rapid growth and versatility in root extension so as to absorb water efficiently; during water shortages, effective means of storing water, diminishing water loss and efficiency of water use; relative tolerance of limited water supply; limitation of other plants by pre-emption of soil volumes by roots, interception by canopy and litter
carbon dioxide, oxygen	present as a reservoir in virtually unlimited amounts in atmosphere and, to some extent, dissolved in water; limitations of O_2 supply occur in waterlogged soil conditions and plants submerged by flooding may be oversupplied with CO_2	leaves and roots	CO_2 is probably rarely limiting except occasionally beneath canopies in very dense vegetation, and in litter and soil where seedlings are germinating; O_2 may be limiting in similar conditions and in waterlogged soils; apart from morphological–physiological efficiency in enhancing absorption and diffusion the ways in which plants can modify conditions are by providing confined canopy or litter conditions; some plants are versatile in waterlogged–dry conditions

* Some authors list living space as a resource for which plants compete. The requirement for space seems to be mainly related to the other resources which are integrated by it. Plants may also compete for certain biotic resources, e.g. pollinators and dispersal agents.

† Limitations on resource use by plants include basic effects of plant structures and physiological processes which provide bottlenecks for resource use. The most efficient plant species in these respects will be successful in competitive contest, other things being equal. The physiological processes of plants are integrated so that competitive limitation of any resource will affect the capacity of the plant to compete for other resources. Weak root growth of seedlings, because food supplies in the seed are limited, will affect subsequent shoot growth, for example.

light for subcanopy plants (Sydes & Grime 1981). The deposition of acid litter by some tree species, which prevents other species having adequate supplies of cations, is another example.

The exudation by plants of toxic compounds which harm others (or even themselves; E. O. Wilson 1969, Webb *et al.* 1967, Gliessman & Muller 1978, A. P. Smith 1979) has been suspected for a long time, but it is very difficult to demonstrate, unequivocally, that such effects occur in nature. Rice (1979) has been an enthusiastic supporter of the view that toxic chemical compounds play an important role in competitive interrelationships. The phenomenon is often termed allelopathy, but **phytotoxicity** is used as an alternative here. Other authors (e.g. Harper 1977) maintain that apparent toxic effects can be explained by direct competition for resources such as nutrients or light. Present understanding (e.g. Kaminsky 1981) is that, in some apparent cases of phytotoxicity, plant detritus encourages soil micro-organisms which produce inhibitory toxic compounds (cf. Ch. 7). Makepeace *et al.* (1985) describe an apparently clear-cut case of phytotoxicity.

Some ecologists (Boughey *et al.* 1964, Rice & Parenti 1967, Rice 1974) have reported toxic effects of plant root exudates on nitrifying bacteria. The inhibition of the bacteria modifies the nitrogen economy in favour of ammonium ion accumulation and nitrogen conservation. Antibiotic effects of plant chemical exudates appear to have complex effects on the biology of soil organisms, and there probably are various feedback effects on plants (Harborne 1982).

The plant species which usually form most of the biomass on a site make most demands on the resources of the site and control site conditions for other species are regarded as the **dominant** plants. They are strong competitors, often limiting or totally excluding other species which, potentially, could occupy a site. The dominants are a major part of the environment of many less-prominent (**subordinate**) species, which contrast with them in the ways just noted. The dominants, and other relatively large plants in forest or scrub, may provide sites for the growth of epiphytes and support for weak-stemmed plants. Their presence will also favour species which require shaded conditions or deep litter layers for their survival.

Wide ecological amplitude, in relation to exogenous factors, is not necessarily correlated with strong competitive power. Some species with wide tolerance for general habitat conditions cannot cope with strongly competitive situations (cf. Burrows 1973). Such species are restricted to disturbed or marginal habitats that are temporarily or permanently unoccupied by stronger competitors.

Enhancement

Enhancement of conditions by plants that favour other plants (the opposite of competition) is less well-documented than competition. Enhanced environmental conditions for plant species which result from the presence of other organisms are noted in Table 3.5 under the headings of commensalism, mutualism and benefaction. Self-enhancement can occur, simply through shading and amelioration of temperature and moisture conditions which assist seedling establishment. Examples of self-enhancement by the improvement of nutrient conditions occur in stands of plants with nitrogen-fixing or mycorrhizal associations. Others are noted in Chapter 12 (Miles 1981, Nye & Greenland 1960). In other instances root exudates of plants stimulate their own growth (Turkington *et al.* 1977, Aarsen *et al.* 1979, Newman & Rovira 1975). Obligate mutually positive or positive–neutral relationships shown in the table are mainly between organisms of different type or different trophic levels, or both. Commensal epiphytes are restricted to certain hosts because of specific conditions met there. The benefits to the host may be substantial (Pike 1978). The facultative relationships, shown in Table 3.5 under the heading of benefaction, are more or less accidental. Mutual benefaction between vascular plants is not very easy to demonstrate. One possible instance is the provision of microhabitat conditions by species *A* for seedling establishment of species *B*, which, as an adult provides shade for species *A*.

Autogenic and other effects due to plant behaviour

The term **autogenic** was first used by Tansley (1929) for vegetation change which is caused by the modifying effects of the plants on habitat and the consequent impact of the modified habitat on the plant populations. Nowadays it is used for any effect of plants on environment which, in turn, influences plants (i.e. operates through feedback). Some autogenic phenomena are short-term or reversible, or both (competition, enhancement effects; shading of space beneath leaf canopies affecting light intensity and quality, temperature, air humidity and soil moisture). Others are longer-term and virtually irreversible, e.g. the production of acid compounds in leaf litter which cause weathering of minerals, thus influencing soil development; or filling of a lake with organic sediment.

Competition causes stress. In cases of autotoxicity it is 'suicidal', otherwise it can be 'murderous'. The monocarpic habit of dying after flowering is also 'suicidal'. There are also some autogenic disturbance effects ('self-mutilation', 'suicide' or 'murder'). Some tree species (e.g. *Salix fragilis* and *Fraxinus excelsior*), have weak branches which snap easily. Trees or shrubs with shallow root systems (e.g. *Fremontodendron californica*) collapse relatively easily. Dry-climate trees and shrubs in Australia, many gymnosperm trees and the Mediterranean-type scrub in various parts of the world have foliage, wood and litter that are rich in volatile oils and resins which make them extremely vulnerable to fire. Sometimes parasitic plants, e.g. the dwarf mistletoes of gymnosperm tree species in the Rocky Mountains, kill their hosts (and commit 'suicide' at the same time). Some tropical trees are made more prone to windthrow, and topple to their death because of the heavy load of epiphytes and vines which cover them.

Plants and their environment are so closely coupled that it is sometimes difficult to distinguish cause from effect. This applies particularly to nutrient–plant interactions. Thus, the distinction between allogenic and autogenic causes in plant ecology is not always clear. It is also difficult to decide whether the invasion of a stand of vegetation and the displacement of other species by a plant species that was not previously present should be regarded as allogenic or autogenic.

93

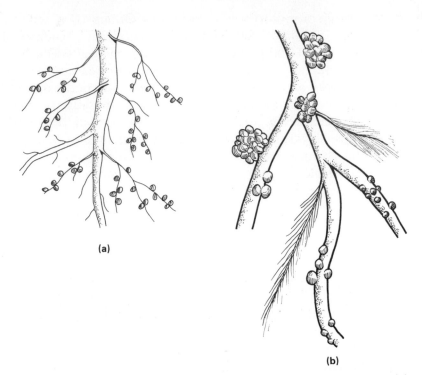

(a)

(b)

Figure 3.8 Associations of nitrogen-fixing organisms with plants. (a) Root nodules containing the bacterium *Rhizobium* on a herbaceous legume, *Trifolium pratense* (red clover). (b) Root nodules containing the actinobacterium *Frankia* on roots of the tree *Alnus glutinosa* (alder).

Nitrogen fixation

Nitrogen fixation is a process which is about as important for life on Earth as photosynthesis is (Subba Rao 1980). Nitrogen is an essential nutrient element, needed by all plants in relatively large quantities. It is a major component of proteins and nucleic acids. None is available from the rocks. Although vast amounts are present in gaseous form in the atmosphere (about 80 per cent of the atmosphere is composed of it), this gas is inert and little is available to higher plants growing on new land surfaces, unless it is converted into nitrate or ammonium by the activity of certain micro-organisms. The nitrogen-fixing micro-organisms are free-living bacteria and cyanobacteria (blue-green algae); or bacteria, cyanobacteria and actino-bacteria (actinomycetes) in symbiotic association with higher plants; or cyanobacteria present in lichens and plants from a few other groups (Fig. 3.8) (W. D. P Stewart 1967, Postgate 1980). Table 3.7 lists the known nitrogen-fixing genera of micro-organisms and the symbiotic associations. Free-living micro-organisms are the most important sources of nitrogen supply to plants in some aquatic and terrestrial situations where symbiotic associations are lacking. They can fix nitrogen at rates of 10–50 kg ha^{-1} year^{-1} (A. W. Moore 1966).

Nitrogen-fixing symbioses take several forms. Some are rather loose associations of bacteria with root (rhizosphere) and leaf (phyllosphere) surfaces of higher plants with production of varying amounts of surplus nitrogen that might be used by the

94

Table 3.7 Nitrogen-fixing moneran genera and symbiotic associations with plants and fungi.

Bacteria

I. free-living
 (a) aerobic
 Azomonas, Azotobacter, Azotococcus, Beijerinckia, Derxia, Methanosinus, etc.
 (b) microaerobic
 Azospirillum, Rhizobia (some), *Spirillum, Thiobacillus, Xanthobacter*
 (c) anaerobic
 Chlorobium, Chromatium, Clostridium, Desulfotomaculum, Desulfovibrio, Ectothiospira, Propionobacterium, Thiopedia, Rhodopseudomonas, Rhodospirillum
 (facultative nitrogen-fixers include:
 Bacillus, Citrobacter, Enterobacter, Klebsiella)

II. symbiotic
 (a) close associations (with angiosperms)
 Rhizobium in legume root nodules (most Papilionaceae, Mimosaceae are nodulated; about 30% of Caesalpiniaceae are nodulated)
 Rhizobium in root nodules of **Parasponia** (= **Trema**), a tropical small tree and possibly in root nodules of a grass, **Calamogrostis**
 Klebsiella in leaf glands of **Psychotria**.
 (b) looser associations (with angiosperms)
 Azotobacter, Beijerinckia, Spirillum on rhizosphere of **Digitaria, Paspalum, Saccharum, Sorghum** and other tropical grasses, weeds; also rice (**Oryza**), mangroves (**Avicennia, Rhizophora**), water hyacinth (**Eichhornia**)
 Azotobacter, Beijerinckia on leaves *may* supply nitrogen

Actinobacteria (actinomycetes)

symbiotic
close associations (with angiosperms and fern) – numbers of known nodulated species shown in parentheses
Frankia in root nodules of a fern, **Comptonia** (1) and angiosperms **Alnus** (32), **Casuarina** (24), **Ceanothus** (31), **Cercocarpus** (4), **Colletia** (1), **Coriaria** (13), **Cowania** (1), **Datisca** (2), **Discaria** (2), **Dryas** (3), **Eleagnus** (16), **Hippophae** (1), **Myrica** (26), **Purshia** (2), **Rubus** (1), **Shepherdia** (2), **Trevoa** (1)

Cyanobacteria (blue-green algae)

I. free-living
 (a) aerobic
 Anabaena, Anabaenopsis, Aulosira, Calothrix, Chlorogloea, Cylindrospermum, Dichothrix, Fischerella, Gloeocapsa (some strains), *Hapalosiphon, Mastigocladus, Nostoc, Scytonema, Stigonema, Tolypothrix, Westiellopsis*
 all heterocystous species (the order Nostocales, except Oscillatoriaceae and the order Stigonematales) are presumed to fix nitrogen
 (b) microaerobic
 Gloeocapsa, Lyngbya, Oscillatoria, Phormidium, Plectonema, Trichodesmium

II. symbiotic
 close associations
 Nostoc in **Geosiphon** (a Phycomycete fungus)
 Calothrix in **Enteromorpha** (green alga)
 (*Richelia* (= **Microchaete**) in **Rhizoselenia** (diatom))
 Calothrix, Dichothrix, Nostoc[a], *Scytonema*[b], *Stigonema*
 in **Collema**[a], **Heppia**[b], **Lobaria**[a], **Nephroma, Peltigera**[a], **Solorina, Stereocaulon**, etc. (lichens)
 about 5% of lichen species are known to have a nitrogen-fixing association
 Nostoc in **Anthoceros, Dendroceros** (hornworts) and in **Blasia, Cavicularia** (liverworts)
 Hapalosiphon in **Sphagnum** (moss)
 Anabaena in **Azolla** fronds (aquatic fern)
 Nostoc, Anabaena in root nodules of **Bowenia, Ceratozamia, Cycas, Dioon, Encephalartos, Macrozamia, Stangeria, Zamia** (cycads)
 Nostoc in stem glands of **Gunnera** and rarely in a species of **Trifolium** (angiosperms)

'host' (which secretes compounds useful to the micro-organisms). Otherwise cyano-bacteria live inside the thalli of about 5 per cent of lichens or inside the leaves, stems or specialized root nodules of a few species of higher plants. Root nodules containing bacteria (*Rhizobium* spp.) are associated with the legume families of flowering plants and a few species from other families. Actinobacteria (*Frankia* spp.) are found in the root nodules of many other groups of flowering plants. The endophytic micro-organisms are somehow induced to fix much more nitrogen than they, themselves, need; the surplus is released to the host. Quantities of 50–225 kg ha^{-1} year^{-1} may be fixed by the symbiotic microbes.

Once the host plants incorporate nitrogen into their own systems it can be made available to other plants by the death and decay of parts of the original beneficiary. Nitrogen is quite labile and some will be lost back to the atmosphere, but most will be recycled (see Fig. 3.9, the nitrogen cycle). Some of the saprophytic bacteria which participate in decay processes can themselves fix atmospheric nitrogen.

Sources of nitrogen which plants can use include relatively small amounts of nitrate, formed in the upper atmosphere by lightning discharge, or aerosols of unknown origin. Industrial effluent may contribute to this. This nitrogen descends in the rain. Quantities of 1–10 kg ha^{-1} year^{-1} have been measured (Egner & Eriksson 1955) and these will be important for plant nutrition wherever nitrogen is otherwise in short supply.

Other sources of nitrogen (and other nutrients) are secondary – from plant and animal detritus, e.g. sea-bird excreta, plant and animal debris washed up on shore as jetsam, or blown over land surfaces by the wind. On land humus forms a vast nutrient reservoir. Fossil peat, wood or coal may also be used as nitrogen sources by plants in some circumstances.

Mycorrhizas and plant nutrition

Mycorrhizas are symbiotic associations of fungi with plant roots or rhizoids (Fig. 3.10) (Trappe & Fogel 1977, Mosse 1978). Ectomycorrhizas are found sheathing the roots of many tree species (Table 3.8). Swollen roots containing endomycorrhizas occur in some liverworts, ferns, some gymnosperms and probably about 80 per cent of angiosperms, especially the woody ones. They are found also in many herbaceous species, notably in grasses. Root nodule mycorrhizas occur in the Podocarpaceae (podocarp) family of gymnosperms, widespread in the Southern Hemisphere, and also in the Tropics, subtropics and as far north as Japan and Ethiopia. They also occur in the Southern Hemisphere Araucariaceae family of gymnosperms. Ericoid mycorrhizas are confined to the heath families, the ericas which are widespread on generally very infertile soils, mainly in the Northern Hemisphere, South America, and East and South Africa and the epacrids on similar soils, mainly in the Southern Hemisphere. Orchid-type mycorrhizas are confined to the orchid family.

Mycorrhizas are of very considerable importance to their host plants; they assist in the uptake by these plants of phosphorus in organic and inorganic forms (cf. Fig. 3.11, the phosphorus cycle). This can be especially important in soils where chemi-cally bound (occluded), forms (cf. T. W. Walker & Syers 1976) otherwise make

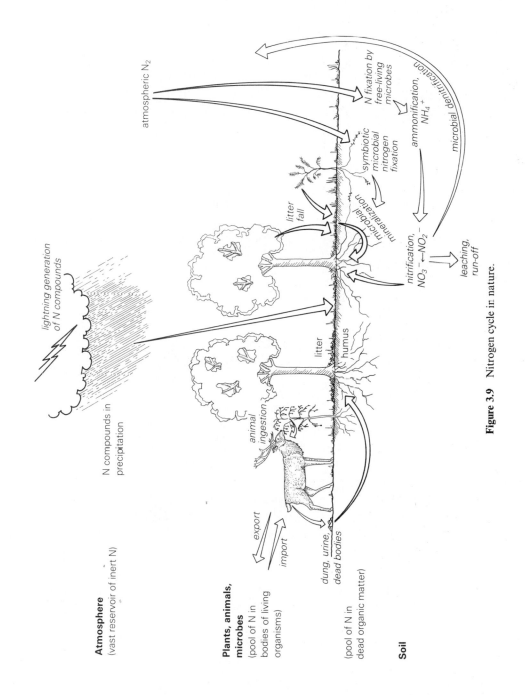

Figure 3.9 Nitrogen cycle in nature.

Atmosphere
(vast reservoir of inert N)

lightning generation
of N compounds

N compounds in
precipitation

**Plants, animals,
microbes**
(pool of N in
bodies of living
organisms)

animal
ingestion

export

import

(pool of N in
dead organic matter)

dung, urine,
dead bodies

Soil

atmospheric N_2

litter
fall

litter

humus

N fixation by
free-living
microbes

symbiotic
microbial
nitrogen
fixation

microbial
mineralization

ammonification,
NH_4^+

microbial denitrification

nitrification,
$NO_3^- \leftarrow NO_2^-$

leaching,
run-off

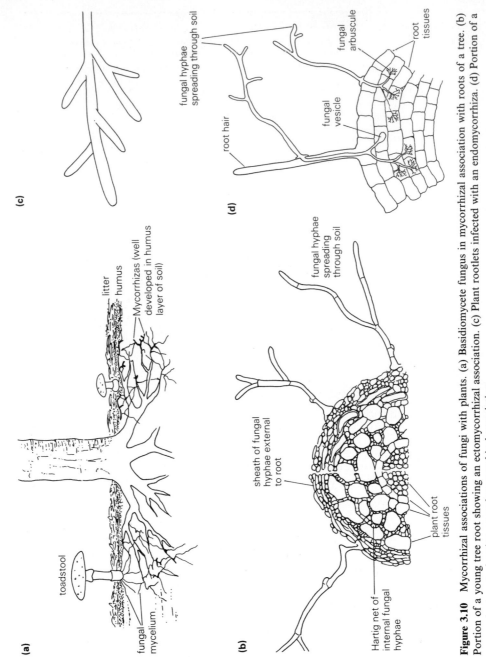

Figure 3.10 Mycorrhizal associations of fungi with plants. (a) Basidiomycete fungus in mycorrhizal association with roots of a tree. (b) Portion of a young tree root showing an ectomycorrhizal association. (c) Plant rootlets infected with an endomycorrhiza. (d) Portion of a young plant root, showing an endomycorrhizal association.

(a)

toadstool

fungal
mycelium

litter

humus

Mycorrhizas (well
developed in humus
layer of soil)

(b)

fungal hyphae
spreading through soil

sheath of fungal
hyphae external
to root

Hartig net of
internal fungal
hyphae

plant root
tissues

(c)

(d)

fungal hyphae
spreading through soil

root hair

fungal
vesicle

fungal
arbuscule

root
tissues

Table 3.8 Mycorrhizal symbiotic associations.

---

1. Ectomycorrhizae: sheath the roots of many woody species from the plant families Fagaceae (beech), Betulaceae (birch), Pinaceae (pine), Cupressaceae (cypress), Tiliaceae* (lime), Salicaceae* (willow), Juglandaceae* (walnut), Myrtaceae* (myrtle), Ulmaceae (elm), Rosaceae (rose), Sapindaceae (sapinda), Accraceae (maple), Dipterocarpaceae (dipterocarp) and the legume family Caesalpinaccac; fungal genera which include mycorrhiza-formers are:

Ascomycetes
Balsamea, Cenococcum, Elaphomyces, Geopora, Helvella, Tuber, etc.

Basidiomycetes
Amanita, Boletus, Cantharellus, Clavaria, Clitocybe, Cortinarius, Entoloma, Hysterangium, Inocybe, Lactarius, Leccinum, Melanogaster, Paxillus, Ramaria, Rhizopogon, Russula, Scleroderma, Suillus, Thelephora, Tricholoma, etc.

2. Endomycorrhizae: develop internally, forming swollen roots on some ferns, some gymnosperms, and a very wide range (probably about 80%) of angiosperms; more especially found in woody species, but also present in many herbaceous species, notably in grasses; also found in some liverworts

septate fungi
occur in heath families Ericaceae (e.g. **Gaultheria**), Epacridaceae (e.g. **Dracophyllum**), (fungus – *Pezizella*, a Discomycete); also occur in Orchidaceae (fungus – '*Rhizoctonia*' – including genera *Ceratobasidium, Sebacina*) and Gentianaceae

non-septate fungi
very widespread in bryophytes (liverworts), pteridophytes (ferns), gymnosperms and many angiosperm families; the fungi are Phycomycetes, *Glomus* and related genera in the Endogonaceae, which form vesicular–arbuscular mycorrhizas; in the gymnosperm families Podocarpaceae (**Podocarpus, Dacrydium, Phyllocladus**, etc.) and Araucariaceae (**Araucaria, Agathis**) there are distinctive mycorrhizal root nodules

---

* Also infected by endomycorrhizas, sometimes both on the same plant. There are also some ecto-endomycorrhizas, with sheathing and intracellular hyphae in Ericaceae (heaths) and related families. They are formed by some of the Basidiomycete fungi listed above.

phosphorus uptake difficult or impossible and in localities where competition for phosphorus is severe (Mosse 1978, Malajczuk & Lamont 1981, Janos 1983). Mycorrhizas have also been shown in some instances to improve their host's uptake of zinc, sulphur, potassium and nitrogen mineralized from organic matter. They have been generally implicated, also, with transfers of carbon compounds and with improved water uptake (Reid & Woods 1969, Trappe & Fogel 1977, K. M. Cooper 1982) as well as deterrence of pathogenic soil fungi and root nematode infections (K. M. Cooper 1982). The fungal components are not necessarily confined to areas immediately adjacent to their host plant; hyphae may anastomose through the soil some distance (up to several metres) away. Through this property, as well as the much increased surface absorptive area, materials can be conducted back to the plant roots from large volumes of soil.

The formation of mycorrhizas is a prerequisite for nodulation of many legume species by nitrogen-fixing bacteria (Mosse 1978, Janos 1980). Orchids have tiny seeds with little or no stored food. Infection with mycorrhizal fungal hyphae, dispersed with the seeds, is essential for the establishment of their seedlings. However, mycorrhizas have sometimes been implicated with inhibition of nitrification (Gadgil & Gadgil 1975) and also with the inhibition of root growth through the production of toxic compounds (D. H. Ashton & Willis 1982).

The Southern Hemisphere proteas (Proteaceae) and restiads (Restionaceae) and the widespread rushes (Juncaceae) and sedges (Cyperaceae) have specialized, non-mycorrhizal roots which assist their nutrition in impoverished soils (Lamont 1981). Non-mycorrhizal species occur in various kinds of vegetation. They seem to have specialized on the production of organic acids which can liberate phosphorus from bound forms (Janos 1983).

Pathogens

Among the harmful organisms affecting plants, and dispersed through the air, through the soil, or sometimes by animal vectors, are many fungi and bacteria (Agrios 1969, Burdon 1982). They harm plants with enzymes and toxins, causing reduced growth, clogging of vascular tissues, necrosis and destruction of cellular tissue. They are usually manifest as rots, wilting and various kinds of necrotic area (as well as by production of sporangiophores). Entry to plant leaves, stems or roots is achieved through natural openings (e.g. stomata and lenticels), as well as through damaged areas or by animal vectors. Some of the fungi are rather indiscriminate pathogens (e.g. Phycomycetes such as *Pythium*, which attacks seeds and seedlings, or *Phytophthora* and the Basidiomycete *Armillaria*, which can kill whole adult plants). Large numbers of rot fungi from the Ascomycete and Basidiomycete groups affect woody tissue, and shorten the life of woody plants. The longevity of trees often depends as much on their resistance to breakage and resistance to insect damage, which allow rot fungi to enter the plants, as on their inherent resistance to the fungi. Contrasting with these general pathogens are many host-specific types, including the often very virulent rust and smut fungi.

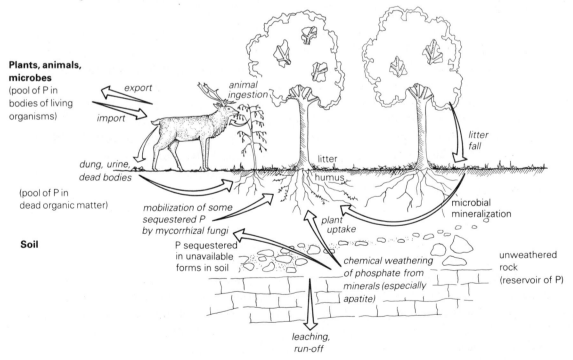

Figure 3.11 Phosphorus cycle in nature.

Table 3.9 Plant defence mechanisms against pathogens.

1   structural defences

   (a)   external barriers

      waxy covering to leaves
      thick, tough cuticle
      thick-walled epidermal cells
      structure and function of natural openings preventing penetration
      sloughing of bark from stems, roots
      attraction of harmless phyllosphere and rhizosphere organisms antagonistic to pathogens

   (b)   internal barriers

      thick, tough cell walls and dense tissue (fibres and stone cells)

   (c)   structures produced in response to pathogenic invasion

      tissues such as cork layers which close off infected areas and prevent further penetration; tyloses; cell-wall swellings, lignitubers; gum deposition; abscission layers and shedding of infected parts

2   biochemical defences

   (a)   pre-formed barriers

      fungal and bacterial inhibiting substances released by plant (e.g. HCN from glucosides)
      intracellular toxic compounds (e.g. phenolics)
      defence through deficiency of substances essential to the pathogen
      defence through denaturing of pathogen enzymes or toxins

   (b)   resistance produced in response to pathogenic attack, injury

      production of inhibitors in tissues surrounding the site of attack (phytoalexins and other kinds of compound)
      induced protein synthesis leading to the development of resistant or immune layer around infection sites
      formation of substrates which resist enzymes of pathogen
      defence through hypersensitivity and rapid tissue death in infected area, thus containing spread of the disease

Viruses are submicroscopic pathogens affecting plants and transferred almost solely through animal vectors, such as aphids. They reproduce only in living tissues, and damage the photosynthetic, transport and general metabolic mechanisms. They can interfere with plant growth substances. Their presence is detected often by chlorotic leaves and necrotic patches. Viruses are often host-specific.

Plants possess a wide range of defence mechanisms against pathogens (Table 3.9). As well as the mechanical and chemical first lines of defence, second-line defences, both mechanical and chemical, are created in response to a pathogenic attack.

Saprophages

Many invertebrate animals (Fig. 3.12) are detritivores which, in feeding on plant and animal remains, begin the process of decay by mechanical comminution and enzymic digestion (Dickinson & Pugh 1974, Mason 1976). The fungi and bacteria which cause the decay of detritus are very important to plants. Mycorrhizal fungi are among them. Some plants (saprophytes) (Table 3.10), which lack chlorophyll, are totally dependent on mycorrhizal connections with decaying plant litter. Some saprophytic

Table 3.10 Saprophytic flowering plant groups* (after J. Hutchinson 1959).

Family (no. of saprophytic genera)	Genera		Distribution of saprophytic representatives
monocotyledonous angiosperms			
Orchidaceae (approx.15)	*Corallorhiza*	coral 'root' orchid	northern temperate to India,
	Neottia	bird's nest orchis	SE Asia, Indonesia, New
	Gastrodia		Guinea, Australasia, South
	Galeola		America
	Hexalectris		
Burmanniaceae (approx.8)	*Burmannia*		Tropics
	Gymnosiphon		
Thismiaceae† (approx.10)	*Thismia*		Tropics to south temperate
	Sarcosiphon		zone
Corsiaceae† (2)	*Corsia*		New Guinea, Chile
Triuridaceae† (4)	*Triuris*		Tropics
	Sciaphila		
Petrosaviaceae† (1)	*Petrosavia*		Malaysia, Sabah, China, Japan
dicotyledonous angiosperms			
Monotropaceae† (approx.12)	*Monotropa*	Indian pipe	north temperate zone
	Pterospora	pine drops	
	Cryptophila		
Gentianaceae (approx.6)	*Voyria*		South America
	Bartonia		
	Cotylanthera		
Polygalaceae (1)	*Salomonia*		SE Asia, New Guinea, North America

  * All are perennial herbs. At least 50 species of orchids from about 12 genera are saprophytic. Some are entirely cryptic, remaining beneath leaf litter. There are difficulties in deciding which plants are truly saprophytic. In a sense most plants are saprophytic to some degree, because they obtain nutrients from dead plant and animal detritus. Those listed, having very little or no chlorophyll, have been thought to be obligate or near-obligate saprophytes. They occur in environments with abundant organic matter in the soil or above it (rotting logs, tree stumps, etc.). The dicotyledonous *Salomonia* and Monotropaceae are regarded by some authors as parasites, on the grounds that mycorrhizal connections between 'saprophyte' roots and other 'host' species carry nutrient material back to the 'saprophyte'. As all saprophytes have strong mycorrhizal connections, the same might apply to the members of the Gentianaceae and the monocotyledonous groups which are listed. In any case this type of 'parasitism' differs from that where there are haustorial connections between parasite and host, or where the parasite invades the host tissues. At least to some degree the angiosperms parasitize the mycorrhizal fungus.
  † Family exclusively saprophytic.

microbes are closely associated with plants, e.g. in the root zone. Others proliferating through the detritus, above and below the soil surface, affect plants less directly but no less importantly. It is their activity which produces humus (important for ion exchange) and also reduces humic material to its ultimate molecular or elemental form, thus releasing nutrients needed by plants (Fig. 3.13).

Plant parasites and carnivorous plants

Plants which make their living by invading the tissues of other plants, and extracting their nutriment directly, belong to a limited number of families (Table 3.11). Parasitism is an obligate relationship for the parasite, but there are different degrees of dependence of the parasite on the host. They range from total dependence, where the

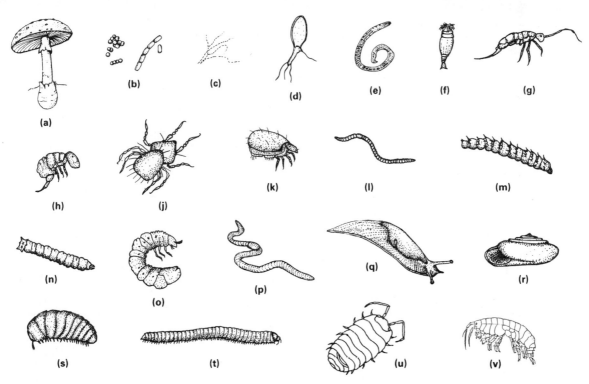

Figure 3.12 Representatives of the main groups of detritivore (decomposer) organisms in soils (a) Fungi. (b) Bacteria. (c) Actinobacteria. (d) Shelled amoeba. (e) Nematodes. (f) Rotifers. (g, h) Collembola. (j, k) Oribatid mites. (l) Enchytraeid worms. (m, n) Dipteran larvae (crane flies). (o) Scarab beetle larvae. (p) Earthworms. (q, r) Molluscs (slugs, snails). (s, t) Millipedes. (u, v) Crustacea (slaters, hoppers). After Packham & Harding (1982).

parasites have no chlorophyll, to hemiparasitism, where the parasites photosynthe-size to a greater or lesser degree (Kuijt 1969). Some parasites are host-specific, but most have more general host-preferences.

A few plant groups, on nitrogen- and phosphorus-deficient soils, although retaining their photosynthetic capacity, have developed means for trapping small animals (Table 3.12). These are digested by the plant enzymes (possibly assisted by bacteria), thus supplementing their nutrient supplies.

Herbivores

In all ecosystems animals interact with plants in many ways. Here the influence of herbivory is examined. **Herbivory** is a term used broadly to mean the removal from live plants and the consumption of any kind of tissue (leaves, stems, flowers, fruit, seeds and roots) by any kind of animal. The definition, thus, encompasses not only leaf-, stem- and root-eating, but also more specialized types of feeding such as seed predation, wood-boring, leaf-mining, the sucking of plant juices by scale insects or aphids, gall formation, etc. Reference will mainly be made here, however, to defoli-ation (grazing). Nectar-taking, pollen and ripe fruit removal, not often harmful, are usually not included within the term.

Plant parts (including seeds) are eaten by a very wide range of animals. Feeding

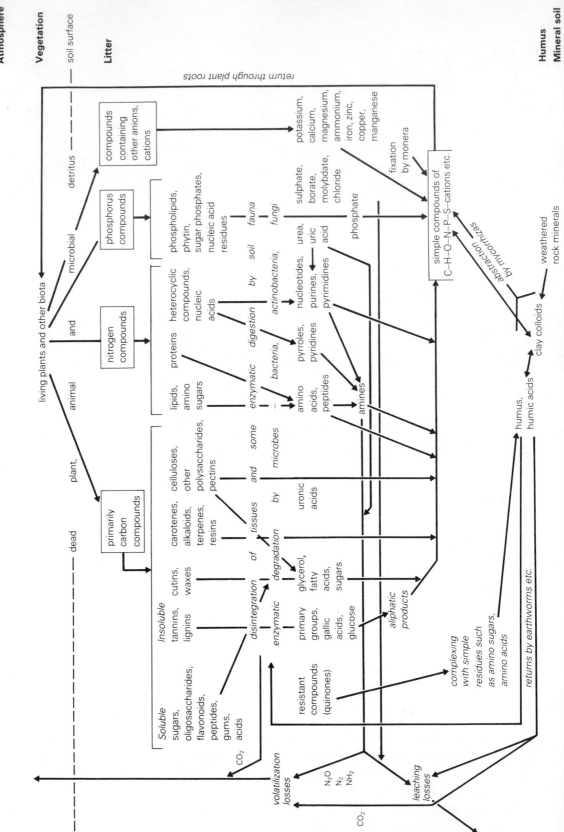

Table 3.11 Parasitic seed plant groups* (after Kuijt 1969).

Plant family (no. of parasitic genera)	Representative genera		Types of plant life form	Degree of parasitism†	Host organs attacked‡	Distribution
Loranthaceae§ (approx.50)	*Ileostylus* *Loranthus* *Nuytsia* *Psittacanthus* *Struthanthus*	mistletoe	trees, shrubs, a few vines	H	mostly s, a few r	some temperate mostly tropical, subtropical
Viscaceae§ (approx.15)	*Arceuthobium* *Phoradendron* *Viscum*	dwarf mistletoe mistletoe	shrubs	H	s	temperate–subtropical tropical
Santalaceae (approx.17)	*Comandra* *Geocaulon* *Exocarpos* *Santalum*	sandalwood	trees, shrubs, herbs	H	r	temperate–subtropical, tropical
Olacaceae (approx.4)	*Olax* *Ximenia*		trees, shrubs, vines	H	r	tropical–subtropical
Myzodendraceae (1)	*Myzodendron*		shrubs	H	s	southern South America
Orobanchaceae§ (approx.10)	*Orobanche* *Epifagus* *Conophilis*	broomrape beech drops squaw root	herbs	F	r	cosmopolitan
Scrophulariaceae (approx.26)	*Bartsia* *Castilleja* *Euphrasia* *Pedicularis* *Striga*	Indian paintbrush eyebright lousewort devil weed	herbs	H (a few F or with weakly developed photosynthetic tissue)	r	mostly temperate, a few subtropical–tropical
Rafflesiaceae§ (approx.8)	*Cytinus* *Rafflesia*		'herbs' (endophyte tissues throughout host)	F	r,s	tropical, a few subtropical, temperate
Hydnoraceae§ (2)	*Hydnora*		'herbs' (no distinct shoot system above ground)	F	r	Africa, South America

Table 3.11 (*cont.*)

Plant family (no. of parasitic genera)	Representative genera	Types of plant life form	Degree of parasitism†	Host organs attacked‡	Distribution
Balanophoraceae§ (approx.20)	*Balanophora Cynomorium Dactylanthus Sarcophyte*	'herbs' (no distinct shoot system above ground)	F	r	tropical, a few subtropical temperate
Convolvulaceae (1)	*Cuscuta* dodder	annual vine	almost completely F, weakly developed photosynthetic tissue	s	temperate to tropical
Lauraceae (1)	*Cassytha*	vine	H	s	temperate to tropical
Lennoaceae§ (3)	*Lennoa*	'herbs' (no distinct shoot system above ground)	F	r	North and South American deserts
Krameriaceae§ (1)	*Krameria*	shrub	H	r	North, Central and South America

* All are dicotyledonous angiosperms. They occur from the subpolar regions to the tropics, but most are subtropical to tropical. Few are strictly confined to one host. One gymnosperm root hemiparasite is known: *Parasitaxus*, a podocarp from New Caledonia.
† F, full parasites, lacking photosynthetic tissue; H, hemiparasites, with photosynthetic tissue.
‡ s, shoot, r, root.
§ Family exclusively parasitic.

Table 3.12 Carnivorous flowering plant groups* (after Lloyd 1942).

Family	Genera		Distribution	Kind of trap
Sarraceniaceae	*Sarracenia*	pitcher plants	E. North America	pitcher-like pitfalls – highly modified leaves, with fluid in the bottom into which prey falls
	Hiliamphora	pitcher plants	South America	as for *Sarracenia*
	Darlingtonia	pitcher plants	W. North America	as for *Sarracenia*
Nepenthaceae	*Nepenthes*	pitcher vines	Madagascar, Ceylon, SE Asia, Indonesia, New Guinea, Australia	as for *Sarracenia*, but 'pitchers' are on elongated, pendulous, tendril-like organs
Droseraceae	*Drosera*	sundews	cosmopolitan	prey adheres to sticky, glandular hairs on leaves; hairs and leaves curl over prey
	Dionaea	Venus fly trap	SE North America	leaf 'spring' trap with marginal teeth; the trap snaps shut when prey touches sensitive hairs
	Aldrovandra		Europe, Africa, India, Japan, Australia	like *Dionaea*, but aquatic, with whorls of trap leaves
	Drosophyllum	fly trap	Iberia, North Africa	sticky glandular hairs and digestive glands on leaves; hairs and leaves remain passive
Byblidaceae	*Byblis*		Australia	as for *Drosophyllum*
Cephalotaceae	*Cephalotus*		Australia	like *Sarracenia*, but 'pitcher' more elaborate
Lentibulariaceae	*Pinguicula*	butterwort	Northern Hemisphere	sticky leaves to which prey adheres; the leaves then curl over the prey
	Utricularia	bladderworts	cosmopolitan	rootless, aquatic; small bladders on stem have trapdoors which spring open when prey touches sensitive hairs; prey is carried in by rush of water
	Biovularia	bladderworts	Cuba, South America	as for *Utricularia*
	Polypompholyx	bladderworts	Australia	as for *Utricularia*
	Glenlisea		Africa, South America	aquatic; highly specialized leaves have spiral organ with deflexed stiff hairs which, once prey enters, force it to continue deep into trap

* The members of the six families (about 450 species), are exclusively carnivorous, mainly on small arthropods. They are all perennial, dicotyledonous angiosperms. Most are herbaceous, but some are woody, or with woody rootstock. Most grow in acidic, low-fertility freshwater wetlands; a few grow on dry sites.

modes are also extremely varied (Table 3.13). Some animal species are very specific in their requirements (e.g. restricted to one host species or one type of organ). Others are relatively indiscriminate (for host species or types of organ). Some plant species are palatable to herbivores, others are not (Wratten *et al.* 1981).

Animals such as locusts can devastate vast areas of vegetation, but often herbivory of adult plants is not so damaging and it is, in some situations, distinctly beneficial (Mattson & Addy 1975). Most plants have powers of regeneration of their parts which allow them to overcome the damaging effects of light-to-moderate or infrequent grazing (Mooney & Gulman 1982). They may even survive complete

Table 3.13 Land animals that feed on living plants* (Imms 1957, Walker 1964).

		Family	Representatives†	Distribution
I.	*Vertebrates*			
A.	Mammals			
	Marsupalia	Didelphidae	American opossums 1,6	Americas
		Phalangeridae	Australian phalangers, koala, etc. 1,5,6	Australia, New Guinea
		Phascolomidae	wombats 1,4	Australia
		Macropodidae	kangaroos, wallabies, etc. 1,2	Australia, New Guinea
	Carnivora	Procyonidae	coatis, raccoons X 1	Americas
		Mustelidae	badgers X	Northern Hemisphere, South America
		Ursidae	bears X	Northern Hemisphere, South America
	Perissodactyla	Rhinocerotidae	rhinoceros 1,2	Africa, India
		Tapiridae	tapirs 1,2	South America, SE Asia, Indonesia
		Equidae	horses, zebras 1,2	Africa, Northern Hemisphere, except America
	Artiodactyla	Suidae	pigs X 4,6	Africa, Northern Hemisphere, except America
		Tayassuidae	peccaries X	Americas
		Hippopotamidae	hippopotamus 1 (emerge from water to feed)	Africa
		Camelidae	camels, llamas 1,2	Central Asia, Arabia, etc., South America
		Cervidae	deer, moose 1,2	Northern Hemisphere, South America
		Giraffidae	giraffe 1,2	Africa
		Bovidae	cattle, bison, buffalo 1,2	Northern Hemisphere, Africa
			musk ox, sheep, goats 1,2	Northern Hemisphere
			chamois 1,2	Europe
			wildebeest, hartebeest, roan, sable, waterbuck, impala, springbok, gazelles, etc. 1,2	Africa
	Proboscidea	Elephantidae	elephants 1,2	Africa, South Asia
	Hyracoidea	Procaviidae	hyraxes X 4	Africa, Middle East
	Lagomorpha	Leporidae	rabbits, hares 1,2	Northern Hemisphere, Africa
		Ochotonidae	pikas 1	North Asia, North America
	Rodentia	Sciuridae	squirrels, marmots, prairie dogs, chipmunks; some 1,3,4,6	Northern Hemisphere, South America, Africa
		Geomyidae	pocket gophers 1,6	North America
		Heteromyidae	pocket mice, kangaroo rats 1,6	North America
		Castoridae	beavers 3	Northern Hemisphere
		Muridae	Old World rats, mice 1	Northern Hemisphere, except America
		Microtidae	voles, lemmings 1	Northern Hemisphere
		Cricetidae	hamsters, gerbils, American mice, rats, 1,6	Northern Hemisphere, Africa
		Dipodidae	jerboas X 1	Africa, Asia

Table 3.13 (*cont.*)

		Hystricidae	Old World porcupines 2,3	Eurasia
		Erethizontidae	American porcupines 2,3	North America, South America
		Myocastoridae	coypu 1,4 (emerge from water to feed)	South America
		Chinchillidae	chinchillas 1	South America
		Dasyproctidae	agoutis 1,2	South, Central America
		Hydrochoeridae	capybara 1 (emerge from water to feed)	South America
		Caviidae	guinea-pigs 1	South America
		Various other small groups of herbivorous rodents.		
	Edentata	Bradypodidae	sloths X 1	South America
	Primata	Cebidae	American monkeys, some X, some 1,6	South, Central America
		Cercopithecidae	Old World monkeys, baboons, some X, some 1,6	Northern Hemisphere, Africa
		Hylobatidae	gibbons X	South Asia, Indonesia
		Pongidae	gorilla, orang-utan, chimpanzee X	Africa, south Asia, Sabah, Indonesia
B.	Birds			
	Passeriformes	Fringillidae	finches, grosbeaks, etc. 5,6	Northern Hemisphere, South America
		Ploceidae	weaver birds X, 1,6	Europe, Africa, Asia
		Sittidae	nuthatches 6	Europe, Asia, Australia, North America
		Paridae	tits 6	Northern Hemisphere
		Corvidae	jays, rooks, crows X, 6	cosmopolitan
		Sturnidae	starlings, mynahs X 6	Northern Hemisphere, Africa
	Gruiformes	Otidae	bustards 6	Africa, Europe, Asia
		Rallidae	rails 1,4,6,	cosmopolitan
	Columbiformes	Columbidae	pigeons X 1,6	cosmopolitan
	Galliformes	Phasianidae	fowls, quail, grouse, partridges, turkeys 1,2,6,	Northern Hemisphere, Africa, Australia
		Megapodidae	brush turkeys 4,6	Australia, New Guinea
	Anseriformes	Anatidae	ducks, geese, swans 1,4,6	cosmopolitan
	Psittaciformes	Psittacidae	parrots X 1,5,6	Australasia, North and South America
	Struthioni-formes	Struthionidae and related families	ostrich, emu, etc. X 1,2,6	Southern Hemisphere, Africa
II.	Invertebrates			
A.	Arthropoda			
(i)	*Insecta* (L = larvae)			
	Collembola	Isotomidae	springtails 1 (seedlings)	cosmopolitan
	Orthoptera	Gryllidae	crickets 1	cosmopolitan
		Acrididae	locusts, short-horn grasshoppers 1	cosmopolitan
		Tettigonidae	long-horn grasshoppers 1	cosmopolitan
	Phasmida	Phasmidae	stick insects 1	tropics, subtropics, warm temperate
		Phyllidae	leaf insects 1	tropics
	Isoptera	Termitidae	termites 3,4	tropics, subtropics, warm temperate

Table 3.13 *(cont.)*

Hemiptera	Cicadidae	cicadas 1,2,4(L)	cosmopolitan
	Membracidae	tree hoppers 1,2	"
	Jassidae	jassid bugs	"
	Cercopidae	cuckoo spit bugs 1	"
	Cixiidae	cixid bugs 1,4	"
	Psyllidae	jumping plant lice 1,2,5	"
	Aleurodidae	white flies 1	"
	Aphididae	aphids 1,2,5	"
	Margarodidae	mealy bugs 1,2,3,4	"
	Coccididae	scales 1,2,3	"
	Pseudococcididae	scales 1,2,3	"
	Miridae	capsid bugs 1	"
	Lygaeidae	lygaeid bugs 1,6	"
	Coreidae	squash bugs 1	"
	Tingidae	lace bugs 1,5	"
	Pentatomidae and many other families	pentatomid bugs 1	"
Thysanoptera	Thripidae	thrips 5	cosmopolitan
Lepidoptera	Hepialidae	swift moths 3,4(L)	"
	Gelechidae	grain moths 1,2,4,5,6,(L)	"
	Cossidae	carpenter moths 3 (L)	"
	Psychidae	bag moths 1,2(L)	"
	Eucosmidae	fruit moths 1,2,6(L)	"
	Tortricidae	tortrix moths 1,2(L)	"
	Crambidae	grass moths 1(L)	"
	Pyraustidae	web moths 1,2(L)	"
	Saturniidae	emperor moths, silk worms 1,2(L)	"
	Nymphalidae	butterflies 1(L)	"
	Lycaenidae	butterflies 1(L)	"
	Pieridae	butterflies 1(L)	"
	Papilionidae	butterflies 1(L)	"
	Hesperiidae	butterflies 1(L)	"
	Geometridae	looper moths 1(L)	"
	Sphingidae	hawk moths 1(L)	"
	Noctuidae	cutworms 1,2,6(L)	"
	Arctiidae	tigermoths 1,5,6(L)	"
	Lymantridae and many other families	gypsy moths 1,2(L)	"
Diptera	Tipulidae	crane flies 4(L)	cosmopolitan
	Cecidomyidae	gall midges 1,2,3,4,5,6(L)	"
	Bibionidae	bibionid flies 4(L)	"
	Trypetidae	fruit flies 1,2,5,6,(L)	"
	Agromyzidae	miner flies 1(L)	"
	Psilidae and various other fly families	root flies 1,4(L)	Northern Hemisphere
Hymenoptera	Siricidae	wood wasps 3(L)	cosmopolitan
	Cephidae	stem saw flies 3(L)	Northern Hemisphere
	Tenthredinidae	sawflies 1,2,5(L)	cosmopolitan
	Diprionidae	conifer wasps 1(L)	Northern Hemisphere
	Cynipidae	gall wasps 1,2,3(L)	cosmopolitan
	Eurytomidae	chalcid gall wasps 1,2(L)	"
	Torymidae and various other wasp families	chalcid wasps 6(L)	"

Table 3.13 *(cont.)*

		Scarabaeidae	chafers 1,4,6(L)	cosmopolitan
	Coleoptera	Dascillidae	dascillid beetles 4(L)	"
		Buprestidae	bark beetles 1,3,4(L)	"
		Elateridae	click beetles 3,4(L)	"
		Cerambycidae	longhorn beetles 3,4(L)	"
		Bruchidae	seed beetles 6(L)	"
		Chrysomelidae	tortoise beetles 1,2,4(L)	"
		Scolytidae	bark beetles 2,3,4(L)	"
		Platypodidae	bark beetles 3(L)	"
		Curculionidae and many other families	weevils 1,3,5,6 (adults & L)	"
(ii)	*Myriapoda* Diplopoda	Julidae and other families	millipedes 3,4	cosmopolitan
(iii)	*Crustacea* Isopoda	Porcellionidae and other families	slaters 3,4	cosmopolitan
(iv)	*Arachnida* *Acarina*	Tetranychidae	spider mites 1	cosmopolitan
		Pyemotidae	mites 1	"
		Tarsenemidae	mites 1	"
		Eriophyidae	gall mites, rust mites 1	"
B.	Mollusca *Gastropoda*	Arionidae	slugs 1	Northern Hemisphere
		Limacidae	slugs 1	" "
		Helicidae	snails 1	" "
		Veronicellidae and other groups of snails	slugs 1	tropical
C.	Nematoda *Rhabditoidea*	Tylenchidae	round worms 4	cosmopolitan

* Animals which feed on nectar, pollen and fleshy fruit and generally have a positive influence are not listed.

† Parts of plant eaten: 1, leaves; 2, twigs; 3, bark or stems; 4, roots; 5, flowers; 6, developing fruit, seeds. X, Omnivores which feed on substantial amounts of plant material. The rest of the animals are obligate herbivores, or nearly completely so.

defoliation. Nevertheless, foliage herbivory may have profound effects on general plant growth, seed production and competitiveness (Crawley 1985).

Beneficial, as well as adverse, effects of grazing on grasslands and other systems are noted in Chapters 7 and 9. Several authors (Stebbins 1981, Owen 1980, Chew 1974) consider that a close coevolutionary relationship has developed between grasses and grazers. Many grasses and other grazed plants are stimulated to more vigorous growth by removal of some leaf tissue, although vegetative growth may be at the expense of flower and seed production. Some evidence suggests that stimulation for growth is provided by compounds in the saliva of the grazers (Reardon *et al.* 1972). Coevolutionary relationships between many other kinds of plants and herbivores are also well-known (Gilbert & Raven 1975, Tilman 1978, Rosenthal & Janzen 1979).

Pollen and nectar are usually removed from flowers by design on the part of the plant. Animals which 'steal' nectar or pollen without carrying out effective pollination

Table 3.14 Features of plants which enable them to resist, evade or recover from grazing by animals.[*]

---

1. resistance

   (a) physical deterrence before attack or reaction when attacked

   stinging hairs; spines, thorns; leaf hairiness; plant form impenetrable to grazers
   impenetrable, unpalatable and indigestible tissues (wax, tough cuticles, tough seed coats, cork; abundant fibre, abundant cortical tissue, stone cells, hard wood; abundant cellulose, lignin, oxalic acid crystals; concentration of silica)
   entrapment with sticky substances (resins, latex and others)
   attraction of organisms (especially ants) antagonistic to grazers
   sloughing of bark or roots, shoots

   (b) chemical deterrence before attack or reaction when attacked

   unattractiveness through low nutrient content
   repellents (taste, odour) – bitter, sour, pungent, disgusting compounds (terpenes, coumarin, alkaloids, tannins, mustard oils and other essential oils, gums)
   toxic substances – monofluoracetate, cyanogenic glycosides, alkaloids, strychnine, nicotine, derris, pyrethrum, quassia; concentration of selenium and other metallic toxins
   interference with animal digestion by production of compounds that reduce digestibility of plant tissue or affect animal digestive process (tannins, mustard oils, proteinase inhibitors)
   interference with animal (particularly insect) growth processes by production of compounds that assume the role of hormones or hormone inhibitors, preventing completion of life-cycles

2. evasion (in space or time)

   meristems maintained close to ground
   absence or reduction of leaves
   underground organs well-developed
   dormancy, leaf-shedding when conditions for growth are unfavourable
   leaf production at times when animals are not active
   leaf-flush episodes (satiation of grazers)
   mimicry of unpalatable plant leaves, or dead tissues
   egg mimicry to deter oviposition by insects

3. features facilitating recovery

   abundant reserves
   brittleness or other properties of tissues ensuring breakage without extreme damage
   rapid growth response to damage and, in some cases (especially grasses) particular stimulation of growth response by grazing; rapid wound healing
   abundant dormant buds activated after grazing; resprouting from stems, stem bases, rhizomes, stolons, roots, etc.
   regrowth of detached portions

---

[*] By grazers are meant any animals which eat any part of living plants, thus causing them direct harm.

are restricting opportunities for their host, however. Removal of fruit and seeds by animals is also usually by the design of the plants, but again harm is done if the seeds are destroyed.

Aphids, feeding on the phloem stream, affect the vitality of plants and introduce harmful viruses. Root feeders, cambium feeders and feeders on young buds, flowers or seed may also do considerable harm to individual plants or plant populations. Feeding on seedlings is usually detrimental to the population.

Plants have evolved a very wide range of defences against herbivory (Rosenthal & Janzen 1979), including a vast array of toxic chemicals (Table 3.14). Some kinds of plant form or function are as beneficial against herbivory as they are against other kinds of hazard. For example, winter leaflessness protects plants against grazing as well as against cold. The sprouting habit of many woody plants is a response both to grazing and to fire or other kinds of damage.

Other animal influences

Animal activities affecting plants include the very important function of seed dispersal. Seeds may be carried internally for a time after animals feed on fruit (by birds, mammals and some invertebrates) and then voided. Some animals (ants, birds, small mammals) carry seeds externally in a deliberate fashion to form food hoards, which are not all subsequently recovered. Otherwise seeds are carried externally, accidentally, adhering to feathers or fur (birds and mammals) and are subsequently removed during grooming.

Heavy trampling and substrate upheaval by animals are beneficial to plant species which tolerate or require frequent disturbance, but detrimental to other plants. Large mammals, such as elephants or deer, often maintain relatively bare ground around their herding areas and along their trails. A few highly specialized, trampling-resistant plant species may be restricted to such sites. Excavations of subsurface living areas by animals such as ants, moles, prairie dogs, ground squirrels or rabbits, provide limited areas of fresh soil, usually inhabited by species different from those on adjacent undisturbed sites. The termite mounds of subtropical and tropical regions also create distinctive plant habitats.

In a striking symbiotic relationship, some plants harbour fierce ants which protect them from herbivores or the establishment of epiphytes. The ants live in specialized parts of leaves or stem. One of the most bizarre plants of this kind is the epiphytic *Myrmecodia* of New Guinea, with its bulbous stem elaborated into intricate ant galleries. Animal–plant relationships are discussed further in Chapter 10.

Dung and urine deposition by mammals and birds benefits some plants. Plants which require high nitrogen and phosphorus levels are often present in herding or roosting areas of these animals, or along the margins of animal trails. Seabird roosts have particularly distinctive flora and vegetation in response to high phosphorus and nitrogen levels. Figure 3.14 depicts some of the interactions of plants and other organisms in a European forest ecosystem. Fig. 3.15 shows the generalized energy flow pattern in an ecosystem.

PLANT SENESCENCE

The death of many individual plants can be attributed to environmental hazards such as diseases, pests or herbivores, and climatic extremes such as drought or cold. **Senescence** (gradual, endogenously determined, cellular decline) is another possible cause of death. The term senescence implies a programmed and inevitable lifespan.

Some plant organs which reach the end of their useful life senesce and die, while the rest of the plant is healthy and vigorous. Foliage or other above-ground parts of

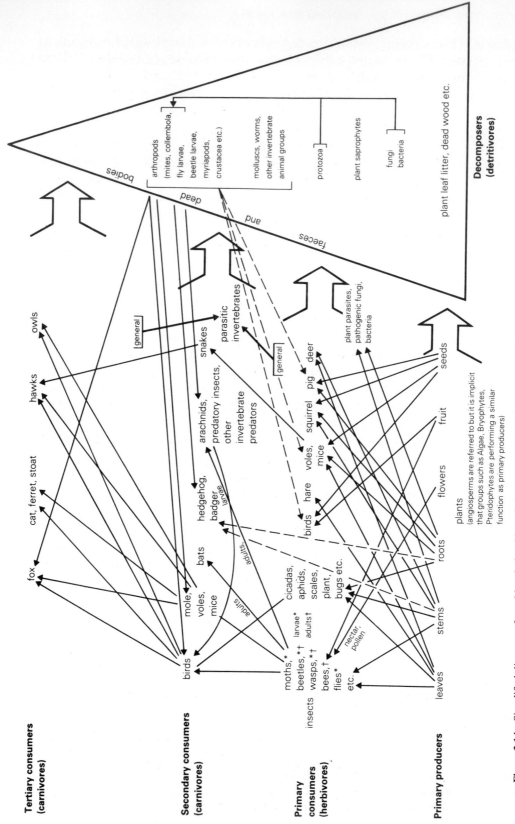

Figure 3.14 Simplified diagram of trophic relationships of plants, animals, monera and fungi in European oak and beech forest ecosystems.

many plants senesce and die with the onset of unfavourable conditions (winter cold or summer drought) while the rest of the plant survives in a quiescent state. Whole individuals of monocarpic (i.e. once-flowering) plants (annuals, biennials and some longer-lived species) senesce and die after flowering. These phenomena are physiologically explicable in terms of altered balance of growth substances in plants, in response to particular stimuli (cf. M. Black & Edelman 1970).

Although many species have characteristic lifespans (Leopold 1961, 1980), it is not clear whether other whole, relatively long-lived and polycarpic plants undergo senescence. It is difficult to disentangle the possible senescence of trees, for example, from the influences of pests, pathogens and climatic factors (see Ch. 9). Longevity may be related to factors such as strength and hardness of wood, distastefulness to insects, resistance to fungal attack, or limitation of pests and diseases by climatic factors. However, there are some indications, considered in Chapters 7 and 9, that certain perennial plant species experience programmed senescence.

ECOLOGICAL NICHES

One way of attempting to portray, in relatively simple terms, the complexities of plant–environment relationships has been by means of the ecological niche concept (G. E. Hutchinson 1958, 1965, May 1981, Newman 1982, Shugart 1984). The **fundamental niche** of a species (i.e. its ecological amplitude) is defined by its limits (the 'space' that it occupies) within the multidimensional complex of environmental gradients which affect it (excluding the effects of competing plant species) (Fig. 3.16).

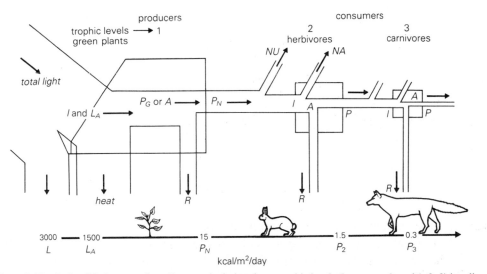

Figure 3.15 A simplified energy flow diagram depicting three trophic levels (boxes numbered 1, 2, 3) in a linear food chain. Standard notations for successive energy flows are as follow: I, total energy input; L_A, light absorbed by plant cover; P_G, gross primary production; A, total assimilation; P_N, net primary production; P, secondary (consumer) production; NU, energy not used (stored or exported); NA, energy not assimilated by consumers (egested); R, respiration. Bottom line in the diagram shows the order of magnitude of energy losses expected at major transfer points, starting with a solar input of 3000 kcal m^{-2} day^{-1} (after Odum 1971).

115

(1) Fundamental niches for each of six plant species (E–F) evident in the absence of competition from other species: the constraints are the conditions of the physical and chemical environment

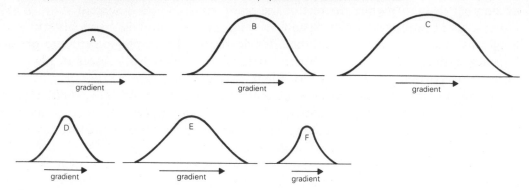

(2) Occurrence of the same six species in a community, showing their realized niches – the expression of each when subjected to competitive pressure from neighbours

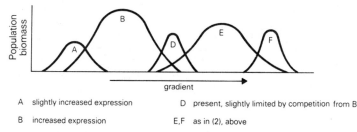

A   a weak competitor, considerably restricted by niche overlap with B, C and thus confined to marginal conditions

B   some restriction by niche overlap with C

C   only slight restriction of this strongly competitive species by B

D   a weak competitor excluded because of complete niche overlap with C

E   little or no influence by competitors, as this species functions at a different season from the others

F   only slight influence on this specialist species by competitors – niche overlap is minimal and the species is well suited to marginal conditions

(3) Expression of the other species when species C is absent from the community

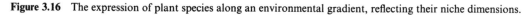

A   slightly increased expression

B   increased expression

D   present, slightly limited by competition from B

E,F   as in (2), above

Figure 3.16   The expression of plant species along an environmental gradient, reflecting their niche dimensions.

The fundamental niches of individual members of an interbreeding population are identical; those of all members of a species are usually held to differ only slightly. We have seen that within-plant, and within- and between-population genetic variation causes this to be not quite true.

The niches of different species are each distinct in various ways. In a general sense some are 'narrow', some 'broad' and some intermediate between these extremes. However, the niches of different species in the same location usually overlap to some degree. This means that, in any region, many of the species present have the potential

116

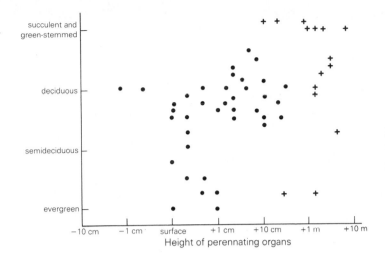

Figure 3.17 Niche differentiation of plant species in semi-desert, Arizona, expressed by type and behaviour of their shoot systems (vertical axis) and stature in relation to the ground surface (horizontal scale). Crosses indicate plants with spines (after Whittaker & Niering 1965).

to occupy the same piece of ground. Close neighbours with a degree of niche overlap constrain each individual plant, so that the **realized niches** of species, the 'spaces' which they really occupy in vegetation, are narrowed.

A species whose fundamental niche overlaps with the niches of other species may or may not be a strong competitor, with repercussions which are expressed in Figure 3.16. Individual members of a species will be very strong competitors with one another, since their fundamental niches overlap almost perfectly. Competition will determine the spacing of individuals.

The degree to which each species is expressed in stands of vegetation is determined, thus, by its fundamental niche 'breadth' and relative competitive ability. The realized niche of any species is variable, depending on the particular array of competitors present and their relative competitive strengths.

The idea of the ecological niche is difficult to translate from abstract into concrete terms. A simple illustration is that for growth behaviour and growth forms of plants in the Sonoran desert of the southwestern USA (Whittaker & Niering 1965) (Fig. 3.17). Niche differences are also evident from data on differences of form and the timing of plant activity in the Belgian oak–beech forest (Ch. 2). To gauge the full nature of niche differences, however, long-term experimentation with single and mixed species populations is necessary.

Species living in the same stand can avoid competition by:

(1) Functioning at different times: species have their growth flushes, flowering, etc., at intervals through the year, so that there is not a total overlap in the demands for nutrients and light.

(2) Functioning in different space, e.g. mosaics of short and tall species in a grassland, with roots also stratified at different depths.

(3) Possession of marked physiologic or functional difference so that demands for resources are different, e.g. different requirements of 'sun' and 'shade' species for light; different absolute requirements for nutrients such as phosphorus and nitrogen, or for water; different dependences on pollinators; different modes of seed dispersal; etc.

Table 3.15 Some variable attributes of seeds of different plant species which influence their ecological behaviour

Attribute	Categories*				Number of categories listed
shape, affecting relationships with soil irregularities, moisture	elongate–cylindric	oval–flattened	spherical	multifacial	4
type of stored reserves	oil	protein	starch		3
amount of stored reserves	none	small (μg)	medium (mg)	large (g)	4
dispersal mode	gravity	animal internal	animal external	wind; expulsion by plant	5
dispersal distance	short (a few m)	intermediate (many m)	long (hundreds of m)		3
viability retention	a few weeks	3–6 months	12–18–24 months	periods of many years	4
delay in germination	none	short after-ripening requirement	overwintering requirement	delay of many years: impenetrable seed coat or biochemical blocking	5
germination requirement for light	none required	light required but indifference to wavelength	white light requirement	far-red light requirement	4
germination position	epigeal	hypogeal			2
	total number of permutations if all categories shown are independent				115 200

* Categories are shown as being discrete. In rich floras the species will display an almost continuous range of properties such as amount of stored reserves, dispersal distance and delay in germination. Some properties for some seed species will not necessarily be independent; syndromes of attributes will occur, reducing the number of possible permutations. However, in some species there will be polymorphic variability in seed properties such as delay in germination, dispersal mode, increasing the number of possible permutations. Furthermore, various other attributes could have been listed.

In fact, in relation to the third point, there may be quite considerable heterogeneity of microhabitats in some stands. As a result, competition between seedlings and some smaller plants may not be severe.

The potential for enormous complexity of niche dimensions can be gauged from consideration of some seed characters alone. In a flora, among other properties, seeds may possess a range of characteristics as shown in Table 3.15. For the characteristics listed there are 115 200 possible permutations. Numbers of the characteristics may be grouped in particular syndromes for any one seed type; different plant species may possess similar arrays of seed characters. Nevertheless, a large flora may carry a very large number of permutations of seed characteristics. Considering all other attributes which may determine the ecological behaviour of species in a rich flora, the numbers of potential permutations (and thus the numbers of sets of niche properties which differ in at least some respects) are huge.

As Whittaker (1965, 1969b) points out, competition must be a very strong selective force; species which habitually occur together will have evolved so that their niches differ, at least to some degree. They are only partial competitors (cf. Fig. 3.16). Although they will overlap in their requirements for temperature, water, nutrients, etc., they can be expected to vary in other vital attributes, including those related to the establishment of seedlings (Grubb 1977, Angevine & Chabot 1979, I. R. Noble & Slatyer 1980). More-elaborate data illustrating the effects of niche differences in plant communities are found in studies by Whittaker (1956), Sagar & Harper (1961), Whittaker & Niering (1965), Putwain & Harper (1970), Werner & Platt (1976), Parrish & Bazzaz (1982a, b) and Newman (1982). A full account of niche theory is outlined in Giller (1984).

Mathematical descriptions of niches have been attempted by various authors (e.g. Colwell & Futuyma (1971), Pielou (1975) (see summaries by Garbutt & Zangerl 1983, Shugart 1984). The most satisfactory empirical approaches to the measurement of niche dimensions for plant species require the manipulation of stands of vegetation so as to alter the range of competing species present and also the growth of single plant species and mixtures in carefully controlled environments.

The relative importance of the component plant species in communities (i.e. their relative numerical abundance or biomass) seems to be best expressed by two hypotheses (McNaughton & Wolf 1970, Whittaker 1975b) (Fig 3.18):

(1) Niche pre-emption, where there is a geometric series of degrees of relative dominance (i.e. one main dominant and successive partitioning of the remaining habitat in a geometric fashion by the other species).
(2) Lognormal distribution, where there are few very important species, few very rare species and many species of intermediate importance.

Communities with relatively small numbers of species tend to fit the niche pre-emption model. Communities that are rich in species fit the lognormal distribution model.

Since niches are difficult to define, it may be more realistic to refer to **lifestyles** for individual plant species. Lifestyles are simply defined as the sets of readily recognized

119

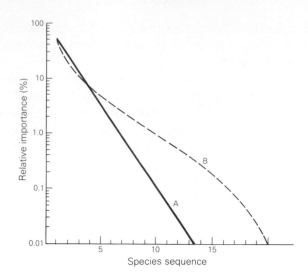

Figure 3.18 Hypotheses explaining the relative importance of plant species in communities. Importance value curves computed for a hypothetical sample of 20 species. (A) *Geometric series* ($c = 0.5$) or *niche-pre-emption hypothesis,* computed from

$$n_i = Nk(1 - k)^{i-1} = n_1 c^{i-1}$$
$$c = 1 - k, \quad N \approx n_1/k$$

where N is the total of importance values for all species in the sample and n_i the importance value for species i in the sequence from the most important to the least important species; c is the ratio of the importance value of a species to that of its predecessor in the sequence, and n_1 is the importance value of the first and most important species. (B) *Lognormal distribution,* or *independent variable hypothesis*

$$s_r = s_0 e^{(aR)2}$$
$$\Sigma s_r = S = s_0 \sqrt{\pi/a}$$

where s_r is the number of species in an octave R octaves distant from the modal octave, which contains s_0 species; a is a constant that often approximates 0.2. After Whittaker (1975b).

properties (stature, life-form, periodic behaviour, life-history, etc.) which characterize the way in which the species fit into a local or regional vegetation complex. For example, among a group of temperate zone weedy herbs some of the lifestyles are as follows.

(1) Ephemerals: annuals which germinate and flower over a brief period when conditions are specifically favourable.
(2) Winter annuals, which germinate in autumn and flower in early summer.
(3) Summer annuals, which germinate in spring and flower in late summer and autumn.
(4) Biennials, which germinate in spring and flower in summer of the following year.
(5) Perennials with different lifespans; differing vegetative and flowering periods; differing statures; differing growth habits (tufts, rosettes, loose mats, climbers, rhizomatous turf formers, etc.).

CONCLUSIONS

In Chapters 2 and 3 we have seen that the distribution and performance of plant species (manifest in the degree to which they are expressed in the vegetation) are determined by a vast multiplicity of interacting factors, many of which operate at random and, thus, are unpredictable. The factors that are independent of plants give rise to an almost infinitely variable environmental complex. The intensity of expression of factors is unevenly distributed in space. In this way spatial discontinuities occur, but, in abstract, the environmental complex is continuously variable. Plant species themselves vary in their capacity to react to the external environment, and this creates another dimension of variables superimposed on the external environmental complex. These are not the only variable components to be considered, because the vegetation modifies the impact of the external environment, in turn influencing plants by various feedback processes. Furthermore, plant species interact with one another, as well as with other biota, in various positive and negative ways. Therefore, the full elucidation of the causal ecology of any vegetation phenomenon usually becomes a very difficult and complex matter.

To help ecologists to try to understand the dynamics of the situations which they are observing in nature, some quantitative or semiquantitative model methods have been developed. These include, for example, loop analysis, ecosystem energy flow models, discriminant analysis, computer simulation models and the Lotka–Volterra equations (outlined in later chapters). All such aids assist in understanding and in the prediction of likely trends, but they all fall short of comprehensive explanation, simply because the natural systems are so complex. Perhaps, in due course, computers will be available that are sufficiently powerful to handle the enormous amount of program data which would be needed to allow the computer simulations to approach reality. However, the chaotic effects of the operation of vast numbers of random variables may defeat the explanation of ecosystem dynamics through modelling.

Chapters 2 and 3 only scratch the surface in attempting to describe the intricacies of plant–environment relationships. Other texts such as Odum (1971), Whittaker (1975b), Etherington (1975), Harper (1977), McNaughton & Wolf (1979), Grime (1979), Barbour *et al.* (1980), Larcher (1980), Fitter & Hay (1981) give fuller treatments. Tables 2.1 and 3.1 summarize some of the most significant points about plant–environment relationships.

One very important point which Chapter 3 attempts to convey is that plant species are probably all uniquely different in their responses to their external environment and their neighbours. It is relatively easy to demonstrate that there are important differences between species in a stand; it is implicit that different species which can inhabit the same site also have much in common. Differences in response to environment and neighbours may be more significant at the levels of germinating seeds and developing juveniles than they are in mature populations. Success in competition in youth will determine which species actually inhabit any site (Grubb 1977). Nevertheless, differences in vigour and longevity and reproductive success of mature plants will determine other aspects of regeneration, and thus the overall patterns of species expression in the vegetation.

It is difficult to express in any simple and easily assimilable way the concept of the ecological niche, which attempts to define how plant species differ from one another in all their dimensions. Nevertheless, this concept is useful, for it enables us to understand how plants fit into the vegetation according to their ecological preferences and tolerances. The niche concept is invoked in later chapters, not only to explain spatial distribution patterns in vegetation, but also to explain the analogous temporal patterns arising during the processes of vegetation change. The term *lifestyle* is used where morphological and functional differences between plant species are apparent in a local or regional context.

4

Vegetation development on volcanic ejecta

The periodic eruption of volcanoes, with emission and deposition of rock materials which originated, molten and sterile, in the depths of the Earth, provides new sites which plants colonize. Patterns of vegetation change on new land surfaces have been less thoroughly investigated than the changes that occur after disturbance of established vegetation (considered at greater length in Chs 7, 9 & 10). However, the development of vegetation on new volcanic debris is an appropriate starting point for the examination of vegetation change processes.

Volcanoes take different forms. Some are the familiar, somewhat symmetrical cones, with a crater on top. Their slopes are built up from layer after layer of ejected debris thrown out by explosive eruptions. Some volcanoes are domes, formed by the welling-out of fluid lava. Others are simply fissures (Francis 1976).

Many different forms of rock material are discharged from volcanoes. From time to time very hot, molten lava streams may simply pour from vents and flow down the sides of a volcano until they solidify. Such eruptions occur often at volcanoes such as Mauna Loa and Kilauea in the Hawaiian Islands, USA, or (giving rise to sheets of lava), from the Laki fissure in Iceland. The solidified lava may be relatively smooth (**pahoehoe**) or very irregular and rough (**aa**) (from the Hawaiian words meaning these forms of lava).

Eruptions at volcanoes such as Mt St Helens in Washington, USA, Aggung in Java and Vesuvius in Italy are often much more violent, throwing out hot rocks ranging from huge boulders to broken angular fragments of various sizes: slag-like scoria, small, pebble-like lapilli or fine particles of tephra (volcanic ash). The larger fragments, including scoria and lapilli, cooled by their course through the air, are deposited as layers on the flanks of the volcano. Much of the ash may rise high into the atmosphere, be transported for considerable distances away by wind and fall as a layer on the surrounding terrain. Some ash settles in compacted layers near the volcano, forming tuff or, mixed with larger fragments, volcanic breccia.

Another type of eruption is the very destructive *nuée ardente* or glowing ash cloud. Incandescent clouds of solid particles, mobilized by gas, rapidly traverse the surrounding terrain and terminate many kilometres from the vent. A large eruption of this kind was that of Katmai in Alaska in 1912, when 10 km^3 of relatively unconsolidated pumice (porous volcanic glass) and ash were deposited. Vegetation adjustments are still occurring after the huge *nuée ardente* Taupo eruptions of about AD 150 which deposited the soft rock ignimbrite (compacted ash) over at least 20 000 km^2 in New Zealand.

Near any volcano or group of volcanoes the range of landform surfaces, as well as the chemistry of the rock materials, is often very diverse. Usually there are fields of very rough scoria or sheets of pumice, blocky (aa) lava fields, solid lava sheets with irregular surfaces, screes of finer lapilli, sand and ash on the slopes, and plains of similar material on level sites. The rocks include dark **basalts**, with a low content of the mineral silica (quartz), but containing feldspar, pyroxene, olivine and oxides, relatively high in bases (calcium, magnesium, potassium and sodium). At the other extreme are **rhyolites**, with large amounts of silica and some feldspar, mica and amphibole, acidic and low in bases. Intermediate in chemistry are **andesites**, containing much silica, as well as feldspar, amphibole, pyroxene and mica (Francis 1976, Dietrich & Skinner 1979). Solid lavas are usually basaltic and, although they may be hard and slow to weather, they are better-supplied with plant nutrients than the rhyolitic or andesitic rocks ejected during explosive eruptions. Pumice and ignimbrite are rhyolitic. Elements or compounds toxic to plants (e.g. fluorine and sulphuric acid) may escape from volcanic rocks.

Two approaches to examining the vegetation development on volcanic deposits are possible. The first is to study the vegetation sequence which develops after one particular eruption. The other is to examine, at one time, the development of soils and vegetation on a series of deposits from several eruptions. The latter approach, even if all conditions of **chronosequence** criteria are met (i.e. time is variable, and other factors are uniform when each surface is exposed to the weather and vegetation development), will nevertheless form an interrupted chronosequence, since deposition is not continuous. The variability of terrain created by different volcanic events and the difficulty of dating the older deposits have limited the use of the second method. Most of this account of vegetation change on volcanic debris will refer to sequences which have been observed since individual, large volcanic events.

SURTSEY, ICELAND

Out of the sea at the southwestern end of the Vestmannaeyjar Island chain, off the south coast of Iceland, on 15 November 1963, emerged the steaming, hot rocks of a new island, since named Surtsey (63° 18′N, 20° 37′W) (Fig. 4.1). After a series of violent explosive eruptions which threw out large quantities of broken debris, the activity changed to emission of flowing lava. By May 1965 Surtsey was about 2.45 km^2 in area. It had also received tephra material, from the eruptions of two adjacent, small, new volcanic islands, which have since been eroded away by the sea. More lava poured out and, in June 1967, the island, with two small volcanic cones forming a summit 160 m high at the northern end, was about 2.8 km^2 in area (Einarsson 1967).

The environment is severe. The island is windswept, with heavy coastal wave action and spray splash during storms. Climatic records for Heimaey, the largest of the Vestmann Islands, 21.6 km distant, to the north-east (mean annual temperature 5.4°C, mean January temperature 1.4°C, mean July temperature 10.3°C, mean annual precipitation 139.7 cm) indicate the general conditions experienced at Surtsey.

The volcanic activity has settled down, but there are areas of hot rock and numbers

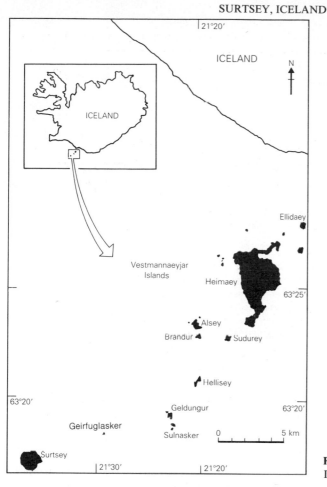

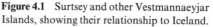

Figure 4.1 Surtsey and other Vestmannaeyjar Islands, showing their relationship to Iceland.

of fumaroles from which hot steam emerges. Round the shore are lava cliffs and bouldery or sandy beaches. The irregular and rough surface of the island is either hard, basaltic lava or unconsolidated, unstable tephra which blows about, abrasively. Both kinds of substrate are generally free-draining. The pH of the substrate was originally 4 – 6.8, but is now 5.6 –8.5 (resulting from the abundant bird excrement). Pools on the surface are saline.

The colonization of Surtsey by organisms has been closely monitored by Icelandic biologists (Einarsson 1967, Fridriksson 1975, 1982). Table 4.1 summarizes the main details of the timing of the advent of plants and some aspects of their ecology on the island; Table 4.2 lists the statistics for vascular plants up to 1980. Figure 4.2 shows the distribution of plants in 1971 and 1980.

There has been development of only a scattered plant cover on Surtsey. Land plants, animals and land-dwelling microbes have had to be dispersed across the ocean by natural means. The nearest land, 5.1 km to the north-east, is the small island Geirfuglasker (0.02 km² in area). Others of the Vestmann chain lie between it and Heimaey (16 km² in area), and the mainland of Iceland is 32 km away.

All of the angiosperms, lichens and algae and about half of the mosses and liverworts could have come from the other Vestmann islands. The other bryophytes

Table 4.1 History of colonization of Surtsey by land plants (after Fridriksson 1970, 1975, 1982, 1987).

November	1963	eruption begins
May	1964	aerial trapping shows constant, relatively sparse, (in numbers and kinds) microbial dispersal to the island; bacteria in rotting seaweed, bird carcasses
	1965	first angiosperm, *Cakile edentula*[*3] in tephra and decaying seaweed
	1966	*Elymus arenarius*[*3] arrives; it and *Cakile* wiped out 1966, re-established 1967
	1967	first mosses *Funaria* and *Bryum*; angiosperms *Honkenya peploides*[*3] and *Mertensia maritima*[*3]
	1968	8 cyanobacteria including nitrogen-fixers *Anabaena* and 3 *Nostoc* spp., but these are not abundant; 100 species of other algae including 74 diatom species; mosses *Leptobryum*, *Pohlia* and *Ceratodon*
	1969	first non-littoral angiosperm *Cochlearia officinalis*[1], transported by birds
	1970	many species of bacteria, widely distributed in the substrate among plant roots and rhizoids; nitrifying and denitrifying species present; fungi include moulds, *Penicillium* and *Phoma*; first lichens *Stereocaulon*, *Trapelia* and *Placopsis*; 16 moss species including *Racomitrium canescens* and *R. lanuginosum*; angiosperm *Stellaria media*, bird transported[2]
	1971	36 species of mosses; *R. canescens* is most common, forming scattered small tufts; 4 liverworts *Marchantia*, *Cephaloziella*, *Scapania* and *Solenostoma*
	1973	colonies of coccoid green algae; 5 lichen species; 69 species of mosses now inhabiting almost all readily available habitats; *Racomitrium lanuginosum* producing capsules; 11 angiosperm species (additions are *Puccinellia retroflexa*[*1], *Tripleurospermum maritimum*[*1], *Festuca rubra*[1], *Angelica archangelica*[3] and *Carex maritima*[*3]; 1 fern *Cystopteris fragilis*[3]
	1981	[incomplete information] bacteria *Azotobacter* and *Beggiatoa* and denitrifier *Thiobacillus* present; nitrification taking place by *Nitrosomonas* and *Nitrobacter*, which are very common; a few more mosses present covering most of the suitable terrain, *Racomitrium lanuginosum* the most common; *Bryum* and *Funaria* also abundant; *Equisetum arvense* present 19 angiosperm species, including *Cerastium fontanum*; *Honkenya* the most common (about 100 000 plants); it and some others flowering and seeding; *Archangelica* and *Stellaria* now absent, *Cakile* rare; many small animals; 112 insect spp., 24 arachnid spp., some are breeding; 60 species of birds have been recorded, some are breeding
	1986	[incomplete information] 23 vascular plants present; innumerable *Honkenya*; *Rumex acetosella* and *Mertensia* slowly increasing; number and size of *Elymus* colonies (second-most-abundant plant species) increasing; additional angiosperms *Sagina saginoides*, *Poa pratensis* and *Armeria vulgaris*; *Cochlearia* declining, now rare; *Cerastium fontanum* slowly declining; *Cakile* and *Cystopteris* now absent

* Obligate littoral plant.
Nearest sources: [1]Geirfuglasker (5 km); [2]Geldungur, Sulnasker (10 km); [3]Heimaey and adjacent islands (20–25 km).

came from Iceland. The littoral (coastal) species of angiosperms have seeds which float in the sea. Many seabirds inhabit Surtsey, and some migratory land birds have also been observed there. Some land plant species apparently were brought by birds (probably adhering to plumage). The fern and the mosses and other non-vascular plants have tiny spores carried by wind.

Mosses established themselves on the lava, but the first angiosperm colonists became established in sandy substrates on the beaches. These are hazardous habitats, liable to be disturbed during storms. In summer 1971 there were 83 individual angiosperm plants on Surtsey, of which only 49 survived the following winter. In 1971 (and in subsequent years) some angiosperm individuals were growing on more-secure sites in sand-filled hollows in the lava (Fig. 4.3). By 1973 the number of individual plants was 1273. In 1973 *Honkenya peploides* and *Cochlearia officinalis*, both perennials, were the most vigorous and abundant, but *Honkenya* is now the most numerous. However, their spread is slow, both vegetatively and, more recently, by seed. *Cakile maritima* (an annual) has not yet become permanently established – conditions are too rigorous for it to flower and seed.

Mosses are hardier plants than the angiosperms. Once established they form tight cushions, maintaining their own tiny microclimates close to the ground surface. The cushions expand relatively quickly by vegetative proliferation. The mosses (especially *Racomitrium lanuginosum*) have become much more widespread on Surtsey than the

Table 4.2 Total number of vascular plants on Surtsey, 1965–1985 (after Fridriksson 1987).

Species	1965	1966	1967	1968	1969	1970	1975	1980	1985
Cakile edentula (= *arctica*)	23	1	22		2		5	1	
Elymus arenarius		4	4	6	5	4	12	5	50
Honkenya peploides			24	103	52	63	428	50 000	∞
Mertensia maritima			1	4			11	7	100
Cochlearia officinalis					4	30	863	75	4
Stellaria media						4			
Cystopteris fragilis							2	5	
Carex maritima							3	1	1
Puccinellia retroflexa							8	7	2
Tripleurospermum maritimum							2	1	1
Festuca rubra							2	3	1
Cerastium fontanum							106	150	20
Equisetum arvensis							2		
Silene vulgaris							1		
Sagina sp.							1		
Juncus sp.							1		
Rumex acetosella								40	50
Cardaminopsis petraea								8	
unidentified plants				1			2		
total:	23	5	51	114	63	101	1449	50 000 +	∞

Other species present in the same period were *Angelica archangelica* (2 plants in 1972, 1973) and *Atriplex patula* (?) (1 plant in 1977). Population numbers of other species have fluctuated considerably. Peak numbers of *Cakile edentula* were 33 (1973), of *Elymus arenarius* 66 (1973), of *Rumex acetosella* 124 (1978).

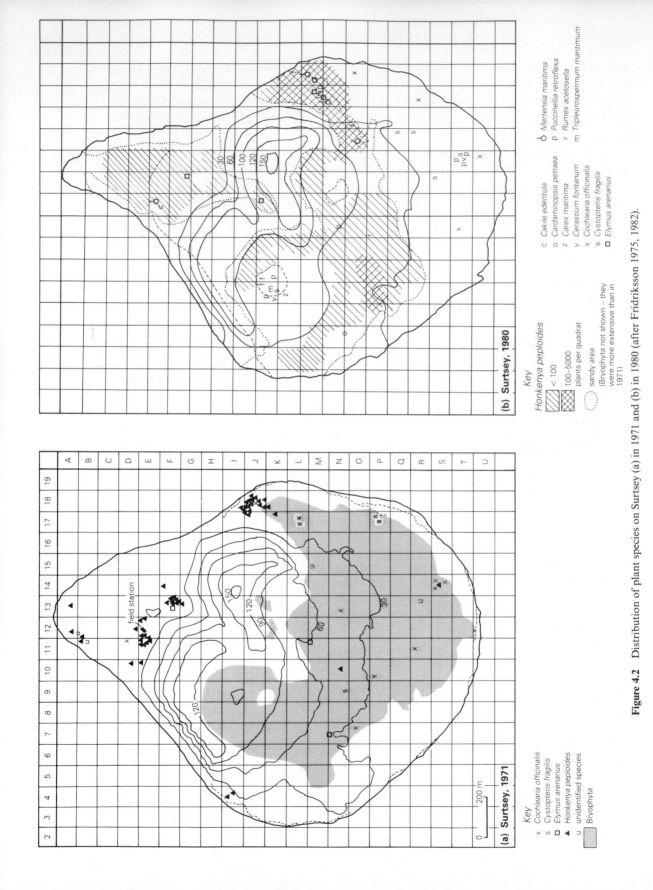

(a) Surtsey, 1971

Key
x Cochlearia officinalis
s Cystopteris fragilis
□ Elymus arenarius
▲ Honkenya peploides
u unidentified species
▓ Bryophyta

0 ___ 200 m

(b) Surtsey, 1980

Key
Honkenya peploides
▨ < 100
▩ 100–5000 plants per quadrat

░ sandy area

◌ (Bryophyta not shown – they were more extensive than in 1971)

x Cakile edentula   c
o Cardaminopsis petraea
z Carex maritima
v Cerastium fontanum
x Cochlearia officinalis
s Cystopteris fragilis
□ Elymus arenarius

♂ Mertensia maritima
p Puccinellia retroflexa
r Rumex acetosella
m Tripleurospermum maritimum

Figure 4.2 Distribution of plant species on Surtsey (a) in 1971 and (b) in 1980 (after Fridriksson 1975, 1982).

Figure 4.3 Surtsey. (a) The moss *Racomitrium lanuginosum* colonizing lava. (b) Patches of *Honkenya peploides* on scoria (photos H. Kristinsson).

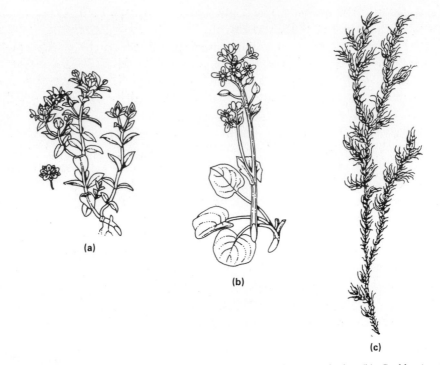

Figure 4.4 Important colonist plant species on Surtsey. (a) *Honkenya peploides*. (b) *Cochlearia officinalis*. (c) *Racomitrium lanuginosum*.

angiosperms, and they form a more-continuous cover in suitable habitats (Figs. 4.3 & 4). Often one angiosperm individual will be associated with a few moss species and one or more species of algae, moulds and bacteria (Fridriksson 1975, 1982). Some nutrient for the plants is obtained from the abundant bird excrement, which contains nitrogen, phosphorus and cations. Nutrients will also be slowly weathering from the rocks, which have good supplies of calcium, magnesium and sodium but are very low in available phosphorus.

The future development of vegetation on Surtsey will be of great interest; careful, frequent measurements of its progress by Icelandic scientists, since the origin of the island, make this a very important place for the study of dispersal to and establishment of plants on new terrain, and the development of ecosystems there.

KRAKATAU, INDONESIA

The island group Krakatau (Fig. 4.5) in the Sunda Straits between Java and Sumatra (6° 06′S, 105° 25′E) is the remnant of an ancient volcano. A volcanic vent on the main island erupted very violently in 1883, blasting away about 15 km^3 of debris and reducing the island from 10 km long and 5 km wide, to 5 km × 4.5 km. All that remained was the 813 m peak of Rakata, split down the middle by the eruption, leaving a very steep cliff on its northwestern side. Rakata has gentler slopes on its east, south and west. The other islands were slightly increased in size. The luxuriant

130

higher plant life and the animals which previously had inhabited Krakatau and its neighbouring two smaller islands, Sertung and Rakata Kecil, were totally destroyed (D. H. Campbell 1909, Docters van Leeuwen 1936, Francis 1976). It is possible that some microscopic life survived. The name Krakatau is a byword, for this eruption was the most catastrophic recorded in written history. Thousands of people on surrounding land, or on boats at sea, were overwhelmed by waves initiated by the blast.

Moderate volcanic activity resumed in 1927 in the sea just north of Rakata, on the site of the vanished north end of Krakatau, until in 1930 a small island, Anak Krakatau ('child of Krakatau'), was built up. Small eruptions occurred on Anak Krakatau in the 1940s, 1952 and 1979, destroying the developing vegetation each time. This island is now about 2 km × 2 km in size and rises 180 m above the sea.

Rakata is composed of older volcanic ejecta overlain by debris from the 1883 eruption. Basaltic lava, scoria, breccia and rhyolitic tuff and pumice form a very rough, bouldery cover over any relatively level or gently sloping areas. Much of the southern side of the island, however, is dissected by deep gullies cut in the tuff by streams, often with sharp-crested intervening ridges. Coastal erosion by the sea is vigorous, and landslides are common on the steepened slopes.

There are no long-term climate records for the islands. Those for Labuhan, about

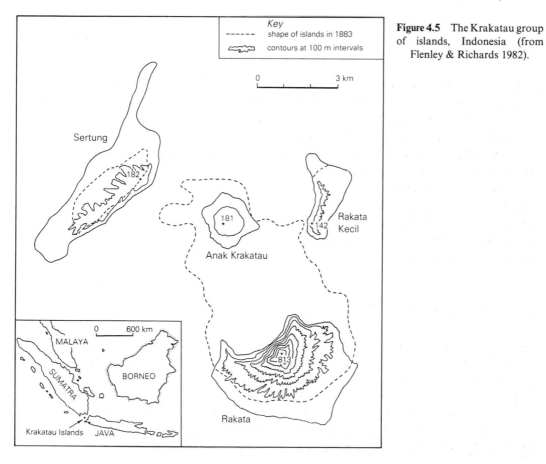

Figure 4.5 The Krakatau group of islands, Indonesia (from Flenley & Richards 1982).

Key
------ shape of islands in 1883
〰〰 contours at 100 m intervals

0   3 km

Sertung

182

181

Anak Krakatau

142 Rakata Kecil

813

Rakata

0   600 km

MALAYA

SUMATRA

BORNEO

Krakatau Islands   JAVA

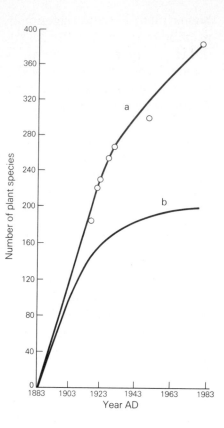

Figure 4.6 Approximate numbers of plant species recorded on Rakata 1883–1979. (a) Cumulative numbers (including species not collected on subsequent visits). (b) Actual numbers of species collected (after Flenley & Richards 1982).

70 km to the south are: mean annual temperature 18.9°C, mean January temperature 18.9°C, mean July temperature 18.5°C, mean annual rainfall 355 cm. There is a marked wet season in December–January–February.

The precipitation is greatest and evaporation least on the peak of Rakata, where there is frequently a cloud cover. Water runs off rapidly and the substrate is generally well-drained. Before it became vegetated the rocky ground would have experienced strong solar heating and shade would have been minimal. Some areas are volcanically heated.

Since the 1883 eruption biologists have visited the Krakatau group from time to time, beginning with M. Treub in 1886, 1897, and 1905. Other visits which have resulted in substantial contributions to the information about vegetation development were in 1929 and 1934 (Docters van Leeuwen 1934, which also summarizes the earlier work); 1951 (van Borssum Walkes 1960) 1979 and 1983 (Flenley 1980, Flenley & Richards 1982, Whittaker *et al.* 1989). Much of the botanical study of Krakatau has consisted of the collection of the plant species present, with brief descriptions of aspects of the vegetation and its ecology. The 1979 work initiated a fuller study of ecosystem development.

Recolonization of the islands by plants, animals and microbes took place from Sumatra and Java (35 and 45 km away, respectively, at the nearest) and possibly from some nearer small islands. Dispersal of plants to Krakatau, rapid in the first 40 years, has since slowed (Fig. 4.6; Table 4.3). Whittaker *et al.* (1989) give full lists of the

Table 4.3 Numbers of vascular plant species recorded* on the three main islands of Krakatau (various sources, including Flenley & Richards 1982)

	Year						Total known to have occurred
	1886	1897	1908	1928	1934	1979	
total number of species	26	64	115	214	271	196	387
species typically littoral in the region	10	30	60	68	70	75	n.d.
angiosperms							
dicotyledons	11	35	85	123	180	114	206
monocotyledons	4	16	19	42	81	37	88
ferns	11	12	9	38	55	41	83
others†		1	3	8	8	4	10

n.d., not determined.

* Numbers are only approximate, as there are various problems of compiling comparative lists, discussed by Flenley & Richards (1982). Some species recorded on one or more early visit have not been found subsequently. In some instances they may have been lost from the flora, but otherwise it is possible that further search would have discovered them. The 1979 list is from Rakata only.

† In 1979 these included 0 (2) *Psilophytes*, 0 (1) *Equisetiphyte, 2 (5) Lycophytes*, 1 (1) *Cycadophyte* and 1 (1) *Gnetophyte* (figures in parentheses are the total number known to have occurred).

developing flora for each island. Up to 1908 about half of the vascular species present were typical inhabitants of coastal sites in the region, although some of them lived inland on Rakata. Normally their seed is transported from island to island by flotation in the sea. Species whose propagules are transported by wind (ferns, terrestrial algae and some bryophytes, as well as bacteria, fungi, lichens and small animals) were also common in 1908. Subsequently species presumed to be dispersed by birds and bats have become more common. In 1934 41 per cent of the vascular plants (including Asteraceae, orchids and some other groups) were presumed to be wind-dispersed, 28 per cent by sea, 25 per cent by animals and the rest by human agency. Flenley (1980) notes the slowing down of the increment of new colonists since 1934 and suggests, first, that this may be due to the difficulty experienced by propagules in crossing a wide stretch of sea. Secondly, it may be because the cutting of much lowland rainforest in west Java and south Sumatra has reduced the number of species which could contribute to the seed supply. Many mature forest species also have seeds which disperse relatively short distances from their parents, so opportunities for them to reach Krakatau must be rare. Another possibility is that the present dense, vigorous vegetation cover may inhibit further influx of species, at least for a time.

Vegetation development on Krakatau may be presumed to have started in a sterile environment. Backer (1909, 1929) believed that some plants or seeds could have survived, but other authors maintain that this is impossible, because of the intense heat and the violence of the eruption (Docters van Leeuwen 1936, P. W. Richards 1964). P. W. Richards drew up a model for vegetation development on Rakata, based on information from earlier papers. Figure 4.7 is a modified, up-to-date version of this. Figure 4.8 depicts some of the important plant species. Similar patterns of vegetation development have been observed on volcanoes in New Guinea (B. W. Taylor 1957).

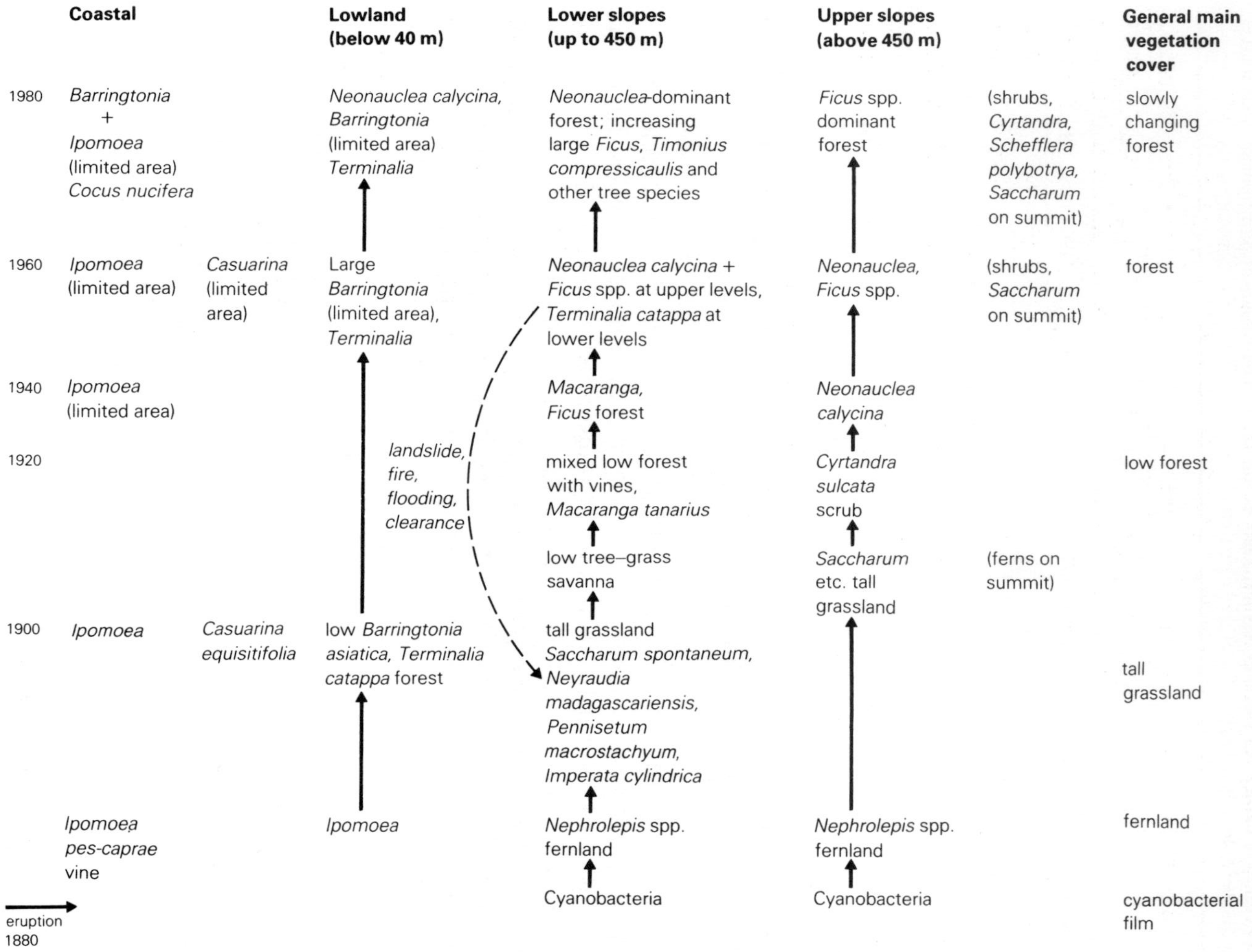

Figure 4.7 Vegetation development trends on Rakata.

	Coastal	Lowland (below 40 m)	Lower slopes (up to 450 m)	Upper slopes (above 450 m)		General main vegetation cover
1980	Barringtonia + Ipomoea (limited area) Cocus nucifera	Neonauclea calycina, Barringtonia (limited area) Terminalia	Neonauclea-dominant forest; increasing large Ficus, Timonius compressicaulis and other tree species ←	Ficus spp. dominant forest ←	(shrubs, Cyrtandra, Schefflera polybotrya, Saccharum on summit)	slowly changing forest
1960	Casuarina (limited area) Ipomoea (limited area)	Large Barringtonia (limited area), Terminalia ←	Neonauclea calycina + Ficus spp. at upper levels, Terminalia catappa at lower levels ←	Neonauclea, Ficus spp. ←	(shrubs, Saccharum on summit)	forest
1940	Ipomoea (limited area)		Macaranga, Ficus forest ←	Neonauclea calycina ←		
1920		landslide, fire, flooding, clearance	mixed low forest with vines, Macaranga tanarius	Cyrtandra sulcata scrub ←		low forest
			low tree–grass savanna ←	Saccharum etc. tall grassland ←	(ferns on summit)	
1900	Casuarina equisitifolia Ipomoea	low Barringtonia asiatica, Terminalia catappa forest ←	tall grassland Saccharum spontaneum, Neyraudia madagascariensis, Pennisetum macrostachyum, Imperata cylindrica			tall grassland
	Ipomoea	Ipomoea ←	Nephrolepis spp. fernland ←	Nephrolepis spp. fernland ←		fernland
			Cyanobacteria	Cyanobacteria ←		cyanobacterial film
eruption 1880	Ipomoea pes-caprae vine					

Nitrogen sources and vegetation development on Krakatau

An important early observation made in 1886 (Treub 1888) was that, as well as the spectacular growth of ferns which covered much of the island at the time (species of *Dryopteris, Nephrolepis* and *Pteris*), much of the ground of Rakata was covered by a thin layer of cyanobacteria (blue-green algae) with a gelatinous and hygroscopic coat. Docters van Leeuwen (1936) noticed that, as tuff cliffs were exposed by landslides, the raw surfaces soon became covered with a greenish film of cyanobacteria. Among the algal genera present were *Schizothrix, Lyngbya, Symploca, Anabaena* and *Tolypothrix. Anabaena* and *Tolypothrix* fix atmospheric nitrogen in aerobic conditions. Treub's observations showed that the cyanobacterial film allowed fern spores to germinate and that, subsequently, the adult ferns, by provision of shade, encouraged flowering plants to become established. He was not fully aware, then, of the significance of his observations in relation to nitrogen fixation.

Ernst (1908) and D. H. Campbell (1909) found aerobic nitrogen-fixing bacteria in the Krakatau soils and believed them to be important in the provision of initial

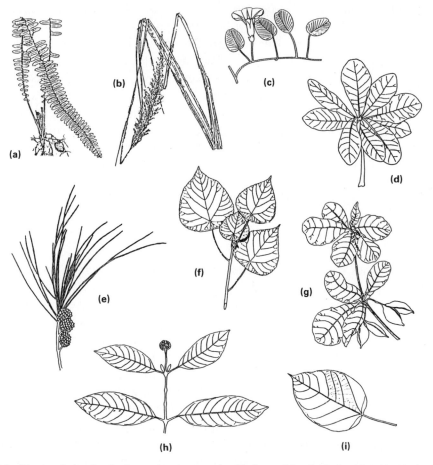

Figure 4.8 Plants of the developing vegetation on the Krakatau Islands, Indonesia. (a) *Nephrolepis* sp. (b) *Saccharum spontaneum*. (c) *Ipomoea pes-caprae*. (d) *Barringtonia asiatica*. (e) *Casuarina equisetifolia*. (f) *Macaranga tanarius*. (g) *Terminalia catappa*. (h) *Neonauclea calycina*. (i) *Ficus* sp.

nitrogen supplies. The amounts of nitrogen available from sources such as the rainfall, free-living bacteria or extra-system animal or plant sources, in the early days of colonization of the bare rocks of Krakatau, may not have been enough to support large vascular plants. However, the cyanobacteria fix substantial amounts of nitrogen, which can be passed on to plants (Singh 1961, Postgate 1980). Once this occurs it can be retained in the biomass of the ecosystem by recycling.

No measurements of soil nutrients were made in the early days of study of Krakatau and its vegetation and soils, but we may presume that the growth of the cyanobacteria film was a key step in ecosystem development, permitting rapid colonization of the island by ferns and other plants. Other things being favourable, the rate of vegetation development in all systems will depend on the rate of incorporation of nitrogen into biomass.

On Krakatau the cyanobacterial source of nitrogen was soon supplemented by supplies from symbiotic organisms living in the root nodules of certain of the vascular plants which colonized the island. These are *Casuarina* (with *Frankia* nodules) and members of the *Rhizobium*-carrying legume genera such as *Erythrina*, *Canavalia*, *Derris*, *Phaseolus*, *Indigofera*, *Caesalpinia*, *Albizzia* and *Sophora*, and *Parasponia* (= *Trema*), a non-legume. In 1905 there were at least 13 legume species in the flora. Corresponding figures for 1934 are 21; for 1951, nine; and for 1979, 11 or 12. A total of 24 legume species has occurred in the flora. Some lichens containing cyanobacteria also inhabit the island. It is very likely, also, that bacteria in the rhizosphere of the tall grass *Saccharum* were a source of nitrogen. The role of such plants as vigorously growing pioneers on new or disturbed ground is understood more easily when it is realized that they carry their own nitrogen-supplying equipment with them. They are usually supplanted by other species which make use of recycled nitrogen from the pioneers and then pre-empt their living space. Some of the original shore plants would also have had supplies of nitrogen from the organic debris washed up along the beaches.

Patterns of vegetation change on Krakatau

Figure 4.7 shows that some communities (dominated by *Barringtonia* and *Ipomoea*), usually strictly littoral in established vegetation systems in Indonesia, were widespread on the lower slopes of Rakata in the early decades after the eruption, but have since become restricted as the burgeoning land vegetation crowded them into the coastal margin. Formerly common species have been stifled by more-efficient competitors. *Casuarina*, once dominant in some uncrowded habitats, now has a very limited distribution on Rakata. Continued disturbance by intermittent eruptions on Anak Krakatau maintains habitat open enough for *Casuarina* which, in 1979, was common there, along with much *Ipomoea* and *Saccharum* (Fig. 4.9).

The forest of Rakata is now lush, floristically moderately diverse and, at lower altitude, dominated by the tall *Neonauclea* trees (Fig. 4.10). The early establishment of *Neonauclea* at altitudes above 400 m and its subsequent invasion of the *Macaranga–Ficus* forest at lower altitude is probably due to, first, the moisture conditions on the higher slopes being suitable for its seedlings to establish. Secondly,

Figure 4.9 Colonizing vegetation on Anak Krakatau, 1983: scattered tussocks of the grass *Saccharum spontaneum* and a stand of the tree *Casuarina equisetifolia*. Forested Rakata Kecil Island in background (photo R. Whittaker).

Figure 4.10 View of the forest on the main Krakatau island, Rakata, from its southeastern shore, 1983 (photo J. Flenley).

the lower slope conditions were unsuitable for its seedlings until the water-holding capacity had been improved by the build-up of soil humus. *Ficus* spp. have replaced *Neonauclea* as dominants in higher-level forests, and may eventually become more-prominent at lower levels, also.

By now the most mature forest vegetation has a continuous canopy of broad-leaved trees up to about 20 m high, some with large stems (1 m or more in diameter) with buttressed bases. They are festooned with creepers, vines and epiphytes. The understorey consists of lower trees, treeferns and shrubs. On the ground is a discontinuous cover of ferns, some forbs and creepers, and an almost complete cover of leaf litter.

Disturbance of the developed inland forest vegetation on Rakata results in a return to earlier phases of vegetation development (as noted, the cyanobacterial phase, usually followed by a telescoping of fern and tall grassland phases, before return through scrubby forest to *Neonauclea*). The most frequent disturbances are land-slides, but there was a fire in 1919 on the western side of the island, and from time to time there is some human disturbance – clearance of plots for coconut groves (van Borssum-Walkes 1960).

Some preliminary studies of soil formation on Rakata have been carried out only recently (Flenley 1979, Flenley & Richards 1982, Whittaker *et al.* 1989). The substratum is unconsolidated volcanic ash and pumice. The soil profiles are very immature. Shallow organic accumulations formed from plant litter rest on only slightly altered parent material which shows some signs of clay formation and organic staining. The soil is relatively rich in the bases calcium and magnesium, and is neutral to slightly acid.

MOUNT TARAWERA, NEW ZEALAND

There are no quantitative data on the rate of nitrogen accumulation in the living systems on Surtsey and Krakatau. Some values for nitrogen accumulation in other developing ecosystems are discussed in later chapters. The role of one plant species and its associated nitrogen-fixing symbiont in the colonization of volcanic debris on a New Zealand volcano is mentioned here because it illustrates a common pattern of vegetation development on new surfaces.

Mount Tarawera (1110 m), in North Island, erupted violently in 1886, destroy-ing all of the forest vegetation on its upper slopes and burying the remains under thick layers of scoria and lapilli. Studies by Aston (1916), Turner (1928), Nicholls (1959), Burke (1974), Druce & Ogle (1978), Clarkson (1980) and Timmins (1981) have determined the main patterns of re-establishment of the vegetation on the raw scoria of the devastated upper part of the mountain (Fig. 4.11). Tall shrubs and low, short-lived trees are now well-established over most of the slopes, and the juveniles of taller, longer-lived trees are beginning to be apparent. A rich flora from which colonists could be drawn was nearby.

Although the rainfall on Tarawera is moderately high and evenly spread (annual average rainfall for a nearby site 142 cm, mean temperature 11.1°C, mean January

Approximate timescale	10 years	15 years	25–30 years	50–70 years	100 years	c. 200 years (predicted future)
Main growth forms	mat herb-tufted grass	low heath shrub	tall shrub (to 5 m)	low broad-leaved angiosperm forest (to 15 m)	moderately tall broad-leaved angiosperm forest (to 22 m)	tall broad-leaved angiosperm forest (to 26 m) with emergent podocarps
Prominent species	*Racomitrium lanuginosum*, *Cladonia* spp. → *Raoulia* spp., *Muehlenbeckia axillaris*, *Rytidosperma* sp., *Luzula* sp., mosses, lichens	*Gaultheria* spp., *Cyathodes* spp., *Dracophyllum* spp., mosses, lichens	*Leptospermum scoparium*, *Olearia arborescens*, *O. furfuracea*, *Coprosma robusta*, *Gaultheria* spp., *Cyathodes* spp., *Hebe salicifolia*, some ferns	*Weinmannia racemosa*, *Griselinia littoralis*, *Aristotelia serrata*, *Melicytus ramiflorus*, *Carpodetus serratus*, *Brachyglottis repanda*, *Pseudopanax* spp., *Coprosma* spp., many ferns	*Weinmannia*, *Knightia excelsa*, *Beilschmiedia tawa*, *Elaeocarpus dentatus*, *Myrsine australis*, *M. salicina*, pole *Podocarpus hallii*, treeferns, many ground ferns	*Weinmannia*, *Beilschmiedia*, *Elaeocarpus*, *Podocarpus hallii*, *P. ferrugineus*, *Dacrycarpus dacrydioides*, pole *Dacrydium cupressinum*, tree ferns, ground ferns, vascular epiphytes, vines

Coriaria arborea (short circuits timing of tall shrub development by about 15–20 years)

*Initial vegetation development to this stage has occurred at progressively later dates with increasing altitude and exposure: it is still occurring on parts of the summit (i.e. 100 years after the eruption)

Figure 4.11 Vegetation development trends on eastern side of Mt Tarawera, North Island, New Zealand, above 600 m altitude.

Table 4.4 Modes of dispersal of fruit, seeds and spores of vascular plant species above 600 m altitude on Mount Tarawera.

	Birds (fleshy fruit)	Wind	Mammals or birds (hooked or barbed fruit)	Indeterminate	Total
trees	19	2	–	2	23
shrubs	22	8	–	2	32
vines	2	2	–	–	4
herbs	8	44*	5	24†	81
ferns } lycopods	–	48	–	–	48
total	51	104	5	28	188

* 11 Orchidaceae (orchids), 10 Asteraceae (daisies), 13 *Epilobium* spp.
† 11 Poaceae (grasses), probably by wind or on birds, or both.

temperature 12.2°C, mean July temperature 5.5°C), the thick scoria and lapilli deposits on summit and slopes form a very freely drained substrate. For this reason the establishment of plants is difficult. Among the early colonists are lichens, mosses and mat-forming herbaceous angiosperms. These are followed by the woody species, of which a very high proportion is fleshy-fruited. Their seeds are transported by birds. The minute spores of the lower plants and the plumed or tiny seeds of many herbs are distributed by wind (Table 4.4). Five herb species have hooked or barbed fruit, presumably transported on mammal fur or bird plumage. Some herbs, including 11 grass species and a few woody species, have seeds or fruit with indeterminate means of dispersal. Some may be dispersed by wind or by animals.

One of the fleshy-fruited, bird-dispersed species, which is rapidly entering the open communities of low shrubs, moss, lichens and herbs near Tarawera's summit at present, is *Coriaria arborea* (Fig. 4.12). This broad-leaved shrub can grow to 5 m tall or more, and can have a canopy spread of 5 m or more. *Coriaria* was the first woody species to re-invade many sites soon after the eruption (Aston 1916), but it has been

Figure 4.12 (a) *Coriaria arborea* (Poole & Adams 1963). (b) *Metrosideros polymorpha*.

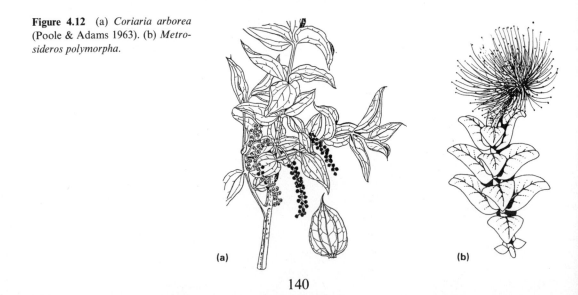

(a) (b)

140

slower to occupy the upper slopes. Timmins (1981) suggests that this is because of difficulty in establishing a connection with its root nodule symbiont *Frankia*, which is not borne on its seeds. The shrub prefers sites that are moderately to well-supplied with moisture, and this may also account in part for its slowness to colonize the whole mountain.

The *Coriaria–Frankia* association rapidly boosts the nitrogen input to developing ecosystems. Silvester (1969, 1977) has shown that the symbiosis can fix up to $192\,kg\,ha^{-1}$ of nitrogen annually – a very substantial amount. On Tarawera no other known symbiotic nitrogen-fixing species are apparent in the bare debris colonized by mosses, lichens and herbs, so they may be nourished by nitrogen from rainfall or free-living organisms. Below the 600 m level on Tarawera the dead wood and plant litter on the old forest soils would have been a source of nitrogen for invading plants. Some logs may also have survived at higher altitude, but only in isolated locations. There, tree roots may ultimately penetrate the scoria to the old buried soils, however. Once *Coriaria* is established, other tall, shrubby species soon follow. In the absence of *Coriaria* several hardy heath-shrubs (*Leptospermum, Dracophyllum* and *Gaultheria*) colonize the low herbaceous plant and moss mats, and the rate of vegetation change is slower.

As the shrub communities grow taller, the initially very dense populations thin out, the canopy becomes somewhat more open, and shade-tolerant young plants of tree species, established within the shrubbery, begin to overtop the shrub canopy. Fern cover (*Asplenium, Blechnum* and *Polystichum*) increases on the ground.

Already a diverse, medium-height forest, dominated by a mixture of broad-leaved angiosperm species, is present near the 600 m level; its height will increase to about 25 m in the next 50 years. The presence of juveniles of potentially taller gymnosperms (*Podocarpus, Dacrycarpus, Prumnopitys* and *Dacrydium*) shows that by the second hundred years the forest will have a canopy of mixed angiosperms with emergent podocarps. In the long term (possibly another 100 years) the slowly growing rimu, *Dacrydium cupressinum*, will probably become important in the vegetation, as it is on surrounding undisturbed sites (Nicholls 1959, Burke 1974).

MAUNA LOA AND KILAUEA, HAWAII, USA

The results of many studies of vegetation development on the basaltic lava flows and other volcanic deposits of the great dome volcanoes Mauna Loa (4171 m) and Kilauea (1247 m) (19° 30′N, 155° 25′W) (Fig. 4.13) were summarized by Smathers & Mueller-Dombois (1974). Their own study was a very detailed observation, over 9 years, of the establishment of plants on a range of different site types, after an eruption from the Kilauea Iki ('Little Kilauea') crater in 1959. About 500 ha of standing forest within a kilometre of the crater rim had been totally destroyed, and most had been buried by scoria and a cinder cone. Further away severely damaged trees and shrubs survived. Some shrubs (*Vaccinium reticulatum* and *Dubautia ciliolata*) sprouted again even when totally buried. *Metrosideros polymorpha* trees within 1–2 km of the vent sprouted from totally defoliated upright stems (some after partial burial).

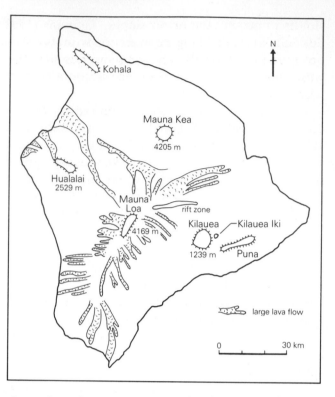

Figure 4.13 The island of Hawaii and its volcanoes.

Measurements of the colonization of surfaces were made in the crater floor of Kilauea Iki (filled with a solid sheet of lava after the eruption) and on the scoria fields outside the crater. The lava and scoria are very difficult sites for plant establishment. Lack of a reliable moisture supply is a main limiting factor. The annual precipitation on Kilauea ranges from about 220 to 330 cm (mean temperature about 18°C, January mean about 17°C, July mean about 20°C) but drainage is very free on the rocky surfaces. The crater floor and cinder cone remained hot for some years after the eruption.

Cyanobacteria, including the known nitrogen-fixers *Hapalosiphon, Scytonema* and *Stigonema*, are always the earliest colonists. Mosses and ferns are also early, though sporadic, colonists. Their first-generation stage, the gametophyte, is very small and not easily noticed, so it is possible that they establish about as early as the algae, germinating in the cyanobacterial mats. The much more prominent sporophytes then develop. A few angiosperms are also early immigrants. Lichens usually follow the algal development; the main one, *Stereocaulon volcani*, contains a cyanobacterial symbiont which probably fixes nitrogen, also. Algae and lichens are able to establish on the surface of the rock, but mosses and vascular plants first settle in crevices or holes. Water conditions are improved in these microsites by the provision of shade and also by the accumulation of fine debris.

In various parts of the world other observations have been made of the prominence of cyanobacteria early in vegetation development on volcanic or other substrates (Doty 1967, Hasselo & Swarbrick 1960, Booth 1941). Not all of these observers have attributed to the cyanobacteria an important role in nitrogen fixation, but Booth was convinced of their very important role in surface stabilization.

142

Other limiting factors for seedling establishment on the bare rocks of the Hawaiian volcanoes are the exposure to strong solar radiation and underground volcanic heating (which slowed the rate of development of the vegetation in the crater and on a cinder cone). Chemicals in the rocks may also inhibit the establishment of lichens.

Smathers & Mueller-Dombois (1974) point out that there is no well-marked sequence of colonist species at Kilauea Iki. Chance factors must determine the first arrivals at a site. After nine years there was still an incomplete cover on the new surfaces.

The woody species *Metrosideros polymorpha* (Fig. 4.12) has tiny, light, wind-transported seed. Its place among the early colonists is interesting because it also is the main dominant in many of the mature Hawaiian forests. In these remote islands a limited number of tree species has evolved to assume various roles. Their ancestors either had to cross wide stretches of the Pacific Ocean to reach Hawaii or are remnants of an ancient, more-extensive flora on a once-larger Hawaiian land mass. *Metrosideros polymorpha* is ecologically and morphologically a very flexible species which can persist in difficult environments as a shrub or stunted tree, but also may be the tall-canopy dominant in old forests. Other features of its versatility are its powers of vegetative regrowth after damage and vegetative proliferation, by production of adventitious roots and development of new stem sprouts on prostrate branches.

Hawaiian chronosequences

Although vegetation changes on some newly formed land surfaces have been observed over one or two human generations, there have been relatively few opportunities for ecologists to study the developmental stages of terrestrial sequences which have been proceeding for longer periods of hundreds or thousands of years. A necessary prerequisite for this is the presence of a true **chronosequence** of land system, soils and vegetation. A chronosequence may be defined as:

> *A sequence of landforms, soils and vegetation, developed on similar terrain and substrates under the influence of constant or ineffectively varying climate, whose differences can be ascribed to the lapses of differing increments of time since the initiation of the sequence.* (Modified from P. R. Stevens & Walker 1970.)

Chronosequences spanning 300–400 years or more on basalt lava flows of Kilauea and Mauna Loa are described by Atkinson (1970) (Fig. 4.14). In his study area mean annual temperature is 23.0°C, mean summer temperature 23.5°C and mean winter temperature 22.0°C. The annual average rainfall is 190–345 cm (more at higher altitude).

Atkinson took care to ensure that his sites fulfilled the requirements for a chronosequence. All of the vegetation sequences which he described are relatively simple. Near the coast, in a moderate precipitation area (178 cm), *M. polymorpha* forest becomes established by about 220 years, and apparently is eventually replaced by the palm-like monocotyledon *Pandanus tectorius*, probably because the latter is more tolerant of salt spray. In places with a summer dry period, mature *M. polymorpha*

Approximate timescale	5 years	15 years	130 years	220 years	c. 360 years	unknown age (? c. 500 years)
Near coast (rainfall 178 cm year⁻¹)	*Stereocaulon, Nephrolepis* →	*Nephrolepis,* scattered *Metrosideros* shrubs →	scattered *Metrosideros* trees/ *Metrosideros* shrubs →	*Metrosideros* forest →	*Metrosideros* with juvenile *Pandanus* →	*Pandanus–Metrosideros* or *Pandanus* forest
Inland (rainfall 200–355 cm year⁻¹)		*Dicranopteris,* scattered *Metrosideros* shrubs →	scattered *Metrosideros* trees/ *Dicranopteris* →	*Metrosideros* forest (*Dicranopteris* may persist in places on smooth lava) →	*Metrosideros* forest →	*Metrosideros* forest

Figure 4.14 Vegetation development trends on lava flows in Hawaii (from Atkinson 1970).

forest is accompanied by the lower *Diospyros ferrea*. Inland, where the rainfall is moderate to high, mosaic patches of the fern *Dicranopteris linearis* and *M. polymorpha* coexist for long periods, probably culminating in pure *M. polymorpha* forest. Where the rainfall is highest, the mature forest contains *M. polymorpha* and *Cibotium* spp. (treeferns).

Nephrolepis, after its early establishment, often inhibits further establishment of *Metrosideros* for a time. It pre-empts nearly all available microsites, but eventually the patches of *Metrosideros* coalesce, forming a canopy over the fern ground cover.

Moisture conditions in particular sites seem to be a very important control of the rate of vegetation change, as well as determining differences in composition of some of the mature vegetation. Scoria and the rough *aa* lava, because of their porosity, are more easily colonized than smooth *pahoehoe* lava. However, unless precipitation is frequent, the free drainage of the porous substrates makes them difficult habitats. Bouldery substrates are less favourable than pebbly or finer substrates, which, as a consequence, support taller *Metrosideros* forest. The amount of cracking and fissuring of the pahoehoe lava determines the suitability of sites for establishment and root extension of *Metrosideros*.

Soil formation is very slow on the hard rock lava flows; very little rock weathering has occurred by 300 years or more (Atkinson 1970). More time is needed for mature soil formation, and presumably for the development of mature vegetation. Eggler (1971) found that fine soils on lava flows often consisted of wind-blown debris originating from distant sources.

SOME CONCLUSIONS

Some important principles arise from the examination of vegetation development on volcanic rock deposits. First, the ability of plant propagules to be transported for considerable distances by natural means is amply illustrated, especially by the data from Surtsey and Krakatau. The order in which organisms arrive at a new site depends on their abundance in the biosphere and the facility with which they are transported, which to a great extent depends on whether they are specialized for carriage over long distances by ocean currents, wind or birds. Micro-organisms (including bacteria, fungi and some algae) are very abundant in nature; they or their spores, invisible to us, are being constantly dispersed into the air and carried about by the wind. Wherever an appropriate substrate is encountered, they settle and go about the business of reproducing more of their kind. The lichens, mosses and, sometimes, ferns are also quite abundant. Their tiny spores are also transported by wind to new sites. Although the angiosperms and some gymnosperms are also abundant, most have relatively large propagules which are not so readily dispersed as those of the organisms already mentioned. Nevertheless, many angiosperms and some gymnosperms have evolved effective means of long-distance dispersal. Empirical studies show that seeds of certain taxa are carried for long distances in the air, in water, or on the fur or feathers of animals, and can become established when they are brought to new sites (cf. Ridley 1930, J. M. B. Smith 1978).

The abundance of seeds (or other propagules) produced by established populations in adjacent areas, and the distance across which the propagules must be transported, interacting with properties of the propagules and the agencies of transport, will determine which species, what numbers and the rates of arrival at a new site. Persistence is assured by establishment and growth to sufficient size for the species to be able to reproduce (either by seeds or by vegetative proliferation). By vegetative spread close to the initial site of colonization, large populations may be built up locally, in relatively favourable sites. They also provide further propagules for colonizing further afield. Seeds or spores will usually be the most effective way of spreading a species widely over the new surfaces, but some vegetatively proliferating plants are also effective in this respect.

When a plant arrives at a brand new site, it usually faces very hazardous environmental conditions. In particular, it must overcome the problems of extreme temperatures, limited water availability and low supplies of certain nutrients that are necessary for growth. Rock and other bare surfaces heat readily when the Sun shines, often to temperatures unbearable to most plants (cf. Richardson 1958). At sundown there is rapid cooling by re-radiation, and very low temperatures may be experienced within a few hours. Unless precipitation is frequent, moisture supplies will be deficient on exposed rock surfaces, not only because of the efficient drainage, but because of strong solar heating. However, pores and crevices provide micro-environments that are both shaded from the worst direct effects of heating and cooling, and buffered from strong evaporation. Condensation at night in holes and pores provides a regular moisture supply (a phenomenon noticed also in well-drained sand dune soils and stony sites; Ranwell 1972).

The third hazard, lack of appropriate nutrients, can be overcome by imports of materials to the site from developed ecosystems in the form of dead plants and animals or animal excreta, imports in the rainfall, and by the activities of the pioneer organisms themselves in fixing nitrogen and abstracting nutrients from the rocks by causing them to disintegrate to their constituent elements (i.e. to weather).

Settlement of new surfaces is usually first done by monerans, lichens and lower plants. These have the hardiness to withstand the climatic hazards, and many also fix their own nitrogen. Their requirements for nutrients are in proportion to their small size. They are enabled, also by their small size and depressed growth habit, to take advantage of any favourable microtopography. Fractions of a millimetre difference in height might spell the difference between life or death in such circumstances. Despite their limited size, it is the lower plants, algae, mosses and lichens which often begin the transformation of substrate and creation of a microclimate and nutrient pool which enhances their environment and insulates them from its worst extremes. By doing so they provide habitat suitable for plants with slightly more-exacting requirements including some higher plants (Lee & Hewitt 1982, T. A. Jackson 1969). The detailed modifications of habitat – changes in raw substrates wrought by small, humble plants – are reviewed by R. Cooper & Rudolph (1953) and Jacks (1965).

As a simple soil is built up, nevertheless containing a complex ecosystem with vast numbers of bacteria, actinobacteria, cyanobacteria, algae, fungi and a microfauna (acarid and oribatid mites, collembola and protozoa), the processes of nutrient cycling are well under way. The heterotrophic organisms rely on exudates from the original photosynthetic colonists, as well as on the dead remains of them and all other organisms present, and the animal faeces. Rock minerals are changed, new secondary minerals are formed and the first clay colloids, together with organic colloids from the humus, provide the soil with effective ion-exchange surfaces. It is into this developed, but small-scale system that higher plants intrude. Rarely, if ever, can they be the first pioneers on new surfaces, but the small size of the real pioneers often causes them to be overlooked or ignored.

The pace quickens with the advent of larger plants with symbiotic nitrogen-fixing associations. The developing systems can become increasingly complex as larger plants gain greater control over the microclimate conditions and nutrients. With their shoots high above the ground, they determine the moisture and light conditions for all smaller organisms. With their roots deep in the substrate and by the fall of their dead leaves, they determine the nutrient regime. The role of smaller organisms, including bacteria and fungi, must not be underemphasized, however. Nor can we regard animals as unimportant in ecosystem development, for many (those listed earlier, plus many kinds of arthropods and worms) participate in processes of litter breakdown and decay. Not only do they reduce litter to sizes that are manageable by the microbes, but also their enzyme systems aid in the decay process, so that they and the microbes perform important functions in the reduction of organic detritus to its ultimate mineral components (Greenfield 1981). The rate of ecosystem development on new ground will depend on many local factors of substrate, general climate and the plants available to colonize sites.

No detailed studies of changing light or moisture conditions have been done in the sequences described here. Except for the Surtsey data and some rather general information from Krakatau, Tarawera and Hawaii, there are no detailed records of changing plant populations in the sequences. Specific information about the changing microbe populations and the fluxes of nutrients is also lacking.

Sketchy as it is, the information about shifts in the species composition of stands in the sequences on volcanic debris tells us quite a lot about the importance of specific plant lifestyles in determining patterns of vegetation development. Each local situation has some unique aspects. The local floras available to participate in vegetation change vary in richness. Also, there are differential patterns resulting from the numbers of propagules being dispersed, the ease with which dispersal occurs and the distances which the propagules must travel. In some localities associations of vascular plants with nitrogen-fixing organisms are important. They are correlated with faster rates of change than in places where nitrogen is supplied in other ways.

The flexible, very broad fundamental (and realized) niche of *Metrosideros polymorpha* in Hawaii is notable. In places with a richer tree flora, narrower realized niches are apparent but fundamental niche dimensions are not so clear. However, on Krakatau the widespread distribution, early in the sequence, of some species which are later confined to colonist or marginal habitat roles, suggests that colonist species often have wide fundamental niches but weak competitive power. Also on Krakatau the behaviour of *Neonauclea* suggests that its fundamental niche is relatively narrow, compared with early colonists. At first *Neonauclea* inhabited only sites where allogenic control of habitat conditions permitted this. Later, the change of soil moisture conditions by autogenic means allowed it to expand its distribution. It seems to be a moderately strong competitor, able to establish and hold its own in crowded habitats, at least for a time. Extreme shade or other factors may account for its subsequent replacement by *Ficus* spp.

Compared with the Hawaiian and Krakatau situations, it is interesting that some New Zealand tree species of *Metrosideros* are colonists on volcanic debris (and in other open sites), but restricted in crowded habitats. For example, *M. excelsa* invades raw volcanic debris, but strong competition from other forest tree species confines it otherwise to seashore, lake shore and cliff habitats.

5

Vegetation development on sand dunes

Coastal sand dunes in Europe (Warming 1891) and lake shore dunes in North America (Cowles 1899a, b, c) were among the first sites of description of relatively long-term vegetation changes. They were also early sites for experimental study of environmental changes correlated with vegetation change. G. D. Fuller, a student of H. C. Cowles, wrote in 1914 of his measurements of evaporation and soil moisture in plant communities believed to form a successional sequence on the Lake Michigan dunes.

More recently several detailed studies of sequences of plant species replacements on dunes have focused on changing soil properties with time, improving our understanding of the nature of long-term ecosystem development (Salisbury 1925, 1952, Burges & Drover 1953, Olson 1958, Syers & Walker 1969a, b, Enright 1978, J. Walker *et al*. 1980, C. H. Thompson 1980). Dunes are among the most useful sites for chronosequence studies, because of the uniformity of the substrate materials.

Dunes are landforms found in many parts of the world where there are both large supplies of sand and winds which pile it up into rather unstable heaps with intervening hollows. They occur in deserts where there may be little or no vegetation and on river floodplains, as well as along the sea coast or the shores of lakes (Bagnold 1971). The discussion here is restricted to coastal and lake shore dunes. There, large supplies of sand, brought on to beaches by currents and wave action, are blown by onshore winds into mounds just beyond the high-water mark. The sand is usually composed of a large amount of the inert mineral quartz but, depending on the source, there will be different proportions of other minerals, some of which can weather to release the various plant nutrient cations and other nutrients such as phosphorus and sulphur. Fresh sand may contain considerable amounts of calcium carbonate, derived from the skeletons of aquatic animals, as well as other organic remains.

The complexity and size of the individual dunes, and the dune system as a whole, depend on a number of factors, including the constancy and quantity of sand supply; the texture (size of individual grains) and the specific gravity of the grains; the regularity of wind force and direction; and the sand-binding plants available to colonize the dunes. The foredunes may or may not be stabilized. Parallel with the prevailing wind direction will be wind-funnel hollows. However, the dunes often lie in ranks parallel to the beach (Figs. 5.1 & 2). Unstable foredunes may form and then march inland as the sand on dune crests is blown forward to fall down the steep lee slope and is then replaced from the windward side by more sand. The plants that inhabit foredunes are often highly specialized species which do not necessarily participate in a vegetation sequence. Other species colonize the rear dunes and fix them more or less firmly in place. Blowouts during heavy windstorms may set stable dunes in motion again, but eventually they become fully stabilized. Often the ranks of

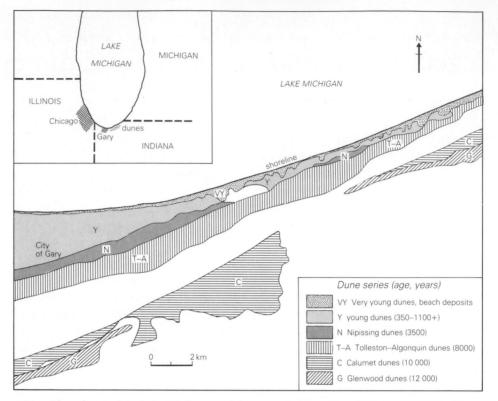

Figure 5.1   Plan of part of the Lake Michigan sand dune complex, near Gary, Indiana, USA (after Olson 1958).

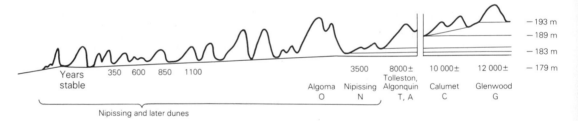

Figure 5.2   Diagrammatic section through the sequence of Lake Michigan sand dunes, near Gary, Indiana, USA (after Olson 1958).

dunes farthest inland are more subdued in height and slope because they have lost sand which has not been replaced. A dune system consists not only of the irregular hills and ridges of sand, but also of intervening hollows which may be sand-flats, or valleys, some of which may carry lakes and marshes (Fig. 5.3). Environmental conditions for plants are variable in such complex terrain (Esler 1970, Cowles 1899a, b, c).

Some geomorphic situations lead to the occurrence of long time-series of dunes. One is the sudden or gradual **tectonic** uplift of a shore (i.e. the upward movement of a block of land relative to lake- or sea-level, impelled by a shift in part of the Earth's crust). Another is a fall of the water level in a lake, due to some change in its regime. In effect these changes shift the shoreline further into the sea or lake, creating the

Figure 5.3 Oblique aerial view of some of the younger dune sets, Manawatu dunes, Tangimoana, North Island, New Zealand. Beach in foreground marked by line of breakers. Young foredunes with grasses, sedges and low shrubs. Rear dune sequence consists of scrub, fern and grass-covered dunes, then, at top right, older dunes with patches of forest. Wet sand flat on left and in centre (photo Manawatu Standard Ltd).

potential for dune formation at distances from the previous shoreline which depend on the amount of vertical change and the general slope of the land surface.

THE INDIANA DUNES, LAKE MICHIGAN, USA

The first studies of soil changes in relation to vegetation development on dunes, in which the dune sequence was reasonably well-dated (spanning about 300 years) were those of Salisbury (1925, 1952), in Britain. Olson (1958), working on the Indiana dunes of Lake Michigan (the scene of Cowles' early dune research), was the first, however, to make a chronologically well-documented study of vegetation on a long-term dune sequence. 'Long-term', for the sake of this discussion, is defined as being 1000 years or more. The dunes (41° 35′N, 87° 20′W) experience a mean annual temperature of 10°C, with a July mean of 23°C, and a January mean of −4°C. Annual average precipitation is 85 cm.

The ages of the dunes were obtained from historical records, maps and photographs, measurements of tree ages and other plant markers of the elapse of time, and

151

from radiocarbon dates for organic material buried in some of the dunes. The oldest dunes were formed about 12 000 years ago (Fig. 5.2).

A note of caution is needed about not confusing **toposequences** (vegetation distributed according to spatial environmental gradients) with chronosequences. Foredunes, although they may be colonized by sand-binding grasses and shrubs, remain as foredunes because they are exposed to the full force of wind and weather. Trees can only live in the more-sheltered sites further inland. Unless the foredune sand migrates inland and becomes stable there (or the water level falls, or the land level rises), the foredune at a particular place remains in the same position and has the same plant cover. This will be so even if some of its constituent sand is moved inland by the wind as new sand is added to it from the beach.

Olson carried out concordant studies of differences in the plant populations present and the soil properties on dunes of successively greater age. He pointed out the great heterogeneity of habitat conditions among the dunes and stressed the difficulty of finding sample areas that are exactly comparable in all significant characters on each age-group of dunes, so that criteria for a chronosequence could be assumed to have been met. He also noted the unlikelihood that a vegetation sequence could be found, extending over such a great length of time as 10 000–12 000 years, which had not experienced some major disturbing events that would override the effects created by the vegetation chronosequence alone. Nevertheless, parity of general site conditions in the sequence of dunes aged 0–8000 years encouraged him to regard the present vegetation series as a close approximation to a chronosequence; all factors except time have been reasonably constant. The two older dune systems, 10 000 and 12 000 years old (Fig. 5.2), have soils with greater than expected amounts of silt and clay, mingled with the sand. This wind-blown material, derived from areas behind the dunes, probably accounts for some unexpected differences in soil chemistry and physical properties, but the older dunes are included with the others to give an indication of the kinds of ecosystem changes which probably occur over that time period.

The most usual trend of vegetation development on these Michigan dunes is shown in Figure 5.4. After early colonization by sand-binding grasses, prostrate shrubs such as sand cherry (*Prunus pumila*) or the low tree, cottonwood (*Populus deltoides*), jack pine (*Pinus banksiana*) often invade. Jack pine dominates from about 90 years onward and lasts for a few centuries. Black oak (*Quercus velutina*) may enter the sequence early, following shrub or cottonwood communities after about 150 years, or it may follow the jack pine (Fig. 5.5). Very commonly black oak, then, is the dominant tree for up to 10 000 or 12 000 years. This is quite contrary to the belief of Cowles about the 'climax' vegetation on the Michigan dunes, and the general view of Clements, that succession ends in the most mesic vegetation for the area. In this region the most mesic and structurally complex vegetation is deciduous angiosperm and conifer forest dominated by beech (*Fagus grandifolia*), maples (*Acer* spp.), hemlock (*Tsuga canadensis*), birches (*Betula* spp.), basswood (*Tilia americana*) and tulip tree (*Liriodendron tulipifera*). Black oak, with its large seeds, which are readily spread by squirrels, is clearly a hardy and versatile species, well-suited to the majority of dune habitats, which have water-limitations for plant growth at times and also

152

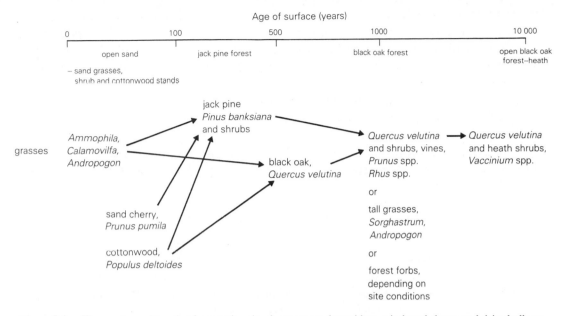

Figure 5.4 The most usual trends of vegetation development on dune ridges, windward slopes and drier hollows, Indiana dunes, Lake Michigan (modified from Olson 1958).

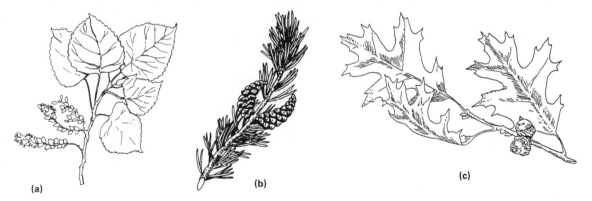

Figure 5.5 Some plants important on the Indiana dunes, Lake Michigan, USA. (a) Cottonwood, *Populus deltoides*; (b) jack pine, *Pinus banksiana*; (c) black oak, *Quercus velutina*.

limited soil fertility. It also survives fire by sprouting from burnt stems. The beech–maple–hemlock forest stands are found among the dunes only on relatively fertile sites that are reliably supplied with moisture, such as some sheltered rear-dunes and dune hollows.

Variations on the relatively simple modal vegetation sequence are found in the wide variety of habitats present in the dunes (Fig. 5.6), but many of the habitats on which alternative sequences occur are of limited extent.

The distribution on the time sequence of some of the more prominent species is indicated in Table 5.1. The individuality of distribution of each species is evident; this is accounted for, in part, by the maintenance, for different lengths of time, of particular sets of environmental conditions, to which different species are suited. The black oak forests are, in fact, quite heterogeneous in their ground conditions, because of the diverse topography. Different sets of understorey and field layer species are associated with particular sets of environmental conditions, with black oak simply being the main, dominant tree overlying this heterogeneity. On dunes that are older than about 1000 years, and more noticeably from about 8000 years onwards, the black oak stands lose some of their vigour. On the older dunes occur rather open stands of trees or mosaics of patches of trees with patches of heath shrubs, especially blueberries (*Vaccinium angustifolia* and *V. vacillans*).

In terms of the establishment and then long dominance of black oak in this system there is little indication, other than the initial binding of sand, of any necessity for preparation of sites by specific precursors. Black oak can enter the shrub stands of sand cherry or cottonwood stands (both early sand-binders), or it can replace jack pine. There is no indication of nitrogen fixation by free-living or symbiotic organisms during the early stages of vegetation development on the dunes. Some of the herbaceous species are known to have a *Rhizobium* association (legumes in the genera *Lathyrus*, *Lespedeza* and *Lupinus*), but they are of only local occurrence. *Ceanothus americanus* has root nodules with *Frankia*, but it is present only on dunes which are 150–850 years old. Some beach debris is a source of nitrogen and other nutrients for the colonists.

Olson (1958) accounted for the amounts of nitrogen present in the system in the early phases of the sequence (Fig. 5.7) (an annual income of about 4.0 kg ha^{-1} in the top 10 cm of soil), in terms of supplies from the rainfall. It is estimated that as much.

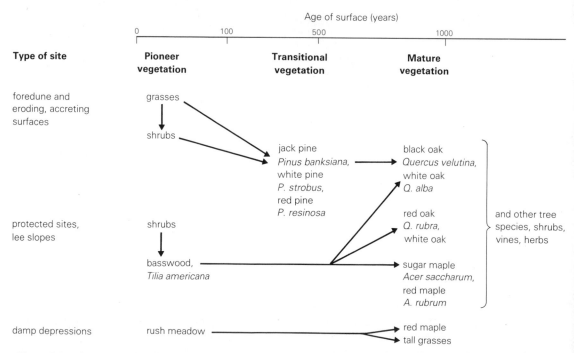

Figure 5.6 Some other trends of vegetation development on the Indiana dunes, Lake Michigan, USA (after Olson 1958).

Table 5.1 Indiana sand dunes, Lake Michigan: relative abundance of the more-prominent plant species on dunes of successively greater age (modified from Olson 1958).

Plant species	Type of plant*	Age of dune system (years)					
		20–40	90–350	600–1100	1700–1800	10 000	12 000
Juniperus virginiana	s,t	2	2	–	–	–	
Cornus stolonifera	s	3	2	–	–	–	
Prunus pumila	s	3	2	–	–	–	
Juniperus communis	s	2	3	–	–	–	
Rhus radicans	v,s	1	3	2–3	2	2	
Vitis riparia	v	2	3	3	3	2	
Prunus virginiana	s	2	3	3	2	2	
Rhus aromatica	s	1–2	2–3	2	–	–	
Artemisia caudata	h	3	2	2	–	1–2	
Calamovilfa longifolia	h(g)	3	2	1	1	–	
Andropogon scoparius	h(g)	3	2	2	2–3	–	
Asclepias syriaca	h	3	2	1	1–2	–	
Lithospermum croceum	h	3	2	2	3	2	
Monarda punctata	h	2	2	2	2	1	
Panicum villosissimum	h(g)	2	2	–	–	–	
Solidago racemosa	h	3	2	–	–	–	
Pinus banksiana	t	1	2–3	2	–	–	
Quercus velutina	t	1	3	3	3	3	
Arctostaphylos uva-ursi	s	–	2–3	2	1	–	
Celastrus scandens	v	–	3	3	–	–	
Parthenocissus quinquefolia	v	1	1	1	2	2–3	
Rosa blanda	v	–	3	3	3	2–3	
Ceanothus americana	s	–	2	2	2	–	
Aster linariifolius	h	–	2	2	2–3	–	
Helianthus divaricatus	h	–	2	2	3	2	
Koeleria cristata	h(g)	1	2	2	2–3	1–2	
Polygonatum pubescens	h	–	2	1	2	3	
Aquilegia canadensis	h	–	1	2	2	–	
Maianthemum canadense	h	–	2	3	2	1–2	
Smilacina stellata	h	–	3	3	2	–	
Solidago caesia	h	–	3	2	2	2	
Gaylussacia baccata	s	–	–	2	2	2–3	
Vaccinium angustifolium	s	–	–	2	2	2	
V. vacillans	s	–	–	3	2	2–3	
Carex pensylvanica	h	–	1	3	3	3	
Polygonatum canaliculatum	h	–	–	2	2	2	
Rudbeckia hirta	h	–	1	–	2	–	
Coreopsis tripteris	h	–	–	–	2	1	
Sorghastrum nutans	h(g)	–	–	1	2	1	
Hamamelis virginiana	s,t	–	–	–	3	3	
Prunus serotina	t	–	–	–	2	2	
Sassafras albidum	t	–	1	–	2	2	
Smilacina racemosa	h	–	–	1	2	2	
Acer rubrum	t	–	–	–	2	2	
Helianthemum canadense	h	–	–	1	2	2	
Pteridium aquilinum	h	–	1	–	2–3	2	

Relative abundance: – absent; 1, rare or local; 2, uncommon–moderately common, sometimes locally abundant; 3, abundant to dominant.

* h, herb; (g) grass; s, shrub; v, vine; t, tree.

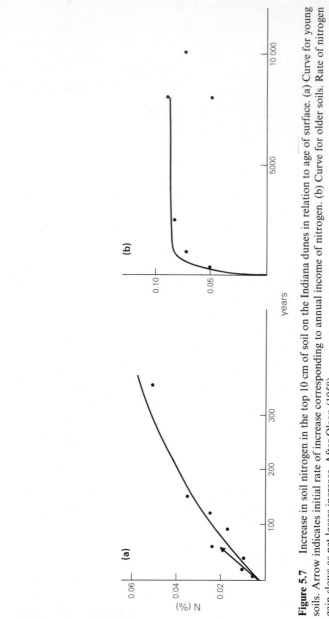

Figure 5.7 Increase in soil nitrogen in the top 10 cm of soil on the Indiana dunes in relation to age of surface. (a) Curve for young soils. Arrow indicates initial rate of increase corresponding to annual income of nitrogen. (b) Curve for older soils. Rate of nitrogen gain slows as net losses increase. After Olson (1958).

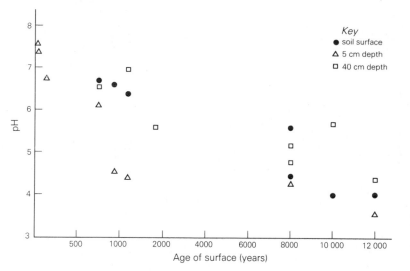

Figure 5.8 Changes in the pH of soils, with time, on the Indiana dunes, Lake Michigan, USA (after Olson 1958).

as 4.6–9.2 kg ha^{-1} of nitrogen accrue in this way to the top 15 cm of soil, in the northeastern USA. (Northeastern Soil Research Committee 1954). However, there may be yet unrecognized free-living or symbiotic nitrogen-fixers associated with the rhizosphere of the early colonist plants.

Chemical analyses of the soils on the dune chronosequence indicate changing conditions which are both caused by the action of plants and influence them in numerous ways (Figs 5.7–10). Some of Olson's analytical methods were not very precise, and his samples, unfortunately, were not taken on a volume–weight basis. Definite *weights* of soil are used to carry out soil tests. To relate these to the actual volumes of soil from which plant roots obtain their nutrients and water, and to be able to calculate the absolute quantities of nutrients in the different soils, definite *volume* samples of soil must be taken and their densities measured. The older soils, with much added organic matter, are less dense than the raw beach sands. Thus, Olson's measurements can be used only as a broad indication of the general trends along the chronosequence in the sites inhabited by black oak. The parameter pH (relative acidity) is independent of the volume of soil.

Calcium carbonate declines to near zero in the top 10 cm of soil by 600 years. By 1200 years none is left in the top 2 m of soil, below the root levels of most plants. Parallel with these changes the pH in the topmost few centimetres of soil declines from 7.6 (distinctly alkaline) until by 8000 years it is only 4.4 (markedly acid) in some sites (Fig. 5.8). At depth 10 cm the pH is 4.4 by 1200 years and as low as 3.6 by 12 000 years. Thus, the upper soil levels, especially those within stands of blueberry, are extremely acid. The acidity of the soil controls the rate of decay of litter and the rate of return of nutrients to plants by cycling (see Ch. 2). In acid soils, also, weathering of minerals from the parent material and leaching of soluble materials, including plant nutrients, are promoted, so that nutrients end up deep in the soil. The free-draining quality of the sandy soil ensures that leaching is rapid. Soils under dune grassland and black oak that have an understorey of mixed, tall, broad-leaved shrubs are less acid than those under blueberry.

157

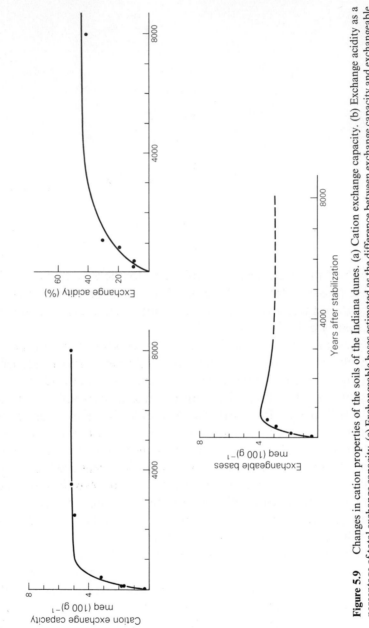

Figure 5.9 Changes in cation properties of the soils of the Indiana dunes. (a) Cation exchange capacity. (b) Exchange acidity as a percentage of total exchange capacity. (c) Exchangeable bases estimated as the difference between exchange capacity and exchangeable acids. After Olson (1958).

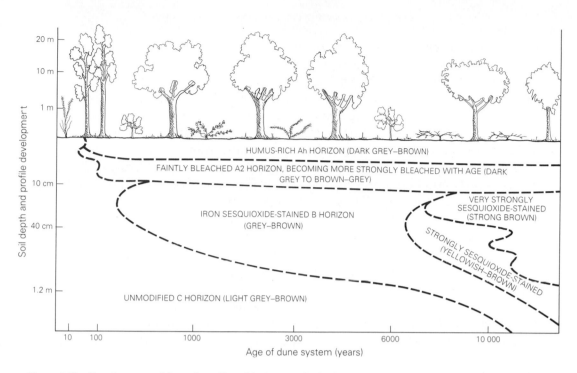

Figure 5.10 Development of the soil profile, with time, on the Indiana dunes, Lake Michigan (after Olson 1958).

The cation exchange capacity (CEC) in the top 10 cm of soil (a measurement of the development in the soils of a colloidal cation-exchange complex, dependent here on the content of humus) rises rapidly at first, then levels off (Fig. 5.9). Soils older than 8000 years, with their content of clay, have higher CEC. Much of the exchange complex in the soils older than about 1000 years is occupied by hydrogen ions (or possibly aluminium ions) as indicated by the pH. The amounts of exchangeable bases (Ca, Mg, K, Na and NH_4) available for use by plants peak at about 1000 years and generally decline thereafter until 8000 years. The total amount of nitrogen in the soil rises sharply up to 1000 years, then levels off. However, the amount available to plant roots depends on the amount of carbon present. If there is too much carbon, in relation to nitrogen, the microbes decaying plant residues use all the released nitrogen. A favourable ratio of carbon:nitrogen is about 12 or less. In the upper soil layers this is met only in marram grass and cottonwood stands. In pine and black oak stands the carbon:nitrogen ratio is unfavourable. Continued supplies of nitrogen from the rainfall and from free-living organisms or symbionts will be important for maintaining or increasing plant productivity.

More-precise measurements are needed to establish, accurately, the details of the nitrogen economy, as well as the relative fertility of sites of different age with respect to other plant nutrients. Although most of the feeding roots of plants are in the top 10–20 cm of soil, comprehensive analyses need to be done at intervals below the surface to the full depth penetrated by roots. Some pH measurements in the upper soil layers on dunes that are 700–1200 years old suggest that materials being brought into the plants from well below the surface are helping to maintain somewhat enhanced nutrient supplies. The litter from the plants is, thus, less acid than might be

159

expected. However, in general the picture is one of soils of only moderate to low initial fertility becoming decreasingly fertile with time.

Coincident with some of the profound changes in plant nutrient budgets in the soil, distinct soil layers begin to be apparent. After about 1000 years the soil profile has a deep organic horizon overlying a thin, faintly bleached A2 horizon. The bleached horizon is found beneath accumulated black oak litter, but especially under litter in blueberry thickets. This indicates that soil acidification, leaching and movement downward of iron and aluminium from the upper mineral soil horizon (the process known as **podsolization**) are occurring. The soil organic component is of **mor** type, acid and slowly decaying. Thus, it has relatively low capacity for recycling of nutrients. By 8000 years the bleached A2 horizon is much more evident and, in the B horizon beneath, strong yellow-brown colours indicate the weathering of minerals and accumulation of ferric sesquioxide (Fig. 5.10). This trend continues so that the altered part of the total soil profile becomes deeper in sites up to 12 000 years old. However, the soils are classed as **podsolic** (i.e. not yet true podsols).

Net nutrient losses to the system occur as a result of acid-litter production and downward percolation of acid solutions. The development of forest understorey *Vaccinium* heath (with its very acid litter) seems to be a cause of the increased rate of decline in soil fertility.

The various physical and chemical changes in the soil are unquestionably autogenic. In the absence of plants they would not occur (or, at least, the rate of those changes which did occur would be very much slower). Differences between soil fertility on the black oak dunes and those with 'richer' communities of species such as basswood (*Tilia americana*), tulip tree (*Liriodendron tulipifera*) and sugar maple (*Acer saccharum*) also demonstrate autogenic processes related to the effects of different species on the recycling of nutrients and their retention within the system. Although the initial parent material is the same, the more-efficient nutrient economy of the 'richer' forest, e.g. in the recycling of calcium, results in them having more-fertile soils. Their soil types are of **mull** type (or **mulloid**), with little accumulation of organic material above the mineral soil. The litter is relatively high in bases, and it decays relatively rapidly. Moist sites seem to be important for the initial establishment and maintenance of this soil–vegetation system.

G. D. Fuller's (1914) determinations of evaporation and soil moisture in stands of vegetation that are thought to be representative of phases in the Michigan dune sequence showed that a site with cottonwood had the best supply of water, compared with sites with pine and black oak. The cottonwood site also lost more water by evaporation than the other two sites (usually about twice as much). It seems likely that cottonwood establishes in sites that are supplied relatively well with water, or which retain water readily. In summer wilting levels (which severely limit plant growth) were often approached, or reached, in the soils of both the pine and black oak dunes. Again, the water economy of this chronosequence needs further study.

Table 5.2   Natural vegetation and soil properties, Manawatu dunes, New Zealand (after Syers & Walker 1969a, b)

	Dune series	Beach sand	Waitarere	Motuiti	Foxton	Kaputaroa
	Approximate age (years)	0	50–100	500–1000	2000–4000	10 000–14 000
	Original natural vegetation	Sedge-grass	Scrub	Low angiosperm forest	Tall mixed forest	Tall mixed forest
	pH A horizon	7.7	6.4	5.4	6.0	4.9
	B horizon	7.8	7.7	6.3	6.3	5.3
in 1 m depth of soil ($\times 10^3$kg ha^{-1})	$CaCO_3$	126	20	nil	nil	nil
	clay	19	28	74	185	723
	oxidizable carbon	5.7	13.7	32.1	77.6	147.6
	total nitrogen	0.8	1.7	3.8	7.5	10.5
	base saturation (%)	–	100	67	61	35
in 1 m depth of soil (ppm)	total phosphorus throughout the profile	347	1956	2241	1711	1483
	organic phosphorus throughout the profile	11	169	546	536	635

THE MANAWATU DUNES, NEW ZEALAND

The essential plant nutrient element phosphorus, not measured by Olson in Indiana, USA, is regarded by some soil scientists as playing a fundamental role in soil–biota relationships (T. W. Walker & Syers 1976). Syers & Walker (1969a, b) described changes of the forms in which phosphorus occurs in a coastal dune chronosequence in the Manawatu region, New Zealand (40° 25′S, 170° 18′E) (Table 5.2).

The Manawatu dunes are in a region with relatively mild, moist climate (mean annual temperature 12.6°C, mean January temperature 17.2°C, mean July temperature 8.7°C, mean annual precipitation 89 cm). They occur in four series spanning the period from the present to about 10 000–14 000 years (Table 5.2). After a period, in the first 500–1000 years, when the total amount of phosphorus in the top metre depth of soil rises to about 4800 kg ha^{-1}, the amount falls to about 4300 kg ha^{-1} by 2000–4000 years and about 3000 kg ha^{-1} by 10 000–15 000 years. The amount of calcium-bound, plant-available phosphorus diminishes from about 2500 kg ha^{-1} in the youngest soils to 1200 kg ha^{-1} by 2000–4000 years and about 50 kg ha^{-1} in 10 000–15 000 years. Other forms of phosphorus, unweathered from minerals (and probably only very slowly ultimately available to plant roots, by weathering), diminishes less dramatically, from just over 1000 kg ha^{-1} in the young soils to about 1000 kg ha^{-1} in the oldest soils. Organically incorporated soil phosphorus, however, rises from less than 1000 kg ha^{-1} in the younger soils to about 1000 kg ha^{-1} by 2000–4000 years and more than 1300 kg ha^{-1} by 10 000–15 000 years. In the older soils an unknown amount of phosphorus is included in the living biomass. It may approximately balance the discrepancy in total amounts of soil phosphorus between young and oldest soils, but effectively it is sequestered in a form (approximately in steady state) which is beyond the reach of plant roots until mineralized from dead plant material. Rates of production of litter and its mineralization are approximately in equilibrium. Losses will occur continually by various forms of export, including leaching to depths unattainable by plant roots.

On the oldest soils phosphorus is only available to the plants by cycling from organic matter or by slow weathering from any remaining minerals. Syers & Walker suggest that, over the 10 000–15 000 year timespan, in which the older dunes have existed, there has been a net loss of phosphorus to the system because the sands (and by implication the organic system) have poor retention properties.

SOME AUSTRALIAN DUNES

On sand deposits near Albany, in Western Australia (34° 57′S, 117° 58′E), Enright (1978) showed how the plants present modified the soil conditions (Table 5.3). Some species are restricted to calcareous sand, because of their relatively high requirements for nutrients such as calcium. In time, as litter is built up, other less-demanding species enter the vegetation. The amounts of soil calcium are reduced and the soils become increasingly strongly acidified until the 'calciphiles' can no longer inhabit them. The foliage of the invading species contains gallic acid, which complexes with iron compounds. The iron compounds are taken up by the plants or leached from the upper mineral soil horizons, or both. The podsolization process is thus induced by the chemistry of the litter of these species. It may be presumed that supplies of phosphorus to plants vary in the same direction as those of calcium, although Enright did not measure this. Apatite (calcium phosphate) is an important source of phosphorus for plants.

The Cooloola dune systems, subtropical Eastern Australia

Vegetation and soils on a sequence of dunes ranging in age from newly formed to, possibly, 400 000 years, were described by J. Walker et al. (1980). In this region (26° S, 153° E) (Fig. 5.11) the climate is mild and moist (mean temperature 24.0°C, mean January temperature 25°C, mean July temperature 15°C, mean annual precipitation 150 cm).

The six separate dune systems extend from the sea coast for 10 km inland and cover 24 000 ha. The topography of each system is similar, but the older dunes are at slightly higher altitude than the youngest ones and are generally more subdued. Dating is imprecise; Table 5.4 shows the estimated approximate ages. In all of the dunes the sand is composed mainly of quartz grains coated with sesquioxides of iron and aluminium.

The climate must have varied in such ways as to influence vegetation considerably over the long spell of time since dune systems 6, 5 (and probably 4) were formed. J. Walker et al. (1980) suggest, however, that they have been forested throughout their history. The species composition of the vegetation may have differed, but structure and biomass would have remained about the same on dunes at particular stages of development. Disturbance by fire is likely in some, if not all, of the dunes,

Table 5.3 Differences in plant species distribution and ability to complex with iron in relation to soil type, soil pH and carbonate content on dunes, Albany, Western Australia (after Enright 1978).

Species	Mean frequency of occurrence in samples (%)			Relative order of complexing ability
	Non-podsol	Transitional	Podsol	
Andersonia simplex	0	8	38	1
Leucopogon reflexus	13	74	70	2
Dasypogon bromeliaefolius	0	2	17	3
Melaleuca acerosa	34	60	53	4
Casuarina humilis	8	42	24	5
Hakea costata	3	12	31	6
Bossiaea rufa	21	60	4	7
Leucopogon revolutus	24	26	3	no complex
Pimelea rosea	31	20	16	no complex
Olearia axillaris	12	3	0	no complex
Hibbertia racemosa	26	24	17	no complex
average soil pH	7.2	6.3	5.8	
average soil carbonate content (%)				
50 cm depth	approx. 4	approx. 0.5	<0.5	
1 m depth	10	5	<1	

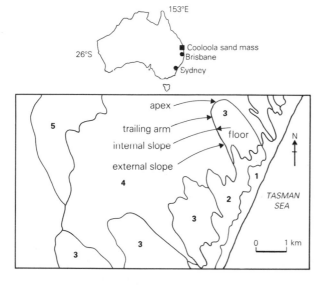

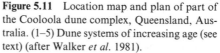

Figure 5.11 Location map and plan of part of the Cooloola dune complex, Queensland, Australia. (1–5) Dune systems of increasing age (see text) (after Walker *et al.* 1981).

and in places this is attested to by charcoal in the soil. Temporal climatic variation is likely to have influenced the vegetation of the older dunes, so strictly it is reasonable to regard only dune systems 1, 2, 3 and possibly 4 as conforming to the requirements for a chronosequence. Dune system 5 is included in the account below, for comparison.

The main features of vegetation and soil changes on the dune systems 1–5 are summarized in Table 5.4 and Figures 5.12 and 5.13. Vegetation, beginning with grasses and shrubs on dune systems 1 and 2, has developed to tall forest on system 4,

Table 5.4 Distribution of plant species on the Cooloola dunes, Queensland, Australia (after Walker *et al.* 1981).

Dune system		1a	1b	2	3	4	5
Approximate age of surface (years)		0–about 100	a few hundred	? < 4000	? > 6000	? > 39 000	? > 100 000
Height above sea level (m)		40	70	100	120–150	240	150
Main vegetation		Pioneer plants on mobile dunes	Shrubby woodland on stabilized dunes	Low grassy forest	Taller grassy forest	Tall forest (rainforest in valleys)	Shrubby woodland
Biomass index*		2	558	780	1400	2300	900
Species	Type of plant†						
Phebalium woombye	S	+	+ ‡				
Pultanaea subternata§	S	+	+				
Banksia integrifolia	T	+	+	+	+	+	
Pteridium esculentum	F	+	+	+	+	+	+
Casuarina littoralis§	T	+	+	+	+	+	
Imperata cylindrica	G	+	+	+	+	+	
Callitris columellaris	T		+				
Casuarina equisitifolia§	T		+				
Exocarpos cupressiformis	S		+				
Acacia leiocalyx§	S		+	+			
Banksia serrata	T		+	+	+		
Banksia integrifolia	T		+	+	+	+	+
Eucalyptus intermedia	T		+	+	+	+	+
Tristania conferta	T		+	+	+	+	+
Eucalyptus signata	T			+			+
Alloteropsis semialata	G			+			
Petalostigma quadriloculari	S			+			
Angophora woodsiana	T			+	+		
Acacia flavescens§	S			+	+	+	+
Themeda australis	G			+	+	+	+
Elaeocarpus reticulatus	S			+	+	+	
Casuarina torulosa§	T			+	+	+	
Xanthorrhoea macronema	'G'			+	+	+	
Leucopogon leptospermoides	S				+		+
Persoonia virgata	S				+	+	
Eucalyptus pilularis	T				+	+	
Leucopogon margoroides	S				+	+	
Leptospermum attenuatum	S					+	+
Banksia aemula	T					+	+
Xanthorrhoea johnsonii	'G'						+
Pimelea linifolia	S						+
Boronia rosmarinifolia	S						+
Acacia ulicifolia§	S						+
Angophora costata	T						+

* The biomass index is derived from crown cover × height of plant.
‡ Boldface indicates that these are prominent plants, characterizing the community.
§ Plant with nitrogen-fixing symbiont.
† S, shrub; T, tree; G, grass; 'G', tufted monocotyledon; F, fern.

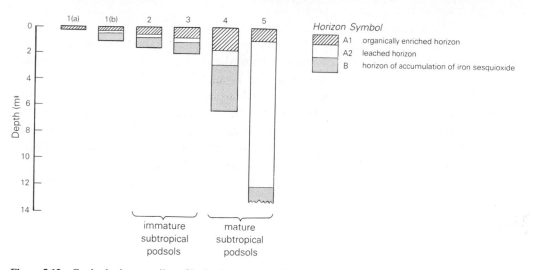

Figure 5.12 Cooloola dunes: soil profile development on the dune chronosequence (dune systems 1–5) (after Walker *et al.* 1981).

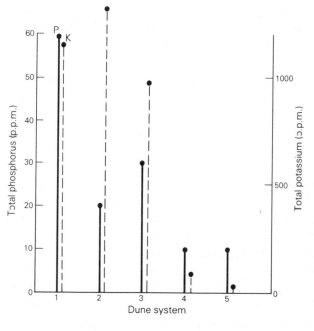

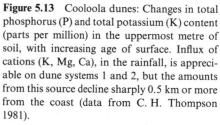

Figure 5.13 Cooloola dunes: Changes in total phosphorus (P) and total potassium (K) content (parts per million) in the uppermost metre of soil, with increasing age of surface. Influx of cations (K, Mg, Ca), in the rainfall, is appreciable on dune systems 1 and 2, but the amounts from this source decline sharply 0.5 km or more from the coast (data from C. H. Thompson 1981).

has begun to decline on system 5, and on system 6 is reduced to scrub or low forest. Plant nutrient supplies in the youngest sands are relatively low, and the amounts of nutrients available to be recycled from organic matter to plants become increasingly important as the system matures. Similar relatively low amounts of nutrients are present in the leaves of individuals of the same species on a range of dune systems, showing that the nutrient demands of each species are similar wherever they occur, and that the return of nutrients to the soil organic matter, in the form of plant litter, will be constant (and relatively small).

Very small amounts of phosphorus are present in the soils of systems 5 and 6,

165

indicating a net loss by leaching and drainage. The vegetation does not retain all recycled nutrients in spite of many of the plants having attributes that are apparently helpful for maintaining efficient phosphorus uptake in conditions of low availability, namely mycorrhizal associations and proteoid roots (Bowen 1980).

Other changes in the soils, concomitant with the changes in nutrient availability, include acidification and podsolization. On all but the very youngest dunes there has been sufficient time for the formation of deep podsols, with thin to deep, bleached A2 horizons and B horizons containing accumulated organic, iron and aluminium compounds. Mature podsols are present on system 4. These changes are similar to those described for the Michigan dunes by Olson (1958), and similar conclusions may be drawn.

SOME CONCLUSIONS

As dune systems age, in localities where precipitation exceeds evapotranspiration and drainage is free, changes in soil organic matter content and water-holding capacity occur. However, probably the main changes affecting plant species are the consequent acidification and losses of nutrients. T. W. Walker & Syers (1976) reviewed the general significance of changes in phosphorus for soil–vegetation interrelationships on chronosequences (Fig. 5.14). Phosphorus is a strictly limited resource in natural systems, since all that is present in a system which receives no additions of new materials is derived, ultimately, from the minerals of the soil parent material. Very little or none comes from the atmosphere, in the rainfall. If 'leakages' of phosphorus occur in a system they cannot be recouped unless new materials are added. In natural systems this could happen by the deposition of fresh, unweathered rock material (alluvium, sand, volcanic ash, etc) by erosion of the site down to unweathered bedrock or other unweathered parent material, or by the addition of relatively large amounts of animal faeces. Phosphorus is one of the most important limiting factors for plant growth, and its availability in soils seems likely to be a key factor controlling the rate and even the direction of long-term vegetation change.

Other nutrients are important, also. Although data on the sources of the initial inputs of nitrogen of the Michigan dunes are not fully conclusive and nothing is available which throws much light on sources of supply of nitrogen in dune systems for the other sequences described above, Silvester (1977) points out that numbers of the known species of vascular plant which have non-rhizobial, nitrogen-fixing root nodules, are important sand dune stabilizers throughout the world. They include the shrub *Hippophäe rhamnoides* in Western and Eastern Europe. It precedes pines (*Pinus*) or sycamore maple (*Acer*) in vegetation sequences and stimulates the growth of nitrophile plants such as *Urtica* (stinging nettle). *Eleagnus* spp. stabilize dunes in Europe and Japan. *Casuarina* occurs on dunes in Australia and India. *Myrica cordifolia* is prominent on dunes in South Africa. Some legumes are also important on dunes. One is *Lupinus arboreus*, a widespread introduced sand dune plant in New Zealand which is also found on dunes in its native habitat, California, USA. On New Zealand dunes foresters use it as a 'nursery' cover to prepare the dunes for planting with *Pinus radiata*, thus providing the nitrogen inputs for an artificial 'succession' (Ritchie 1968).

The species that are present early in dune sequences require relatively high supplies of cations and phosphorus and basic-to-neutral conditions. Most presumably cannot tolerate impoverished, acid soils. As acidification and leaching of soils proceed, the plant populations change. Species that are tolerant of the impoverished soils replace those requiring more-basic, nutrient-rich conditions. Many of the species that are typical of sites with infertile conditions cannot tolerate the richer soils. Ozanne & Specht (1981), for example, report toxicity of relatively high phosphorus levels to heathland plants in South Australia. Other problems with which plants of poor soils must cope are the excess of hydrogen ions, the presence of aluminium ions in toxic amounts and very low quantities of nitrogen (Woolhouse 1981).

The storage of nutrients in organic matter (living and dead) is very important for the maintenance of pools of nutrients in ageing dune systems, but the rate of return of nutrients by mineralization from the litter is essentially constant (J. Walker *et al.* 1980). The plant species inhabiting the old, poor dune soils seem to be very efficient at recycling (from dying tissues and litter) the small amounts of phosphorus and calcium that are available to them (Groves 1981a). Nevertheless, losses occur by exports (in drainage, or by removal of biomass), leaching to depths below root levels and sequestration in unattainable forms (T. W. Walker & Syers 1976) (Fig. 5.14). These changes are autogenic, undoubtedly both the result of the presence of the vegetation and the cause of continuing vegetation change (although they require the allogenic influence of moderate-to-high rainfall). Exogenous disturbing events, such

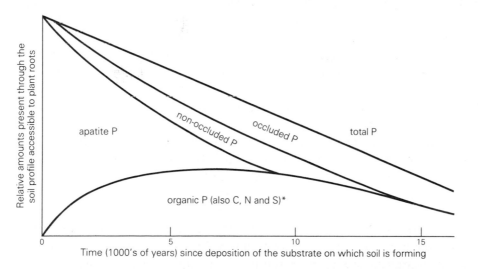

Figure 5.14 Model for long-term changes in amounts and states of soil phosphorus within the zone accessible to plant roots in ecosystems in humid environments. Apatite P, phosphorus still unweathered from rock minerals; occluded P, phosphorus tightly bound to aluminium and iron compounds; some of this P may be released slowly by mycorrhizas; non-occluded P, phosphorus lightly bound on soil colloids and in the soil solution and available to plants; organic P, phosphorus in organic matter, released by microbial activity. C, carbon, N, nitrogen; S, sulphur; Total P, the total amount of phosphorus in the soil profile. The rate of change with time depends on factors such as the initial amounts of apatite present and the rate of soil weathering (faster in high rainfall and warmer climatic regimes) (after T. W. Walker & Syers 1976).

as fire, logging or grazing, can exaggerate or accelerate the process of nutrient pool degradation. The resultant vegetation, on soils with depleted nutrient supplies, is low in stature, vigour and biomass compared with that on younger surfaces. It is composed of plants with relatively low nutrient demands. The species diversity of the vegetation on old soils need not be less than that of more-fertile, younger soils, but the species composition is very different.

The distributions of species on the Michigan dunes demonstrate very individual responses of the species to the prevailing environmental gradients. It may be assumed that, apart from the need for open sites by early colonists, nutrient availability and related soil factors are the main determinants of plant distributions, although soil water-holding capacity may also be important for some species. Some species (e.g. *Quercus velutina*) tolerate a wide range of conditions while other species such as *Celastrus scandens, Aquilegia canadensis* and *Coreopsis tripteris* are more narrowly restricted.

One thing lacking for all of the dune sequences described here is detailed information about changes in the plant species populations with time. It is reasonably certain that in the Michigan dunes only one generation of most of the initial herbaceous sand-binders occurs on any site. *Prunus pumila* and *Populus deltoides* may also last for only one generation, but their vegetative habit of reproduction (along with that of the herbaceous sand-binders) makes it difficult to determine the length of a generation. *Pinus banksiana* stands last for more than one generation.

Beyond that it is not even possible to speculate on the duration of the various species in the sequence in terms of numbers of generations which recur on specific sites. The long timespan over which some species are present (including, of course, *Quercus velutina*), suggests that these species persist for many generations. If there is more than one generation of each of the species in a community on a site, then those species have achieved a relative equilibrium with their environment and their neighbours – at least a state of temporary stability.

The early parts of the dune sequences, with plant species populations apparently being replaced by others after one generation of occupancy of a site, resemble the sequence from abandoned field to oak–hickory forest in North Carolina, USA (outlined in Ch. 11). In spite of long periods of dominance by particular mature vegetation types (black oak, oak–hickory, etc.), the 'steady state' appears to be only a temporary condition in places where leaching losses of nutrients occur, even in the unlikely event of freedom from exogenous disturbance. The gradual compositional changes resulting from the autogenically caused decline of nutrient stocks should perhaps be regarded as differing only in rate (and not, essentially, in kind) from the more-rapid changes occurring as vegetation colonizes new or recently disturbed sites. However, biological factors determining the actual detailed pattern of species-by-species replacement seem easier to identify in the colonizing situation.

6

Vegetation development on glacial deposits

Some long vegetation–soil development sequences occur in mountain areas on deposits left by receding glaciers. Many glaciers take the form of ice tongues projecting into valleys from **névés** (snow accumulation areas high in the mountains) (Fig. 6.1). Others are sheets of ice of different sizes lying on gentler terrain in cold regions.

During cool periods, when snowfall is heavy in a névé, the ice formed from the compacted snow gradually slides and flows to lower levels. Vegetation in its path may be destroyed. Within its ice and on its surface the glacier carries rock debris ranging in size from silt to huge boulders. This is material which has fallen on to the ice surface or which the glacier has scraped from its rocky channel. At the margins of the glacier this debris falls from the ice surface or is released by melting and forms bouldery ridges (**moraines**). Meltwater streams emerge from the glacier, carrying large amounts of alluvial **outwash** (gravel, sand and silt) which is deposited to form plains and terraces in the area downstream. The material emplaced by ice is known as **till**. The rock materials in the till and outwash alluvium vary according to the different types of bedrock forming the terrain traversed by the glacier.

The terminus of an advancing glacier, nourished by abundant snowfall in its névé, reaches a point down-valley where melting balances the continuing supply of ice, and the terminus halts. Large moraines are often formed then. If, during warmer periods, the snowfall in the névé diminishes, the terminal position begins to retreat, sometimes intermittently, leaving minor moraines and outwash plains, till deposited under the ice and areas of smoothed bedrock. These sterile, bare surfaces are colonized by plant propagules derived from the vegetation of nearby unglaciated sites and transported to the new sites by wind, water and animals.

In the past few thousand years periods of glacier advance which left moraine systems are recorded from all of the world's mountain regions with sufficient precipitation and climates cool enough for ice to form. Most recently, after marked advances in the 17th century AD, there has been retreat followed by intermittent, small re-advances. The recession of many glaciers has been accelerated in the 20th century, broken by only a few minor advances.

This chapter considers vegetation change sequences beginning on new moraines and outwash in some temperate localities. Some of the information comes from direct observation of the sequential changes. Sequences longer than about 70 years are reconstructed from chronosequence studies. In one Alaskan locality, where glacier retreat has been almost continuous for nearly 200 years, the chronosequence is similarly continuous. Otherwise the chronosequences are interrupted ones on moraines formed at different times.

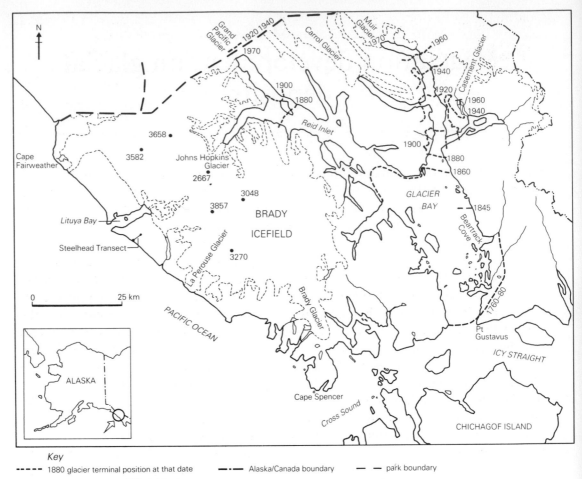

Figure 6.1 Glacier Bay, southeastern Alaska, showing terminal positions of the receding glaciers at intervals from about AD 1760 to 1970 (after Ugolini & Mann 1979, Mirsky 1966).

GLACIER BAY, ALASKA, USA

Glacier Bay, on the coast of southeast Alaska (58–59° N, 135° 30′–136° 30′W) is one of the best-known localities for research on vegetation development on new land surfaces. Its particular value for this arises from the very extensive and almost continuous retreat of the glaciers since 1794, the well-defined site chronology and the numbers of ecological observations, beginning in 1916. W. S. Cooper, of the University of Minnesota, made his first visit there in 1916 and returned in 1921, 1929 and 1935 (W. S. Cooper 1923a, b, c, 1931, 1937, 1939, 1942). D. B. Lawrence and associates, also from the University of Minnesota, made visits in 1941, 1949, 1950, 1955, 1967, 1972, 1982 and 1988 (Lawrence 1950, 1951, 1953, 1958, 1979, Lawrence & Hulbert 1960, Lawrence *et al.* 1967, Schoenike 1957, Reiners *et al.* 1971, Worley 1973, P. Wardle, pers. comm. 1985, D. B. Lawrence pers. comm. 1988). R. P.

Goldthwait and associates, from the University of Ohio, worked there in 1958 and again in 1966 (Goldthwait 1963, Goldthwait *et al.* in Mirsky 1966).

Glacier Bay is a large, complex fjord (Fig. 6.1), with rugged, high mountains, rising to 3800 m, in its steep, rock-walled upper reaches. On the mountains and plateaux surrounding the fjord are snowfields and large glaciers. In the late 18th century AD the glaciers coalesced, forming a very large ice tongue which extended to the outlet of the fjord in Icy Strait. From a position well back in the fjord the glacier had undergone a major advance, possibly beginning in the 17th century and culminating by about AD 1760 (first record in 1794 by the navigator George Vancouver). By then it had expanded over much of the lowland on either side of the present fjord outlet and the islands which dot the fjord, destroying much mature forest (W. S. Cooper 1923a, Goldthwait 1963, Field 1947, 1975). The surface of the glacier was at about 800 m above present sea level in the upper fjord and not far above present sea level 26 km behind the terminus.

Since 1794 there has been steady retreat of the terminus, and downward shrinkage of the glacier surface. By 1880 the former great glacier had separated into many individual ice tongues, which have withdrawn up the separate arms of the fjord. By 1975 there had been more than 105 km of recession in Reid Inlet, the main (northwest) arm of the fjord and more than 80 km in Muir Inlet. Some glaciers became stranded on land and melted *in situ*. The largest glaciers now terminate as ice cliffs in tidewater at the heads of the main branches of the fjord (Fig. 6.2).

The climate at sea level at Glacier Bay (Table 6.1) is cool, cloudy and relatively moist (Crocker & Major 1955, Loewe 1966). Snow often falls on the mountains in summer and at lower altitudes in winter.

The complex of terrain exposed by glacier retreat and developed on thick, old alluvium, old till, younger till and bedrock, includes extensive areas of flat to gently rolling country in the south-east, with some stream-channels, depressions with ponds and ice-melt pits, terraces, ridges of moraine, outwash flats and some low bedrock hills. Land with similar topography is present outside the 18th century ice limit. Islands in the fjord are covered by till. Some strand plains are slowly emerging from the sea by isostatic rebound (uplift after release from ice loading). Further up the fjord are narrow borders of low-lying outwash and moraine and ice-sculptured bedrock hills. Valleys with low floors penetrate the mountains and bedrock islands in the fjord. Ice-worn rock and steep cliffs predominate there.

The substrates available for colonization by plants after ice retreat, thus, include freshwater pools, bare rock, till and alluvium. In the last two of these, textures vary from boulders to cobbles, pebbles, sand, silt and some clay. Rock types are argillite, slate, quartzite, chert, dolomitic limestone, marble, diorite and quartz diorite; different proportions of these are present at any one site, with the igneous and argillaceous rock types predominating (cf. Dietrich & Skinner 1979).

The vegetation development sequence

The chronology for the exposure of localities in Glacier Bay by ice retreat was first worked out by W. S. Cooper (1923a, b, c, 1931, 1939) from historic records, surveys

Figure 6.2 Glacier Bay, southeastern Alaska. View of Muir Inlet (foreground and middle left) and Casement Glacier (background left), 1982, from the slopes of Sebree Hill, opposite the Muir cabin (photo P. Wardle).

Table 6.1 Climate data for localities in southeastern Alaska, near Glacier Bay (after Wernstedt 1972).

	Mean annual precipitation (cm)	Mean annual temperature (°C)	Mean January temperature (°C)	Mean July temperature (°C)
Gustavus 58°12′N, 136°38′W	136.5	−1.9	−18	13.9
Juneau 58°22′N, 134°35′W	230.4	1.4	−10.2	12.6
Cape Spencer 58°25′N, 135°42′W	276.0	−1.5	−20.1	14.1

and tree ages. Further detail was added by Field (1947), Lawrence (1958, 1979) and Goldthwait (1963, 1966) (Fig. 6.1).

Cooper recorded the colonizing pattern of the vegetation in several ways.

1 Observation, description and partial plant lists of several sites at various stages of vegetation development.
2 One-metre square chart quadrats of nine sites, made at intervals, a total of four observations, over 19 years.
3 In the same quadrats, counts of discrete individual plants, records of the occurrence of all species and measurement of the area covered by mat plants.
4 Photographs of numbers of localities, including the permanent quadrats.

Since 1935 the quadrats have been measured again a further six times, up to 1988 (spanning 72 years) (Lawrence 1979, P. Wardle pers. comm. 1985, D. B. Lawrence pers. comm. 1988), the sites being photographed each time. Some results are summarized in Figure 6.3 and Table 6.2. One site has been lost by erosion. The small size of the quadrats limits the value of these records, but the photographic record is very valuable.

Other observations on sites of known age include larger temporary quadrats and other numerical samples, to record the composition of the vegetation (Decker 1966, Reiners *et al.* 1971). Soil studies parallel to the vegetation sequence were done by Crocker & Major (1955) and Ugolini (1966). There have also been studies of the changes in animal communities (Table 6.3) (E. E. Good 1966, Trautman 1966) and some fungal studies (Sprague & Lawrence 1960).

W. S. Cooper (1923b) divided the perceived time series of vegetation on terrestrial sites into three stages: pioneer community; willow (*Salix*)–alder (*Alnus*) thicket; spruce (*Picea*) forest, ' . . . with the understanding that these are mere cross-sections of a continuous stream of development, made with reference to certain points in this stream where change is most evident'. Aquatic sequences in ponds were also briefly described.

Additional phases in the vegetation sequence were recognized by the later workers (Table 6.4) and studies of other aspects of the biology of the plants were initiated. One such study was the examination of the sources of plants colonizing the bare ground and their dispersal capabilities (W. S. Cooper 1942, Lawrence 1979). The extensive 18th-century ice cover and rapid ice recession at Glacier Bay meant that the initial immigrant plants had to come either from the area covered by forest, bog or shore vegetation to the south, outside the former ice limits, or from the alpine tundra with scrub and some scattered trees, above the ice limit on the surrounding mountains (Fig. 6.1). It seems likely that propagules of most of the initial herbaceous pioneers were derived from tundra or from small, disturbed sites such as minor landslides and stream margins, in the forest zone. Shrubs and trees were also probably derived from the varied habitats present in the forest zone. Many of the colonists are well-suited to wind dispersal by possession of spores, small seeds, or plumed or winged seeds or fruits. Others have fleshy fruits, some with large seeds transported by birds, mammals and water. Apparently unspecialized seeds and fruits are usually carried by wind, water or on the fur of mammals.

Earlier part of the sequence

The general development of the vegetation on gravelly substrates, according to Decker (1966), Reiners *et al.* (1971) and Lawrence (1979), is as follows (using Decker's time-framework of eight intergrading stages) (Figs 6.4–7). A lag deposit of pebbles often forms on the ground surface on new till as a result of rain-wash and frost action. Plant seedlings and sporelings establish between or on the pebbles. On silty and sandy surfaces a 'black crust' (actually dark green) develops. This has varying amounts of a tiny liverwort, *Lophozia*, cyanobacteria including *Anabaena*, *Nostoc*, *Scytonema*, *Gloeocapsa* and *Oscillatoria*, and lichens (*Lempholemma* and an

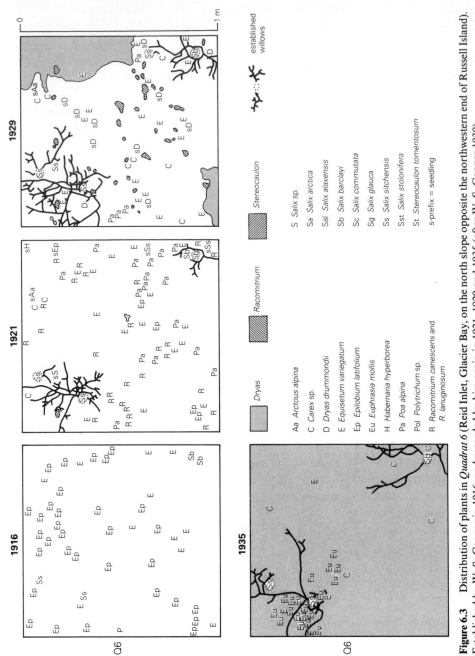

Figure 6.3 Distribution of plants in *Quadrat 6* (Reid Inlet, Glacier Bay, on the north slope opposite the northwestern end of Russell Island). Established by W. S. Cooper in 1916, and recorded by him again in 1921, 1929 and 1935 (after W. S. Cooper 1939).

Table 6.2 Summary of data from W. S. Cooper's Quadrat 1, Reid Inlet, Glacier Bay (north shore, opposite middle of Russell Island; 1 m² quadrat on site covered by ice in 1879, but ice-free very soon after) (after W. S. Cooper 1939, Lawrence, 1979, 1988, D. B. Lawrence, unpublished data).

Date of recording (time ice-free)	Main cover	Individual shrubs and trees	Herbs
1916 (37 years)	30% *Racomitrium* moss mats; the rest bare	4 prostrate *Salix barclayi* 1 *S. arctica*, many seedlings	100 *Equisetum variegatum*, 2 *Carex*, 1 tuft *Stereocaulon* lichen
1935 (56 years)	Over 50% plot still bare; *Salix*, *Racomitrium*	14 *Salix*, forming a mat <30 cm high	*Equisetum* constant, *Carex* much increased
1955 (76 years)	*Alnus crispa**, 50% *Shepherdia canadensis** (rooted outside plot)	1 *Alnus* 53 cm tall, numerous *Alnus* seedlings; willows sending up erect, shoots 30 cm high	*Equisetum*, *Racomitrium* still present
1967 (88 years)	98% *Alnus* canopy cover, over-hung by *Populus trichocarpa* tree (rooted outside plot); deep litter accumulating.	*Alnus* 4.6 m tall; all *Racomitrium*, low-statured willows and *Shepherdia* dead	*Pyrola asarifolia* abundant, *Equisetum* still present but rather attenuated
1982 (103 years)	98% *Alnus* canopy cover, over-hung by *P. trichocarpa* and *Salix sitchensis* (both rooted outside plot); dense *Alnus* leaf and twig litter	one *Alnus* stem emerging from plot; another forms low cover from outside plot	about 25 living stems of *Equisetum*, 10 cm tall, without strobili; about 5 cm² moss, 5 cm² lichens, on *Alnus* twig litter; all *Pyrola* dead
1988 (109 years)	As above	the adjacent *P. trichocarpa* has 30 cm basal diameter and *S. sitchensis* 15 cm basal diameter; all *Alnus* low cover dead	about 10 living stems of *Equisetum*, maximum 7 cm tall and very weak; mosses on exposed stem, litter

* with nitrogen-fixing symbiont.

Table 6.3 Vertebrate animals associated with the vegetation sequence at Glacier Bay (after E. E. Good 1966, Trautman 1966).

Estimated age of site (years) Vegetation phase (Decker 1966)		0–5 I	5–20 II,III	20–30 IV,V	30–50 VI	50–90+ VII	90 VIII
birds							
Acanthis flammea	redpoll	+					
Dendroica petechia	yellow warbler	+	+	+			
Erolia minutilla	least sandpiper	+					
Hyloichla guttata	hermit thrush	+	+	+			
Junco oreganus	Oregon junco	+	+	+			
Lagopus lagopus	willow ptarmigan	+	+	+			
L. mutus	rock ptarmigan	+					
Leucosticta tephrocotis	rosy finch	+					
Passerculus sandwichensis	Savannah sparrow	+					
Passerella iliaca	fox sparrow	+	+	+			
Plectrophenax nivalis	snow bunting	+					
Vermivora celata	orange-crowned warbler	+					
Wilsonia pusilla	Wilson warbler	+	+	+			
Hylocichla minima	gray-cheeked thrush			+			
Stercorarius parasiticus	parasitic jaeger			+			
Dendroica coronata	myrtle warbler				+		
Ixoreus naevius	varied thrush				+		
Corvus caurinus	north-western crow					+	+
C. corax	common raven					+	+
Dendragapus obscurus	blue grouse					+	+
Parus pubescens	chestnut-backed chickadee					+	+
Pinicola enucleator	pine grosbeak					+	+
Regalus calendula	ruby-crowned kinglet					+	+
small mammals							
Microtus oeconomus	tundra vole		+	+	+	+	
Peromyscus maniculatus	deer mouse		+	+	+	+	
Sorex vagrans	wandering shrew		+	+	+	+	
Clethrionomys rutilis	red-backed vole						+
Sorex cinerus	masked shrew						+

unidentified species). The crust stabilizes the substrate. Worley (1973) thought the crust was not an essential precursor for later successional communities. However, Lawrence (1979) pointed out, first, the nitrogen-fixing properties of cyanobacteria, both free-living and symbiotic in *Lempholemma*, and, secondly, that the 'black crust' may absorb heat and warm the substrate, making microsites more suitable for colonization. Crocker & Major (1955) reported increased amounts of nitrogen under the cyanobacteria crust, compared with barren sites. On the pebble pavement the colonists are the moss *Racomitrium canescens*; the lichen *Stereocaulon* sp. (also with a cyanobacteria symbiont) and the angiosperm herb *Epilobium latifolium*. They are unable to endure crowding and shading. It is also possible that they and other initial colonists require some special chemical, physical or microbial conditions of the open habitats.

Table 6.4 Representation of plant species in samples from the developmental sequence, Glacier Bay, southeastern Alaska (after Reiners et al. 1971, Decker 1966, Ugolini 1968).

Estimated age of site (years)	Growth form	0–5	5–20	20–30	30–50	50–90+	c. 250	Possibly 800–1500
Phase (Decker 1966)		I	II,III	IV,V	VI	VII	VIII	
Dominant vegetation cover		Pioneer herb	Dryas–low willow–cottonwood scrub	Willow–alder scrub	Alder–willow–cottonwood–spruce forest transition	Spruce forest	Spruce–hemlock forest	Muskeg bog
ground cover herbs; low shrubs; juvenile trees, tall shrubs								
Epilobium latifolium	h	+						
Equisetum variegatum	h(horsetail)	+	+	+	+			
Salix arctica	ls	+	+	1				
S. sitchensis	js	+	1	2				
S. barclayi	js	+	+	1				
S. stolonifera	ls		+					
S. commutata	js		+	1				
Populus trichocarpa	jt		1					
Dryas drummondii	h		1–5	4				
Alnus crispa	jt		+	3				
Athyrium filix-femina	h(fern)				+			
Dryopteris dilatata	h(fern)				+			
Actaea rubra	h				+			
Aruncus sylvester	h				1			
Picea sitchensis	jt				+	+	+	
Tsuga heterophylla	jt				+			
Listera cordata	h					1		
Goodyera oblongifolia	h					+		
Vaccinium ovalifolium	ls					2	3	
Cornus canadensis	ls						1	
Menziesia ferruginea	ls						1	
Lysichiton americana	h						1	1
Vaccinium vitis-idaea	ls						+	1
Carex spp.	h							5
Eriophorum angustifolium	h							1

Table 6.4 (cont.)

Species	Form	1	2	3	4	5	6
Trichophorum caespitosum	h						+
Podagrostis aequivalvis	h						1
Cornus suecica	ls						2
Empetrum nigrum	ls						2
Kalmia polifolium	ls						1
Ledum decumbens	ls						1
Vaccinium oxycoccus	ls						2
Drosera rotundifolia	h						1
lichens, mosses, liverworts							
Bryum sp.	m	+					
Racomitrium canescens	m	+	1–2	2			
Stereocaulon alpinum	li		+	+			
Drepanocladus uncinatus	m		1	3	+		
Brachythecium sp.	m				+		
Rhytidiadelphus squarrosus	m				2		
R. triquetrus	m				2	+	+
R. loreus	m				2	4	4
Mnium insigne	m				1	+	
M. glabrescens	m				1	1	
Hylocomium splendens	m				1	4	1
Calopogeia mulleriana	lv				+	1	
Plagiochila asplenioides	lv				+	1	
Dicranum majus	m				+	1	
Plagiothecium undulatum	lv				+	1	+
Pleurozium schreberi	m					1	2
Ptilidium crista-castrensis	lv				+	+	+
Scapania bolanderi	lv					1	
Sphagnum girgensohnii	m					1	
Aulocomnion palustre	m						
Sphagnum fuscum	m						1
S. magellanicum	m						2
S. palustre	m						3
S. papillosum	m						1
S. recurvum	m						+
S. tenellum	m						1
S. warnstorfianum	m						3

h, herb; ls, low shrub; jt, juvenile tree; js, juvenile tall shrub; li, lichen; m, moss; lv, liverwort.
Cover values: + = <1%; 1 = 1–4%; 2 = 5–9%; 3 = 10–25%; 4 = 25–50%; 5 = >50%.

Trees and tall shrubs

	5–20+		20–30+		30–50+		50–90+		possibly c. 350		possibly 800–1500	
Estimated age of site (years)	II,III		IV,V		VI		VII		VIII			
Phase (Decker 1966)	Dryas–low willow–cottonwood		Willow–alder scrub scrub		Alder–willow–cottonwood spruce forest transition		Spruce forest		Hemlock forest		Muskeg bog	
Dominant vegetation cover	D	BA	D	BA	D	BA	D	BA	D	BA	D	BA
trees												
Salix alaxensis					28	0.4						
Populus trichocarpa					56	1.2						
Picea sitchensis					444	29.0	562	75.2	104	34.1		
Tsuga heterophylla							229	7.3	542	41.2		
Pinus contorta											222	7.5
total					528	30.6	791	82.5	646	75.3	222	7.5
tall shrubs												
Salix alaxensis	83	0.02	167	0.07	111	0.13						
Populus trichocarpa	125	0.07			278	1.96						
Alnus crispa	42	0.03	17 167	8.45	2889	8.31	42	0.09				
Salix commutata			250	0.06	56	0.02						
S. sitchensis			750	0.24	333	1.22						
S. barclayi			750	0.23	222	0.19						
Picea sitchensis			83	0.03	1278	2.12	42	0.34	125	0.29		
Tsuga heterophylla							250	0.83	708	2.88		
Menziesia ferruginea									2083	0.72		
Vaccinium ovalifolium									42	0.00		
Pinus contorta											444	2.64
Tsuga mertensiana											56	0.04
total	250	0.12	19 167	9.10	5107	13.95	334	1.26	2958	3.89	500	2.68

D, density (stems ha^{-1}); BA, basal area (m^2 ha^{-1}).

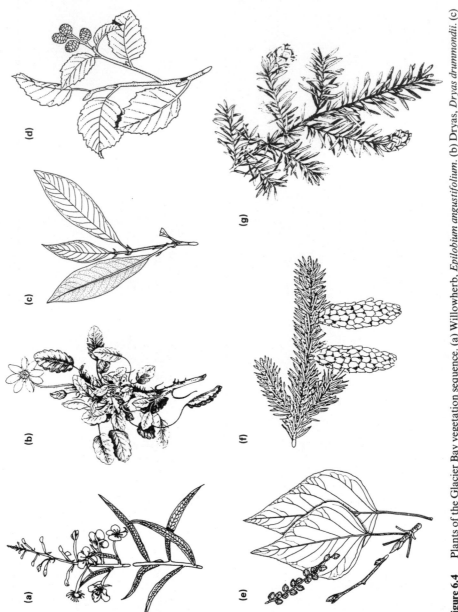

Figure 6.4 Plants of the Glacier Bay vegetation sequence. (a) Willowherb, *Epilobium angustifolium*. (b) Dryas, *Dryas drummondii*. (c) Alaskan willow, *Salix alaxensis*. (d) Sitka alder, *Alnus crispa*. (e) Cottonwood, *Populus trichocarpa*. (f) Sitka spruce, *Picea sitchensis*. (g) Western hemlock, *Tsuga heterophylla*.

Figure 6.5 Goose Cove, Muir Inlet, looking north toward the receding Muir Glacier, 1941. *Dryas* spp. and other herbs and *Salix* (willow) shrublets in foreground; clumps of *Salix, Alnus crispa* (Sitka alder) and *Populus trichocarpa* (cottonwood) saplings in middle. About 20 years after deglaciation (photo D. B. Lawrence).

Figure 6.6 From Russell Island, looking towards Reid Inlet, Glacier Bay, 1982. Dense *Alnus* thicket with emergent *Populus* and *Picea sitchensis* (Sitka spruce). About 50 years after deglaciation (photo P. Wardle).

Figure 6.7 Interior of first generation *Picea sitchensis* forest, Bartlett Cove, Glacier Bay, 1968. Ground and logs, tree stumps densely covered by *Hylocomium splendens* moss. About 200 years after deglaciation (photo D. B. Lawrence).

The most prominent early vascular plant pioneers of Phase I are *Dryas drummondii*, *Salix* spp., *Epilobium latifolium* and (mainly in moist spots) *Equisetum variegatum* (cf. Table 6.3), but Lawrence (1979) also recorded milk vetches *Hedysaryum alpinum*, *Astragalus alpinus* and soap berry (*Shepherdia canadensis*). Scattered cottonwoods (*Populus trichocarpa*) and sitka spruce (*Picea sitchensis*) plants colonize open sites from Phase I onwards.

Phase II is dominated by large *Dryas* mats (only a few centimetres high) which form by coalescence of the original colonies and eventually cover as much as 90 per cent of the terrain. This phase grades into late pioneer scrub, Phase III, in which young sitka alder *Alnus crispa* and cottonwoods are prominent among shrub willows, *Salix* spp. The alder colonies increase in number and size and, as they coalesce, exclude *Dryas*, eventually forming clumps 2 m or more high (Phase IV). In a few decades the alders often dominate the landscape, forming a very dense thicket, with decumbent basal branches (Phase V). On the ground are some ferns and many herb species. Scattered willow bushes and cottonwoods are present; the latter begin to emerge above the 4–5 m high alder canopy (Phase VI).

Soil changes paralleling the vegetation changes

The role of plants and other organisms in bringing about change in soil characteristics is stressed by Crocker & Major (1955), Lawrence (1979) and Ugolini (1966,

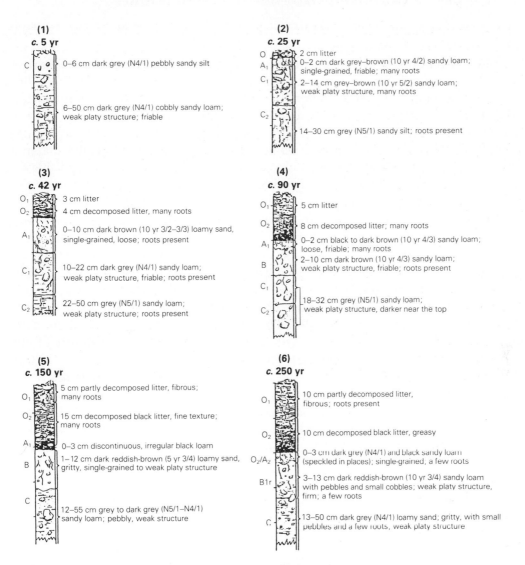

(1)
c. 5 yr

C — 0–6 cm dark grey (N4/1) pebbly sandy silt

6–50 cm dark grey (N4/1) cobbly sandy loam; weak platy structure; friable

(2)
c. 25 yr

O — 2 cm litter
A₁ — 0–2 cm dark grey–brown (10 yr 4/2) sandy loam; single-grained, friable; many roots
C₁ — 2–14 cm grey–brown (10 yr 5/2) sandy loam; weak platy structure, many roots

C₂ — 14–30 cm grey (N5/1) sandy silt; roots present

(3)
c. 42 yr

O₁ — 3 cm litter
O₂ — 4 cm decomposed litter, many roots

A₁ — 0–10 cm dark brown (10 yr 3/2–3/3) loamy sand, single-grained, loose; roots present

C₁ — 10–22 cm dark grey (N4/1) sandy loam; weak platy structure, friable; roots present

C₂ — 22–50 cm grey (N5/1) sandy loam; weak platy structure; roots present

(4)
c. 90 yr

O₁ — 5 cm litter

O₂ — 8 cm decomposed litter; many roots

A₁ — 0–2 cm black to dark brown (10 yr 4/3) sandy loam; loose, friable; many roots
B — 2–10 cm dark brown (10 yr 4/3) sandy loam; weak platy structure, friable; roots present

C₁

C₂ — 18–32 cm grey (N5/1) sandy loam; weak platy structure, darker near the top

(5)
c. 150 yr

O₁ — 5 cm partly decomposed litter, fibrous; many roots

O₂ — 15 cm decomposed black litter, fine texture; many roots

A₁ — 0–3 cm discontinuous, irregular black loam
B — 1–12 cm dark reddish-brown (5 yr 3/4) loamy sand, gritty, single-grained to weak platy structure

C — 12–55 cm grey to dark grey (N5/1–N4/1) sandy loam; pebbly, weak structure

(6)
c. 250 yr

O₁ — 10 cm partly decomposed litter, fibrous; roots present

O₂ — 10 cm decomposed black litter, greasy

O₂/A₂ — 0–3 cm dark grey (N4/1) and black sandy loam (speckled in places); single-grained, a few roots

B1r — 3–13 cm dark reddish-brown (10 yr 3/4) sandy loam with pebbles and small cobbles; weak platy structure, firm; a few roots

C — 13–50 cm dark grey (N4/1) loamy sand; gritty, with small pebbles and a few roots, weak platy structure

Figure 6.8 Soil profile development, with time, on glacial till, Glacier Bay, Alaska (after Ugolini 1968).

1968) (Fig. 6.8). Among the nitrogen-fixers present are the cyanobacteria mentioned earlier, *Rhizobium* in symbiotic association with legumes (*Astragalus* and *Hedysaryum*) and *Frankia* in the root nodules of *Dryas*, *Alnus* and *Shepherdia*. Free-living nitrogen-fixing bacteria are probably also present. Supplies of nitrogen in rainfall can thus be considerably augmented from these many sources. Lawrence (pers. comm. 1967) also pointed out that animal faeces and carcasses, plant litter derived from established vegetation and fossil plant horizons (exposed in some places) are nitrogen and phosphorus sources in some localized sites. *Dryas* and *Alnus*, with *Frankia*, are the two most important vascular symbiotic nitrogen-fixing associations.

Lawrence (1979) notes that any pioneer plants at Glacier Bay except those with the nitrogen-fixing association are, at first, sickly and yellowish, with slow growth rate and prostrate form. These are symptoms of nitrogen deficiency. This applies even to willows, cottonwoods, spruce and hemlocks, which are erect later in life. Lawrence (1958) experimented with applications of ammonium nitrate around such stunted cottonwoods, and achieved spectacular responses. If they grow alongside *Dryas* or *Alnus* the nitrogen deficiency in these plants is overcome (Lawrence & Hulbert 1960).

In the studies by Crocker & Major (1955) it was shown that concomitant with the changes in plant species composition, and continuing up to 180 years after initiation of the sequence there were: (1) a rapid build-up of carbon in the litter residues in the soil; (2) a substantial build-up of organic nitrogen in the soil especially under the influence of alder; and (3) a marked increase in soil acidity (Figs 6.9–11, Table 6.5). After 50 years about 3800 kg ha^{-1} of carbon and 300 kg ha^{-1} of organic nitrogen had accumulated in the upper soil layers under alder. No measurement has been made of the total amounts of nitrogen accumulating with time in the living part of the system, but the substantial quantities present in the litter demonstrate that the alder symbionts are very efficient nitrogen-fixers. The *Alnus rugosa* symbiont has been shown to fix nitrogen at rates in excess of 815 kg ha^{-1} per annum in eastern Canada (Daly 1966) and near Fairbanks, Alaska, USA, 20-year-old stands of *A. incana*, on alluvium, contained 274 kg ha^{-1} of nitrogen, 24 per cent of it in the leaves and the remainder in the branches and boles. The soil beneath this stand, to 70 cm depth, contained 2300 kg ha^{-1} of nitrogen (Van Cleve *et al.* 1970). The distribution of other chemical elements (and microclimate) were studied by Viereck (1970) and Van Cleve & Viereck (1972, 1981) in vegetation sequences on alluvium near Fairbanks.

The soil nitrogen is available for recycling to other plants. It is notable during Phase VI at Glacier Bay that cottonwood plants, hitherto inconspicuous, begin to overtop the alders. Apparently it takes at least 30–50 years for the recycling process to have its effect on *Populus*. The *Dryas* symbiont is not as spectacular at fixing

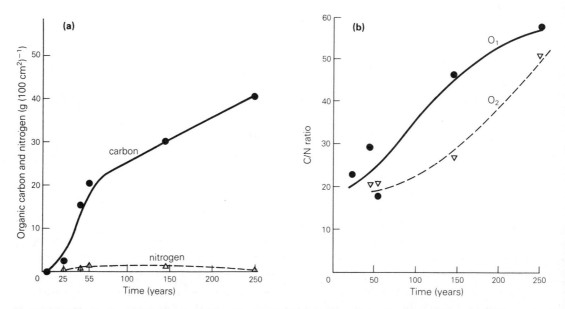

Figure 6.9 Glacier Bay. (a) Carbon and nitrogen accumulation in the soil as a function of time. Surface organic layer. (b) C/N ratio for O1 and O2 horizons as a function of time. After Ugolini (1968).

184

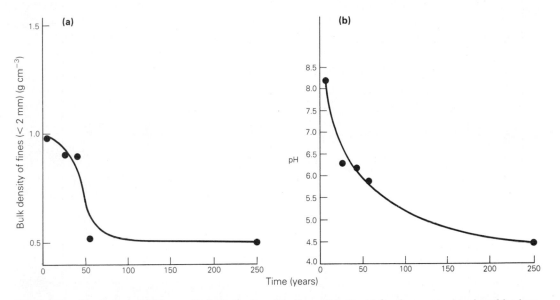

Figure 6.10 Glacier Bay. (a) Changes in bulk density of the fine soil (<2 mm) for the uppermost mineral horizon, with time. (b) Changes in pH for the uppermost mineral soil horizon, with time. After Ugolini (1969).

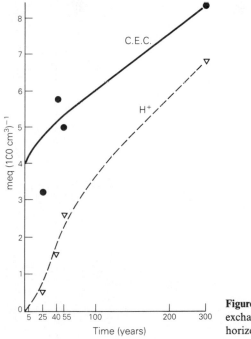

Figure 6.11 Glacier Bay. Changes in cation exchange capacity and exchangeable hydrogen with time for the uppermost mineral soil horizon (after Ugolini 1968).

nitrogen as is that of *Alnus* (cf. Schoenike 1957, Lawrence *et al*. 1967), but its effects are appreciable (Crocker & Major 1955).

There is no doubt that the rate of vegetation change is substantially increased by the activity of the *Alnus* symbiont as a nitrogen-fixer. *Alnus* also brings about marked acidification of the upper soil layers. It is likely that *Dryas*, which prefers calcareous

Table 6.5 Soil chemical properties associated with the vegetation sequence at Glacier Bay (after Ugolini 1968).

Vegetation phase and age of surface (years)	Horizon and depth (cm)		pH	Nitrogen (%)	Exchangeable cations (m eq per 100 g)		Cation exchange capacity (m eq per 100 g)
					K^+	Ca^{2+}	
I–II (5)		0–6	8.2	–	1.3	4.0	5.0
		6–15	8.4	–	1.3	4.0	5.0
		15–22	8.4	–	2.0	3.3	5.0
III–IV (24)	O1	0–4	5.2	2.12			
	A1	0–2	6.3	0.04	0.13	2.2	4.0
	C1	2–14	8.2	–	1.3	3.0	5.0
	C2	14–18	8.2	–	2.0	3.2	5.0
VI (41)	O1	0–4	4.9	1.81			
	O2	0–5	4.6	1.17			
	A1	0–10	6.2	0.03	0.7	3.0	6.4
	C1	10–22	7.4	0.02	0.7	3.0	6.4
	C2	22–23	8.0	0.01	1.3	3.1	6.0
VI (55)	O1	0–4	4.4	2.33	–	–	–
	O2	0–6	5.1	2.54	–	–	–
	A1	0–6	5.9	0.09	0.13	3.8	10.1
	A/B	6–11	7.4	0.01	0.7	3.0	4.4
	C1	11–16	8.0	0.02	1.2	3.0	4.0
	C2	16–26	8.1	0.01	1.3	3.0	4.0
VII (150)	O1	0–5	4.0	1.09	–	–	–
	O2	5–20	4.4	1.00	–	–	–
	B	0–12	5.9	0.05	0.1	4.0	7.8
	C	12–55	7.7	0.02	0.1	5.0	6.0
VIII (c. 250)	O1	5–20	4.0	0.95	–	–	–
	O2	5–20	3.7	0.96	–	–	–
	O2/A2	0–5	4.4	0.37	0.47	1.3	16.6
	B1r	5–10	4.8	0.05	0.17	0.9	12.7
	C	10–20	5.8	0.03	0.1	2.0	6.3

soil conditions, is ousted as much by this acidification as by shading. However, in general, plant-by-plant replacements can be accounted for by the great height growth potential and vigorous spread of *Alnus* which simply shades out earlier colonists.

Later vegetation and soil changes

Phase VII is the development of the young *Picea sitchensis* forest, by 50–60 years, or more. Isolated trees established on bare till grow slowly. Others, established more densely in the alder thicket, grow quickly, then rise above the alders and suppress them. Some cottonwoods persist in the spruce forest. As the spruce trees mature (increasing to > 1 m in diameter and 30 m tall) signs of the previous alder stand disappear and the ground is covered by a dense mat of mosses and many herb and small shrub species (Table 6.4).

A later phase (Phase VIII) is mature spruce–western hemlock forest, about 35 m tall or more, with *Tsuga heterophylla* gradually becoming increasingly prominent with time, until it dominates. Sometimes yellow cedar (*Chamaecyparis nootkaensis*) and mountain hemlock (*Tsuga mertensiana*) are present. Figure 6.12 shows the 1965 distribution, in part of Muir Inlet, of vegetation cover classed into the categories Phases I–VII (Decker 1966).

It is estimated that, at Glacier Bay, the sequence from bare ground to the beginning of establishment of western hemlock as a prominent component of the forest takes at least 90–150 years (Decker 1966, Reiners *et al.* 1971, Ugolini 1968). In the period from about 40 to 90 years, first under alder, followed by alder–spruce, then spruce, the immature soil profile gradually develops a coloured B horizon and is classed as a brown podsolic soil. Subsequently under spruce, then spruce and hemlock, the soil acquires more of the characteristics of a podsol (with a markedly leached A2 horizon). From 55 to 250 years the pH of the litter and soil is lowered from about 5 to less than 4. Other changes include a decline in exchangeable potassium, and an increase in exchangeable hydrogen and cation exchange capacity, due mostly to an increase in humic colloids (Ugolini 1968). Both Ugolini (1968) and Crocker & Dickson (1957) showed that there is a net loss of nitrogen in the upper soil layers after about 50 years. This is nitrogen which has been taken up by the spruce trees and incorporated into their biomass. There will be slow cycling of nitrogen and other nutrients from litter to trees in the spruce, spruce–hemlock and hemlock forests. In these conditions some saprophytic nitrogen-fixing bacteria (*Clostridium*) may be contributing nitrogen to the system, but in the form of ammonium rather than nitrate. By 100–250 years after ice retreat a relatively species-rich forest is evident (Reiners *et al.* 1971, M. Noble & Sandgren 1976) and numerous fungi are present (Sprague & Lawrence 1960).

There have been no measurements of changes in light characteristics or microclimate within the vegetation as the sequence proceeds. Microclimatic differences between sites within the vegetation are likely to be as great in magnitude (or greater) than in sites lacking vegetation.

Some of the vertebrate animal life associated with the vegetation development sequence is listed in Table 6.3 (F. E. Good 1966, Trautman 1966). A fuller study of

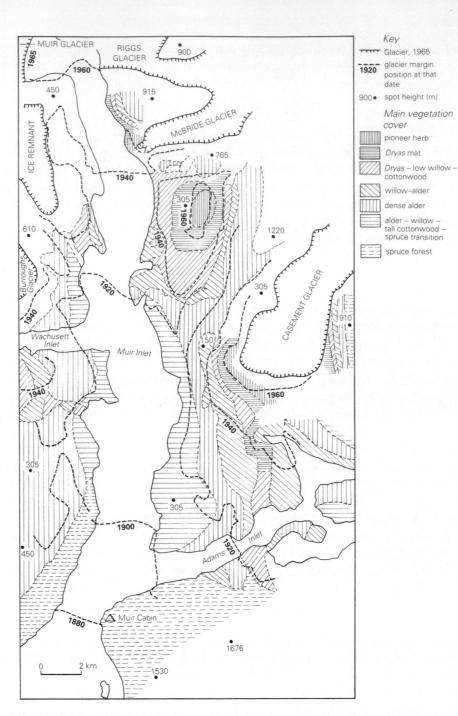

Figure 6.12 Distribution of vegetation types at various stages of development, in part of Muir Inlet, Glacier Bay, south-west Alaska, 1965 (after Mirsky 1966).

the fauna, including the many small inhabitants of the soil, is needed to elucidate the complex biological interactions, as the sequence proceeds.

Degradation of soil nutrient stocks and its effects on vegetation

Extensive bogs, known locally as **muskegs**, cover level or gently sloping sites on old moraines beyond the 17th–18th century ice limits at Glacier Bay (Fig. 6.13). The bogs are believed to originate as a result of establishment of *Sphagnum* sponge moss colonies within hemlock forest (Lawrence 1958). Gradually the *Sphagnum* spreads and, by its very great capacity for water retention, causes waterlogging of the ground and then peat accumulation. The hemlocks become stunted, die and are replaced by muskeg bog communities dominated by *Sphagnum* spp., heath shrubs and sedges, with scattered lodgepole pine (*Pinus contorta*) and some mountain hemlock (Table 6.4). Western hemlock–yellow cedar forest remains only on slopes or other sites where drainage is free. Evidence for sequences of change in the bog communities themselves is outlined by Lawrence (1958), but will not be considered further here. The reduced stature of the muskeg vegetation reflects the low carrying capacity of sites with respect to nutrients and as a result of waterlogging. Near Glacier Bay the evidence for this is only circumstantial. Near Fairbanks, Alaska, however, Heilman (1968) showed that declining tree productivity in black spruce (*Picea mariana*) stands, degrading to muskeg, was accompanied by marked decline in phosphorus levels and, to a lesser extent, potassium levels, in the tree foliage. The nutrients are either lost to the system completely or, in the case of nitrogen, sequestered in the deeply accumulating peat, from which nutrients are not cycled.

Estimates for the time required for development of hemlock-dominant forest and muskeg bog at Glacier Bay (c. 250 and 1500 years, respectively) are approximate (Ugolini 1968, Lawrence 1958, Reiners *et al.* 1971). Another estimate for the rates of development of these kinds of vegetation comes from a locality about 60 km west of Glacier Bay, at Lituya Bay, where tectonic uplift (movement of the earth relative to sea level) and eustatic sea-level changes (independent of any earth movements) have resulted in the formation of three beach-sand and gravel-covered terraces, 9, 26 and 80 m above present sea level (Ugolini & Mann 1979) (Fig. 6.14). The terraces have estimated ages of 400, 2000–3000 and 6000–8000 years, respectively. The outer edge of each terrace is younger and slightly better-drained than the inner side. Spruce forest dominates the lower terrace. The middle terrace, on its seaward side carries spruce with a western hemlock understorey and on its inland side western hemlock and yellow cedar. The upper terrace has stunted yellow cedar and mountain hemlock on its seaward side, with a boggy forest floor and, on the inland side muskeg bog dominated by *Sphagnum* spp., sedges and heath scrub.

It is a reasonable inference that each of the sites has undergone a process of vegetation change similar to that at Glacier Bay. Direct evidence for part of the sequence is preserved in the peat bog on the upper terrace, which contains a basal forest layer that is rich in wood and also fossils of a forest carabid beetle. The original muskeg bog would have developed between about 3000 and 8000 years ago.

The middle terrace has, on its outer side, soil profiles like the oldest described from

Figure 6.13 Muskeg (bog) on Pleasant Island, Icy Strait, east of Glacier Bay. *Pinus contorta* (lodgepole pine) scattered on bog which is dominated by sedges (*Eriophorum*), dwarf heath shrubs (*Empetrum*, *Kalmia*, *Ledum*, *Oxycoccus*), herbs, *Sphagnum* spp. and other mosses and various lichens. *Tsuga* (hemlock) forest in background. Ice-free for at least 14 000 years (photo D. B. Lawrence).

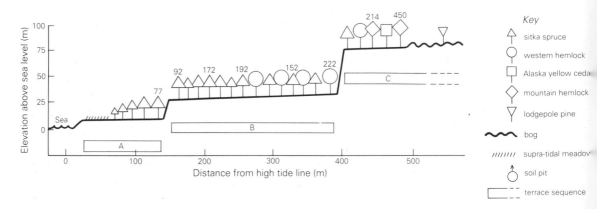

Figure 6.14 Cross-section of a raised terrace sequence (A, B, C) at Steelhead Creek, near Lituya Bay, Alaska (see Fig. 6.1), showing vegetation changes with time. Refer to the text for details (after Ugolini & Mann 1979).

Glacier Bay (Fig. 6.8). However, on its inner side there is a thick organic horizon, and an impervious B horizon, preventing free downward drainage. The upper terrace has a deep, poorly drained peat soil and the inner side an even deeper peat (1.5–2.0 m thick). There is evidence for iron-pan formation at the hemlock forest stage. The water table is held up above it and decomposition of plant material is slowed, resulting in peat accumulation. In different sites waterlogging could proceed either by swamping through spread of *Sphagnum* or by iron-pan formation or by both. This **paludification** process is common in cool, humid environments (Zach 1950, Heinselman 1975, Burrows & Dobson 1972, Burrows *et al.* 1979; see Ch. 8). The nutrient status of the bog communities is very poor, compared with young sites on similar terrain.

Diverse patterns at Glacier Bay

The trends of vegetation change at Glacier Bay are more complex than is indicated by the simple scheme outlined in Table 6.4. All sites vacated by ice at any one time are not simultaneously colonized by plants. Neither does the changing species composition necessarily follow the same path at each site, nor does the succession proceed at the same rate at each site. The vegetation at any locality at any time during the developmental phases is patchy. This applies to the early pioneer herb, *Dryas* and willow phases, and especially to the alder–willow phase, where mosaics of alder and willow are usual. Spruce and even, sometimes, hemlock may be early colonists, although closed spruce forest develops more readily through the alder–willow thicket phase. Cottonwood, though often an earlier colonist than alder, may gradually supplant it. Patchiness is also a feature of the mature vegetation; mosaics of hemlock, spruce and muskeg (itself heterogeneous), and sometimes alder and willow occur, according to different local site conditions. The patchiness is related to heterogeneous initial substrates (bedrock, boulders, gravel, dry ridges and wet hollows) and to subsequently varied environmental conditions, as well as to chance factors that decide which species happen to arrive at a place first and to periodic disturbances which damage the vegetation. The vegetative modes of proliferation of some species (particularly the herbs and some shrubs) contribute to the heterogeneity.

GLACIER MORAINES IN OTHER LOCALITIES IN NORTH AMERICA

Although there are some general patterns that are common to vegetation change on moraine chronosequences, the details differ according to the local circumstances. Near the Mendenhall and Herbert Glaciers, Juneau, Alaska, USA, a series of recessional moraines has a vegetation sequence similar to that at Glacier Bay, including hemlock forest and muskeg bog on the older surfaces (Lutz 1930a, Chandler 1942, Lawrence 1950, 1958, Crocker & Dickson 1957). The rate of development is more rapid because the moraines are close together and seed sources are in close proximity. The alder scrub phase is over by 50 years. Some spruce seedlings establish 3–7 years after ice retreat, and dense spruce forest is present by 100–120 years.

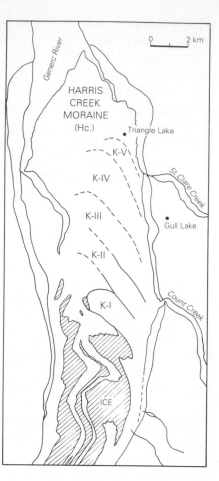

Figure 6.15 Moraines resulting from ice down-wasting after surges affecting the Klutlan Glacier, St Elias Range, Yukon Territory, Canada (after Birks 1980a, b). See text and Table 6.6 for details.

In a different climatic region (drier, colder in winter and warmer in summer) lie the moraines of the Klutlan Glacier on the inland side of the St Elias Range, Yukon Territory, Canada (Fig. 6.15). The moraines were formed by surges of a tributary glacier. The slow wasting away of the ice creates a special milieu for vegetation development (H. J. B. Birks 1980a, Jacobson & Birks 1980, Whiteside *et al.* 1980), repeated at some Alaskan glaciers, especially the Malaspina, in Yakutat Bay (Tarr 1909, Clayton 1965).

The Klutlan moraines consist of a thick mantle of hummocky surface debris lying over stagnant ice, which slowly and irregularly melts. Some areas are undisturbed for long periods; elsewhere there are meltwater streams in channels, ice-cliffs and melt and collapse hollows, some of which contain lakes. On the oldest moraines the surface is stable.

The locality is at 1250 m altitude. A station 90 km north of it has a January mean temperature of −28°C and July mean of 14°C. Annual mean precipitation is 36 cm.

The pioneer herb and shrub cover persists, on stable sites, for 20–40 years (Table 6.6). Unstable sites remain at (or return to) early phases of vegetation development. The harsh climate and lack of *Alnus* probably accounts for the relatively slow rate of development of closed, tall vegetation. By about 100–150 years,

Table 6.6 Representation of common plant species in samples from vegetation on successively older moraines of the Klutlan Glacier, Yukon Territory, Canada (after H. J. B. Birks 1980a).

Estimated age of site (years)		2–30	32–58	58–80	96–178	177–240	615–1220
Moraines on which present		KII–KV	KIII–KIV	KIV	KIV	KV	HC
Dominant vegetation cover		Pioneer herb–dwarf shrub complex	Willow–soapberry scrub	Spruce–willow	Spruce–bearberry	Spruce–labrador tea	Spruce–moss
vascular plants							
Crepis nana	h	0–1					
Dryas drummondii*	h	+ –3	+				
Hedysaryum mackenzii*	h	+ –3	+	+			
Epilobium latifolium	h	+ –2	+	+	+		0
Shepherdia canadensis*	s	+ –1	2	2	1	+	
Arctostaphylos uva-ursi	s	+ –1	2	2	1	1	
Salix glauca	s	+ –3	2	2	2	+	+
S. alaxensis	s	+ –2	1	1	+	+	0
S. arbusculoides	s	0–1	+	+	+		+
Populus balsamifera	s–t	0–1	+	1	+	+	+
Hedysaryum alpinum*	h		+	+	1	+	+
Dryas integrifolia	h		+	1	1	1	+
Pyrola grandiflora	h		1	1	1	1	1
Carex concinna	h		1	+	1	+	+
Betula glandulosa	s		+	1	1	1	
Salix reticulata	s		+	+	1	+	+
Arctostaphylos rubra	s		1	1	1	1	1
Picea glauca	t		+	1	+	+	0
Parnassia palustris	h		+	0	+	0	
Calamogrostis purpurascens	h		+		2	3	3
Ledum palustre	s				+	2	+
Vaccinium uliginosum	s					1	1
V. vitis-idaea	s					1	2
Empetrum nigrum	s					1	2

* Associated with nitrogen-fixing symbiont.
h, herb; s, shrub; t, tree.
0, present rarely; +, present infrequently, low cover–abundance; 1, present frequently, low cover – abundance, 2, present frequently, moderate cover – abundance; 3, present frequently, high cover – abundance.

after phases when willows and the nitrogen-fixing associations with *Hedysaryum alpinum*, *Dryas drummondii* and *Shepherdia canadensis* are important, open white spruce (*Picea glauca*) forest develops, with bear-berry (*Arctostaphylos uva-ursi*) as a ground cover. The first-generation forest becomes denser by about 200 years. On a moraine, probably about 1000 years old, mature spruce forest is present, with widely spaced old trees, stunted for their age. The ground between the trees is covered by heath shrubs, some herbs and abundant mosses and fruticose (erect and shrub-like) lichens. *Racomitrium lanuginosum* and *Sphagnum girgensohnii* occur in the oldest forest stands. Their presence and that of the heath shrubs, as well as the decrepit *Picea glauca* forest, indicate poor nutrient status of the soils, but there is no soil waterlogging, except around ponds and mires. The ecosystem nutrient complement seems to be declining.

Independent checks on the correctness of the assumption that the spatial sequence at the Klutlan Glacier, observed at an instant of time, also represents or is close to the temporal sequence, comes from several pieces of evidence (H. J. B. Birks 1980a, b). The maximum age of woody plants in the successively older plant communities is one indicator. Also, on two of the older moraines (IV and V), beneath spruce forest, spruce litter layers overlie humus deposits, with macrofossils of the species from earlier developmental phases, *Dryas drummondii*, *Salix glauca* and *Arctostaphylos uva-ursi*. In pollen analyses from sediments in lakes on the oldest moraine, also, the sequence of pollen assemblages from bottom (oldest) to top (youngest) is *Salix–Shepherdia* → *Picea–Salix* → *Picea*.

DIRECT COLONIZATION OF MORAINES BY TREES

Observations by W. S. Cooper (1916) on the phenomenon of direct colonization of talus and moraine at Mt Robson in the Canadian Rockies, by tree species which occur in the mature forest of adjacent slopes, are of some interest in relation to hypotheses about the causes of sequential vegetation change. W. S. Cooper visited the site in 1914 and described three communities on moraines, dominated, respectively, by *Dryas octopetala–Arctostophylos rubra*, *Salix* spp.–*Betula glandulosa* and engelmann spruce (*Pinus engelmannii*). The spruce, along with mountain fir (*Abies lasiocarpa*) and whitebark pine (*Pinus albicaulis*) dominates the mature forest. All three communities may establish on bare moraines; the herb and shrub communities also persist on moraines which have progressed to predominantly forest cover. W. S. Cooper pointed out that modifications of the habitat by earlier plant populations did not seem to be a necessary precondition for the appearance of the later populations (although vegetation change in some sites proceeds through the sequence herbs → shrubs → trees).

Tisdale *et al.* (1966) continued studies on this moraine sequence in 1963. The older five moraines then ranged in age from 50 to 180 years. Four younger moraines had formed since then, but carried sparse vegetation. Table 6.7 shows that there is a temporal vegetation trend, but that pioneers tend to persist to a late stage in the rather open forest. Some spruce trees occur on moraine 5, but they are not abundant. There was no sign of nitrogen deficiency in any plants on the younger moraines and no differences in nitrogen content were evident in foliar samples from plants on the three moraines. Although the forest on the older moraines is superficially like that of mature forest in the region, it differs floristically and in structure so that the situation is more complex than W. S. Cooper had indicated.

It was found by Sigafoos & Hendricks (1969), on the moraines of the Nisqually and other glaciers, on Mt Rainier, Washington, USA, that tree seedlings (*Picea engelmannii*, *Pinus contorta* and others) became established 2–13 years after sites had been vacated by ice. Willows, alder and heath shrubs form mosaics with the developing trees. Sigafoos & Hendricks (1969) suggested that no real, regular vegetation sequence was occurring. Change in vegetation through time, which might be inferred from differences in vegetation on surfaces of different age, they proposed, seems to be

Table 6.7 Vegetation composition on successively older moraines of Robson Glacier, Canadian Rockies (after Tisdale *et al.* 1966).

Moraine		Moraine 5 (ice-free AD 1914)	Moraine 3 (ice-free c. AD 1890)	Moraine 1 (ice-free c. AD 1800)
Main vegetation type		Herb–scrub	Herb–scrub–scattered trees	Moss–lichen–herb–scrub–open forest
			Canopy cover for plant species (%)	
mosses, lichens		3.6	11.7	34.7
*Dryas drummondii**	h	7.2	4.9	0.7
*D. octopetala**	h	1.9	21.5	7.4
*Hedysaryum mackenzii**	h	46.8	52.0	14.6
Castilleja pallida	h	6.8	–	–
Arctostaphylos rubra	s	0.8	5.4	23.8
Salix brachycarpa	s	21.1	19.8	8.5
S. glauca	s	4.9	11.6	16.9
S. reticulata	s	0.7	3.8	1.5
Betula glandulosa	s	–	1.7	14.9
*Shepherdia canadensis**	s	–	–	3.8
Picea engelmannii	t	–	13.1	19.6†
Pyrola chlorantha	h	–	–	1.0
Carex spp.	h	–	0.1	12.8

h, herb; s, shrub; t, tree.

* Associated with nitrogen-fixing organism.

† NB: In mature forest nearby *Picea engelmannii* dominates, *Cornus canadensis*, *Pyrola* spp. and *Phyllodoce* spp. are the main understorey species. *Dryas*, *Hedysaryum*, *Arctostaphylos* and *Salix* are absent.

related to the death of shorter-lived species and the continued growth of the trees. They give no detailed floristic analyses of the vegetation, but plant lists suggest that the floristic composition of the young forest differs from that of the old forests on surrounding slopes. In this locality alder forms dense thickets in places on young moraines where coniferous trees are small and widely scattered. Observations on the Mt Rainier moraines by Dr Curt Wiberg (pers. comm. 1980) also show that herbs, trees (lodgepole pine) and shrubs (including alders) colonize bare areas at the same time. Lodgepole pine is especially important. Silver fir (*Abies amabilis*), douglas fir (*Pseudotsuga menziesii*) and western hemlock (*Tsuga heterophylla*) make their appearance later.

Thus, in some morainic sites, a sequence of vegetation beginning with herbaceous pioneers and ending with trees is not strongly developed. Sufficient nitrogen must accrue to the sites to permit vigorous growth of the young trees. It may come down in the rain or be derived from plant litter from established vegetation on adjacent slopes. The openness of the tree communities may permit smaller 'pioneers' (including species with nitrogen-fixing associations, *Dryas*, *Hedysaryum* and *Shepherdia*) to co-occupy sites and thus provide the trees with their nitrogen requirements. The trees which can colonize moraines are versatile species, with efficient means of dispersal. The seed sources are very close. Their ability to function very effectively as early colonists in some places must be determined by their broad tolerance in relation to their microsite requirements for establishment and subsequent growth.

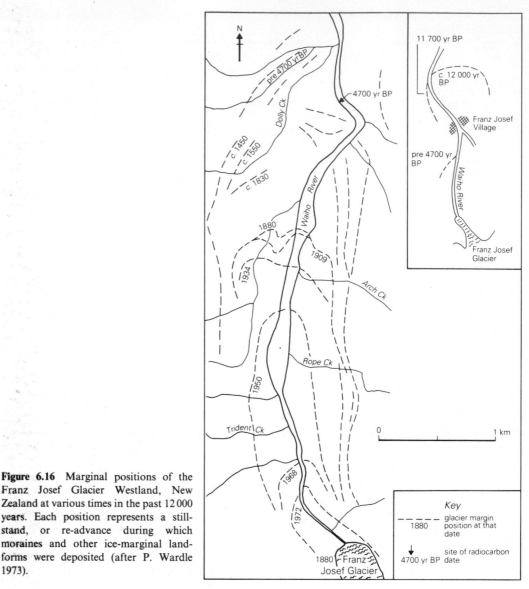

Figure 6.16 Marginal positions of the Franz Josef Glacier Westland, New Zealand at various times in the past 12 000 years. Each position represents a still-stand, or re-advance during which moraines and other ice-marginal landforms were deposited (after P. Wardle 1973).

FRANZ JOSEF GLACIER, NEW ZEALAND

Moraines of the Franz Josef Glacier, in a mild, humid part of New Zealand (43° 23′S., 170° 11′E; mean temperature 14°C, mean January temperature 18.5°C, mean July temperature 9°C, mean annual rainfall 500 cm) (Fig. 6.16) carry a sequence of vegetation and soil development spanning about 12 000 years (Stevens 1963, 1968, P. Wardle 1977, 1980, Burrows unpubl. data 1982).

The bouldery moraines consist of hard sandstone and schist debris. They are assumed to conform to the conditions for a chronosequence. Some of the outwash surfaces probably do not do so because they contain some alluvial material deposited by valley-wall streams.

Figure 6.17 outlines the main vegetation sequence on surfaces up to about 12 000 years old. First colonists are angiosperm herbs and mosses, but then there is rapid development of dense scrub, followed by low forest, before stands of tall, long-lived rata (*Metrosideros umbellata*) develop. On moraines older than about 600 years *Metrosideros* is replaced by the more shade-tolerant kamahi (*Weinmannia racemosa*), with gradual increase in the populations of shade-tolerant gymnosperm species. The gymnosperms miro (*Prumnopitys ferruginea*), totara (*Podocarpus hallii*) and occasional rimu (*Dacrydium cupressinum*) are present on surfaces up to several thousand years old. *Dacrydium cupressinum* then becomes more abundant until it is co-dominant with *Weinmannia*. *Prumnopitys ferruginea* and *Podocarpus hallii* also occur, together with a range of other subcanopy tree species. This kind of forest still persists on surfaces about 11 500 years old, although understorey species have changed.

The development of soil profiles (Fig. 6.17) and the changes in soil chemistry were examined on outwash surfaces by Stevens (1963, 1968) but not on moraines, which because of their often very bouldery nature, are extremely difficult to sample for soil analysis. The results of the chemical analyses are shown in Figure 6.18. The data may be taken to be a rough approximation to the soil changes occurring on moraines.

Starting at time zero, after retreat of the ice, the unaltered parent material, (containing rock fragments of various sizes) soon accumulates a surface layer of organic matter; the pH in the upper soil horizons falls from 8 to 5.5 in 55 years. Subsequently it falls to less than 5 by 2000 years and to 4 by 10 000 years. Organic carbon in the soil builds up to about $4000\,kg\,ha^{-1}$ in about 300 years, then is steady. Soil nitrogen rises to nearly $3\,kg\,ha^{-1}$ in 300 years, falls slightly, rises to nearly $4\,kg\,ha^{-1}$ by 2000 years, then is steady. Stevens (1963) attributed the initial rapid rise in nitrogen to the role of the legume shrub *Carmichaelia grandiflora*. *Coriaria arborea* is also implicated on most moraines. An often unconsidered source of nitrogen (and other nutrients), plant litter from outside the system, is also likely to make some contributions. Nitrogen in the precipitation amounts to about $1–2\,kg\,ha^{-1}\,year^{-1}$.

Organic phosphorus rises to about $0.5\,kg\,ha^{-1}$ in the first 2000 years, then levels off. Cation exchange capacity is still rising until about 10 000 years, but the value for total exchangeable bases is at its highest by 300 years (a peak mainly accounted for by exchangeable calcium). It then falls and is steady from 2000 to 10 000 years. Meanwhile, the soil profile develops (Fig. 6.17) with iron sesquioxide accumulations and clay mineral formation (Mokma *et al.* 1973). Shallow, weathered horizons form in the first 1000 years, and a gleyed A2 horizon some time after 2000 years. By 10 000 years the soil is a well-developed gley-podsol, with a weathered yellow-brown B horizon and relatively little unweathered parent material remaining in the zone accessible to roots. On older surfaces (more than about 13 000 years old) the soils contain easily crumbled stones with a minimal residue of unweathered minerals. The plants must then rely almost totally on recycling from the organic component of the system. Exports of nutrients probably exceed imports.

Estimated age of surface (yr)

	5	10	50	150	500	1000	2000	5000	12 000
Vegetation phase	pioneer →	early transitional scrub	→ early transitional low forest	→	transitional rata forest	→ transitional to mature kamahi forest	→ kamahi – mixed podocarp forest	→ mature rimu–kamahi forest	
Prominent plant species	bare rock debris with *Epilobium* spp., h *Raoulia* spp., h *Poa novae-zelandiae*, hg *Racomitrium crispulum*, hm *Stereocaulon* sp., l	*Carmichaelia grandiflora*, s* *Olearia avicenniaefolia*, s *Coprosma rugosa*, s *Coriaria arborea*, s*	*Olearia*, s *Coriaria*, s* *Schefflera digitata*, t *Melicytus ramiflorus*, t *Pseudopanax colensoi*, s	*Aristotelia serrata*, t *Fuchsia excorticata*, t *Griselinia littoralis*, t *Hoheria glabrata*, t	*Metrosideros umbellata*, t *Weinmannia racemosa*, t *Carpodetus serratus*, t	*Weinmannia*, t *Prumnopitys ferruginea*, tp *Podocarpus hallii*, tp *Dacrydium cupressinum*, tp other angiosperm trees, *Cyathea* sp. tf		*Dacrydium cupressinum*, tp *Weinmannia*, t *Quintinia acutifolia*, t other podocarps and angiosperm trees	
Rate of change	very rapid	rapid	rapid but slowing		slow	slow		very slow	
Soil sequence		lithosol LFH weak A1 C1 C2		weakly podsolic LFH A1 AB C1 C2	podsolic LFH A1 AB weak B1 C1 C2	yellow–brown podsolic LFH A1 B1 B2 C		yellow–brown gley podsolic thick LFH weak A2 B1 B2G BG C	

Key

h herb (g = grass, m = moss)   l lichen   s shrub (*with nitrogen-fixing symbiont)   t tree   tf treefern   tp podocarp (gymnosperm)

Figure 6.17 Vegetation development on moraines of the Franz Josef Glacier, Westland, New Zealand (adapted from Stevens 1963, 1968, P. Wardle 1980).

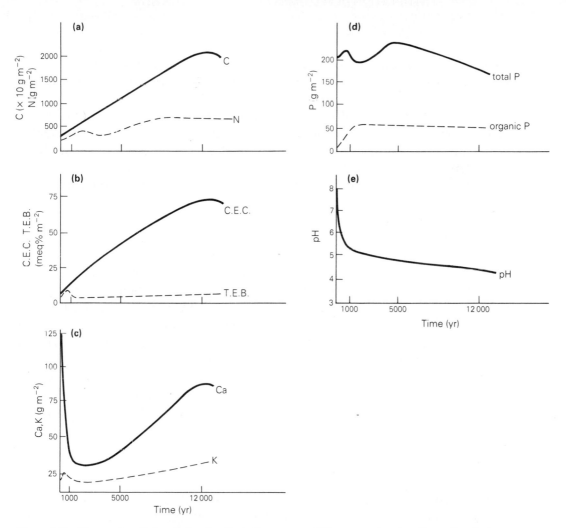

Figure 6.18 Changes in soil chemical properties in the uppermost 33 cm, on outwash surfaces after glacier recession Franz Josef Glacier, Westland, New Zealand. (a) Carbon $g\,m^{-2} \times 10$; nitrogen $g\,m^{-2}$. (b) Cation exchange capacity and total exchangeable bases, milli-equivalents $\%\,m^{-2}$. (c) Calcium and potassium, $g\,m^{-2}$. (d) Total and organic phosphorus $g\,m^{-2}$. (e) pH. After Stevens (1963, 1968).

SUGGESTIONS ABOUT CAUSES OF THE VEGETATION CHANGES

Can any of the observed changes in vegetation species composition up to 12 000 years at the Franz Josef Glacier, or on moraines elsewhere, be attributed to the soil changes? At present some specific correlations can be made, although our understanding of the causal phenomena is imperfect. Interactions of biological and soil phenomena are implicated. As Runge (1973) points out, the separation of cause from effect is difficult in slowly changing systems under continuous change. The greatest lack is information about the quantities of nutrients in the soil as well as those in the total biomass. Some data which are relevant to this are those gathered by foresters in uniform, planted forests of different age, on uniform

199

substrates, (e.g. Ovington 1959, Ritchie 1968). The kinds of effects noticed are:

1   Large, progressive additions of organic matter in the form of litter. This conserves water and contains considerable amounts of carbon, nitrogen, phosphorus and cations.
2   In young stands, removal of nutrients (e.g. calcium and potassium) from near-surface horizons and their incorporation into biomass exceeds return, but eventually the upper soil nutrient levels are restored as plants draw nutrients from greater depths and recycle them.
3   In time most nutrient elements within plant root depths become concentrated in the upper horizons.

The natural ecosystem on the Franz Josef moraines and outwash surfaces is much more complex than the plantation forests, and the time sequence there spans a much longer period, but analogies between them can be drawn. Properties of plants influencing the main processes inferred to be occurring in the sequence of changes on the Franz Josef moraines up to about 550 years old are summarized in Table 6.8. The pioneers are undemanding with respect to nutrients and moisture; there is enough nitrogen from free-living nitrogen-fixers, rainfall and from wind-blown debris to support them (and *Carmichaelia*, *Coriaria* and *Gunnera* have nitrogen-fixing associations). Significant habitat change is effected by *Carmichaelia* and *Coriaria* in this way, but otherwise changes occur through provision of a wider range of microhabitat conditions which later colonists can exploit. A few seedlings of some versatile tree species (e.g. *Weinmannia* and *Aristotelia*) appear on young moraines, but seem to be suppressed later by shrubs.

The shrubs *Coriaria* and *Olearia* rapidly supplant the early pioneers and modify the light, moisture and surface soil environments. They have relatively high requirements for calcium, nitrogen and phosphorus. *Coriaria* and *Carmichaelia* build up the soil nitrogen; there is a large accumulation of carbon and a decline in pH. Tree seedlings, which germinate abundantly below the shrub canopy, tend to be suppressed by *Olearia* competition but, as the shrub vigour declines with age, the canopy opens. The decline of *Olearia* appears to be increased by abundant epiphyte growth on its leaning stems.

Reduced pH levels and calcium supplies and takeover of organic nitrogen and phosphorus, released by recycling, permit quickly growing angiosperm trees to overtop the shrubs, which die out. The trees tap deeper levels of unweathered soil and incorporate much nitrogen, phosphorus and other nutrients into their biomass; the upper soil layers may be depleted of nutrients for a time.

Progressive increase in the range of microsites for seedling establishment enhances the possibilities for development of diverse vegetation. The two species which now begin to invade the low forest have some different characteristics, although both have tiny, wind-dispersed seeds, and both are slower-growing than the species they supplant. *Metrosideros* requires relatively well-lit sites; it can establish on bouldery substrates under a shrub canopy, provided that they are sufficiently moist. It is probably moderate in its nutrient demands. *Weinmannia* is shade-tolerant and

Table 6.8 Salient points about the vegetation sequence on 0–550-year-old moraines of the Franz Josef Glacier, New Zealand.

Approximate age of surface (years)	Main cover	Other features	Plant influences on the vegetation sequence	
10	bare cobbles, boulders c.50% bare sand, pebbles c.30% Racomitrium crispulum (on boulders, cobbles) 10%	propagules are derived from steep, vegetated slopes above	mat plants trap silt, sand, increasing soil depth	F†
	mat plants Raoulia spp., Gunnera sp.*	A few Aristotelia serrata, Weinmannia racemosa (tree juveniles) are present in sheltered crevices among boulders (which contain wind-blown litter, sand, silt)	Coriaria, Gunnera, Carmichaelia release N from litter to soil (and P, cations etc.)	F
	shrubs up to 50 cm tall Coriaria arborea*, Carmichaelia grandiflora*, Olearia avicenniaefolia 10% (all on finer-textured material).		young trees are mainly suppressed by shrubs, but some may survive	I T
30	dense shrubbery. Olearia up to 2.5 m tall 50% Coriaria 2–2.5 m tall 20%	ground well-covered by litter, a few open areas by huge boulders or boulder heaps a few young Aristotelia and Weinmannia saplings ferns common on ground	shrubs have pre-empted site persistent young trees	I T
50	dense tall shrubbery Olearia, Schefflera digitata up to 6 m or more tall; also some Coriaria, Melicytus ramiflorus, Pseudopanax colensoi, Aristotelia trees	All Olearia adults have epiphytic bryophytes, ferns and orchids on their leaning stems sapling Melicytus, Griselinia littoralis, Weinmannia and Metrosideros umbellata are present ground ferns present	Coriaria and Olearia decline – being suppressed by shading by taller plants epiphyte growth dependent on host plant shaded site and presumably nutrient cycling favour the more shade-tolerant species (e.g. Griselinia, Weinmannia, Melicytus)	F F (and T?)

Table 6.8 (*cont.*)

	Vegetation	Process	
150	dense low forest *Melicytus*, old *Olearia* 7–8 m *Weinmannia*, a few *Metrosideros* to 9 m; *Griselinia*, *Schefflera*, *Coprosma lucida*, *Cyathea smithii* (tree fern)	ground cover is litter and (about 30%) fern	fern cover favoured by canopy shading and pre-empts ground space — F
		root suckering of *Weinmannia*, *Griselinia*, *Metrosideros*	maintenance of potentially long-lived tree species — I, T
		Weinmannia and *Metrosideros* saplings established on leaning stems and rotting logs of other species	provision of sites for tree establishment — F
		abundant bryophyte, lichen, fern, orchid epiphytes on leaning stems	epiphyte growth on *Olearia* may hasten its death, but *Olearia* is doomed mainly by its inability to grow tall or establish seedlings in the shade
400	tall *Metrosideros* forest 18–20 m, trunk diameter 40–90 cm; also some *Weinmannia*; subcanopy *Weinmannia*, *Carpodetus*; abundant saplings, shrubs *Weinmannia*, *Griselinia*, *Coprosma lucida*, *Schefflera*, *Cyathea*; a few saplings of *Podocarpus hallii*, *Prumnopitys ferruginea*	ground cover includes about 30% fern	*Metrosideros* pre-empts the site — I
		multi-stemmed *Metrosideros* seem to have coalesced from several individuals	*Weinmannia* and other shade-tolerant woody species, including podocarps, persist — T
		vines *Ripogonum scandens*, *Rubus cissoides* are common	vines extend to canopy, scrambling on tree stems. — F
550	tall *Metrosideros* forest 25m; canopy spread of individuals up to 18 m; *Weinmannia* subdominant; subcanopy *Weinmannia*, *Melicytus*; smaller trees, shrubs include *Schefflera*, *Pseudopanax simplex*, *Griselinia*, *Cyathea*, *Coprosma* spp.; saplings of *Prumnopitys ferruginea* and (rarely) *Dacrydium cupressinum*	ground well-covered by forest litter; ground cover (mainly ferns, herbaceous angiosperms) about 70%	epiphytes dependent on tree sites — F
		epiphytes very abundant on trees – ferns, lycopods, psilophyte and some tree and shrub species	*Weinmannia* (with tiny seeds) has little chance of establishing on the ground, but treefern stems are very suitable sites — I, F
		Weinmannia commonly epiphytic on treeferns	*Cyathea* pre-empts gaps for some time until, on its death, shade-tolerant species replace it — I

| a few big trees of *Metrosideros*, *Weinmannia* have fallen; in the gaps are dense *Cyathea*, *Schefflera* or young *Weinmannia*

some fallen *Metrosideros* have developed upright stems

vines include *Ripogonum*, *Rubus* and *Clematis paniculata* | podocarps continue to grow slowly | T |

---

* Plants with nitrogen-fixing symbiont.

† F = **facilitation** – a process of species replacement arising from changes wrought by the plants resident on a site.

I = **inhibition** – a process of limitation of other species by occupation of a site by a species and pre-emption of resources there, at least for a time.

T = **tolerance** – the existence of species which are very tolerant of difficult conditions and persist until they grow (very slowly) to adulthood. It is unlikely that they are not affected at some time in their life by the facilitation or inhibition phenomena.

tolerant of a rather wide range of nutrient conditions. It has the capacity to establish epiphytically, as well as on the ground. Once *Metrosideros* is well-established, its large, dense canopy tends to inhibit the growth of most other trees, except some *Weinmannia*. *Metrosideros* is relatively long-lived (perhaps to 600 years or more), so it can hold its position for a long time, but it only occupies the sites for one generation, even if there are disturbances. Fallen trees are usually replaced by *Weinmannia* or quickly growing treeferns or other tree species (which are relatively more shade-tolerant than *Metrosideros*). Patches of treeferns inhibit the development of tree species by their dense canopies and heavy leaf-shed.

Weinmannia, a versatile species, may then dominate the forest for a time, but populations of the slower-growing podocarps eventually build up on surfaces that are older than about 1000 years. They tolerate the increasingly low pH of the upper soil horizons and lowered soil fertility. They also seem to have specific microsite requirements for establishment that are not met in earlier phases of the sequence, and their juveniles can cope with deep shade. They are not seen on moraine surfaces that are younger than about 400 years, although adults are present not far away. Forest dominated by *Prumnopitys*, *Podocarpus* and *Weinmannia* persists, probably for several thousand years, before giving way to dominance by *Dacrydium cupressinum*, *Weinmannia* and *Quintinia acutifolia*. Various other tree species and many shrubs and ferns are common.

The mature forest phases are diverse, with complex structure. The many more kinds of microsite for seedling establishment include the rotting leaf litter, tree bases, fallen logs, root masses of fallen trees, treefern stems and the stems, branches and crotches of old trees. The environment is so moist and litter so abundant that organic soil collects on such sites, providing habitat not only for obligate epiphytes such as the many fern and orchid species, but also for other species which customarily are rooted on the ground. Some of these eventually send roots down to the ground.

The soil on surfaces that are older than about 2000 years is progressively more gleyed (waterlogged) and becoming poorer in nutrient content. Much of the nutrient pool is retained in large, long-lived, slowly growing trees. The litter and upper soil layers, which also contain a pool of nutrients, mineralizing slowly, are very acid. Mycorrhizal fungi are prominent in the often deep mor humus layers. Tree and shrub species that are typical of the earlier phases would be unable to grow in these conditions, even if there was sufficient light for them to do so.

Because of the potential for the *Dacrydium–Weinmannia–Quintinia* forest to persist for thousands of years it may be regarded as 'most-mature' or 'steady-state' vegetation. However, strong leaching continues and, if sufficient time were available, the forest structure and species composition would probably gradually decline, in parallel with the diminished nutrient status of the soil and increasing degree of gleying (enhanced by the development of an impervious iron-pan below the B horizon). Lower-statured forest, with understorey species that are tolerant of very low soil fertility and waterlogging, is present nearby on surfaces that are older than 12 000 years.

SOME CONCLUSIONS

The data from glacier moraine and outwash chronosequences provide further insights into the complex interactions of biotic, soil, climatic and other components in ecosystems developing over long periods. Although the early phases of vegetation development on moraines (in the first few hundred years) often resemble the sequences of forest recovery after disturbance (cf. Ch. 9), the long-term sequences are dissimilar to short-term 'secondary succession'. The differences are not due only to the relatively long time required for soils to develop on new surfaces. Differences are inherent also because soil conditions continue to change in the long sequences by a decline in nutrient stocks and a change in recognizable physical and chemical soil properties, evident from the soil profiles and chemistry.

Under mor-forming conditions the plants induce soil acidity, which in turn affects the weathering and leaching of nutrients and changes in the physicochemical soil properties. Then there are feedback effects on the populations of plants from the changed soil conditions. Relatively abundant supplies of nutrients occur in young systems, because nutrients are being released from the rocks by weathering and nitrogen is being fixed abundantly by symbiotic associations. In maturing systems there is an increasing trend of incorporation of nutrients into living biomass and soil organic matter, so that plants depend, increasingly, on nutrient cycling and the intercession of mycorrhizal fungi for their nutrient supplies.

The rate of change is slow in undisturbed systems, but eventually the vegetation and soil reach an equilibrium when no more nutrients can be weathered from the rocks in the soil depth that is accessible to roots, and net losses of nutrients occur. In this way the total nutrient pool of old systems can be depleted. The diminished nutrient status and the altered soil physical conditions on any site are unfavourable to earlier inhabitants. The vegetation stature may eventually decline, with replacement of the initial 'equilibrium' species populations by others in which the species are smaller, more-slowly growing and tolerant of very low nutrient supplies. They may also have to be tolerant of waterlogged soils, although waterlogging is not an inevitable concomitant of long-term soil development.

Such patterns of change in soil–vegetation systems, as a consequence of modifications to various compartments of the total nutrient pool (see Ch. 2) were outlined by Billings (1941) in an early review, generally neglected by plant ecologists. The diminution of nutrient stocks and degradation of old systems in leaching environments, under mor-forming plants, are noted in Chapter 5 and are met again in relation to peat formation in Chapter 8 and in tropical forests in Chapter 10. The phenomenon is quite widespread in the Tropics and moist climates elsewhere (Stark 1978, Zach 1950, Heilman 1966, 1968, Groves 1981b).

Changes in states of phosphorus have been implicated as a critical determinant of vegetation change on chronosequences (Stevens & Walker 1970, T. W. Walker & Syers 1976, Runge 1973; Ch. 5). With possible exceptions noted later, once an ecosystem begins to develop, virtually no phosphorus is added to it and despite conservation by the biota, losses ultimately occur. Smeck (1973) pointed out that phosphorus from calcium phosphate is most readily available to plants at pH of 7. At

205

pH levels above that its solubility decreases. At pH values below 7 the solubility is governed by complexes with iron and aluminium and decreased rapidly with declining pH. In mature, clay-rich soils phosphorus is also tightly held (Bieleski 1973). Bieleski also describes other aspects of the phosphorus economy of soils important in relation to availability to plants.

As a system ages the forms of phosphorus in the soil system change as shown in Figure 5.14 (from T. W. Walker & Syers 1976). Young soils are relatively fertile with respect to phosphorus, and old soils less fertile. The time taken for weathering of all the apatite (the main primary source of phosphorus in the rocks) after colonization of new sites by plants will be of the order of approximately 15 000 years. Once it is all weathered only small amounts of phosphorus can be made available to plants by the mineralization of organic matter (cf. Harrison 1979, Duvigneaud & Denaeyer-de Smet 1970, Gorham et al. 1979). However, the mycorrhizal associations enable plants to abstract phosphorus from the occluded as well as the organic forms.

The disturbance of an ecosystem by fire or overgrazing will be likely to accelerate losses of phosphorus and other nutrients. Accretions of fresh materials can rejuvenate the nutrient stocks of a system. This can occur through siltation by rivers, loess accumulation, volcanic ash showers, erosion to unweathered rock or additions by animals (e.g. seabird faeces in roosting areas).

There seem to be strong correlations between the long-term soil and vegetation changes on glacial moraines. However, superimposed on these are changes controlled by the properties of individual plant species which determine their tolerance ranges in relation to the prevailing environmental gradients and their ability to compete successfully with their neighbours. Light and moisture relationships, as well as nutrients, are implicated.

Other significant matters governing the patterns of colonization and species replacements are the range of microsites available for establishment of young plants and the degree of control over the environment by species or groups of species. Prominent species usually dominate for a time, during vegetation development, pre-empting sites and preventing potential successors from taking them over. Eventually, because the members of the prevailing population begin to senesce, or the successors gradually become favoured by changes in habitat conditions, or both, the prevailing population is supplanted.

In the early phases of vegetation development on moraines the species appearing in sequences are often not competitively versatile, so that they last for only one generation (or, at least, for a short time). In later phases the species are competitively more versatile and last for a few to many generations. Exceptions are known where direct colonization of the fresh surfaces is by species that are versatile both in their microsite requirements for establishment and in competition.

7

Influences of strong environmental pressures

Not all of the Earth's surface is favourable for habitation by plants. Among the places where the vegetation is sparse or non-existent over extensive areas are: hot, dry deserts; cold, dry polar deserts and icefields, summits of high mountains and glaciers; large areas of the sea and large freshwater lakes (Fig. 7.1). At intervals on a small local scale, or seasonally on a wider scale, other places which, for at least part of the time, are quite favourable for plant growth, experience limiting conditions. They are caused by climatic variation, by other specific site factors or by disturbance. Under these conditions plant cover is continually or seasonally absent, scattered or reduced in stature, according to the capacity of the plants for coping with the particular habitat conditions.

Some of the difficult habitats (dry or cold sites, sites with hard substrates, or infertile or toxic soils) are physiologically very stressful for plants. Places where destructive forces such as fire, grazing and wind occur frequently, or where there are unstable substrates, soil erosion or debris deposition are also difficult plant habitats. Most disturbances tend to be episodic, but in some localities they are virtually incessant (e.g. wave action on sea shores, soil turbation by frost in high alpine regions). Often several strong environmental pressures affect plants at the same time.

DESERTS OF THE WARM TEMPERATE TO SUBTROPIC ZONES

Deserts are places where the potential evapotranspiration is two or more times the annual supply of water. Precipitation occurs irregularly, if at all. Consequently, the occurrence of vascular plants in deserts is sparse and limited to places where soil moisture is available from time to time (Hillel & Tadmor 1962, N. E. West 1983) (Fig. 7.2).

The desert terrain is varied, depending on the locality. It may include mountain ranges and intervening basins, stony alluvial fans and plains, large areas of rock pavement, dune systems, dry stream courses and lake beds. Some desert climate data are summarized in Table 7.1 and Figure 7.3.

As well as the almost perpetual drought, other harsh environmental conditions prevail in deserts. These include extreme diurnal temperature ranges – very hot in the daytime, cold at night. Bare rock, or stony or sandy surfaces heat very strongly in the daytime. The light intensity is also very high. The relative humidity is very low during the day, although dew may form at night.

Deserts are usually windy; loose surface sand and fine gravel, driven by the wind,

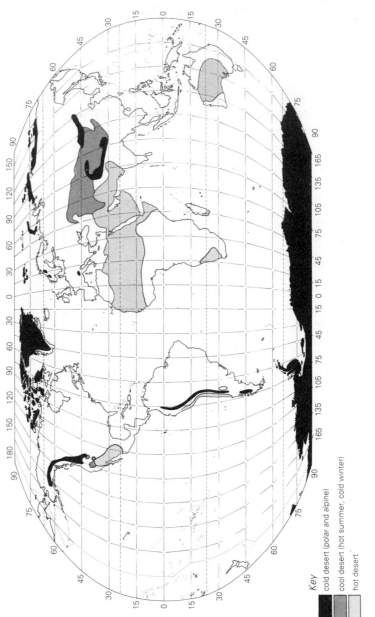

Figure 7.1   The main barren or sparsely vegetated areas of the world (many small areas of high mountains or other mainly barren terrain are not shown).

Figure 7.2 Desert, Monument Valley, Utah, USA, with scattered *Chrysothamnus* (rabbit brush) shrubs (photo C. J. Burrows).

Table 7.1 Climate data for North American desert and Great Plains grassland localities (after Wernstedt 1972).

Locality	Mean annual precipitation (cm)	Mean annual temperature (°C)	Mean January temperature (°C)	Mean July temperature (°C)
desert				
Richfield, Utah 38°46′N, 112°05′W	20.6	10.3	− 3.0	24.0
Beowawe, Nevada 40°36′N, 116°29′W	16.4	19.2	8.1	31.5
Chandler, Arizona 33°18′N, 111°50′W	18.7	21.1	10.7	32.2
Great Plains grasslands				
Scottsbluff, Nebraska 41°52′N, 107°36′W	38.9	10.8	− 4.8	25.3
Lincoln, Nebraska 40°51′N, 96°37′W	69.8	9.5	− 3.1	23.0
Fairfield, Iowa 41°01′N, 91°57′W	90.7	8.1	− 7.7	22.8
Amarillo, Texas 35°14′N, 101°42′W	52.9	20.1	11.8	28.1

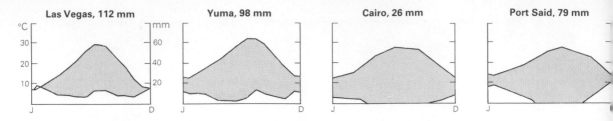

Figure 7.3 The annual march of temperature (upper curve) and precipitation (lower curve) for desert localities in North America (Las Vegas, Nevada and Yuma, Arizona) and North Africa (Cairo, Egypt and Port Said, Egypt) (from Wernstedt 1972).

may abrade objects that project above the surface. Soil surfaces are wind-scoured, and seeds and plant litter are widely dispersed. Paradoxically, scattered, severe, periodic thunderstorms may occur, with downpours of rain and local flash-flooding which causes severe surface erosion and the formation of shallow, ephemeral pools and lakes, but the water rapidly evaporates. Soluble salts accumulate in the soils and lake beds in quantities that are toxic to most plants. The salts usually include sodium and magnesium chlorides, sodium and calcium sulphates, and calcium carbonate. They often form cemented masses in the soil, or a white crust over the soil surface.

Desert soils (**aridisols**) are highly variable in texture. Most contain unweathered rock fragments and sand, as well as silt and clay. Organic matter is very sparse and, although structures and some horizon differentiation are evident, soil profiles are weakly developed (W. H. Fuller 1974).

Although plant nutrient cations and anions are usually abundant in desert soils, the unavailability of nitrogen and phosphorus can be major limiting factors. Phosphorus is usually plentiful, but it is in calcium-bound forms, which are difficult for plants to use under dry, alkaline conditions. Mycorrhizas occur, but little is known of their effects. The low amount of organic matter present means that nitrogen reserves are inherently low. Some legumes (with *Rhizobium*) are present in the floras, as are species with *Frankia* associations. Some of the symbiotic lichen associations which occupy rock and soil surface layers, and free-living cyanobacteria, may form nitrogen-rich soil crusts (McCleary 1968). Free-living nitrogen-fixing bacteria seem to be uncommon (W. H. Fuller 1974, Friedman & Galun 1974).

In spite of the difficult environment, many plants have taken up life in deserts. The flora varies in richness according to the moisture supply. In southwestern North American deserts (Fig. 7.4) poor floras, with from one to three species in a stand, occur in the driest localities. Up to 12 species or more may occur in stands in moister localities (Shreve 1942). Large stretches of ground may carry no vascular plants, and even in those places where there is enough moisture to maintain perennial plants the shoot systems are usually widely spaced.

After a rainstorm (possibly occurring at intervals of many years) dense stands of short-lived annuals may appear, dying down when the water supply is exhausted (Went 1948, 1949, Freas & Kemp 1983). They survive in the form of dormant seed populations. In the southwestern North American deserts annual plants fall into two distinct groups. Summer annuals appear after a summer rainstorm, flower and die down in 4–6 weeks. Winter annuals appear in autumn and winter and flower in

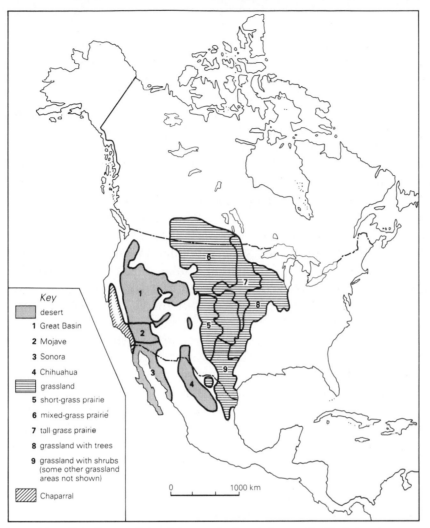

Figure 7.4 Deserts, Great Plains grasslands and chaparral in North America.

Key
- desert
  1 Great Basin
  2 Mojave
  3 Sonora
  4 Chihuahua
- grassland
  5 short-grass prairie
  6 mixed-grass prairie
  7 tall grass prairie
  8 grassland with trees
  9 grassland with shrubs
    (some other grassland
    areas not shown)
- Chaparral

0   1000 km

spring. They live for three months or more. Summer annuals form a smaller group and the plants are often very small (Shreve 1942).

The most common vascular plant life-forms in the North American deserts (simplified from Shreve 1942) are:

1   Winter annuals (ephemerals).
2   Short-lived herbaceous perennials.
3   Perennials with various types of underground resting organs.
4   Succulent-stemmed perennials.
5   Succulent-leaved perennials.
6   Tough-leaved perennials with very short stems.
7   Perennial shrubs with different degrees of leaflessness, small leaves or leaf shedding in dry periods, or both.

211

In deserts elsewhere other life-forms are also prominent (Barker & Greenslade 1982, Evenari *et al.* 1986, McCleary 1968). Many of the perennials are very long-lived. The annuals evade water stress by means of their dormant seeds; they have precise moisture and temperature requirements for breaking seed dormancy. Only a proportion of the seeds germinate at any one time (Went 1948, 1949, Beatley 1974). The perennials are all drought-tolerant to some degree (i.e. their protoplasm resists dehydration), but some also avoid water stress by devices such as general plant dormancy, stomatal closure or leaf-shedding. Many morphological and physiological specializations enable the plants to persist in their extreme environment (McCleary 1968, Hadley 1973, Rhoades 1977, D. J. Anderson 1982, Ehleringer 1985). The C4 and CAM photosynthetic modes, both advantageous for water conservation in dry habitats, are common in desert plants (Caldwell *et al.* 1977, Ting & Gibbs 1982).

Some species have extensive root systems which permit them to utilize water on a wide radius whenever rain occurs. Some desert plants have extremely long roots, descending to reliable subsurface water supplies (Walter & Stadelmann 1974). However, many species have quite small and shallow root systems (Barbour 1973). Competition for water is believed by some authors to account for the spacing of desert perennials (King & Woodell 1984), but the causes may be more complex (Silvertown 1985, Fowler 1986).

Water is the main control over desert vegetation development; general substrate conditions are important with respect to water supply when rainfall occurs, with stony and sandy soils providing better short-term reservoirs than impervious clays (Walter & Stadelmann 1974). The salt content of the soil is also important; some species are restricted to sites that are relatively low in total soluble salts or soluble sodium chloride, whereas others tolerate high salt concentrations, although they can grow on sites with less salt (Gates *et al.* 1956).

Desert vegetation and vegetation change

Examples from the southwestern North American deserts may be cited to illustrate patterns of vegetation composition, structure and change phenomena (Fig. 7.5). In the drier deserts large areas are dominated by stands of the perennial shrubs creosote bush (*Larrea tridentata*) or sagebrush (*Artemisia tridentata*), each up to 1.5 m tall. Communities are monotonously uniform, with only two or three other species present. Plants may cover up to 15 per cent of the ground, and open spaces are wider than the canopy spread of the shrubs. The associated species have different root depths and different seasonal leaf behaviour, so competition between them for water is probably reduced (Shreve 1942).

Seedlings are very scarce in undisturbed creosote bush or sagebrush stands. If the old plants are removed, or if the soil surface is disturbed, large numbers of seedlings of the same species appear after the next rainfall, but thin out until the stand density is similar to that of the original stand (Shreve 1942). Shreve (1917) and Shreve & Hinckley (1937) followed the course of establishment of seedlings of perennial species over seven and 30 years, respectively, and found that, although large numbers of seedlings may germinate, very few survive (Table 7.2). C. H. Muller (1940) described

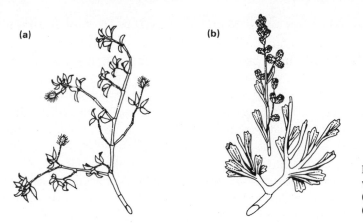

(a)

(b)

Figure 7.5 Two plants important in western North American deserts. (a) Creosote bush, *Larrea tridentata*. (b) Sagebrush, *Artemisia tridentata*.

Table 7.2 Composition of desert vegetation in a quadrat* near Tucson, Arizona, 1910–1936 (after Shreve & Hinckley 1937).

	Numbers of individuals		
	1910	1928	1936
Acacia paucispina	10	11	12
Carnegiea gigantea LC[†]	2	3	2
Cercidium microphyllum	22	21	21
Ferocactus wislizeni C	1	2	2
Fouquieria splendens	6	6	5
Jatropha cardiophylla	0	1	1
Krameria canescens	2	2	2
Larrea tridentata	24	22	26
Lycium berlandieri	3	5	10
Opuntia phaeacantha C	3	5	5
O. versicolor C	9	8	8

* Quadrat a slightly irregular rectangle 557 m^2.
† C, cactus; LC, large cactus.

vegetation changes in a community dominated by creosote bush and tar bush (*Flourensia cernua*). The community is best developed on rubbly soil which is sufficiently deep to retain moisture when rain occurs. Erosion of the soil down to impervious clay reduced the community to a poor cover with scattered, stunted individuals of the main species, or shallow-rooted grasses. In the most extreme cases no plants survive. Accumulation of rubble, which builds up on plains, leads to a simple and slowly changing sequence, beginning with the grasses and ending with the shrubs, all of which form part of the final community. This simple pattern of vegetation replacement was referred to by C. H. Muller as autosuccession. Hereafter it will be termed **direct replacement** in order to avoid confusion with other terminology.

Vasek (1980a) described a simple sequence of change in a grossly disturbed site in the Mojave Desert, but Beatley (1980) noted in the same region that species composition and relative proportions did not change over at least a decade. *Larrea tridentata* has occupied some locations for very long periods. Vasek (1980b)

213

described clones which he thought were about 11 000 years old. P. H. Zedler (1981) noted that *Larrea–Ambrosia–Encelia–Opuntia* stands in Southern California were stable, with no seedling establishment, until after a major rainfall, when seedlings of some of the resident species, as well as species which were not previously present, become established. Seeds of the latter presumably had been transported by flood water. P. H. Zedler suggests that extreme events (rainstorms) may determine the species composition and stand structure of perennial desert vegetation, arising from the brief periods of abundant seedling establishment. There seems to be little information on whether annual communities undergo repetitive development with the same composition on the same patches of ground. Work by Venable & Lawlor (1980) suggests that this is not the case.

As the plant cover is sparse and simple in structure, autogenic influences in most desert systems are not as complex as in closed vegetation of well-watered habitats. Spinous cacti or shrubs protect smaller plants beneath them from predation (Fowler 1986). Litter and windblown soil accumulate under some shrubs with prostrate branch habit (W. H. Muller & Muller 1956). Particular species of annual herb grow there also, presumably benefiting from recycled nutrients, slightly increased moisture and shade. However, other shrubs have upright branch systems which do not trap litter. Some of the desert shrubs (e.g. *Encelia farinosa*) are known to produce leaf exudates which are very toxic to other plants under laboratory conditions. However, neither chemotoxic nor root competition effects have been very convincingly demonstrated as reasons for the spacing of desert plants (Barbour 1973). *Larrea, Prosopis* and *Cercidium* are known to excrete chemicals which cause the development of a hydrophobic crust in the soil. This impedes water infiltration and apparently inhibits the growth of other plant species, thus diminishing competition for water (Adams *et al.* 1970, Mabry *et al.* 1977).

The binding of sand particles by rootlets may influence other species, as also may nitrogen fixation by symbiotic associations. Green algae are present in many places and bind the soil, as do cyanobacteria, and some lichens and fungi. Otherwise, the soil structure and chemistry is altered little by the plant cover.

Vegetation changes of the sequential (successional) type, familiar from regions with moister climates and denser vegetation, are not usually part of the desert scene. There the floristic composition of the vegetation in any locality is so restricted and the plants so well suited to (and constrained by) their difficult environment that the vegetation is simply replaced, if disturbed. If a new land surface is formed, it is gradually colonized by the species which comprise the ultimate and apparently stable vegetation there. Very slow, unpredictable changes occur in some communities in response to sporadic climatic events (P. H. Zedler 1981). In effect they are fluctuations. Otherwise change in most desert vegetation is imperceptible, except for the ephemeral annual communities.

Figure 7.6 Short grass prairie along the Platte River, Great Plains, western Nebraska, USA. The prominent plants include blue grama grass (*Bouteloua gracilis*), wheatgrass (*Agropyron smithii*) and a cactus (*Opuntia* sp.) (photo W. D. Billings).

GRASSLANDS AND DROUGHT

Natural grasslands are extensive in semi-arid climates. Grasslands may be prominent, also, in some well-watered natural or semi-natural habitats where other environmental pressures (such as periodic flooding, frequent fire or cold climate) favour their presence, at the expense of woody plants. Grass plants are particularly well-suited to live in semi-arid conditions by several of their properties, as shown below.

The western Great Plains grasslands of North America (Fig. 7.4 & 6) in Nebraska, South Dakota, western Kansas and Colorado, USA, will serve as a model to exemplify the patterns of vegetation change during oscillations from relatively moist to drier than usual weather. Detailed studies on the grasslands have been done by Weaver & Albertson (1940), Coupland (1958, 1974), Albertson & Tomanek (1965), Weaver (1968), Wali (1975), Ellis (1977) and Risser *et al.* (1981).

Landforms in the western Great Plains region include minor hills and mesas, areas of sandhills, plains with a loess cover and shallow valleys with terraces cut by the rivers and streams. There is a gradual rise from about 500 m in the east, to about 1500 m at the base of the Rocky Mountains, in Colorado and Wyoming.

The climate varies on gradients from west to east and from north to south. Annual average precipitation is about 25 cm at the foot of the Rockies rising to about 65 cm

215

further east in Kansas. The winters are cold, with strong wind, snow and extreme low temperatures down to − 35°C in the north. Summers are hot, with extreme highs up to 40°C or more in the south. There is a dry period with very little rain from mid- to late summer. Some rainfall occurs in autumn, but most is at intervals through spring and early summer. Evapotranspiration in summer considerably exceeds the precipitation and is exaggerated by strong, drying winds. The growing season averages about 150 days.

From west to east the soil types are: *brown* and *chestnut* (mainly under short grass, 30–55 cm annual average precipitation and shallow, with lime accumulation) and *chernozem* and *prairie* (mixed grass in west, tall grass in east, > 55 cm up to about 100 cm precipitation and deeper, with a deep, organic A horizon, some clay formation in the B horizon and a zone of lime accumulation if the precipitation is < 65 cm) (Fig. 7.7). The soils, especially those in the more-easterly mixed and tall grass areas, are very fertile. The dark organic A horizon has developed by long-term incorporation of dead leaf and root material. Nutrients are retained there and recycled efficiently by the grasses. Rainfall wets the upper soil layers to a limited depth. Calcium carbonate (lime) accumulates at this depth.

The weathered upper part of the solum is vulnerable to erosion by wind and runoff if it is physically disturbed, or if the grass cover is removed, but the undisturbed grassland sod conserves both soil and nutrients. There is no runoff. Even heavy rainfall simply soaks into the ground or evaporates.

Much of the grassland area of the Great Plains has been totally modified by modern agricultural activity. Particularly towards the western side, however, large areas are managed for rangeland cattle grazing. The vegetation characteristics of some of these areas appear to be not grossly different from those which prevailed before European settlement. Foreign plants have failed to invade the relatively undisturbed grasslands, except for bluegrass (*Poa pratensis*), a rhizomatous European species which has spread in some places. Much of our information about the grasslands has been gathered from reserves of various kinds and sizes. The discussion below applies to areas that still remain in more or less their natural state.

During the spring and early summer, perennial grasses form a close green cover of variable height, according to the rainfall distribution. Perennial forbs are dotted throughout, but woody plants are excluded by the dense sward.

Grasses

The nature of grass plants suits them well for life in semi-arid climates. They also possess properties which allow them to withstand grazing and fire, as will be seen later in the chapter. Perennial grasses proliferate vegetatively and some species spread widely in this way. The perennial grass plant has basal stem axes from which emerge **tillers** (branches), each having a cluster of leaves enfolding one another and successively younger towards the centre of the cluster (Fig. 7.8). Many tillers are terminated, eventually, by a flower stem. Some grasses have a series of very short axes on which the tillers are clustered together to form a tussock or bunch grass. In other species there are somewhat longer axes, arranged more laxly so that the plant forms a

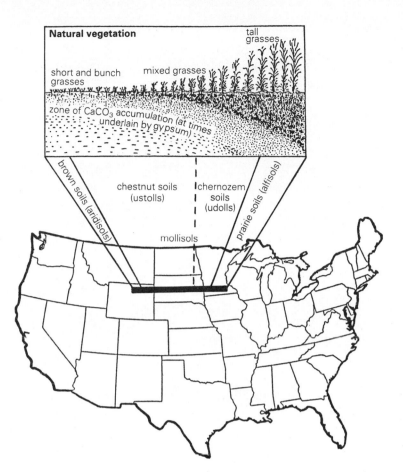

Figure 7.7 The correlation between grassland type and soil type on a transect from Wyoming east to Minnesota, USA. On the same transect mean annual rainfall increases from about 30 to about 90 cm (after Buckman & Brady 1969).

loose patch or tuft. Other grasses have long, horizontal, overground axes (stolons) with roots and clusters of leaves at the nodes; these can spread relatively rapidly. Species with short or long horizontal underground stem axes (rhizomes) have short side-branches with shoots that ascend vertically so that their tillers form a turf or sod. Those with long rhizomes also spread rapidly. All of these forms are represented in the Great Plains grasslands (Fig. 7.9).

The prominent grasses of the Great Plains grasslands are listed in Table 7.3 and Figure 7.9. They are perennial species with stem axes and roots that live for at least 3–4 years. The drought resistance of the grasses is well-known (Julander 1945, Whalley & Davidson 1968). They die down and become dormant in dry seasons (and also during the winter cold period).

The main flush of growth and flowering of the grasses is in spring and early summer (generally in order of short to tall species), but some grass and forb species flower at intervals through the summer and autumn. Seeds of the grasses and many forbs may fall near their parents, but wind can distribute them widely. Another feature which fits the grasses (and many forbs) to withstand summer drought and winter cold is the

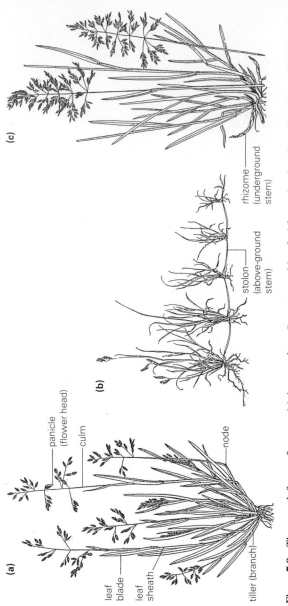

Figure 7.8 The growth forms of grasses. (a) A grass plant, *Poa annua*, with tufted form, showing the various parts. The meristems (areas of cell division) are at the base of stems and leaves and (during elongation), at each node (after Lambrechtsen 1972). (b) *Buchloe dactyloides* (buffalo grass), showing its stolons (after Gleason 1968). (c) *Poa pratensis* (blue grass), showing its rhizomes (after Lambrechtsen 1972). Bunch grasses (or tussocks), like *Poa annua* in general form, but with many tillers packed close together, are shown in Figure 7.9 (e.g. *Stipa spartea* (needle grass); *Aristida purpurea* (purple three-awn)).

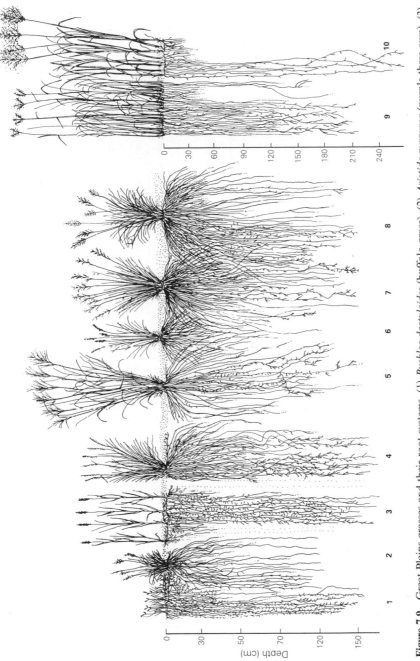

Figure 7.9 Great Plains grasses and their root systems. (1) *Buchloe dactyloides* (buffalo grass). (2) *Aristida purpurea* (purple three-awn). (3) *Agropyron smithii* (wheat grass). (4) *Bouteloua curtipendula* (side-oats grama). (5) *Stipa spartea* (needle grass). (6) *Koeleria cristata* (june grass). (7) *Andropogon scoparius* (little bluestem). (8) *Sporobolus heterolepis* (prairie dropseed). (9) *Andropogon gerardi* (big bluestem). (10) *Panicum virgatum* (switch grass). Note change of scale for (9) and (10). After Weaver (1968).

Table 7.3 Grasses of the Great Plains (from Weaver 1958, 1968).

	Usual height of foliage (cm)	Usual height of flower stem (cm)	Main flowering time	Usual depth of root system (m)	Growth habit	Grassland type in which species occurs/size classes†
Agropyron smithii wheat grass	30–50	60–80	June–Aug.	1.6	long, slender rhizomes	M,S/m*
Andropogon gerardii big blue-stem	90	180–250	Aug.–Sept.	2.1	short, branched rhizomes, forming dense mat	T/t
A.(= Schizachyrium) scoparius little blue-stem	40–50	60–100	July–Aug.	1.5	very short, branched rhizomes, bunchgrass form, but can form turf	T,M/m
Aristida longiseta red three-awn	10–20	30–40	July–Sept.		bunchgrass	S/s
A. purpurea purple three-awn	20–30	30–50	May–Aug.	1.5	bunchgrass	M,S/s
Bouteloua curtipendula side-oats grama	40	60	June–Sept.	1.6	slender, short rhizomes	T,M/m
B. gracilis blue grama	10–20	20–40	June–Sept.	1.5	short rhizomes, bunchgrass form	M,S/s
B. hirsuta hairy grama	15–20	30–50	July–Sept.		bunchgrass	T,M/s
Buchloe dactyloides buffalo grass	10–20	15–20	May–June	1.5	long stolons and short rhizomes	S/s
Elymus canadensis Canada wild rye	50–60	60–150	June–July	0.75	short rhizomes, tufted habit	T/m
Koeleria cristata june grass	30	60	May–June	0.6	bunchgrass	M/s
Panicum virgatum switch grass	120	180–240	July–Sept.	3.0	long, branched rhizomes, forming open mat	T/t
Poa pratensis blue grass	25–30	50–80	May–July	0.9	long, slender rhizomes	T/m
Sorghastrum nutans Indian grass	90	180–250	July–Sept.	1.8	short to long branched rhizomes	T/t
Sporobolus cryptandrus sand dropseed	20–40	30–90	July–Sept.	1.2	tufted	M/m
S. heterolepis prairie dropseed	40–45	60–65	Aug.–Sept.	1.5	bunchgrass	T,M/m
Stipa comata needle and thread	25–30	30–60	June–Aug.	1.5	bunchgrass	M,S/m*
S. spartea needle grass	45	60–90	June	1.3	bunchgrass	T,M/m

† Grassland type: T, tall grass prairie; M, mixed grass prairie; S, short grass prairie. Grass size class: t, tall; m, mid; s, short; *short in hard sites.

widespread occurrence of seed dormancy (Mueller 1971, Iverson & Wali 1982). A reservoir of dormant seeds is ready to sprout as soon as conditions are favourable. Many of the grasses have dispersal units with awns which move in response to changes in moisture, and thus bury the seed below the soil surface.

The grassland vegetation types are differentiated mainly on the stature of the prominent grasses (Table 7.3). The species composition varies according to local site differences, including soil water regime, soil texture and grazing pressure. The many forbs present are also mainly perennials; annual species comprise only a few per cent of the total flora (Oosting 1956, Weaver 1968, Risser *et al.* 1981).

Drought

There is a long history of recurrent drought on the Great Plains, e.g. in 1890, 1894, 1910, 1913, 1917, 1924 (Dix 1964). Severe drought occurred in the 1930s; in the years just before that (1927–1930) rainfall was above average (Dix 1964). Rainfall in the 1940s was average or higher (except for 1943 and 1946). Then in 1952–1955 there was another drought, not quite as severe as the earlier one. Up to 1961 the weather was again wetter (Fig. 7.10). Weaver (1950, 1968), Weaver & Albertson (1956), Albertson *et al.* (1957) and Albertson & Tomanek (1965) studied the effects of these climatic oscillations on the composition and ground cover of the grasslands. They made records in permanent study sites (Figs. 7.10 & 11) as well as over the grasslands in general.

There were very profound effects on the grasslands in the 1930s, when annual precipitation was reduced to 75 per cent or less of the average amount and evaporation increased by about 20 per cent. Wind velocity also increased and much topsoil blew away. There was a great decrease in plant cover (generally to at most 50 per cent of that in 1930 and locally to 10 per cent or less). Some areas became completely bare. The tall grass and mixed grass prairie belts were each displaced towards the east by about 160 km or more, and were replaced by the grassland type which normally occurred further west. There were complex changes in species composition (Fig. 7.10). For example, in short grass prairie, mixed buffalo grass (*Buchloe dactyloides*) and blue grama (*Bouteloua gracilis*) swards were reduced to the latter only; mixed grass prairie dominated by little bluestem (*Andropogon scoparium*) changed to stands of blue grama and sideoats grama (*B. curtipendula*) only; big bluestem (*A. gerardii*) took over dominance in tall grass prairie where it previously had been co-dominant with little bluestem. Certain foreign annual forbs increased on the greatly extended bare ground. The most severe changes were those where grazing compounded the effects of drought.

Yet the grasslands were resilient, for in the wetter early 1940s the ground cover increased substantially, and species which apparently had been eliminated on particular sites reappeared. Some species had not returned to the cover values of the late 1920s before the 1950s drought struck, but little bare ground remained. This trend of improvement in cover continued in the 1960s (Fig. 7.10).

Details of the rapid recovery in 1941–1943, after the 1930s drought, are enlightening (Weaver 1942, 1950, Coupland 1974). Although to some extent the plants

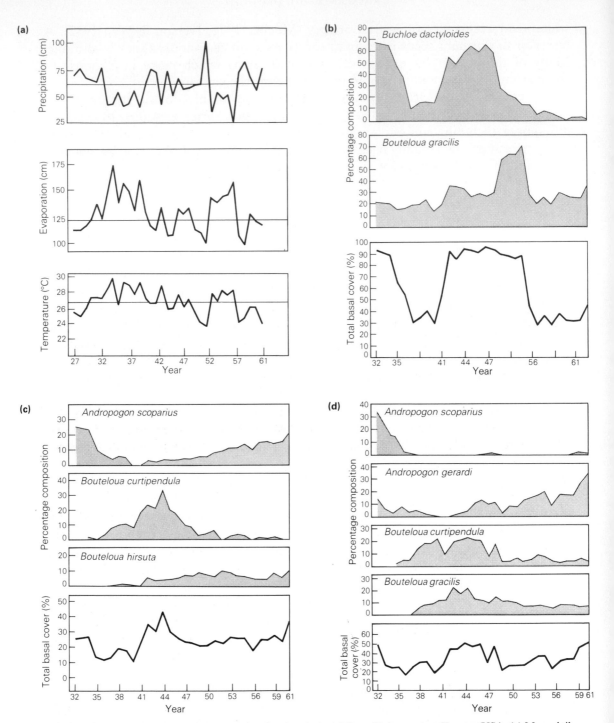

Figure 7.10 The effects of a drought cycle on the abundance of Great Plains grasses, Kansas, USA. (a) Mean daily temperature and evaporation during the growing season and total annual precipitation, at Hays, Kansas, from 1927 to 1961. Changes in total basal cover (per cent) (t.b.c.) and percentage composition of dominant grasses: (b) Short grass prairie. (c) Mixed grass prairie. (d) Tall grass prairie. After Albertson & Tomanek (1965).

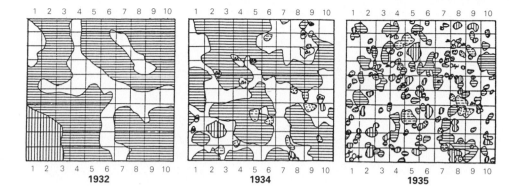

Composition of basal cover in percentages

Species	1932	1934	1935
Andropogon scoparius (horizontal hatch)	67.5	51.7	14.4
Bare ground (unmarked)	24.8	42.5	74.0
Andropogon gerardi (vertical hatch)	7.7	1.5	7.7
Bouteloua gracilis (dotted)	0.0	1.2	2.8
Bouteloua hirsuta (Bh)	0.0	1.1	0.5
Bouteloua curtipendula (broken horizontal hatch)	0.0	2.0	0.6

Figure 7.11 Changes in cover and composition of grassland in a square metre quadrat in Nebraska, USA, as drought intensified over the period 1932–1935 (after Weaver & Albertson 1956).

(especially those of short grass prairie) are drought-tolerant, avoidance and evasion are the main modes by which both the grasses and forbs cope with drought. During prolonged drought the plants die down to dormant axes. Similarly, their seeds lie dormant in the soil until favourable moisture conditions prevail. In fact, seedling survival in the closed, dense grassland during moist periods is poor; open ground is most favourable for good establishment. The root systems of the plants also enable them to resist drought to some degree. Most of the roots are in the top 25 cm of soil, but some species have roots more than 1.5 m deep in the soil and the roots of some forbs go deeper (Fig. 7.12).

The means of revegetation of bare, black soil by grasses in the early 1940s were:

1   By establishment of seedlings from dormant seeds.
2   By establishment from seeds dispersed from the few stands which survived in sites that were not so badly affected by the drought. After the drought ended, grasses flowered very profusely, probably because of the reduced root competition.
3   By vegetative proliferation from previously dormant stem axes.

Some forbs (which had also been dormant) increased in size, but forb seedlings were rare until 1944. As the sward became denser, foreign weedy species were ousted. Buffalo grass, because of its stoloniferous habit, increased more rapidly than blue grama. Species of grass with deep root systems (wheatgrass (*Agropyron smithii*) and big bluestem) quickly replaced their shorter, more drought-resistant competitors. The very drought-sensitive little bluestem was slow to recover its place as one of the most important dominants.

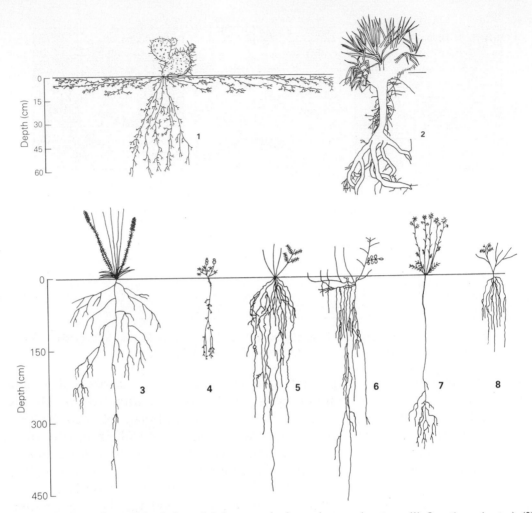

Figure 7.12 Some Great Plains forbs and their root and other underground systems. (1) *Opuntia* sp. (cactus). (2) *Yucca glauca* (soapweed). (3) *Liatris punctata* (snakeroot). (4) *Malvastrum* sp. (globemallow). (5) *Amorpha canescens* (lead plant). (6) *Rosa suffulta* (prairie rose). (7) *Silphium laciniatum* (compass plant). (8) *Astragalus crassicarpus* (ground plum). Note different scales for (1), (2) and (3–8). After Epstein (1973).

A careful analysis of the data on the growth and reproductive behaviour of the grasses during drought and recovery periods shows that each species responded differently to the environmental fluctuations. On some areas new species combinations developed in the 1940s, quite different from those present before the 1930s drought (Weaver 1950, 1968).

Patterns of change in the grasslands

In terms of species composition on any patch of ground the Great Plains grasslands are in a continual state of flux in response to the climatic fluctuations which affect the area. Grazing and periodic fire effects are superimposed on the climatic effects (Ellison 1960). The plants migrate readily, over long distances by seed dispersal, and over short distances by vegetative means. Seed banks are present to replenish the vegetation stands when bare ground in created. Such changes are best described as fluctuations. Fluctuations can hardly be defined as variation about a mean (cf. Miles 1979), because there is no guarantee that, after the stress is relaxed, the vegetation on any particular area will return to species composition similar to that which prevailed before the stressful period. Persistence of a species on the same piece of ground or maintenance of the same assemblage of species there for long periods may or may not occur. The whole grassland system seems to be resilient and stable in the sense that, in general, over large areas, the grasses and forbs cope with the difficult and variable environment and rapidly recover when adverse times are over.

Vegetation change sequences occur on disturbed areas in the grasslands, such as ploughed land, or sites where prairie dogs and badgers burrow and create low mounds of loose soil patches. De Vos (1969) and Vogl (1974) point out that the grassland soils are constantly being turned over and mixed, both by vertebrates and by invertebrates. On the open soil created in this way, seedlings of grasses and forbs can get established. Recolonization of badger-disturbed patches in tall grass prairie was described by Platt (1975) and Platt & Weis (1977). Many of the colonists are perennials, not common in the undisturbed grassland. The patterns of revegetation are similar to gap-filling in forests, and eventually the grassland vegetation takes over each badger patch.

On ploughed fields in South Dakota, USA, after abandonment, colonization by annual and perennial forbs took place. Ten years later a cover of wheat grass, needle and thread (*Stipa comata*) and red three-awn (*Aristida longiseta*) was established. The short grass prairie (with blue grama) may take up to 20 years to develop (Tolstead 1941).

Sequential replacements in the grasslands are explained by the greater niche differentiation (i.e. narrower and more-diverse niches) among the dense grassland species than among colonist species, (evident from the experiments by Parrish & Bazzaz 1976, 1979, 1982a, b; outlined in Ch. 11). Competition between root systems of adults for water and nutrients is probably very important. The deep and extensive root systems of the tall grasses help them to suppress short grasses; so do shading and litter deposition by tall grasses, which limit some of the short grasses and prevent their seedlings from becoming established. Production of toxic compounds by some species may also exclude others (Curtis & Cottam 1950, R. E. Wilson 1970).

ROCK OUTCROPS

Rock outcrops with little or no vascular plant cover are present in many parts of the world They are especially prominent in the alpine, sub-polar and arid regions, but they are also common below the tree limit in regions with broken terrain, but abundant rainfall where otherwise the land surfaces are generally well-vegetated.

The difficulty of rock outcrops as habitats for vascular plants is mainly connected with the slowness of rock weathering and often with steep slope angles. The result is a lack of rooting medium, a shortage of water and strong surface heating in sunny weather. Some plant nutrients may also be in short supply. Exposed rock surfaces are often almost totally covered by a thin crust composed of several to many species of lichen, usually with sparser patches of mosses. Together they may form the only cover.

Lichens have often been implicated in the weathering of rock surfaces (Hale 1967), but the process is slow and may be equalled by the rate of removal of the loosened material. On some concave or flat surfaces, ledges and crevices, small amounts of sand, silt and organic matter accumulate. These are derived mainly from sources external to the particular site, and are blown or washed into place. It is in these sites that vascular plants get established. The roots of the plants penetrate deeply into crevices, and may be able to tap more-reliable sources of water there. The thin, immature soils are high in organic matter content.

Some vascular plant species are virtually restricted to rock outcrop sites. Freedom from crowding is probably the main reason, although there may be other special features of the habitat (e.g. specific minerals in different types of rock) which the plants require or tolerate. Among the rock plants are herbs and shrubs, some of which form mats or cushions which hug the rock tightly. Rock outcrop plant communities also include species which occur more widely in adjacent vegetation on deeper soils, often as pioneers.

In southern Illinois, USA (Winterringer & Vestal 1956), the first colonists of cliff-top sandstone rock outcrops are large patches of the semi-foliose lichen *Parmelia conspersa* as well as smaller thalli of the crustose lichens *Rinodina* and *Acarospora*. Large cushions of the moss *Hedwigia ciliata* and smaller colonies of the moss *Grimmia olneyi* also occur. The moss cushions can absorb much water, but dry out after a few days without rain. Sites with a thin soil, washed from above, may be colonized by a low cover of the fruticose lichens *Cladonia* spp., and annual forbs (species of *Sedum* and *Diodia*), some small annual grasses and the cactus *Opuntia rafinesquii*. On deeper soils, in hollows and on ledges, shrubs (especially *Vaccinium arboreum*) form extensive patches. Crevices may be colonized by shrubs and small trees (*V. arboreum, Juniperus virginiana, Celtis pumila* and *Amelanchier arborea*), as well as grasses and some forbs and the fleshy *Opuntia* and *Agave virginica*.

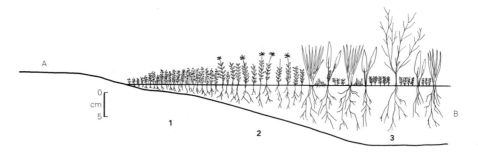

Figure 7.13 Section at the edge of a vegetated hollow in a gneiss outcrop in Georgia, USA. (A) Hard rock surface with thin cover of mosses and lichens. (B) Hollow filled with sandy organic soil. (1) *Sedum smallii* (winter annual) zone in very shallow soil. (2) *Minuartia uniflora* (winter annual) zone in shallow soil. (3) Densely vegetated innermost zone, with perennials and annuals on deepest soil (*Agrostis, Hypericum, Viguera, Senecio*, lichens, etc.). After Sharitz & McCormick (1973).

226

The communities mentioned do not seem to be developmentally related, although it is possible on some stable sites that moss and herb stands are invaded by shrubs. Drought and erosion by wind and water during storms tend to remove the cover in places, and thus set back the development of soil and vegetation.

On gneiss outcrops in Georgia, USA, islands of vegetation a few metres in diameter occur where shallow soil accumulates in hollows on top of the rock (Sharitz & McCormick 1973). The plants form a concentric pattern (Fig. 7.13) with the outer two zones dominated, respectively, by the winter annuals *Sedum smallii* and *Minuartia uniflora*. The innermost zone has taller annual and perennial species *Viguera porteri*, *Hypericum gentianoides*, *Senecio tomentosus* and *Agrostis elliotiana*, lichens (*Cladonia* spp.) and the moss *Grimmia laevigata*.

By experiment, it was shown that the vegetation pattern was controlled by the depth of soil in relation to the water content and the competitive power of the respective species. The *Sedum* tolerates low moisture levels in soils less than 4 cm deep, produces vast amounts of seeds, then dies down during the summer drought. *Minuartia* is a superior competitor under more-favourable moisture levels where the soil is 4–10 cm deep. It also seeds heavily and dies down in summer. The larger annuals and perennials are superior competitors at higher moisture levels where the soil is up to 15 cm deep.

If soil continues to accumulate in the hollows and there are spatial shifts in the zones, the zoned vegetation may be regarded as representing the phases in a small-scale (but very slow) sequence of change. However, it is likely that a balance of soil loss over soil accumulation has been reached, in which case the zonation pattern is static, except for fluctuations which will occur as a result of year-to-year variation in precipitation.

Rock outcrops on granite in Victoria, Australia, carry complex mosaics of vegetation, including eucalypt forest and shrub stands, which are rich in species (D. H. Ashton & Webb 1977). The discussion here is restricted to open outcrops of hard rock which, at the most, have a low shrub cover. Soil accumulation on flat sites is partly by rock weathering *in situ*, and partly by trapping of mineral and organic debris washed or blown from adjacent slopes.

On sites below 300 m of altitude, on north exposures with less than about 20° slope, three adjacent vegetation zones are present.

1   Lichen zone. (*Parmelia conspersa*, *P. palla*, *Verrucaria* sp. and the liverwort *Andrewsianthus* sp.) A few millimetres thickness of sand and dust are found beneath the *Parmelia* colonies.
2   Moss zone. (Dense mats of *Campylopus bicolor* with the succulent annuals *Crassula sieberiana* and *Calandrinia* sp., *Drosera* sp. and orchids.) A thin veneer of soil – humus, sand and gravel 1–2 cm thick – is trapped by the moss. On cooler, moister south exposures other moss species are present.
3   Shrub zone. (*Kunzea ambigua* dominant, with *Epacris impressa*, *Pultanea daphnoides*, *Drosera* and orchids.) The soil is shallow and humus-rich. Roots of the *Kunzea* extend out beneath the moss mat and seem to be responsible for holding much of the moss apron in place. This seems to be a good example of mutual benefaction (cf. Ch. 3).

227

Slopes greater than 10° are subject to disturbing rain wash. The moss mat is often removed on slopes greater than 20°. North-facing slopes greater than 30° bear only crustose lichens. South facing steep rock may have *Parmelia* and some mosses. At higher altitudes on the outcrop similar patterns are found but, because of the higher precipitation, the shrubs differ and additional moss species are present.

Sequential invasion patterns are evident in some places. The central raised parts of large *Parmelia* mats may contain small groups of mosses or annual angiosperms. Vigorous moss colonies occasionally overgrow the lichen. Mosses usually colonize bare rock, however. Herb and shrub seedlings often establish in the moss mat in winter and spring, but normally none survives in summer. Drought, disturbance by birds and grazing by small mammals usually remove them. Occasionally shrubs become established in crevices and trap debris, thus building depths of soil up to 30 cm or more. They outshade the mosses and lichens in these sites. However, generally the stringent environment prevents sequential development and, apart from small fluctuations as the climate varies, or removal of patches of the cover by disturbance and subsequent regrowth, the zonation patterns are static.

Invasion of the *Kunzea* scrub by *Eucalyptus* trees is rare. *Eucalyptus* forest with a shrub understorey occurs more extensively in much-jointed areas of rock with deeper soil, a different landform system from the hard rock outcrops. More-rapid weathering and more-extensive microsite availability in the many crevices account for the differences.

Often vegetation development on rock may proceed no further than the initial sparse lichen or moss cover. Sequences of the lichens (Hale 1967) or of the mosses may themselves occur as a result of autogenic alteration of rock surface chemistry, but are not considered further here. The lower plants only rarely participate in a sequence as precursors of vascular plant communities. If sequential change of vascular plant populations occurs on rock it is extremely slow, in parallel with the slow weathering of rock surfaces and slow soil accumulation. Some rock surfaces can never be inhabited by vascular plants.

Soils on rock outcrops form partly as a result of accretion of debris within plant mats (an autogenic process), but the debris is often derived from outside the immediate system. Soils and vegetation are often destroyed by disturbing events, initiating slow revegetation of the site. Another autogenic effect, which may harm the plants causing it, is the weakening of some of the outermost blocks of rock by woody-plant root expansion in cracks. Sometimes pieces of the weak rock collapse, exposing roots and killing the plants. Although fluctuations in vegetation composition are likely to occur, especially according to oscillations in precipitation regimes, it is realistic to regard rock outcrop vegetation as being relatively stable and self-sustaining. Some communities have probably existed with little change for many thousands of years. Only substantial climate shifts, severe grazing or catastrophes such as collapse of part of the rock face are likely to cause major changes.

Figure 7.14 Spring scene at 3300 m on Niwot Ridge, Front Range, Rocky Mountains, Colorado, USA, looking towards Indian Peaks. Alpine tundra dominated by sedge turf (*Kobresia* and *Carex* spp.), with forbs, mosses and lichens (photo C. J. Burrows).

TEMPERATE ALPINE REGIONS

The alpine zone on temperate high mountains (up to 3000–4000 m above sea level) lies above the limit of tree growth (Fig. 7.14). The environment is markedly seasonal and strongly affected by the temperature gradient (mean temperature is about 0.6°C lower for every 100 m rise in altitude). Associated with this gradient are increases in frost effects and snowfall (Table 7.4). Rise in altitude is also accompanied by increased storminess, windiness and often increased precipitation and decreased evaporation. The amounts of precipitation, however, depend on the location of the mountain range. Oceanic localities are much wetter than continental localities, where total amounts of moisture in the air are low and water loss by sublimation may be very strong.

Snow is a special feature, dividing the year into two seasons.

1 The snow season, when an insulating blanket protects plants from the severest cold and wind effects, but when little or no growth is possible and soil biological activity, including nutrient cycling, is minimal.
2 The snow-free season, when plant growth and microbial activity is possible but frost and wind effects may be adverse at times.

In winter the wind tends to redistribute the snow after it falls. Because of the irregular terrain and the wide range of slopes and aspects on mountains, the snow

229

Table 7.4 Biologically important climatic and other effects correlated with increasing altitude on high mountains of the temperate zone.

1. air pressure decreases (affecting atmospheric density and moisture properties)
2. ultraviolet light intensity increases
3. general storminess increases; storms are more frequent and often more violent at higher altitudes
4. windiness increases, especially over ridges
5. cloudiness increases
6. air temperature declines (at a rate of about 0.6°C for every 100 m ascended); the rate varies diurnally, seasonally and on different aspects
7. overall precipitation usually increases (but drought can occur); substrates are often well-drained
8. snowfall frequency and length of snow cover increase (a long winter snow season and a summer snow-free season are special features of temperate mountains); snowfalls may occur in summer; wind redistributes snow (off exposed slopes and on to lee slopes and hollows); snow insulates the ground in winter, but shortens the growing season; abundant water is present in the thaw period
9. soil frost effects increase; their disturbing effects on soil are most severe on bare, snow-free sites; frost may occur in every month; frost wedging is an important rock-weathering mechanism
10. other erosive forces are important, including fluvial and mass-movement processes on slopes
11. microbial activity is slowed, nutrient cycling is slow
12. soil chemical weathering is slowed, but high precipitation may increase the rate
13. soils tend to be shallow, stony and immature; as a consequence some nutrients (cations) may be well-supplied, others (nitrogen) may be in poor supply

distribution is very uneven; greatest in hollows and on shady lee slopes, and least on wind-exposed ridges and sunny, convex slopes. Exposed areas, with little or no snow in winter, experience extremely low temperatures, frequent freeze-and-thaw (which, because of the development of ice, has a disturbing effect on the soil), desiccation of frozen plant tissues and abrasion by wind-driven snow. These are the factors which determine the limits of plant growth on exposed sites.

The depth of snow cover over alpine vegetation controls the length of the growing season (Fig. 7.15). During the thaw period, in early summer, slopes at lower altitude and sunny, convex slopes and ridges lose their snow first. Then the snow retreats, in sequence, uphill and on to the most shaded and concave sites.

Vegetation cover thins out with increasing altitude, giving way to stony ground, screes and bare rock outcrops. The uppermost limit of continuous plant cover is correlated with very late melting of the snow cover or with the lack of snow on exposed sites in winter. It is determined as much by the shortened growing season in late snow sites, and ground disturbance by frost-heaving in exposed snow-free sites as by low temperature *per se*. The plants that are capable of living in late snow areas have their growing season reduced to about two months or less (Burrows 1977).

The vegetation immediately above the timberline on temperate mountains often consists of a narrow shrub zone, grading into a wider zone of tundra composed of grasses, sedges, rushes, forbs and dwarf shrubs, usually with some mosses, liverworts and lichens. Further up-slope the tundra gives way to stony ground where only a few vascular plants occur in sheltered positions. Above this, in the zone of permanent snow, even fewer plants are present, on rock outcrops or debris fields (Fig. 7.14). Mosses and lichens become more prominent where vascular plants thin out.

Tropical high mountains (4000–6000 m above sea level) differ from temperate mountains in having less-pronounced seasonal effects, although there may be a wet

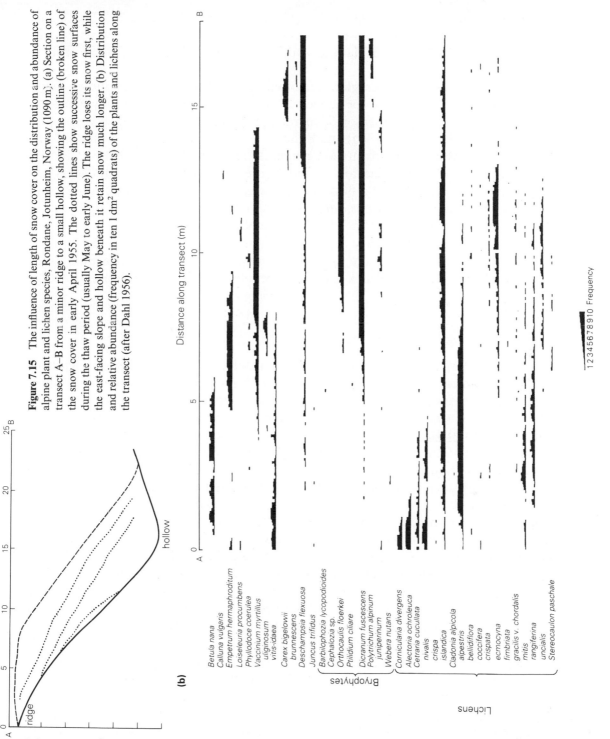

Figure 7.15 The influence of length of snow cover on the distribution and abundance of alpine plant and lichen species, Rondane, Jotunheim, Norway (1090 m). (a) Section on a transect A–B from a minor ridge to a small hollow, showing the outline (broken line) of the snow cover in early April 1955. The dotted lines show successive snow surfaces during the thaw period (usually May to early June). The ridge loses its snow first, while the east-facing slope and hollow beneath it retain snow much longer. (b) Distribution and relative abundance (frequency in ten 1 dm² quadrats) of the plants and lichens along the transect (after Dahl 1956).

season with more snowfall and a dry season with less (Troll 1978). The higher mountain ranges may have frost every night and snowfall whenever it is cloudy. Otherwise the conditions for plant growth are essentially similar to those on temperate mountains.

The plant species of alpine regions are often small. Their physiology and modes of reproduction and growth are specialized to enable them to cope with low temperatures and the short growing season. General alpine plant ecophysiology is covered by Larcher (1980), Bliss (1962, 1985) and Billings (1974b); cold hardiness is discussed by Levitt (1978). Most of the plants are perennial and long-lived; they have efficient means of vegetative proliferation. Annuals are rare because the climate is too uncertain for successful annual flowering and seed set to be assured. Common plant forms are prostrate dwarf shrubs; tuft and rhizomatous herbs (grasses, sedges, rushes and forbs). Underground storage organs are usually well-developed and bulbs, corms or other underground storage organs are common. Some shrubs are winter-deciduous, and many herbs die down to ground level in winter. However, some shrubs and herbs are evergreen, a feature in keeping with the need to conserve energy (Bliss 1962).

Vegetation change patterns and processes

Dynamic patterns of vegetation change in patches of dwarf *Calluna* (heather), *Calluna–Arctostaphylos* and *Empetrum–Racomitrium* heath are found on open, exposed, slopes on the Cairngorm Mountains, Scotland (A. S. Watt 1947; Fig. 7.16). In the simplest case, that of strips of *Calluna* separated by strips of bare, stony, wind-eroded soil, the heather plant stems lie side by side. The stem tips spread out on to the bare soil in the shelter created by the plant and eventually develop adventitious roots. Their old parts die away behind. Some strips contain *Arctostaphylos*, which acts as the advance guard, continually extending itself forward, while being overgrown behind by *Calluna*. The dead and dying *Calluna* at the rear of each patch contains some *Cladonia* lichen. At higher altitude a similar pattern in evident in *Empetrum–Racomitrium* heath. The old dead stems of the shrubs, in the bare patches behind the trailing edge, prove that the patches migrate slowly across the landscape. Ages of *Empetrum* stems show that the rate of advance is about 1 m in 50 years.

Strong wind from one direction is the forcing factor. It kills the rear part of each patch by abrasion with driven sand and snow particles. Wind-blown soil accumulates in the advancing part so that the patches build up a little above the local level. The older part of each patch is thus made susceptible to erosion. A. S. Watt (1947) regarded the sequence as cyclic in nature. He termed different phases of the continuous sequence of changes **pioneer**, **building**, **mature** and **degenerate**.

Similar phenomena are known from other mountain areas (e.g. *Epacris petrophila* heaths in the Snowy Mountains of Australia (Barrow *et al.* 1968), *Dryas octopetala* communities in Iceland (D. J. Anderson 1967)). They also occur in some other windy places such as dune blowout areas and deserts.

In a wet meadow in the Medicine Bow Mountains of Wyoming, USA, Billings & Mooney (1959) found another cyclic pattern. Hummocks of turfy vegetation develop,

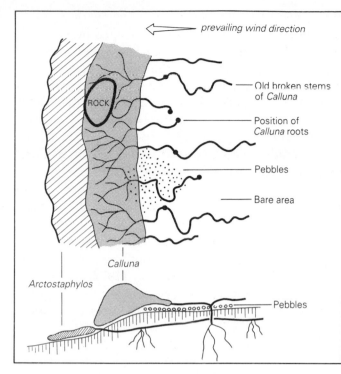

Figure 7.16 Band of vegetation (one of a series of parallel bands that alternate with strips of bare ground), migrating slowly across a mountain plateau, Cairngorm Mountains, Scotland (760 m). Migration is impelled by the blasting effect of wind-driven sand and snow particles which erode the windward side of the vegetation band, while on the sheltered leeward side bear-berry (*Arctostaphylos*) invades bare ground and is overtaken and suppressed by advancing dwarf heather (*Calluna*). The remains of heather and bearberry are evident in the bare windward areas (after A. S. Watt 1947).

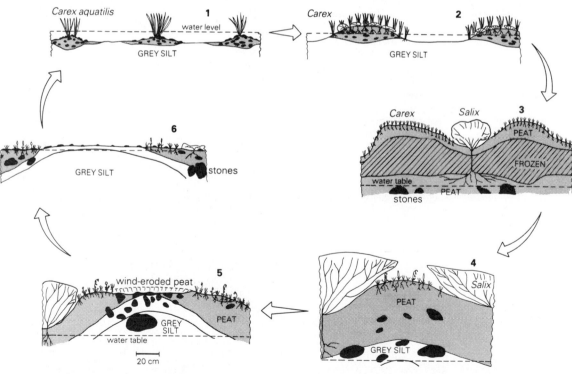

Figure 7.17 Hummock–soil polygon–erosion scar cycle, Medicine Bow Mountains, Wyoming, USA (after Billings & Mooney 1959).

containing *Geum turbinatum, Polygonum viviparum* and *Carex* spp. with marginal dwarf *Salix* shrubs (Fig. 7.17). Beneath each hummock is peat (frozen in winter) in which the plants are rooted and which accumulates from their dead remains. When a hummock is sufficiently high to project above the snow, the turf on its summit is killed by wind-driven snow particles during periods of strong wind. The plants die back and the peat erodes down to near the inorganic, stony silt substrate. Frost-thrusting pushes rocks and silt out through the scar and the rocks form a garland around the margin. The hummock is reduced to the local base-level, which is often about at water level. Building of the mounds begins again with development of *Carex aquatilis* or other *Carex* species on the stone garlands. Species of *Salix, Sedum* and mosses accompany the *Carex* as the mound slowly begins to develop. It is not known how widespread cyclic sequences like this are in alpine regions, but they also occur in the arctic tundra (Hanson 1953, Hopkins & Sigafoos 1951).

Alpine vegetation change on new surfaces resembles the sequences occurring in vegetation at lower altitudes. For example, on the recessional moraines at the Storbreen, a small glacier in Jotunheim, Norway, Matthews (1978, 1979 a, b) described a chronosequence spanning 250 years, using very detailed sampling and ordination techniques (Fig. 7.18 & 19). High and lower altitude series are evident.

Figure 7.18 Vegetation pattern on a chronosequence of alpine glacier moraines, Storbreen, Jotunheim, Norway, in relation to time since deglaciation (——) and contour (– – –); each symbol represents a 16 m² quadrat. Plant species assemblage types are: (×) *Poa alpina–Cerastium* (mainly on surfaces 2–118 years old); (□) *Salix lanata–Cassiope* (mainly on surfaces 62–220 years old; (▽) *Cassiope–Phyllodoce*; (○) *Pinguicula–Tofieldia* (mainly on surfaces 118–220 years old; (▼) *Salix herbacea–Sibbaldia*, (▲) *Betula nana–Juniperus*; (●) *Arctostaphylos–Loiseleuria*; (■) *Vaccinium myrtillus–Salix glauca* (mainly on surfaces >220 years old). After Matthews (1978c).

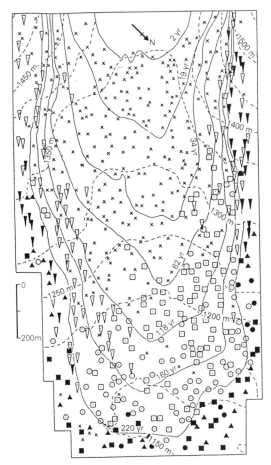

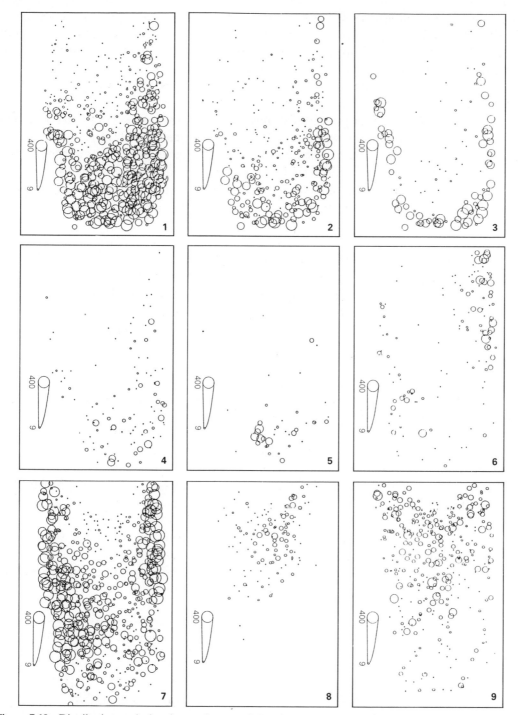

Figure 7.19 Distribution and abundance of some of the plant species on a chronosequence of alpine glacier moraines, Storbreen, Jotunheim, Norway. The individualistic behaviour of each of these species is manifest. (1) *Empetrum hermaphroditum.* (2) *Vaccinium uliginosum.* (3) *V. vitis–idaea.* (4) *Arctostaphylos alpina.* (5) *Tofieldia pusilla.* (6) *Gnaphalium supinum.* (7) *Salix herbacea.* (8) *Arabis alpina.* (9) *Poa alpina.* Each circle represents one 16 m² quadrat. Size of circle proportional to frequency of occurrence in 20 cm × 20 cm subdivisions of the quadrat; largest circle 400, smallest circle 9, <9 indicated with a cross. After Matthews (1978).

The distribution patterns of species on the surfaces of different age, and on different slopes and aspects, tend to be distinctive, indicating individualistic behaviour. Certain species-combinations are apparent in each of the pioneer, transitional and most mature vegetation, but there is considerable overlap of species composition in stands throughout the time-series. Some species have very wide distribution, others are more restricted.

In the alpine tundra, as in grasslands, there is less control of microclimate and light conditions by large dominant species than is found in forests. Different species can be prominent over small areas, according to local site differences; for example, variations in substrate texture, soil moisture, nutrient availability or exposure to wind and frost. The vegetative proliferation habit of many alpine species enhances this kind of effect. This makes for great heterogeneity of the mature vegetation.

SUBPOLAR REGIONS – THE ARCTIC TUNDRA

In localities near the poles, low temperature also is the master factor determining potentialities for plant growth. The winter period is long, and is marked by snow cover and biting cold (Table 7.5). The summer, snow-free season is correspondingly short.

There are similarities between the flora and the vegetation of alpine and arctic tundra in North America, Asia and Scandinavia. In the Arctic the tundra is, however, present on lower, relatively gentle terrain as well as on mountains. It is also often underlain by **permafrost** (permanently frozen ground) (Fig. 7.20).

Permafrost is not found in the alpine tundra of the temperate zone, but it is a dominating phenomenon in the high Arctic (Pewe 1966). There, the frozen soil, snow-covered in winter, thaws in summer only to a depth of about 20–50 cm. Further south the permafrost table may lie several metres below the surface in summer. The depth of the permafrost layer in summer limits plant growth, as plant roots can only grow in unfrozen ground. It also forms an impervious layer which prevents drainage. Large areas of arctic tundra soils are often waterlogged and peaty for this reason (Hanson 1953). Other, more-elevated sites are well-drained.

Habitat diversity is also created by the interaction of small topographic differences

Table 7.5 Climate data for tundra, Northern Canada.

	Mean annual precipitation (cm)	Mean annual temperature (°C)	Mean January temperature (°C)	Mean July temperature (°C)
Fort McPherson, North West Territory 67°29′N, 134°50′W	13.5	−6.2	−28.1	15.5
Fort Norman, North West Territory 64°55′N, 125°29′W	19.8	−4.0	−26.8	16.0

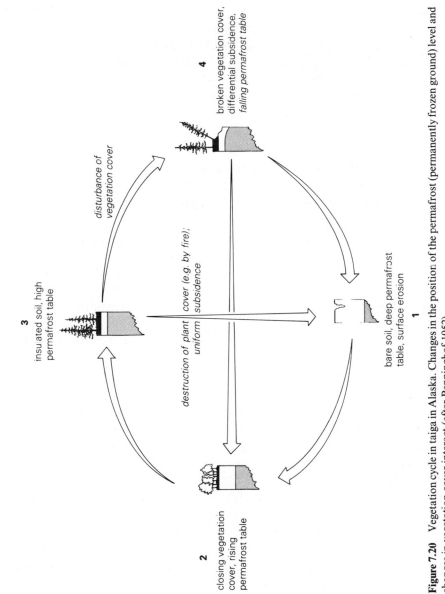

3
insulated soil, high
permafrost table

*disturbance of
vegetation cover*

*destruction of plant cover (e.g. by fire);
uniform subsidence*

4
broken vegetation cover,
differential subsidence,
falling permafrost table

2
closing vegetation
cover, rising
permafrost table

1
bare soil, deep permafrost
table, surface erosion

Figure 7.20 Vegetation cycle in taiga in Alaska. Changes in the position of the permafrost (permanently frozen ground) level and changes in vegetation cover interact (after Benninghof 1952).

with the low solar beam angle, during the brief summer. This influences the temperature and snow melt regimes of the different slopes and aspects. At a finer scale, habitat diversity is created by small irregularities of the substratum such as stones, or cracks in the soil. Such differences are enough to influence potentialities for seedling establishment (Bliss *et al.* 1984, Sohlberg & Bliss 1984).

There are many kinds of interaction of plants in Alaska with soil frost disturbances and other dynamic geomorphic processes (Raup 1951, 1971, Sigafoos 1952, Polunin 1955). For example, on patterned ground, caused by the disturbing effects of frequent freezing and thawing, polygon margins with dwarf birch, *Betula nana*, heath shrubs and *Sphagnum* moss, on thick peat, surround silty centres subject to frost boiling. The frequently disturbed centres are covered by tussocks of *Eriphorum*, the bog-cotton sedge.

Solifluction lobes in the same region are bulges of vegetated, wet soil which slowly creep downhill when the ground thaws. The herbs and dwarf shrubs of areas such as these must adjust to substrate movement. Their root systems are very flexible. The impression gained from these studies is that the tundra vegetation is dynamic — constantly adjusting to site disturbance, although the constraints of a brief growing season cause slow plant responses.

Interactions of vegetation and permafrost near the Arctic timber line in Canada and Alaska, USA, are known. Benninghof (1952) described a cycle beginning with bare soil. Summer warming of the bare soil causes the permafrost table to retreat. *Salix* (willow) shrub vegetation becomes established. Eventually it is invaded by black spruce (*Picea mariana*), which forms a low forest. The build-up of litter and a thick, peaty soil layer beneath the vegetation, and the shading caused by the trees, insulates the ground so well that the permafrost level rises. This continues until the tree roots are affected and the trees begin to die. The peat erodes and trees are tilted by differential thawing in summer. The forest stand dies out, and the cycle begins again (Fig. 7.20).

Other vegetation sequences in the same region have been described by Drury (1956) (e.g. filling of ponds by *Sphagnum* and sedge peat, leading to invasion by heath and birch shrubs, and eventually tamarack (*Larix laricina*) or black spruce, or both). Viereck (1966) also described sequential development of herbaceous and shrub vegetation on glacier outwash plains in Alaska.

Specific vegetation change patterns in the tundra of a hilly Arctic area in northern Scandinavia depend on the response of the plant species present to the relative severity of the site and to relative differences in crowding (C. H. Muller 1952). Three vegetation zones occur on the slopes above the treeline (Fig. 7.21). In the lowermost zone vegetation recovery after disturbance is through a sequence beginning with pioneers (which are species absent from undisturbed tundra in this zone, but present in the higher-altitude zones) and ending with cover dominated by species normally present in the undisturbed tundra. Similar sequences are found in the area studied by Matthews (1978) further south in Norway, and in Alaska (Sigafoos 1952).

In the more-severe middle- and high-altitude zones in northern Scandinavia, Muller noted that species from the original cover serve as colonists in disturbed areas; there is no sequential replacement. Similar direct replacement patterns were observed

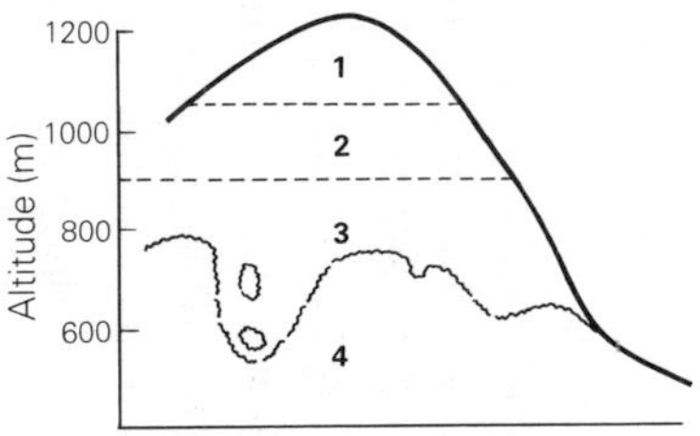

Vegetation zone	Vegetation type and prominent plant species	First colonists on disturbed areas	
1	Feldmark: scattered lichens, mosses, Ranunculus glacialis, Salix herbacea; S. reticulata, Luzula spp., Saxifraga oppositifolia, Astragalus alpinus, Cerastium lapponicum, Silene acaulis on predominantly bare, stony ground	Salix herbacea and other species of the feldmark community	
2	Low heath tundra: Cassiope tetragona, Dryas octopetala, Vaccinium vitis-idaea, V. uliginosum, Salix herbacea, Phyllodoce caerulea, Empetrum hermaphrodium, Carex bigelowii, grasses	Salix herbacea, Carex bigelowii	rhizomatous heath species
3	Dwarf birch–heath shrub–willow tundra: Betula nana, Empetrum hermaphrodium, Vaccinium myrtillus, Salix spp.	Saxifraga aizoides, S. oppositifolia, Polygonum viviparum, Carex bigelowii, Cerastium lapponicum, Astragalus alpinus, Silene acaulis, Dryas octopetala, Betula nana, Arctostaphylos alpina	heath species
4	Birch woodland: Betula pubescens/Heath species and B. nana as in the zone immediately above	Saxifraga aizoides, Carex spp., Astragalus alpinus, Equisetum palustre, Tofieldia pusilla, Carex spp.	heath species

Figure 7.21 Vegetation zones and the colonizing species on disturbed areas, Torne Lappmark, northern Sweden (after C. H. Muller 1952).

in the high Arctic tundra of Canada on King Christian Island (Sohlberg & Bliss 1984) and Baffin Island (Briton 1966). Although some species are better suited than others to occupy open, disturbed sites quickly, microsite differentiation is important for seedling establishment. Direct replacement also takes place in less-stringent conditions than those of the high Arctic or desert, as outlined elsewhere in this chapter.

The plant species of severe (but relatively uncrowded) Arctic sites, which participate in direct replacement, are highly specialized for survival in their harsh environment. None is particularly specialized to be an early colonist only (as is found in the floras of less-stringent environments). They are quite versatile enough to be colonists when the need arises, but competition squeezes them out of the denser vegetation of less-extreme sites. Some Arctic and alpine species of marginal, but uncrowded habitats have very wide fundamental niches (possibly associated with ecotypic variation). They can live in a variety of extreme conditions (for example, *Salix herbacea* on exposed, windswept ridges, snow-free in winter, in late snowbeds and on rock outcrops below the alpine zone). Such species are not found in adjacent closed vegetation because they lack competitive power. The Arctic tundra species of less-extreme locations are more-specialized in competitive ability, a necessity for living in dense, closed communities.

The lability of the Arctic and alpine tundra vegetation has led some ecologists to consider that it is not particularly stable (Raup 1951, 1971, Hopkins & Sigafoos 1951). In the sense that the whole complex is resilient in the face of severe conditions and frequent disturbance, it must be regarded as stable, but in a state of dynamic equilibrium (Churchill & Hanson 1958).

PROTECTED COASTS AND ESTUARIES SUBJECT TO TIDAL INFLUENCES

Gently sloping surfaces of gravel, sand, silt and mud occur between low- and high-tide levels on many coasts in temperate regions. In localities protected from the violence of breaking waves (on the lee of islands and spits, and in bays and estuaries) a limited number of salt-tolerant vascular plants form salt-marshes above the low-water mark (V. J. Chapman 1966, Ranwell 1972, Jefferies & Davy 1979). Embayed and estuarine areas have finer-textured substrates than the more-exposed areas. More or less strong freshwater influences are usual in estuaries.

Two of the main habitat conditions affecting the plants are the tidal flooding cycles and salinity. The lower marsh is submerged and emerges twice-daily. A gradient of decreasing coverage by the tide affects the progressively higher surfaces. The uppermost marsh is flooded only during spring tides or during storms when the wind whips up unusually high waves. Tidal action and storm waves can cause considerable scouring of the soil. The substratum is often sandy and the upper marsh tends to be well-drained, but silt and clay are common and substratum conditions are spatially variable.

The plants must have a highly specialized physiology to cope with alternation of salt- and freshwater influences, with twice-daily fluctuations of salinity (Dainty

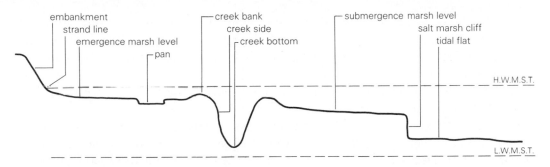

Figure 7.22 Tidal salt marsh landforms which provide a variety of habitat conditions for plants (after Ranwell 1972).

1979). Increased salinity is often caused in the upper marshes by strong summer heating and drying. In winter deep cold is often experienced, and salt may be leached because of the increased rainfall. The short vegetation cover ensures that salt marshes are poorly buffered from climatic extremes (Corré 1979).

Nutrients are usually in adequate supply in salt marshes except for nitrogen which is generally unavailable (G. R. Stewart *et al.* 1979). Cyanobacteria, some bacteria and organic detritus are the main sources.

There are often intricate drainage systems in the marshes, resulting from the inward and outward tidal flows. The salt-marsh landforms are subdued (Fig. 7.22). Minor height differences, however, create large differences in the physical site factors determined by the amount of flooding experienced through the daily and monthly tidal cycles. This means that there are sharp differences from site to site in the degree of drying out at low tide and in salinity. Substrate types, also, tend to vary in space over short distances (Ranwell 1972). Habitat conditions arising from the interaction of the many variables are, thus, complex and heterogeneous.

European salt marshes

Vegetation development trends are well-known in European salt marshes (Ranwell 1972, Brereton 1971). Strictly spatial variations in vegetation composition with floristic overlap from place to place – actually zonations determined by different habitat conditions – have, however, often been confused with sequential series (Beeftink 1979). The vegetation sequences are good examples of land-building, autogenic processes.

Bare mud-flats in Britain and western Europe, exposed only at lowest tide, are inhabited by algae and marine angiosperms such as *Zostera marina* (eelgrass). These plants and colonies of animals, such as bivalved molluscs, trap sediment which is being moved about by the tide. Once the level of sediment has been built up by a few centimetres, salt-marsh angiosperms can become established. The habitat is difficult for seedlings, because of the drag of the tide and waves when the site is submerged. The pioneers include relatively small, annual species of glasswort (*Salicornia*) (which establish very rapidly from seed) and, in mud, the taller perennial cord grasses, *Spartina* spp. *Spartina* plants establish from seed or broken-off rhizome fragments.

241

They soon increase into large colonies by vegetative proliferation. These ultimately coalesce, although large open areas without plants may persist for long periods. Sediment accretion rates in *Spartina* stands can be as high as 5–10 cm vertically in a year, but usually they are less (Ranwell 1972). Accretion can occur when other smaller species colonize, but the rate is much less. If there is no accretion, open 'colonist' stands may persist for long periods.

Raising of the salt-marsh substrate level causes changes in conditions of moisture and salinity. Once environmental thresholds are passed, the reproductive capacity of the original species is depressed and other species invade the site (Ranwell 1972). Examples of common replacement series are shown in Figure 7.23. The passing of particular thresholds can occur quite rapidly, and almost simultaneously, over relatively extensive areas. Some of the perennial, rhizomatous species bind the substrate to form a firm turf.

Competitive replacement of *Spartina* by other grasses and sedges occurs when its growth and reproduction are suppressed by the accumulation of its own litter. The other species take advantage of the less-saline and drier conditions caused by sediment build-up. Their height and vegetative spread further limit the *Spartina* (Ranwell 1972).

The vegetation stands resulting from such shifts in composition may have a relatively stable species structure for decades, until further vertical changes in the substrate level occur. If erosion removes sediment, the surface may be lowered and the vegetation on the new surface consists of a different range of species. They may or may not recreate the earlier sequence on the same place.

In fact, many of the salt-marsh species have wide tolerance ranges. Species such as *Aster tripolium, Puccinellia maritima, Salicornia europaea* or *Plantago maritima* (Fig. 7.24) can behave as pioneers or as members of closed mixed communities. This is partly because of the occurrence of genetically distinct populations in the pioneer and the closed communities, a feature well-studied in salt-marsh species (Gray *et al.* 1979).

Salt marshes in some localities may experience enough climatic variability to cause fluctuation in the plant populations. Corré (1979) recorded substantial changes in populations of *A. tripolium, Limonium vulgare, S. fruticosa, S. radicans, Anthrocnemum glaucum* and other species, in the Camargue in southern France, during 1966–1975. These resulted partly from climatic variations during the alternation from hot, dry summers to cool, moist winters, but mainly from year-to-year climatic differences. The communities of plants, as a consequence, are often transient. After a sharp environmental shift, fluctuations in the abundance of different species may occur over several years, until a new, relatively stable equilibrium is reached. The community composition is usually in a state of flux, however. In some New Zealand salt marshes a considerable degree of persistence of vegetation composition and structure has been noticed, even in stands of annuals such as *Hordeum marinum* (Burrows 1988, unpublished data).

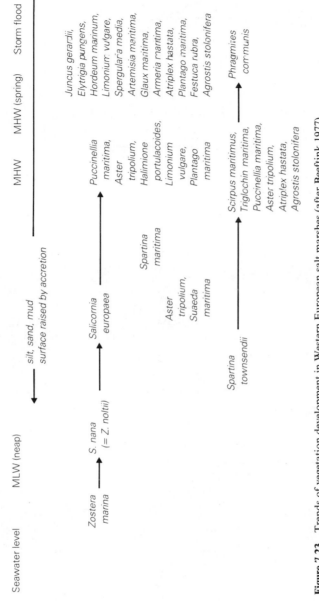

Figure 7.23 Trends of vegetation development in Western European salt marshes (after Beeftink 1977).

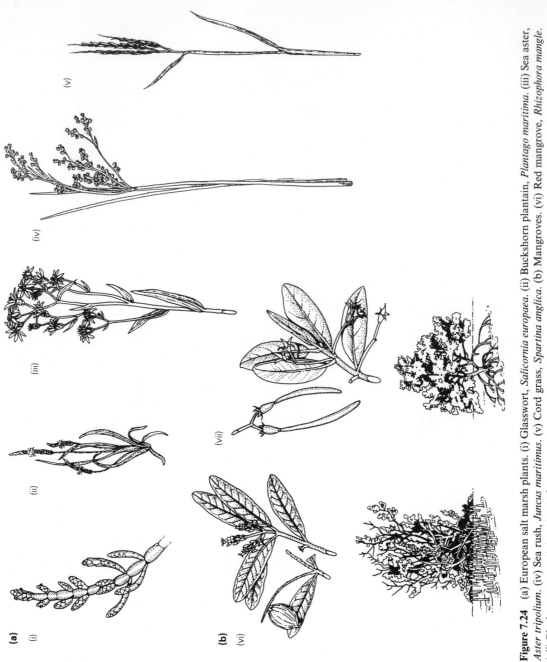

Figure 7.24 (a) European salt marsh plants. (i) Glasswort, *Salicornia europaea*. (ii) Buckshorn plantain, *Plantago maritima*. (iii) Sea aster, *Aster tripolium*. (iv) Sea rush, *Juncus maritimus*. (v) Cord grass, *Spartina anglica*. (b) Mangroves. (vi) Red mangrove, *Rhizophora mangle*. (vii) Black mangrove, *Avicennia nitida*.

Mangroves

Coastal areas in the Tropics and subtropics, in similar geomorphic situations to the temperate salt marshes, are inhabited by the highly specialized evergreen shrubs and trees called **mangroves**. Mangroves belong to about 20 plant families and many genera (V. J. Chapman 1977). The two most characteristic genera are *Rhizophora* and *Avicennia*. Mangroves form a shrubby or low-forest vegetation rooted in the sediment in tidal areas. Among their outstanding features are prop roots, pneumatophores (specialized 'breathing' organs projecting above water level at low tide) and viviparous seeds which drop, already sprouted, into the mud.

Mangrove vegetation, in some circumstances, entraps sediment and builds up the substrate level, resulting in a seaward extension of the vegetation stand (J. H. Davis 1940, Steers 1977, Reimold 1977, Bird & Barson 1982). Rates of up to 1.6 cm per year of upward accretion have been measured under *Avicennia* colonizing mid-tide mudflats in eastern Australia (Bird & Barson 1982). One or more species may be present. If the flora includes several species, then sequential change can occur (Table 7.6) and zonation develops. Each zone is dominated by different species or species mixtures. However, zones are not always indicative of the sequential pattern (Thom 1967, Lugo & Snedaker 1974).

The vegetation may remain static for long periods or, if the substrate is eroded by strong wave action or the vegetation is damaged by hurricanes, the stand may recede shorewards. In fact, in Caribbean hurricane areas the structure and extent of mangrove vegetation is determined by the storm frequency and intensity (Steers 1977).

OTHER EXTREME SOIL CONDITIONS

In many localities species-poor vegetation is induced by toxic soil conditions (e.g. limestone or serpentine soils), with similar effects on vegetation change patterns to those found in saltmarshes or deserts. Waterlogged soils (Ch. 8) and soils that are very low in nutrients (Chs 5, 6, 8 & 10) tend to have vegetation that is richer in species

Table 7.6 Sequential change in mangrove vegetation on sites where sediment is accumulating (data from Lugo & Snedaker 1974, Bird & Barson 1982).

	colonist on tidal flats	invader after rise in substrate through sedimentation	invader after further rise in substrate
	stems submerged at high tide; plants grow well at seawater salinity or in hypersaline soil	ground submerged less frequently and water and substrate less saline	ground rarely or never flooded by sea water
Florida USA	*Rhizophora mangle* (red mangrove)	*Avicennia nitida* (black mangrove)	*Conocarpus erectus* (button mangrove)
Cairns, Queensland, Australia	*Avicennia marina, Sonneratia alba*	*Rhizophora* spp., *Bruguiera* spp., *Ceriops* spp., *Lumnitzera* spp.	

and more-continuous vegetation cover. The vegetation change patterns are distinctive in some respects, but generally resemble those of less-extreme sites.

GRAZING

In Chapter 3 the topic of grazing was introduced generally. The term is used here to mean consumption of leaves, stems and sometimes wood and roots by any kind of animal. Many grazing animals are totally dependent on plants for their sustenance, so their population numbers interact in complex ways with their food resources (Crawley 1983). Both plants and animals interact, also, with regular or random climatic events, and with very destructive disturbances such as fire.

The relative palatability to the particular grazers, as well as the density of the animal populations, determines the effects on populations of plant species. Plants are especially vulnerable at the seedling stage and when new leaves are unfolding, but usually they can withstand light grazing. They may suffer some setback from leaf removal, through diminished photosynthetic activity, but, as will be seen, some leaf or stem removal may stimulate plants to greater productivity.

Selective grazing is one of the main ways in which vegetation change can be effected, for palatable species may be eliminated and unpalatable species increase, aided by the release from competition. Heavy, indiscriminate grazing and associated ground disturbance facilitate change by allowing the increase of unpalatable plant species and species which are tolerant of trampling, or which prefer open habitats. Grazing removes nutrients from the plants. They may be deposited within the system in the form of dung and urine, insect frass and dead bodies, or they may be transported elsewhere. Sustained severe grazing is similar to fire, but usually not quite as comprehensive in its destructive effects; one difference is that some grazers damage plant root systems badly (De Vos 1969).

One of the most important features of grasses is their ability to withstand grazing. They do this better than most dicotyledon plants, because they, like other monocotyledons, have meristems (areas which give rise to new tissues) at the bases of their leaves, rather than on the tips of branches (Fig. 7.8). Removal of the distal part of the leaf does not harm the meristem, and the leaf may extend to replace lost tissue. In fact, light grazing stimulates grasses to grow vigorously, and under these conditions they can be more productive than if there was no grazing at all (Alcock 1964, McNaughton 1979). The removal of rank old foliage permits more leaves to be well-lit, for maximum photosynthesis. It also prevents dead foliage from smothering seedlings.

Co-evolutionary relationships between grasses and herbivores, and the possible stimulation of grass growth by grazer saliva, are noted in Chapter 3. Many grass species respond well to grazing and, in fact, seem to require to be grazed (Chew 1974, Owen 1980, Owen & Wiegert 1982).

The ability of some grasslands to withstand grazing is partly a result of their growth cycle. In the Great Plains grasslands of North America, because of the die-down of leaves through late summer and winter, little direct harm is done to the

246

Table 7.7 Climate data for savannah localities.

	Mean annual precipitation (cm)	Mean annual temperature (°C)	Mean January temperature (°C)	Mean July temperature (°C)
Olala, Tanzania 2°32′S, 35°35′E	59.0	23.5	23.4	22.0
Punda Maria, South Africa 22°41′S, 31°01′E	56.5	17.0	22.0	15.3

plants by grazing in those seasons (although the dead leaves provide some protection from winter cold for the dormant axes). For the same reasons fire is not too harmful to these grasslands (Daubenmire 1968a).

Savannah in East Africa

Savannahs are areas of often tall grassland with scattered tall shrubs or trees (Walter 1971). The tree root-systems are in a lower soil layer than those of the grasses, so competition for water is not severe. Climate data for East African savannahs are summarized in Table 7.7. There is a clear-cut alternation of dry (September–April) and wet (March–August) seasons. Rainfall is less than 60 cm per annum. In the dry season the grasses die off and the savannahs experience periodic fires.

Studies by Boughey (1963), D. R. Stewart & Stewart (1971), J. Phillips (1974), McNaughton (1978, 1979), B. H. Walker (1981), Cumming (1982) (and cf. Diamond 1986) show that changes in the savannah vegetation result from very complex interactions of the climate, fire, grazing ungulates, the plants and nutrient stocks in the ecosystem. Trees commonly include species of *Acacia*. The main grasses are tall species of *Themeda*, *Hyparrhenia*, *Panicum*, *Pennisetum* and *Andropogon* (Fig. 7.25). As well as many small herbivores, vast herds of large grazers such as buffalo (*Syncerus caffer*), wildebeest (*Connochaetes taurinus*), zebra (*Equus burchellii*), thompson's gazelle (*Gazella thomsonii*), grant's gazelle (*G. granti*), hartebeest (*Alcelaphus buselaphus*) and impala (*Aepyceros melampus*) graze the savannah, moving over the landscape in seasonal migrations attuned to the search for water and palatable feed. Different species graze at different heights and have selective preferences (Table 7.8). Highly nutritive bunch-grasses (Table 7.9), with rapid leaf production rates, are preferred, in general. Under light to moderate grazing they are maintained because of the immense supply of nutrients returned in dung and urine. Nutrient turnover is rapid under these circumstances (B. H. Walker 1981) with concentrations in the upper soil layers, the grasses and the animals. In effect, the animals serve as a mobile nutrient store.

Heavy grazing, to near the limits of the carrying capacity, leads to the dominance of short, unpalatable species with large underground systems (including storage organs) and rhizomatous or stoloniferous habit (Table 7.9). These are poor competitors

Figure 7.25 Grasses of East African savannah. (a) Bristle grass, *Aristida junciformis*. (b) Buffalo grass, *Cenchrus ciliaris*. (c) Rhodes grass, *Chloris gayana*. (d) Boat grass, *Hyparrhenia cymbaria*. (e) Red grass, *Themeda triandra*. From Chippindall & Crook (1976–1978).

Table 7.8 Grazing preferences of large herbivores in East African savannah

A. *Nairobi National Park, Kenya* (D. R. Stewart & Stewart 1971)
Preferred grass species, determined by cuticle analysis of faeces (figures are scored percentage values).

	Hartebeest	Grant's gazelle	Thompson's gazelle	Wildebeest	Zebra
Pennisetum mezianum	+	2	1	1	+
Themeda triandra	5	4	5	6	5
Cynodon dactylon	1	4	2	1	1
Hyparrhenia lintonii	2	1	+	+	1
Ischaemum afram	2	2	3	2	4

Fragments recorded: + <1; 1, 1–4.9; 2, 5–9.9; 3, 10–19.9; 4, 20–49.9; 5, 50–79.9; 6, 80+.

B. *Umfolozi Game Reserve, Natal* (R. Emslie, *ex* Diamond 1986)
Preferred grass species and form, determined by observation of grazing and grazed areas.

	Buffalo (unselective)	Wildebeest (unselective)	White rhino (unselective)	Impala (selective)	Zebra (selective)
Themeda triandra					
medium-height tillers	+				
short tillers		+			
Panicum coloratum					
short, erect tillers in centre of tuft		+	+		
long, prostrate tillers at edge of tuft				+	+

Table 7.9 Perennial grasses of East African savannah (after Chippindall & Crook 1976–1978).

		Height of plant (cm)	Growth habit	Palatability, etc.
Andropogon gayanus	blue grass	70–200	densely or loosely tufted	good forage
Aristida junciformis	bristle grass	30–40	densely tufted; may have short rhizomes	unpalatable except at beginning of growing season; indicator of overgrazing
Cenchrus ciliaris	buffalo grass	60–100	often large, bushy tufts; short 'woody' rhizomes	good forage
Chloris gayana	rhodes grass	55–150	mat-forming; long stolons	good forage
Cynodon dactylon	quick grass	10–40	mat-forming; stolons and rhizomes	good forage
Eragrostis racemosa	narrow love grass	10–60	densely tufted, sometimes with short rhizomes	moderately palatable
Hyparrhenia cymbaria	boat grass	120–180	tufted, with stolons	palatable before flowering
Ischaemum afrum	turf grass	50–200	tufted; thick rhizomes	good forage
Panicum coloratum	small buffalo grass	20–200	tufted; some forms have stolons	good forage
Pennisetum clandestinum	kikuyu grass	100–120	mat-forming; close, compact turf; rhizomes	palatable
Sporobolus stapfianus	dropseed	5–75	densely tufted	palatable
Themeda triandra	red grass	30–50	tufted	palatable before flowering

with the erect, leafy bunch grasses and, to persist, need at least intermittent periods of heavy grazing. Overgrazing of the palatable grasses may also lead to the development of persistent scrub thickets, because there is insufficient grass to carry fire.

In lightly grazed savannah the limitations for plant growth are nutrients and water. Eventually litter builds up in the moribund growth, increasing the capacity to carry fire. When the grassland burns, herbivores are attracted to the new, palatable flush of growth, and the nutrient status is improved so that nutritive grasses are encouraged. The effect persists for years, until the regrowth after a fire elsewhere attracts the animals away.

Changes in the composition of savannah vegetation, effected by the interactions of the plants with climate regime, grazing animals, nutrients and fire, can be classed as fluctuations. Temporary equilibrium can be reached, but the situation is relatively labile and unpredictable; balance can be relatively easily tipped in a different direction from that which currently prevails. Other factors contribute to the complexity of the vegetation fluctuations. These include long-term changes in the landforms and soils, and episodic events such as drought, frost, rinderpest (a disease which kills off the ungulate herds), and locust irruptions.

Grazing by insects

The biomass and grazing pressure of insects in some natural ecosystems often equals those of mammals. Locust plagues may cause enormous devastation (R. F. Chapman 1976). Defoliating insects with more-specific habits attack most plant species, usually with minor to moderate impact. From time to time severe epidemics occur. In addition to the example cited below in forests of northern and eastern North America, others are outlined in Chapter 9.

Spruce budworm (*Choristoneura fumiferana*) is the larval stage of a moth, endemic on balsam fir (*Abies balsamea*), white and red spruce (*Picea glauca* and *P. rubens*), but also affecting other gymnosperm species. The animal undergoes population irruptions from time to time, which have devastating effects on the forests containing the host species, particularly balsam fir. Dates of budworm epidemics in Quebec and New Brunswick, Canada, are shown in Figure 7.26 (Blais 1965, Mott 1976, Royama 1984).

Defoliation of balsam fir trees for 2–4 years kills them. Lesser defoliation weakens the trees, leaving them susceptible to other pests and diseases. White and red spruce may also be killed, as may pines and black spruce (*P. mariana*), but some individuals of these latter species often survive repeated defoliation. In the 1970s, in eastern Canada, many thousands of hectares of balsam fir, 40 years old or more, were killed by a budworm epidemic.

Balsam fir trees can live for up to 150 years, but most die long before that. Even-aged stands about 40–80 years old are usual. They arise according to the timing of the budworm epidemic (at about 40-year intervals) in relation to the stage of maturity of the stand. In old stands, which have many suppressed saplings and seedlings beneath the canopy, by the fifth year of an epidemic both the mature trees

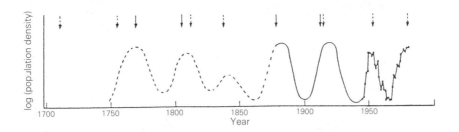

Figure 7.26 Episodes of spruce budworm (*Choristoneura fumiferana*) irruption in balsam fir (*Abies balsamea*) and other conifer forests of eastern Canada in the past two centuries: (–·–·–) from sampling data; (——) from historical records; (– – –) from analysis of growth rings of surviving trees. Arrows indicate the years of first sign of growth ring narrowing. Solid arrows, New Brunswick. Dotted arrows, Quebec (after Royama 1984).

and the saplings are killed. Balsam fir seedlings survive because caterpillars which fall to the ground are voraciously eaten by the many small rodents which are present. A few mature spruce and birches also survive (Figs 7.27, 7.28).

Stands of species of shrubs and small deciduous trees (raspberry (*Rubus*), cherry (*Prunus*), birch (*Betula*) and mountain ash (*Sorbus*)) develop. Dense stands of young balsam fir, derived from the seedlings, grow up through these.

If a budworm attack occurs in young balsam fir stands (which also arise after fire) about half of the trees will survive, but very few seedlings will then be present. The surviving young trees grow on to maturity. Balsam fir tends to be least severely affected by spruce budworm attack in forest stands of mixed composition and age (Fig. 7.27). The effects of spruce budworm resemble those of windthrow and fire in gymnosperm forests. All can result in regeneration of relatively even-aged forest stands of similar character to the previous one (i.e. direct replacement). Spruce budworm epidemics often interact with fire, because the large areas of trees killed by budworm burn very well.

FIRE

Fire is probably the most important destructive phenomenon affecting natural and semi-natural vegetation. In Chapter 9 it is shown that fire has a strong influence in some deciduous forests of eastern North America. It is a much stronger force in the ecology of grasslands (Daubenmire 1968a, Vogl 1974); tropical and subtropical savannahs (B. H. Walker 1981); the Mediterranean scrub vegetation types throughout the world (Naveh 1974, Quezel 1976, Castri *et al.* 1981); much of the vegetation of Australia (A. M. Gill 1975, A. M. Gill *et al.* 1981); heathland types throughout the world (Graebner 1895, 1901, Specht 1979); and most of the gymnosperm forests of North America and Siberia (Heinselman 1981a, b, Ahlgren 1974b, Seitz 1986). The plants in most fire-prone vegetation are evergreen perennials or grasses that die down during a dry season.

A. M. Gill (1975) considers fire to be a unique environmental variable because:

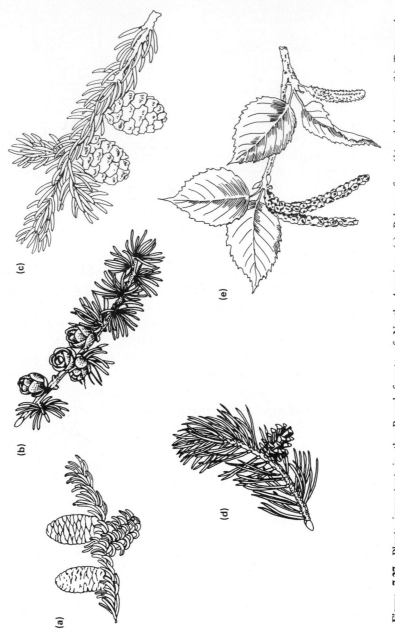

Figure 7.27 Plants important in the Boreal forests of North America. (a) Balsam fir, *Abies balsamea*. (b) Tamarack, *Larix laricina*. (c) Black spruce, *Picea mariana*. (d) Lodgepole pine, *Pinus contorta*. (e) Paper birch, *Betula papyrifera*.

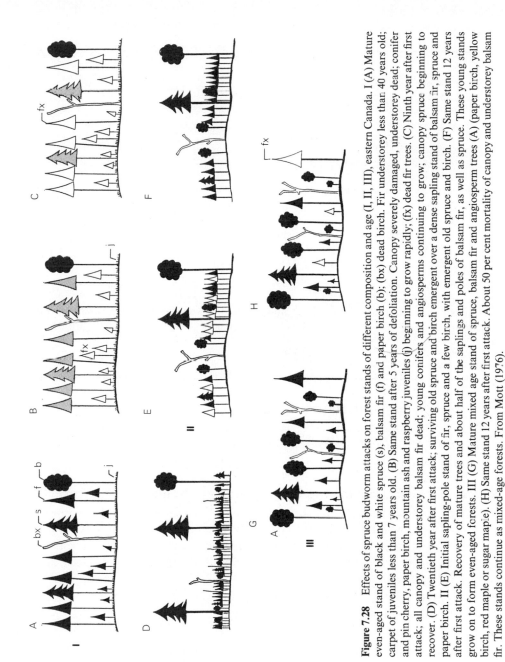

Figure 7.28 Effects of spruce budworm attacks on forest stands of different composition and age (I, II, III), eastern Canada. I (A) Mature even-aged stand of black and white spruce (s), balsam fir (f) and paper birch (b); (bx) dead birch. Fir understorey less than 40 years old; carpet of juveniles less than 7 years old. (B) Same stand after 5 years of defoliation. Canopy severely damaged, understorey dead; conifer and pin cherry, paper birch, mountain ash and raspberry juveniles (j) beginning to grow rapidly; (fx) dead fir trees. (C) Ninth year after first attack; all canopy and understorey balsam fir dead; young conifers and angiosperms continuing to grow; canopy spruce beginning to recover. (D) Twentieth year after first attack; surviving old spruce and birch emergent over a dense sapling stand of balsam fir, spruce and paper birch. II (E) Initial sapling-pole stand of fir, spruce and a few birch, with emergent old spruce and birch. (F) Same stand 12 years after first attack. Recovery of mature trees and about half of the saplings and poles of balsam fir, as well as spruce. These young stands grow on to form even-aged forests. III (G) Mature mixed age stand of spruce, balsam fir and angiosperm trees (A) (paper birch, yellow birch, red maple or sugar maple). (H) Same stand 12 years after first attack. About 50 per cent mortality of canopy and understorey balsam fir. These stands continue as mixed-age forests. From Mott (1976).

Table 7.10 Characteristics enabling plants to resist, evade or recover from fire (N.B. the categories are not all mutually exclusive).

1.   *resistance*

2.   (a) *evasion in space*
      widespread seed dispersal
      restriction to sites which did not carry fire (e.g. wet sites, rocky areas)
      very low growth habit
      tall, clean stems of trees (avoidance of ground fire)
      basal meristems – grasses, etc.
      underground organs including meristems – bulbs, corms (drawn down by contractile roots), rhizomes, lignotubers, etc.
      burial of seeds by mechanical means
      cryptogeal germination (seeds drawn down by pseudo-roots)

   (b) *evasion in time*
      die-down of above-ground organs when likely to be vulnerable – bulbs, corms, grasses
      dormant seeds, seed banks in soil or in canopy of vegetation, in resistant fruit

3.   *recovery*
      abundant reserves
      rapid response to damage
      abundant dormant buds, activated after fire; sprouting from stems, stem bases, lignotubers, rhizomes, roots, etc.
      seed-shed from resistant fruit
      rapid germination of seeds, activated after fire
      stimulation of flowering by fire

(a)   It depends on the vegetation for fuel.
(b)   It is self-propagating.
(c)   It occurs for extremely limited periods.
(d)   It may be devastating.

The immediate effects on the vegetation depend on the intensity of the conflagration. Longer-term effects depend on the frequency of fires and the season in which they occur, in relation to the regeneration of the plants by regrowth or from seeds. To some degree some plants in fire-prone regions may resist fire, but more often they recover from fire by growth or life-cycle responses. Some of the more important of these are listed in Table 7.10.

Mutch (1970) has suggested that the production by plants of flammable, volatile or resinous compounds, or materials such as dry wood, bark and foliage, is an evolutionary development which arose because of the advantages to the plants of being burnt at frequent intervals. Survival is ensured, especially, by sprouting from unburnt organs, or by immediate and abundant germination of seeds after a fire. Some species apparently require fire before they will flower (A. M. Gill 1975, B. H. Walker 1981).

Humans are very frequently responsible for starting fires, but natural fires also occur, usually caused by lightning strike. The evolution of fire-tolerant plant species took place in Australia and North America long before human intervention (D. Walker 1982). The same may be true for fire-tolerant plants elsewhere. However, humans, by increasing fire frequency, may have had the effect of spreading the plants, and the vegetation containing them more widely, in the past 10 000 years or so. Some vegetation (e.g. in the tropics and in wet climates like parts of New Zealand and western Chile) is rather intolerant of fire.

The Californian chaparral

The Californian chaparral, USA (Fig. 7.4) is a scrub vegetation type similar to that of the Mediterranean region (Fig. 7.29). It provides a good example of the interaction of plants with fire (Horton & Kraebel 1955, Sweeney 1956, 1968, C. H. Muller *et al.* 1968, Biswell 1974, P. H. Zedler 1981, Mooney & Miller 1985). The chaparral grows on hill country in a climate with mild, moist, sunny winters and protracted summer drought. Precipitation in the region ranges from < 50 to > 100 cm per annum, and its occurrence (autumn, winter and spring), is very variable in time and space. Mean temperatures are about 16°C, with a mean summer maximum of 28–30°C and a mean winter minimum of 1.7–7.2°C. Maximum summer temperatures reach 40°C and the relative humidity then is often down to 5 per cent, or less. The soils are immature and shallow, with inherently low fertility, low water-holding capacity and a summer water deficit for five months or more.

The chaparral vegetation has variable species composition but similar physiognomy throughout its range. Several hundred plant species are present, and the composition of the shrub dominants and subordinate sub-shrubs and herbs varies according to substrate, rainfall, aspect, proximity to the coast and altitudinal differences. Mature, vigorous chaparral consists of shrubs which are mostly evergreen, with generally small, thick, heavily cutinized (sclerophyll) leaves. Some species have broader leaves. The branches are rigid and the shrubs have flattened canopies with branches of adjoining plants often interlaced. The root systems are large in proportion to the size of the aerial system.

In southern California the most abundant shrub is *Adenostoma fasciculatum.* Species of *Ceanothus, Arctostaphylos, Salvia, Rhus ovata* (winter deciduous), *Pickeringia montana* and *Quercus dumosa* are important, also (Fig. 7.30). In mature chaparral there is little herbaceous ground cover. Table 7.11 gives the important properties of the more-common species.

At intervals of 10–40 years summer fires sweep through the mature chaparral stands, virtually destroying all above-ground plant cover. The fires are very hot (to 700°C). The susceptibility of the plants to fire is enhanced by the low moisture content of their leaves and twigs (< 10 per cent) and by the presence of resin and volatile oils. A fire therefore removes all of the old vegetation cover above ground, initiating a cycle of regrowth which takes decades to complete.

After a fire, redevelopment of the vegetation begins by establishment of herbs and shrubs from seed banks or by sprouting of many of the shrubs from unburnt basal or underground organs. Seeds of herbaceous species may have lain dormant for 40–50 years. Their dormancy is broken by heating and they germinate in the following wet season. Among the herbs are annual and biennial species in genera such as *Antirrhinum, Phacelia, Oenothera, Mimulus* and *Lotus,* as well as bulbous species in the genera *Brodiaea, Chlorogalum, Calochortus* and *Zygadenum.*

Numerous seedlings of many of the dominant shrubs also establish. The adults of

Figure 7.29 (a) Chaparral dominated by chamise (*Adenostoma fasciculatum*), Pinnacles National Monument, California, USA. (b) Burned *Adenostoma* chaparral, Malibu Canyon, California, USA, about 6 months after a fire. Note herb cover on the ground and sprouts from the base of *Adenostoma* shrubs. Photos M. G. Barbour.

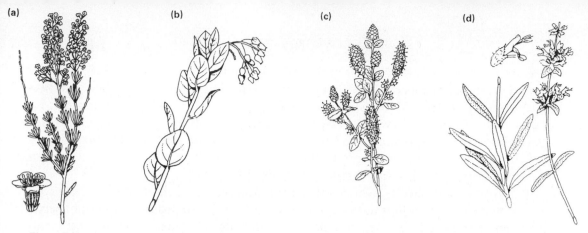

Figure 7.30 Important plants of the California chaparral. (a) Chamise, *Adenostoma fasciculatum*. (b) Manzanita, *Arctostaphylos insularis*. (c) Ceanothus, *Ceanothus tomentosa*. (d) Sage, *Salvia mellifera*.

Table 7.11 Properties of common chaparral species in relation to recovery after fire (after P. H. Zedler 1981).

	Killed by fire	Numbers germinating after fire	Numbers germinating in fire-free interval	Longevity	Numbers germinating in other disturbed sites
1. reproduce only from seed (rarely except after fire)					
Ceanothus tomentosus	A	very high	none	M	none or very few
Cupressus forbesii	A	very high	none–few	L	very few
2. reproduce only from seed, after fire and in fire-free intervals					
Artemisia californica	(C) B	high	moderate–high	S–M	moderate–high
Salvia mellifera	B	high	moderate	S–M	? moderate
3. reproduce by sprouting and from seed (relatively little except after fire)					
Adenostoma fasciculatum	D–C	high	none (moderate)	M–L	few–moderate
Arctostaphylos glandulosa	F–D	low–high	? none	L	? none–few
Pickeringia montana	D–C	low	?	? M	? few–moderate
Rhus laurina	F–D	moderate	? none	M–L	? moderate
4. reproduce by sprouting after fire and rarely or little from seed in fire-free intervals					
Cercocarpus minutiflorus	D–C	none	none–moderate	? L	? few
Quercus dumosa	E	none	? few	L	? none

Mortality after fire: A, very high; B, high; C, medium; D, low; E, very low; F, none. Longevity: L, long (potentially 100 years or more); M, intermediate; S, short (30 years or less).

some of these species are completely killed by fire, but after burning of the aerial parts about half are able to sprout from the base or from roots (Keeley & Zedler 1978) (Table 7.11). Some of the sprouting species (including *Adenostoma* and some of the *Arctostaphylos* and *Ceanothus* spp.) possess lignotubers (swollen basal organs) just below the soil surface. Numerous sprouts develop from the lignotubers. Some very large lignotubers, to 4 m in diameter, are probably hundreds or possibly thousands of years old. Among the sprouting species are some (e.g. *Quercus dumosa*) which have little or no seed germination after fire. Their regeneration from seed appears to depend on the occurrence of wet years from time to time. Some other non-sprouting shrubs (e.g. *Salvia mellifera*) establish seedlings in the intervals between fires (P. H. Zedler 1981). In the first few years after a chaparral burn different herbaceous species appear in sequence, apparently because of different requirements for the breaking of seed dormancy. Some of these are legumes. By the fourth year grasses in the genera *Bromus, Festuca, Elymus* and *Danthonia* are prominent.

By 5–6 years the shrubs, which will be the dominants in the mature vegetation, have grown to be 1 m tall or more. They flower and begin to produce abundant seeds which enter the seed bank each year. Then for the next decade or so the scrub thickens to become an almost continuous cover. The ground cover of herbs and shrub seedlings disappears and some well-grown shrub species (e.g. some *Ceanothus* spp.) also die out. No new species of plant enter the community thereafter.

Features of the habitat which may be involved with the inhibition of plants beneath the chaparral canopy include shading by the dense canopy, the presence of populations of small grazing mammals, litter accumulation and accumulation in the surface soil layers of both phytotoxins and waxy compounds which cause soil particles to be water-repellent. Volatile terpenes from the foliage of *Salvia leucophylla* and phenolic compounds from the foliage of *Adenostoma* and the roots, bark and leaves of *Arctostaphylos* spp. have been thought to inhibit seed germination and seedling growth of plants (C. H. Muller *et al.* 1968, Muller & Del Moral 1971). However, rabbit-grazing and seed predation by small rodents have also been held responsible (Bartholomew 1970, 1971). Halligan (1973) showed that both soil toxins and small mammals limited the growth of the chaparral understorey. Although Christensen & Muller (1975a, b) demonstrated the role of toxins, Kaminsky (1981) found that these were not from the shrub canopy, but rather are derived from soil microbes associated with the litter from chaparral. The toxins inhibit seed germination and the growth of herbs. The inhibition is effective during the time of summer drought, and the inhibitory effect is removed by fire. Shading and root competition for scarce nutrients also limit seedlings to some degree. The water-repellent substances in the soil layers beneath the plants also probably have this effect.

Old chaparral (20–60 years old) tends to be moribund, with little annual production by the shrubs and much dead wood present. The stands cannot regenerate. Fire is the means by which the system can be rejuvenated.

After a burn the abundant ash contains readily available nutrients. The litter and humus with their phytotoxic properties are removed; so are the shade-forming shrubs, grazing animal populations, pathogens and pests. The water-repellent properties of the surface soil layers are removed (they may remain about 10 cm or more below the surface). Seed beds for new plants and suitable conditions for regrowth of sprout-forming shrubs are created. Many of the dormant seeds in the seed bank are stimulated to germinate by the physical effects of the fire which splits seed coats or removes inhibitors, both endogenous and exogenous.

Old, moribund chaparral, if it is not burnt, may eventually open up as senile shrubs die; the spaces then remain open or fill with grasses. Some of the hill-country chaparral stands may be slowly invaded by oaks or pines to form an open woodland, with an understorey of chaparral shrubs (Hanes 1981). If chaparral is burnt more frequently than the ten years or so which is minimum for the 'complete' fire cycle, some shrubs will probably persist, but the grass cover increases.

The question may be asked here whether the main cause of the moribund *Adenostoma* stands is soil toxicity, whether autotoxic effects due to the production of inhibitory chemical compounds is responsible, or whether plant senescence (gradual decline and finally death caused by an inherent ageing process) may be involved.

Senescence has been thought, implicitly, to be a cause of some cyclic patterns of vegetation change (cf. A. S. Watt 1947). Senescence of whole, mature perennial plants, other than monocarpic species, is not understood well (Leopold 1980, Nooden 1980). Nevertheless, the ageing of stands seems to give rise to some replacement patterns (e.g. in the Boreal gymnosperm forests). Some of the cyclic vegetation changes which A. S. Watt (1947, 1955) attributed to ageing (e.g. in bracken stands or bog communities in Britain) have now been shown to arise from causes other than simply the ageing of particular species or communities (Gliessmann & Muller 1978, Barber 1981). Decline in bracken (as in *Adenostoma* chaparral stands) seems to be correlated with the accumulation of toxic compounds, but soil micro-organisms may be involved in each case.

Perhaps the best-documented replacement pattern which seems to result from inherent, programmed decline of a dominant plant species, is the heather (*Calluna vulgaris*) cycle in Britain (A. S. Watt 1947, 1955, Barclay-Estrup & Gimingham 1969, Gimingham *et al.* 1979) (Fig. 7.31), but the effects of soil organisms cannot be ruled out there.

Fire-induced patterns in other regions

Similar fire-induced temporal vegetation patterns are evident in Mediterranean scrub and evergreen oak vegetation (Margaris 1981, Castri *et al.* 1981); in subtropical woodland (B. H. Walker 1981); in heathlands (Specht 1979); and in much of the Australian vegetation, which is dominated by members of the genus *Eucalyptus* (A. M. Gill 1975, A. M. Gill *et al.* 1981, I. R. Noble & Slatyer 1977). Many other genera in the Australian vegetation are fire-tolerant in a variety of ways (Table 7.10).

Mallee eucalypt scrub (with *E. dumosa* and several other eucalypt species) may burn at about ten-yearly intervals (J. C. Noble 1981). The fire cycle period is often determined by the time taken for understorey grass stands to build up sufficient fuel to carry a fire. Some other eucalypt vegetation, however, has very long fire cycles; fire seems to be an obligatory means of ensuring the regeneration of the forests. For example, D. H. Ashton & Willis (1982) showed that forest dominated by tall, long-lived (200–400 years) *E. regnans* requires periodic fire for its replacement. Fire kills the adult trees, but it also removes a number of factors which are inimical to seedling establishment of this species, namely competitors for light, fungal pathogens, seed predators, browsing animals and some soil effects which are the result of long occupancy of the site by the mature trees (acid humus, lipid accumulations which inhibit root hair development and toxic exudates from mycorrhizas). The seed-bank is maintained, not in the soil, but in fire-resistant woody capsules carried

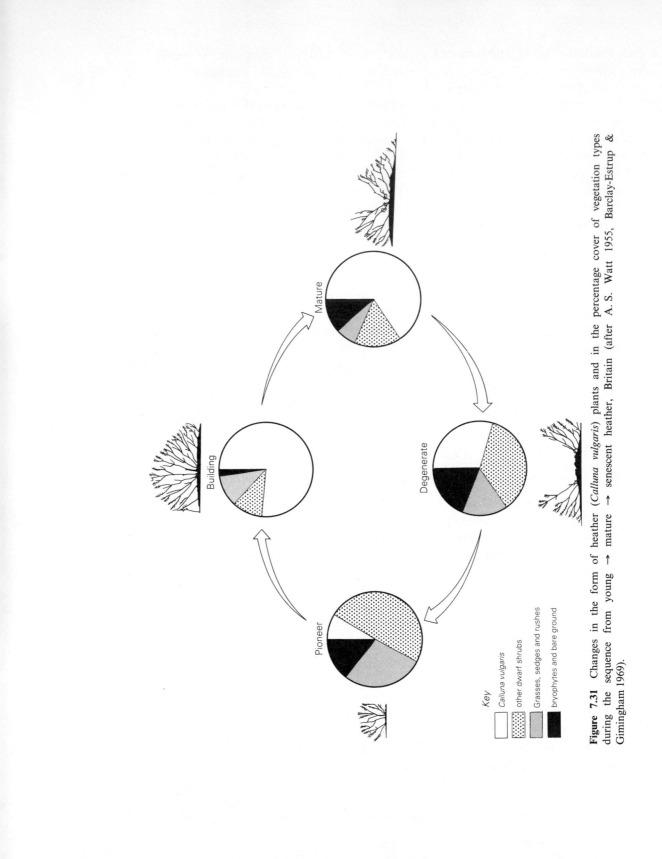

Figure 7.31 Changes in the form of heather (*Calluna vulgaris*) plants and in the percentage cover of vegetation types during the sequence from young → mature → senescent heather, Britain (after A. S. Watt 1955, Barclay-Estrup & Gimingham 1969).

on the upper branches. The young stands which develop on the ash-bed after a fire can begin to produce seed in about 20 years.

Species that are capable of forming lignotubers are common in the Australian flora. So are species with volatile oils and resins which cause them to burn well, and species with woody fruit which must be heated before seeds are shed.

Extreme forms of fire-avoiding woody species occur in sandy soils in the Kalahari Desert and other parts of southern Africa. Species such as *Dichapetalum cymosum*, *Lannea edulis*, *Elephantorrhiza elephantina* and *Combretum platypetalum* are 'underground trees', with extensive underground woody stem systems but only their leaves projecting above the ground (F. White 1976).

North American gymnosperm forests

Fires have affected almost all of the gymnosperm forests of North America within about the past 300 years or less (Ahlgren & Ahlgren 1960, Heinselman 1981a, b). In the Boreal forests the gymnosperm species jack pine, *Pinus banksiana*, lodgepole pine, *P. contorta*, black spruce, white spruce and balsam fir (Fig. 7.27) do not resist fire; their resinous tissues, the large amounts of dry fuel and thick litter and the moss layers present in old forests cause them to burn well during dry spells. The forests are killed over areas which vary from a few tens up to thousands of hectares. Some of the associated deciduous angiosperms, birches (*Betula* spp.), maples (*Acer* spp.) and aspens (*Populus* spp.) are also completely killed, but some regenerate by sprouting, as do understorey shrubs. The gymnosperms, however, must regenerate from seeds. In some species seeds are released from serotinous cones by the fire (i.e. the heat of a fire is needed to melt the resin which seals the cone scales). The specific patterns of forest recovery depend on many factors which include the site type (wet, well drained, etc.), the season, the age of the stand, the availability of propagules and the type and intensity of the fire. Crown fires differ in their effects from those that burn the litter on the ground surface down to mineral soil. The gymnosperm species require bare soil areas for optimum seedling establishment. Given the right conditions, their stands can be immediately replaced. In other instances stands of herbs, shrubs, birches, maples or aspens may regenerate first, but eventually some of the gymnosperms gradually replace them. Figure 7.32 illustrates some replacement patterns in Boreal forests. The pines must be established in the first two years after a burn, but black spruce can establish for up to five years. White spruce and balsam fir can be eliminated locally by burning unless the fire occurs in the autumn of a seed year. The gymnosperm stands (black spruce, jack pine and lodgepole pine) which arise after fire are virtually even-aged and often monotypic (with an understorey of shrubs and a moss field layer). Patches of deciduous angiosperms or mixtures of these with gymnosperms also occur. The resultant forest pattern is a mosaic of stands of different size and shape and different composition, according to the boundaries of fires, the type of fire and its timing with respect to the regeneration cycles of the tree species and, of course, the time which has elapsed since each fire.

The forest communities (except for *Sphagnum* spp., the feather mosses *Hylocomium splendens* and *Pleurozium schreberi* and lichens) are reconstituted quickly. The full

(1) After severe ground fire destroying most of the litter and humus

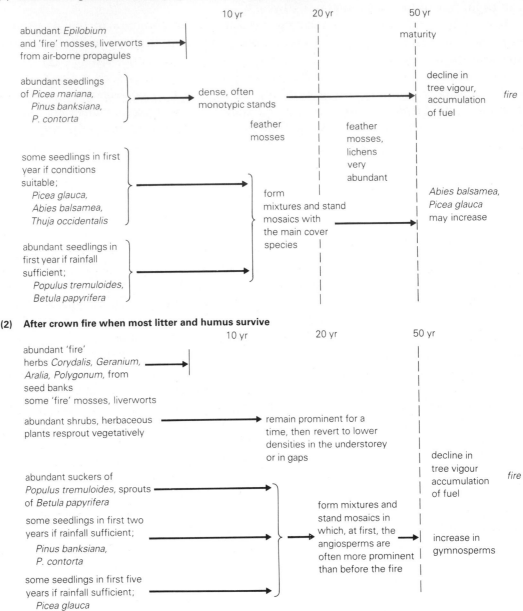

Figure 7.32 Patterns of forest regeneration after fire in North American Boreal forest (from Heinselman 1981a, b).

development of mature forest takes about 50 years. Moribund stands (with many dead standing and fallen trees and deep, dead biomass accumulation on the ground) develop within 120–220 years, but the forests rarely reach this stage because fires usually recur at intervals of 50–120 years. Some of the tree species can live much longer, but do not often get the opportunity, because of disease, pest damage or fire. Old, unburnt black spruce, jack pine or lodgepole pine forests do not regenerate; when they fail the stands open up. Openings are often filled by quaking aspen

(*Populus tremuloides*) or birch (*Betula* spp.) from sprouts, or by white spruce and balsam fir, which can establish on litter or moss layers.

In the Rocky Mountains in lodgepole pine, engelmann spruce (*Picea engelmannii*)–subalpine fir (*Abies lasiocarpa*) and aspen forests there are also similar patterns with different fire-recurrence intervals according to circumstances and species. All of the species that contribute to the mature forest establish in the first year after a fire. The long-leaf pine (*Pinus palustris*) and slash pine (*P. elliottii*) forests of the southeastern USA also regenerate well after burning. Similar fire influences are found in red pine (*P. resinosa*) and white pine (*P. strobus*) forests of the Lake States–St Lawrence region. Mature white pine is somewhat fire-resistant, as its tall, clean stems prevent fire from reaching the canopy.

In all of these instances the dominant species of the mature vegetation are well-suited to cope with periodic fire, which rejuvenates the stands. The fire often benefits the vegetation by killing pests and diseases. There is little or no indication of any regular, linear sequences, leading from simple to complex vegetation with different composition. Most individuals of most species can become established in the first few years after a fire and the stands last for only one generation (usually terminated by another fire). Autogenic processes are involved, however, in bringing the community to a stage where the next fire will be carried most effectively by the vegetation. These cyclic vegetation patterns are predictable, but the timing of the fires is not, because they depend on random factors such as dry years or (in nature) lightning strike.

Various kinds of vegetation, thus, are maintained by fire; like the phoenix, they arise from their own ashes. The course of vegetation change is impelled by fire but strongly dependent on plant characteristics. The interesting paradox of a destructive agency which is also stimulatory (and even necessary to maintain a vegetation pattern), is paralleled by the effects of grazing on grasslands.

In the sense that some of the species which regenerate vegetatively after fire remain on the same spot, there is a degree of stability in fire-prone systems like the Australian mallee or Californian chaparral. The same may be said for gymnosperm forests, which are poor in species, and which regenerate from seeds on the same spot after fire. Stability seems hardly to be the term to apply to systems that are disturbed so regularly, but the overall patterns of vegetation in resilient, fire-prone systems seem to be reasonably stable.

Degradation of nutrient pools by fire

Allogenic vegetation changes initiated by fires include those that result from degradation of ecosystem nutrient pools. The two main ways in which this takes place are by:

1 Losses to the atmosphere (Harwood & Jackson 1975). Carbon, nitrogen and sulphur are most affected.
2 Losses in the run-off (Boerner 1982). Substantial amounts of phosphorus and cations are dissipated.

Loss of nutrients will be most significant in high-rainfall (leaching) environments. The formation of heathlands – areas with very infertile soils which carry vegetation dominated by fire-tolerant and low nutrient-demanding plant species – is attributable to such losses. Permanent changes to former balsam fir forest sites in Newfoundland are one such instance (Damman 1971). The burnt forests are replaced by *Kalmia* heath. Repeated burning causes such substantial losses of nutrients that the effects on the nutrient budget are irreversible. Permanent heathland vegetation in various parts of the world has arisen in the same way (Specht 1979). Most forests, if very frequently burnt, would probably undergo similar changes. Provided that fires are not too frequent, forests seem to be able to replenish their nutrient supplies after burning.

DISTURBANCES IN CONFINED AREAS

In some localities strong disturbances influence plants frequently in the same confined area; adjacent areas remain unaffected. Examples of such situations are snow avalanche areas in temperate mountain areas, river floodplains, ice-push zones around lakes in northern Canada and some wind-exposed sites.

In temperate zone mountains, snow avalanches tend to descend in the same narrow paths. Avalanches start high on mountainsides. When the winter snow pack becomes unstable, large masses of snow thunder down the paths with great destructive power. They may occur several to many times annually, or at less-frequent intervals, depending on the site, the amounts of snow, and the prevailing weather conditions. Some winters have much heavier snowfalls than others. Some sites accumulate unstable snow more readily than others. Some types of weather cause snow instability, others do not.

In forested areas, if avalanches are frequent and sufficiently severe, then no mature trees survive on the avalanche paths. A common pattern of vegetation in such an area in the southern Rocky Mountains in Colorado, USA, is shown in Figure 7.33 (Burrows & Burrows 1976). Seeds from the adjacent mature engelmann spruce–subalpine fir forest fall on the open, avalanche-swept zone after an avalanche, and often dense colonies of young plants become established. The stands may survive for a time, sometimes for decades, before a heavy snow year occurs, setting off severe avalanches which sweep them away. A few damaged survivors may remain. Aspens are common in dense stands along the margins of avalanche paths. If they are snapped off, they regenerate quickly from root sprouts. Otherwise, where avalanches are frequent, the vegetation consists of patches of shrubs (willows, (*Salix* spp.), birches, (*Betula* spp.) and juniper (*Juniperus communis*)) and, on very severely affected sites, grasses and forbs. Many of the herbs and shrubs are also found in the alpine zone above the timber line or on other kinds of disturbed area. Shrub and herb vegetation only survives in these sites because of the frequent disturbance which curtails the forest. Such conditions have probably existed on the same sites for many hundreds, if not thousands, of years.

Most rivers flood from time to time and have distinctive marginal vegetation in the flooded zone. Some rivers in forested mountain areas flood so strongly and

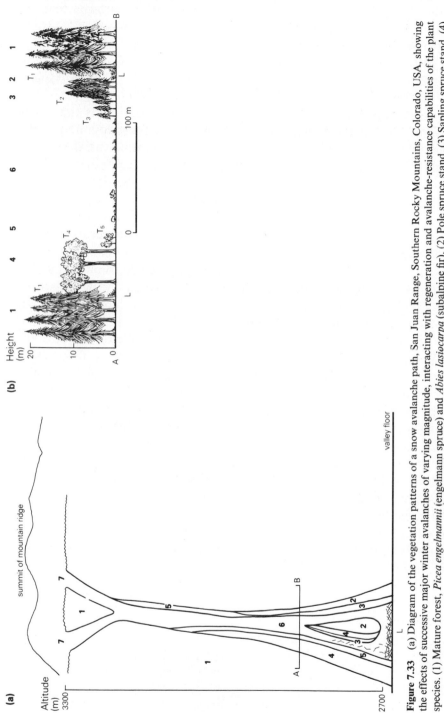

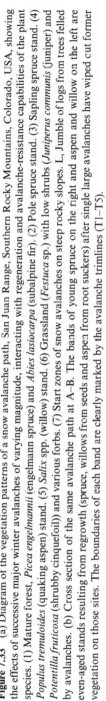

Figure 7.33 (a) Diagram of the vegetation patterns of a snow avalanche path, San Juan Range, Southern Rocky Mountains, Colorado, USA, showing the effects of successive major winter avalanches of varying magnitude, interacting with regeneration and avalanche-resistance capabilities of the plant species. (1) Mature forest, *Picea engelmannii* (engelmann spruce) and *Abies lasiocarpa* (subalpine fir). (2) Pole spruce stand. (3) Sapling spruce stand. (4) *Populus tremuloides* (quaking aspen) stand. (5) *Salix* spp. (willow) stand. (6) Grassland (*Festuca* sp.) with low shrubs (*Juniperus communis* (juniper) and *Potentilla fruticosa* (shrubby cinquefoil) and various forbs. (7) Start zones of snow avalanches on steep rocky slopes. L, Jumble of logs from trees felled by avalanches. (b) Cross section of the same avalanche path at A–B. The bands of young spruce on the right and aspen and willow on the left are even-aged stands resulting from regrowth (spruce, willows from seeds and aspen from root suckers) after single large avalanches have wiped cut former vegetation on those sites. The boundaries of each band are clearly marked by the avalanche trimlines (T1–T5).

frequently that their verges have shrub and herb-rich vegetation patterns that are similar to those on avalanche paths. The force of the flood water is sufficiently strong to remove large woody plants or to prevent them from getting established.

Another kind of river margin system is that found in western Canada, Alaska, Iceland and New Zealand (Nanson & Beach 1977, Burrows 1977). The floodplain consists of wide stretches of relatively unstable alluvial gravel. When the river is low, braided streams occupy only a proportion of its width. During floods the vastly swollen river covers much of the floodplain. The streams shift, eroding and transporting large quantities of alluvium. Areas of floodplain vegetation are destroyed, either by removal or by being covered with debris. During quiescent periods plants again colonize areas that are not covered by water. Areas survive destruction for different lengths of time, so a complex mosaic of vegetation patches of different ages is present. In Canterbury, New Zealand, the most mature vegetation is usually dense grassland or scrub, sometimes with scattered trees. In Alaska it is scrub and low forest (cf. Van Cleve & Viereck 1981). Eventually, during one or another of its floods the shifting river will destroy each vegetated area, reducing the site to bare gravel.

Many Canadian lakes, ice-covered in winter, experience periods during the spring thaw, when the broken-up ice accumulates in vast piles along the shore. Strong winds and wave action shove this ice into the marginal vegetation, physically damaging and often destroying it. Raup (1975) described the results of this process at Lake Athabaska. On sites so affected each year, assemblages of plant species occur which fluctuate in composition from year to year. There are no repeated species groupings closely correlated with particular microsites. The most-specialized plants (halophytes and aquatics) are apparently the least versatile species. Specialization in these directions seems to be at the expense of flexibility in response to disturbance.

Wind may affect vegetation in several ways. Sprugel (1976) described 'regeneration waves' in balsam fir forests in the mountains of Maine, USA. The 'waves' (Fig. 7.34) arise when winter ice storms and desiccation kill exposed old trees along bands at right angles to the prevailing wind direction. In effect, the pattern represents direct replacement, in a cyclic fashion.

Another wind effect is the constant substrate accretion and erosion in the band of foredunes along sandy seacoasts or lake shores (see Ch. 5). The plants present must

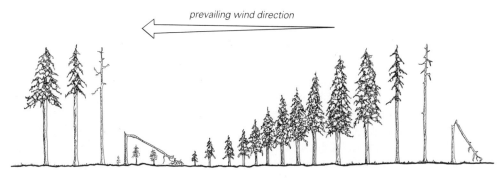

prevailing wind direction

Figure 7.34 Diagrammatic section through a 'regeneration wave' in a balsam fir (*Abies balsamea*) forest in Maine, northeastern USA (after Sprugel 1976).

respond at times to the erosive removal of sand, and at other times to burial. They specialize on rapid extension of stem internodes, and generally on vegetative proliferation.

SOME CONCLUSIONS

Emphasis in this chapter is on the impact on plant species and vegetation of: (a) environmental factors which severely affect the functioning of plants; and (b) profoundly disturbing factors which damage plants. Among the general points which emerge, it can be seen that there are no consistent patterns of vegetation change in some very severe environments with open and sparse plant cover. Presumably this is because autogenic influences and competition are minimal. Direct replacement occurs after the disturbance of vegetation in some severe environments, although it is not necessarily confined to such places. Direct replacement is quite common after fire. Some causes of direct replacement seem to be: (a) a relatively simple flora with species specialized to cope with extremes but versatile as colonists; (b) opportunity for rapid establishment from seeds or by vegetative means; (c) a limited array of relative niche widths; and (d) arising from (a) and (c), less competition than in most less-severe environments (the species are often weak competitors).

Within the limited range of species which inhabit some kinds of severe site, however, some individual species can be recognized as being more-effective early colonists of open sites than others; other species are more-effective inhabitants of relatively crowded sites.

Strong environmental forcing factors interacting with plant growth responses are involved with dynamic vegetation patterns in severe environments (e.g. freeze and thaw, wind, fire and grazing). The interactions of such factors with other environmental variables, with substrate accretion, movement or erosion, and with the growth of resilient plants, are complex. The dynamic interactions are, however, usually more perceptible than in the vegetation of less-severe environments. Autogenic effects of plant development are also often readily perceived.

Fluctuations in response to environmental variation are apparent in the vegetation of some relatively severe environments. Some authors (e.g. Miles 1979) refer to 'variation about a mean', when discussing fluctuation. The implication is that there is a norm of vegetation composition around which species composition varies in response to environmental oscillations. This poses the problem of how the norm is to be defined – any definition must be arbitrary (with respect to space and time) because species composition of vegetation, in the instances of grasslands and some other vegetation types, is in a continual state of flux. The random nature of some environmental factors (such as grazing in savannahs), which force vegetation change, means that the corresponding vegetation changes are unpredictable. If regular environmental oscillations occurred, perhaps fluctuations would resemble cycles, but usually fluctuations seem to be rather irregular in their effects on the species composition on any patch of ground.

Cyclic replacement seems to take various forms. The phenomena to be considered include direct replacements like the balsam fir 'regeneration waves', the alternation of a vegetation patch with bare ground, the heather cycle, the chaparral fire sequence and the Arctic permafrost cycle. Are these similar enough to be classed together? How, if at all, do they differ from gap-filling in forests, or recolonization of disturbances caused by animals in grassland? Or for that matter are they different from sequential vegetation changes in general? Strong forcing factors (as indicated above) are often involved. So are autogenic influences of dominant plant species. Plant senescence and autotoxicity seem to be important in some. Spot-bound plants which survive disturbance are important in other cases. The essence of a cycle is repetition of a sequence of events, so perhaps all such repetitive vegetation phenomena could be classed together as variations of a general theme, the details of which depend on local circumstances, and the behaviour of the plant species concerned.

Sets of local conditions that are physiologically severe or involve strong disturbance often maintain distinctive vegetation in regions where other vegetation is usual. Some of the resilient plants involved are pioneers of sequential replacement series in the more-usual vegetation of the region. Others are extreme specialists for the severe conditions. Often the same conditions have prevailed in the same place for very long periods and the overall vegetation assemblage is stable. Major environmental shifts would be necessary before a change in species composition could take place.

Some vegetation in extreme environments seems to be stable in the sense that the same general composition is maintained on the same area of ground (e.g. desert, or rock outcrop). This seems to arise from the limited niche variety, but relatively wide fundamental niches of species in those habitats. In other instances the vegetation is in a state of flux in terms of local changes (Arctic tundra and fire-prone vegetation), but the overall vegetation complex seems to be stable and resilient in the face of disturbance. There are problems in trying to accommodate fluctuations, cycles, direct replacements and systems maintained by severe disturbance within an orthodox 'succession to climax' model. This point is considered in Chapter 12.

8

Patterns of vegetation change in wetlands

Wetland is a general term for any site where water is the substrate or the dominating influence on plant growth, or both. The water may be above, at, or just below the ground surface and is present continuously or frequently (Fig. 8.1). The plants may be rooted in the ground (consisting of mineral soil or organic sediment) or they may be floating.

Water is a distinctive medium for plants, both physically and chemically. As a result the general ecology of wetland plants contrasts with that of plants in well-drained sites. Arising from the anaerobic conditions which often prevail, the accumulation of relatively undecayed organic sediment is a feature of many aquatic habitats.

As will be seen, certain wetland systems are among the most important places for obtaining a clear record of long-term vegetation changes (over periods from a few hundreds to many thousands of years). This is because of the preservation in their sediments of a record of the history of vegetation of both the wetland itself and adjacent terrestrial sites, in the form of identifiable plant fragments and pollen grains and spores.

Permanent freshwater systems, mainly in temperate regions, are the subject of this chapter (cf. P. D. Moore & Bellamy 1974, Wetzel 1975, Etherington 1983). The wetlands to be considered are **lakes**, and various kinds of **mire** (Fig. 8.1).

LAKES

Lakes are standing water bodies of any size, although small lakes, with an area of less than about a hectare, are often called ponds or tarns. Lake basins have originated in a wide variety of ways (Wetzel 1975), but in each case the essential feature is a ground hollow with an impermeable substratum. The input of water from all sources balances evaporative or other losses for most of the time.

The vegetation of lakes is varied. In temperate regions the waters of lakes usually have, as well as macroscopic and microscopic algal populations, specialized herbaceous **macrophytes** from angiosperm or other higher plant groups. Most are rooted on the lake bottom. Some are totally submerged, and others have floating leaves. In shallow lakes, or deeper lakes with shallow margins, many of the macrophytes are reedswamp species with stems emerging from the water. The sedges (Cyperaceae), rushes (Juncaceae), grasses (Poaceae) and bulrushes (Typhaceae) are especially well-represented. Some woody plants (e.g. in the northern temperate zone willows (*Salix* spp.) and alders (*Alnus* spp.)) may also grow with their roots and stem bases immersed in shallow water. Some highly specialized, aquatic macrophytes may float

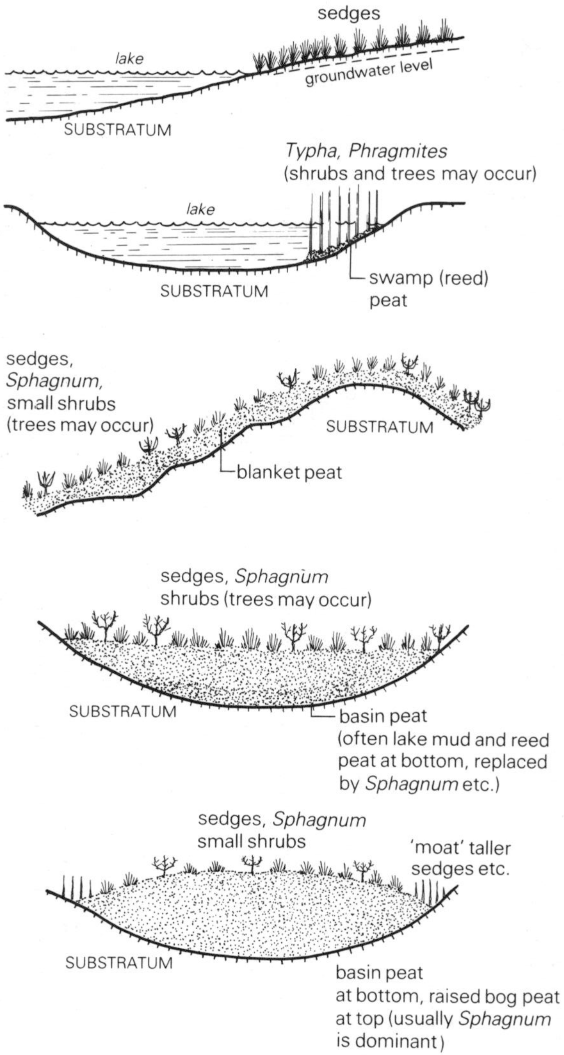

	Water regime	Substrate	Water fertility
(1) Marsh	ground water	often waterlogged mineral soil, little or no peat	usually moderately fertile to fertile
(2) Swamp (fen)	water in more or less permanent lake basin	mineral soil or organic sediment including reed (swamp) peat	usually moderate to strongly fertile
(3) Blanket bog	high rainfall maintains very wet conditions on slopes	acid sedge, shrub or moss peat	usually relatively infertile
(4) Basin bog	water in basin, but maintained by peat (possible influx from land surface)	acid sedge, shrub and moss peat	usually relatively infertile
(5) Raised bog	high rainfall, water maintained by raised peat dome (the only source of water is the rainfall, except at the marginal 'moats')	acid sedge, shrub and moss peat	usually very infertile

Diagram labels:

(1) Marsh — sedges; groundwater level; lake; SUBSTRATUM

(2) Swamp (fen) — *Typha, Phragmites* (shrubs and trees may occur); swamp (reed) peat; lake; SUBSTRATUM

(3) Blanket bog — SUBSTRATUM; blanket peat; sedges, *Sphagnum*, small shrubs (trees may occur)

(4) Basin bog — basin peat (often lake mud and reed peat at bottom, replaced by *Sphagnum* etc.); sedges, *Sphagnum* shrubs (trees may occur); SUBSTRATUM

(5) Raised bog — 'moat' taller sedges etc.; sedges, *Sphagnum* small shrubs; basin peat at bottom, raised bog peat at top (usually *Sphagnum* is dominant); SUBSTRATUM

Figure 8.1 Some wetland types. Other types are described later in the text of this chapter. Various intergrading types occur and some wetlands are dominated by shrubs and trees.

unattached in dense crowds at the surface of still water in small lakes or embayments. This phenomenon is especially well-developed in the Tropics and subtropics. Floating mats of plants, usually attached to the shore, are found around the margins of some small- to medium-sized lakes. The plants are joined together by interlacing rhizome, branch and root systems. The mats are often peaty and may be colonized by woody plants, including trees. Lakes with infertile water or lakes in harsh climates may have little or no macroscopic plant growth.

Oxygen is generally deficient, so all wetland plants must be specialized to allow oxygen to reach their submerged parts. Many aquatic species produce masses of adventitious roots close to the water surface – the zone richest in dissolved oxygen. The leaves, stems and roots of many aquatic macrophytes have extensive development of **aerenchyma** (tissue with large intercell spaces). In emergent plants oxygen diffuses down through these tissues to reach the submerged parts. Plants that are completely submerged use dissolved oxygen. The aerenchyma is mechanically weak, but this does not matter as the plants are supported, in part, by the water.

Various dead remains from the plants living in and around the lake, and from its animal life, sink to the bottom. In summer many deep lakes in the temperate zone have layers of water of contrasting temperature. The colder bottom water is especially poorly oxygenated, and decay of the detritus is minimal. Decay is also slow in infertile, acid water. Fertile, shallow, base-rich water supports highly productive plants, which give rise to abundant debris; decay is usually relatively rapid under these circumstances.

Eventually some lakes become completely infilled by organic sediment. However, other lakes are so nutrient-poor from the outset, or so large, that they have not become infilled many thousands of years after their formation. Many lakes also accumulate inorganic sediments carried by tributary streams or derived from erosion around the shores. There may be complex interrelationships between the organic and inorganic sedimentary components.

MIRES

Wetlands which are essentially vegetated, so that there is relatively little open water, are called **mires**. There are various distinct mire types, according to differences in the substrate type and fertility, but the habitats and vegetation in each tend to form a complex, intergrading continuum.

Marshes are systems developed on sites where the water table lies just above, at or just below the ground surface. The plants which inhabit them are often grasses, rushes and sedges that are rooted in the waterlogged mineral soil. Shrubs and trees may also occur. Marshes may develop on slopes as well as on level sites. The marsh habitat is often marginal to lakes and grades into the fully aquatic lacustrine habitat, but some marshlands have little or no open water.

Swamps (or fens) are sites with shallow, often fluctuating, water levels. The water is fertile and neutral to basic. The plants are rooted in mineral soil or organic sediment. The plants are often 1 m tall or more. Swamps grade into marshes. They also grade into shrub communities and wet-ground woodland; and into the open water of lakes.

Bogs are sites where, although there may be pools and sometimes streams, the plants are mainly rooted in organic sediment. Water is maintained by the sponge-like properties of the sediment (peat), and is acidic and infertile. The plants are often tall to short sedges and mosses, but shrubs and trees may also be present. Quite different suites of species occur in bogs and swamps. As will be seen, the origins and ultimate character of peat mires (swamps and bogs) are quite variable.

WATER CONDITIONS

Physical and chemical properties of the water, which influence the growth of plants in lakes, marshes and mires and which vary from site to site, include:

1   The inherent fertility (i.e. the amounts of essential nutrients such as nitrogen, phosphorus, calcium, potassium and sulphur per unit volume of water, and the flux of nutrients in flowing water).
2   The relative acidity of the water.
3   The amounts of dissolved oxygen present.
4   The water temperature.
5   The penetration of light (in open water only).
6   Toxic compounds.
7   Grazing animals.
8   Wave action (in open water or lake margins).

These will be influenced, in turn, by various features of the physical setting of each site. With respect to lakes, factors such as topography of catchments and drainage systems, size, exposure to wind, steepness of shore, depth, turbidity of water, the kinds of materials forming the substratum, the rock types of their catchment areas, the rates of inflow and outflow of water and whether they freeze in winter will all have important effects on the essential water properties which control plant growth. In fact, an almost infinite array of permutations of site conditions is possible, so each individual wetland tends to have some unique features. Although some mire plant species are widely distributed, others are more local and the proportions of the species in each mire also tend to differ.

Marshes and swamps are influenced by the influx and flow of water through them from groundwater sources, by flooding from stream overflow or by rises in lake levels. The peat mass of many mires maintains large volumes of water by means of its spongy properties. This exerts considerable control over the hydrology, with consequent effects on the vegetation growing on mire surfaces. For example, the lateral movement of water through the peat mass is slow. Mire surfaces are more or less insulated, by the presence of vegetation and peat, from the external influences which can cause rapid day-to-day changes in the open water of lakes. However, spatial and temporal variations affect the hydrology of mires. Streams may flow across their surfaces, and some have large or small pool systems on their surfaces. Others have little or no standing water. Some bogs have raised central domes of peat.

Temperature extremes, drought, grazing and fire may affect marshes, swamps and bogs. The main influences on the ecology of plants in them, however, derive from the fertility conditions of their water. Water which has flowed from soil or rocky substrates is more fertile than that which has only been in contact with acid peat, such as occurs in many mature bogs. Table 8.1 includes some values for nutrients in the water of various types of wetland site and the corresponding vegetation.

The richer (**eutrophic**) waters, which are also usually neutral to basic, have supplies of cations and anions sufficient to support relatively highly productive vegetation (e.g. in North America and Europe reedswamp species such as the reed *Phragmites communis*, bulrush or cat-tail (*Typha latifolia*), sedges (*Carex* spp.) and some species of trees (*Salix* and *Alnus*)).

The poorer (**oligotrophic**) waters are very low in nutrient cations and anions and high in hydrogen ion concentration. They can support only a few herbs such as bog bean (*Menyanthes trifoliata*), relatively slow-growing species of sponge moss (*Sphagnum*) and, on peat, grass-like angiosperm herbs such as species of bog-cotton sedge

Table 8.1 Generalized values for pH and concentrations of major ions in waters of lakes and mires (from P. D. Moore & Bellamy 1974).

				Ionic composition (meq l^{-1})						
	pH	H^+	Ca^{2+}	Mg^{2+}	Na^+	K^+	HCO_3^-	Cl^-	SO_4^{2-}
I. lakes									
A. eutrophic (basic, relatively fertile)									
	8.0	0	24.5	7.3	8.9	2.1	66.3	35.9	14.2
B. oligotrophic (acidic, relatively infertile)									
	4.7	0.01	3.8	1.3	4.5	1.2	0	12.8	8.9
II. mires									
A. mires developed in lakes with stream inflow and outflow (predominant cation Ca^{2+}, anion HCO_3^-)									
1. continuous flowing water which inundates mire surface	7.5	0	4.0	0.6	0.5	0.05	3.9	0.4	0.8
2. continuous flowing water beneath floating vegetation mat	6.9	0	3.2	0.4	0.4	0.08	2.7	0.5	1.0
3. intermittent flowing water which inundates mire surface	6.2	0	1.2	0.4	0.5	0.02	1.0	0.5	0.7
4. intermittent flowing water beneath floating vegetation mat	5.6	0.01	0.7	0.2	0.5	0.04	0.4	0.5	0.5
B. mires developed in lakes with seepage inflow (predominant cation Ca^{2+}, anion SO_4^{2-})									
5. continuous flow of water	4.8	0.03	0.3	0.1	0.3	0.07	0.1	0.3	0.5
6. intermittent flow of water	4.1	0.14	0.2	0.1	0.3	0.04	0	0.4	0.4
C. mires with never any inflow of ground water (predominant cation H^+, anion SO_4^{2-})									
7. sole supply of water is rainfall	3.8	0.16	0.1	0.1	0.2	0.04	0	0.3	0.3

Eriophorum and heath shrubs (cranberry (*Vaccinium*), labrador tea (*Ledum*), leather leaf (*Chamaedaphne*) and crowberry (*Empetrum*)). A few gymnosperm tree species also tolerate these conditions. There are strong deficiencies of nutrients such as nitrogen and phosphorus, and substantial amounts of elements such as aluminium, manganese and iron tend to accumulate. Anaerobic and sulphide-rich reducing conditions are present 10–20 cm below the mire surface (Clymo 1965, P. D. Moore & Bellamy 1974). The plants have a highly specialized physiology to cope with low fertility and concentrations of metallic and sulphide ions which would be toxic to swamp plants (or many terrestrial plants) (Etherington 1983).

In fact, the diversity of the vegetation of wetland complexes usually results from local variations in the fertility of the water. These matters are considered more fully later.

PEAT MIRE STRATIGRAPHY

The development of peat mires has a very large autogenic component. This can be traced by examination of the successive stratified layers of sediments beneath them. Sedimentation has been almost continuous in many mires for many thousands of years. Hiatuses in deposition and signs of disturbance are sometimes evident. Inorganic sedimentation may also complicate the picture. Sands, silts and clays may form distinct horizons or may be intermingled with the organic detritus.

Assuming that organisms colonize a relatively fertile lake soon after it is formed, the first layers of sediment will probably include much inorganic material derived by soil disturbance around the catchment, as well as organic detritus derived from the remains of algae and small animals. Floating-leaved and soft-leaved submerged macrophytes will give rise to somewhat coarser debris in later layers. If the lake is colonized by reedswamp species, then infilling proceeds more rapidly and there will eventually be coarser, fibrous peat deposits overlying the earlier deposits. Woody plants will contribute to coarse peats, also, as the infilling continues up to the local water table. Invasion by *Sphagnum* moss and heath shrubs may complete the sequence. Some peat mires start, not in lakes, but in marshy areas, as a consequence of waterlogging of the ground and peat accumulation.

Raised bogs are the ultimate phase of peat build-up in some mire sites in infilled lakes or marsh sites. Once the water table in the peat dome is raised above that of the local land surface, the only possible source of water on the dome is from precipitation. The only inputs of nutrients are in the rainfall, in various kinds of aerosol and dust, by imports by animals or from the slow decay of peat. The input of nutrients from these sources is almost inevitably less than in sites where water is flowing from mineral soils (Table 8.2). Around the margin of raised bogs, however, usually there is an area with more-fertile water derived both from adjacent land and from the peat. The plant assemblage here differs from that on the peat dome (Fig. 8.2).

The height to which the peat dome may rise is apparently determined by the equilibrium reached between plant growth, peat accumulation, water input and evapotranspiration in the region. Raised bogs are usually found in cool-temperate,

Table 8.2 Nutrient content of the precipitation in some Northern Hemisphere localities.

A   Average concentration of ions in one year's precipitation (in ppm) (after Gorham 1961).

Locality	Ca^{2+}	Mg^{2+}	Na^+	K^+	SO_4^{2-}	Cl^-	NO_3^-	NH_4^+
Newfoundland	0.8	n.d.	5.2	0.3	2.2	8.9	n.d.	n.d.
Wisconsin	1.2	n.d.	0.5	0.2	2.9	0.2	n.d.	n.d.
northern Sweden	1.2	0.2	0.4	0.3	2.5	0.7	n.d.	n.d.

B   Concentration of ions in one year's precipitation (g-equiv ha^{-1}) (after Eriksson 1960).

Norway	515	980	3118	112	959	3465	247	257
Iceland	225	476	1805	58	219	2107	27	32
Netherlands	525	1717	5422	128	1174	5700	298	172
England	309	952	3590	137	576	4190	98	267

n.d., not determined.

Figure 8.2 A raised bog, Dun Moss, in Perthshire, Scotland. Heather (*Calluna vulgaris*) is the dominant plant. Sedges and *Sphagnum* species are also present (photo C. J. Burrows).

humid climates, but they can occur in warmer climates including the tropics (Etherington 1983).

The stratified deposits in mires, thus, are oldest at the bottom and youngest at the top. Each layer will contain more or less well-preserved **macrofossils** – macroscopic plant fragments (seeds, fruit, leaves, stems, roots and wood). There are usually relatively few species in mire vegetation. They tend to grow in dense stands, so that usually the main communities can be identified from the preserved plant remains (although some aquatic species do not preserve well). The sediments will also contain identifiable **microfossils** (pollen grains and spores) which are usually preserved very well. From the macrofossils, supplemented by the microfossil evidence, the vegetation sequence can be reconstructed. Radiocarbon dates can be used to construct a timescale for significant events.

The pollen assemblages in the sediments of a mire include some elements derived from the plants which grew in the mire and others derived from plants which grew on more-distant sites, including well-drained terrestrial ones. If this information is to be used to reconstruct the history of a mire, the component from mire plants must be separated from all others so that a **pollen diagram** can be drawn, representing the changing proportions of plant species through the history of the mire (Fig. 8.3). It is likely that some pollen and spore types, recovered from mire sediments, will represent species not recorded among the macrofossils, and vice versa. The quantitative relationship between pollen spectra and the vegetation types from which they were derived can be assessed, but it is more difficult to quantify the relationship between macrofossils and vegetation. The methodology of pollen analysis is outlined more fully in R. G. West (1971), Faegri & Iversen (1975) and H. H. Birks & Birks (1980).

As will be seen in later examples, each site tends to have a distinct sedimentary sequence, which in some sites is simple, in others more complex. The blanket peat sequences which develop from waterlogged, marshy sites tend to be simpler than those of lake basins.

Following Heinselman (1975), a simple classification of organic sediment types is used here (Table 8.3). More-elaborate classifications (e.g. Faegri & Iversen 1975) employ complex terminology.

SOME BRITISH AND SCANDINAVIAN MIRES

Sedimentary strata in British and Scandinavian mires, developed in hollows formed by glacial processes, reveal complex histories of vegetation development. Similar plant assemblages were present within each region, but the patterns of vegetation change in each mire are distinctive. Changes occurred at widely different rates. These

Table 8.3 Common organic sediment types in wetlands (modified from Heinselman 1975).

Sphagnum peat	
sedge peat	
reed peat	
tree peat	
shrub peat	
organic lake mud	very diverse, depending on source organisms and history – often derived from phytoplankton, zooplankton, aquatic floating, or floating-leaved angiosperms, and debris such as leaves, twigs from outside the system

Combinations of the distinct sediment types may occur, as well as other types not listed here and various kinds of inorganic sediments brought in as solids, or precipitated from solution (e.g. calcium carbonate).

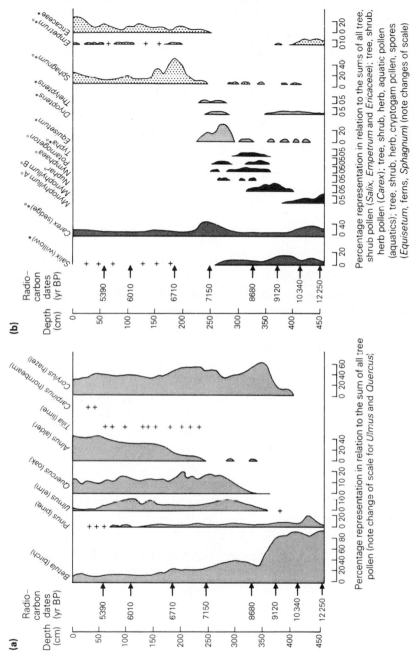

Figure 8.3 Simplified pollen diagrams from the sediments in Din Moss, Cumbria, Britain. (a) The changing tree pollen frequencies. Most of the trees contributing pollen grew in the local region, but some (e.g. *Betula* and *Alnus*) grew around lake or mire margins or sometimes on mire surfaces. (b) The changing pollen frequencies of shrubs and herbs which grew around lake or mire margins or on mire surfaces (*) or were aquatic (○). *Empetrum* was part of the vegetation of the local region early in the sequence. Modified from Hibbert & Switsur (1976).

Table 8.4 Climate data for mire localities (after Wernstedt 1972).

Locality	Mean annual precipitation (cm)	Mean annual temperature (°C)	Mean January temperature (°C)	Mean July temperature (°C)
Carlisle, Cumbria, Britain 54°53′N, 2°57′W	81.6	10.2	4.3	16.3
Emmen, Drenthe, Netherlands 52°47′N, 7°00′E	76.6	10.2	2.0	18.1
Lyngdal, Østfold, Norway 59°54′N, 9°32′E	78.4	6.1	−1.3	15.7

differences can be related to differences in the habitat conditions which have prevailed during the lifetime of the individual mire sites. Four sites are illustrated: two from Cumbria, in Britain, two from southeastern Norway (Figs 8.4–7). In each instance the climate is humid temperate (Table 8.4).

The sites were chosen because significant phases in the mire stratigraphy have been dated. Also, as can be gauged from the internal evidence of the peat types, there has been uninterrupted organic sedimentation with no sign of allogenic interference, except at the top.

Din Moss, in Cumbria, Britain (Hibbert & Switsur 1976), originally a small lake basin with marginal reedswamp, had open water for more than 2000 years (from 12 250 years BP). As organic sediment filled the lake, *Phragmites* reedswamp and **fen** (reedswamp plants growing on peat in shallow, relatively fertile water), with some birch (*Betula*) **carr** (mire vegetation dominated by shrubs or low trees) became predominant in the basin. This phase also lasted for more than 2000 years. It gave way to *Sphagnum* bog which has lasted, with no substantial change, from about 6800 years BP to the present. The change from reedswamp and fen to *Sphagnum* bog resulted from a decline in the water fertility and increased acidity, an outcome of the invasion of the fen by a mat of *Sphagnum*. During this time, as gauged from the terrestrial pollen spectra (Fig. 8.3), the vegetation of the surrounding dry land underwent marked changes presumed to have been caused by climatic variation and (from about 5400 years BP), by human disturbance (Hibbert & Switsur 1976). Subtle changes in the mire vegetation may have resulted from the same causes, but the gross features of the wetland vegetation were unaffected. The plants in their aquatic or semi-aquatic habitat seem to have been insulated from, or relatively insensitive to, these allogenic influences, or both.

A history similar in some respects to that of Din Moss, but different in detail, is recorded by the sediments of the raised peat bog Scaleby Moss, also in Cumbria. Like many other raised mires, its surface has been much disturbed by peat cutting. However, sufficient of the upper peat layers remain to allow the sequential development to be examined (Godwin *et al.* 1957).

278

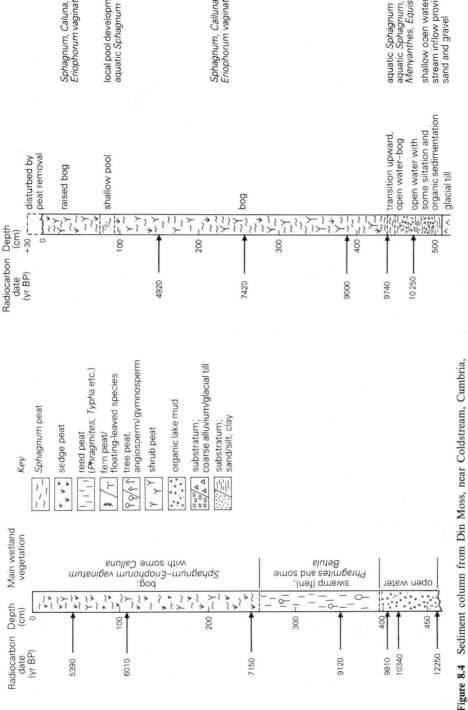

Figure 8.4 Sediment column from Din Moss, near Coldstream, Cumbria, Britain (170m a.s.l.), showing dominant sediment types and the inferred vegetation which prevailed in the wetland during the accumulation of each. N.B. this explanation and the key for sediment types applies to each of Figs 8.5, 6 & 7 and other figures later in the chapter (from Hibbert & Switsur 1976).

Figure 8.5 Sediment column from Scaleby Moss, near Carlisle, Cumbria, Britain (35m a.s.l.) (from Godwin *et al.* 1957).

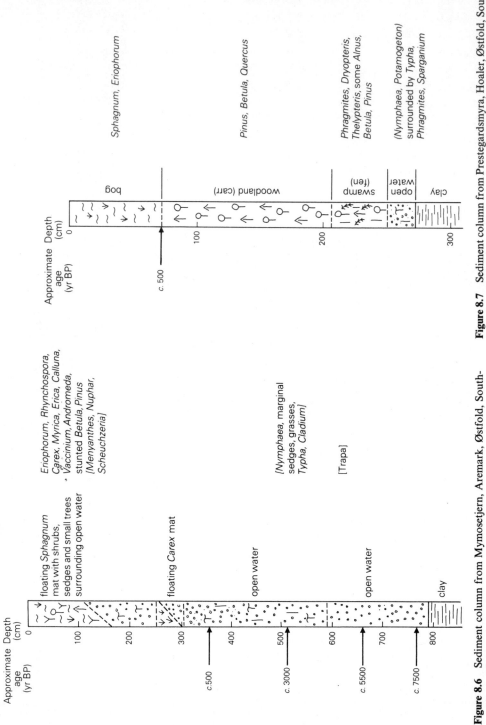

Figure 8.6 Sediment column from Mymosetjern, Aremark, Østfold, Southern Norway (112m a.s.l.) (from Danielson 1970).

Figure 8.7 Sediment column from Prestegardsmyra, Hoaler, Østfold, Southern Norway (16m a.s.l.) (from Danielson 1970).

Originally the site of a shallow lake, the mire occupies a hollow in glacial till. From about 12 000 until about 9600 years BP, some 75 cm of mainly inorganic sediment accumulated. During this period sedges grew around the shores and the organic component of the sediment gradually increased, probably indicating the development of reedswamp. Birch woodland and willow scrub surrounded the lake. Then for a period of a few hundred years the lake was invaded by aquatic *Sphagnum* spp., macrophytic herbs (*Menyanthes* and *Equisetum*) and sedges.

About 9200 years BP (before hazel (*Corylus*) and later elm (*Ulmus*), oak (*Quercus*) and alder (*Alnus*) became prominent in the surrounding forest) the open water finally was replaced by a continuous cover of *Sphagnum*, in which grew the shrub heather (*Calluna vulgaris*) and the sedges *Eriophorum vaginatum*, *E. angustifolia* and *Trichophorum caespitosum*. This phase lasted for at least 7000 years, during which 325 cm of peat accumulated.

About 2000 years BP, possibly after a period of increased precipitation, there occurred a period of vigorous peat development which raised the bog surface by another metre. Several oxidized horizons indicate halts in deposition which were followed by renewed peat accumulation. Before human disturbance in recent centuries, the summit of the bog was a dome of peat standing above the surrounding mire margins.

The site Mymosetjern in southern Norway (Danielson 1970) is a large, flat bog which formed after partial infilling of a lake by gradual accumulation of sediment, over more than 7500 years. The centre of the basin still has open water with floating-leaved aquatic plants and a peripheral, floating *Sphagnum* mat with bog herbs and some heath shrubs. The organic sediment in the mire is derived from the floating-leaved plants and *Sphagnum*. The very simple sedimentary sequence shows that such sites, with relatively infertile water, may take many thousands of years to infill under an organic sedimentation regime. Fossils in the sediments show that fertility has declined and other conditions have changed so that some species have been ousted from the mire.

Prestegardsmyra is a small bog on the site of a completely infilled shallow lake (Danielson 1970). Its history is relatively brief – less than 3000 years, during which time the relatively fertile water was quickly infilled. First there was open water surrounded by *Phragmites* and *Typha*. This was followed by a wooded phase, beginning with *Alnus, Betula* and pines (*Pinus*) and the fern *Dryopteris thelypteris*, then *Pinus* and *Betula* and *Quercus* near the upper part of the wooded phase, which ended about 500 years ago. The growth of *Sphagnum* and sedge has been relatively rapid since then.

MIRES IN THE SOUTHERN GREAT LAKES REGION, NORTH AMERICA

The vegetation history of lakes and mires in the Itasca Park region, Minnesota, on the site of the ancient glacially dammed Lake Agassiz, has been well studied (Buell & Buell 1941, 1975, Lindeman 1941, Conway 1949, Leisman 1953, 1957, McAndrews 1967, Buell *et al.* 1968, Heinselman 1970, 1975). This huge lake drained,

14 000–10 000 years ago, as the ice sheet retreated from the Wisconsin glacial maximum, leaving many small lake basins and wet areas.

Two kinds of site are described below.

1   Those where lake basins are being infilled by organic sediments derived from plants and other organisms living in the lake waters.
2   Those where raised blankets of peat have developed in marshes and spread over surrounding areas.

Not all of the sequences are well dated.

Lake basins

Sediments in the lake basins show that lake infilling has been achieved mainly by sedimentation from pelagic or floating-leaved plants, as well as from the bottom of marginal floating vegetation mats (Figs. 8.7 & 8). The modern patterns of marginal invasion of lakes by vegetation are physiognomically similar to one another, but the floristic composition of the bands of vegetation varies (Buell & Buell 1941, 1975, Conway 1949). The differences are explained in terms of the different fertility and acidity of the water.

The process of invasion begins with floating mats of sedges (or the herbaceous water willow *Decodon verticillatus*) which expand centripetally. The mats are invaded on the shoreward side by *Typha* or shrubs and low trees (*Salix* spp., red alder (*Alnus rugosa*), paper birch (*Betula papyrifera*). etc.), or by heath shrubs such as *Vaccinium* spp., *Ledum groenlandicum* and *Chamaedaphne calyculata* which grow in a matrix of *Sphagnum* moss (Conway 1949). The next colonist is the tamarack (*Larix laricina*) or sometimes black spruce (*Picea mariana*). The *Larix* or *Picea* may remain dominant in some sites, but usually there follows a forest replacement sequence in which they give way to balsam fir (*Abies balsamea*) or white cedar (*Thuja occidentalis*) (Figs. 7.28, 8.9 & 10).

Based on observations of the modern patterns, inferences about these changes in vegetation composition are supported by four kinds of evidence.

(a)   Stratigraphic study of the sediments in the mires showing replacement series, e.g. forest peat on *Sphagnum* peat on sedge peat.
(b)   Study of the age distributions in the populations of trees in the marginal bands, showing replacements such as *Larix* → *Picea* → *Abies*.
(c)   Measurement and later re-measurement of the changing vegetation patterns.
(d)   Measurement of the upward growth of the developing peat bogs.

In some of the older gymnosperm stands near the shore in the forested bogs, some angiosperm trees are present, but there is no sign of the development of forest composition like that on adjacent dry land. Completely infilled, forested lake basins are known, in which sedimentation began more than 10 000 years ago, and in which the forest continues to be dominated by white cedar, black spruce and tamarack

(b)

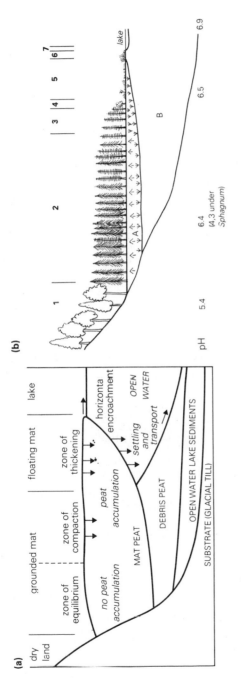

pH   5.4   6.4   6.5   6.9
(4,3 under
Sphagnum)

(a)

dry land | grounded mat | floating mat | lake

zone of equilibrium | zone of compaction | zone of thickening

no peat accumulation | peat accumulation | horizonta encroachment | OPEN WATER | settling and transport

MAT PEAT

DEBRIS PEAT

SUBSTRATE (GLACIAL TILL) | OPEN WATER LAKE SEDIMENTS

Key

1 *Quercus ellipsoidalis* (northern pin oak), *Pinus banksiana* (jack pine) forest on dry land

2 dense *Thuja occidentalis* (eastern white cedar), *Larix laricina* (tamarack) in a band 100 m wide or more; ground cover *Sphagnum magellanicum, Oxycoccus quadripetalus* (cranberry)

3 invading margin of *Larix laricina* 10–15 m wide, with shrub understorey (*Ledum groenlandicum*, labrador tea, *Rhamnus alnifolia, Rhus vernix, Andromeda glaucophylla*)

4 willow shrub zone 2–3 m wide; *Salix petiolaris, S. bebbiana, Larix* juveniles

5 sedge mat zone 20 m wide; *Carex lasiocarpa*, with *Thelypteris palustris* (mire fern), *Muhlenbergia racemosa* (mire grass)

6 Patches of tall 'reeds'; *Typha latifolia* (cat tail), *Eleocharis palustris* (spike sedge)

7 marginal band of *Decodon verticillatus* (water willow) invading open water

A peat derived from surface accumulation of material from the sedge mat and later from the forest which replaces it

B Organic sediment derived from open water organisms and later from debris which falls from the bottom of the floating mat of invading vegetation

Figure 8.8 (a) General model for mire development through autogenic infilling of small lakes with organic sediments, Southern Great Lakes region, North America (after Kratz & de Witt 1986). (b) Section through the vegetation and sediments at the margin of a small lake, Cedar Creek Bog, Minnesota, USA. Other lakes, have a different array of species in the concentric bands of marginal vegetation. The lake water may be much more acidic. *Chamaedaphne calyculata* (leather leaf) is often the species at the margin of open water; different *Carex* species, *Sphagnum* spp. and various small heath shrubs and dwarf birch (*Betula pumila*) occur in the sedge mat; *Larix* is followed by *Picea mariana* (black spruce), then balsam fir (*Abies balsamea*) (after Buell & Buell 1941, Conway 1949).

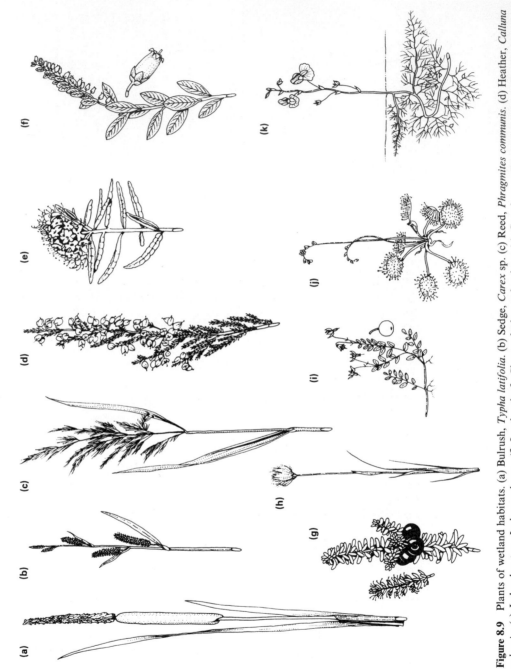

Figure 8.9 Plants of wetland habitats. (a) Bulrush, *Typha latifolia*. (b) Sedge, *Carex* sp. (c) Reed, *Phragmites communis*. (d) Heather, *Calluna vulgaris*. (e) Labrador tea, *Ledum palustre*. (f) Leather leaf, *Chamaedaphne calyculata*. (g) Crowberry, *Empetrum nigrum*. (h) Bog cotton, *Eriophorum vaginatum*. (i) Cranberry, *Oxycoccus palustris*. (j) *Drosera* sp. (sundew). (k) *Utricularia* sp. (bladderwort).

Figure 8.10 A small, acid water lake in Emmet County, Michigan, USA, with invading marginal vegetation. *Chamaedaphne calyculata* (leather leaf) is the primary invader. A narrow *Sphagnum* mat (not visible) lies between it and bog forest dominated by *Picea mariana* (black spruce) and tamarack (*Larix laricina*) (photo S. Planisek *ex* H. Crum).

(Collins *et al.* 1979). This contrasts markedly with angiosperm-dominated forest on well-drained adjacent sites, with species such as sugar maple (*Acer saccharum*), basswood (*Tilia americana*) and red oak (*Quercus rubra*). Some moderately fertile wet sites carry dense stands of angiosperms such as red maple (*Acer rubrum*), yellow birch (*Betula alleghaniensis*) and paper birch, which are less common on well-drained sites.

Blanket peatlands

The history of the very extensive **blanket bogs** of Itasca Park is described by Heinselman (1970, 1975). Blanket peat covers the terrain, regardless of slope. It originated from raised bogs, on infilled lakes and marshes, and spread laterally on to drier ground. Stratigraphic analysis and dating of blanket peat in the Itasca Park region, by McAndrews (1967) and Heinselman (1963, 1975), shows that the swamping of relatively dry land started about 8000 years ago or earlier (Figs. 8.11 & 12).

The peatland topography is semi-independent of underlying topography and is complex, with raised peat domes and ridges having crests up to 3.5 m above their margins, and intervening shallow valleys down which water gradually drains. The raised peat domes are rain-fed. Forested and shrubby areas are present, the former on relatively well-drained sites, the latter in the 'water-tracks' where local drainage

Figure 8.11 Aerial view of a partially forested peatland area near Lac St Jean, Quebec, Canada. The bog surface has a complex pattern of pools, flat *Sphagnum* areas and hummocks with sedges, shrubs and, in places, stunted black spruce (*Picea mariana*) trees. It is similar to the Itasca Park peatlands (photo P. H. Glaser).

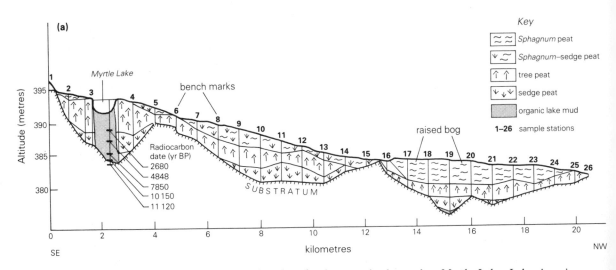

Figure 8.12 (above and opposite) (a) Stratigraphy of a large wetland complex, Myrtle Lake, Lake Agassiz Peatland, Minnesota, USA, showing development of basin mires and blanket peat mires culminating in the formation of raised bog. Note that Myrtle Lake has been in the same general position for more than 10 000 years. (b) Reconstruction of three phases of development of the wetland, based on the changes in peat types (modified from Heinselman 1963).

causes the water to be slightly more fertile than on adjacent ridges. The changed topography created by peat build-up has, in turn, changed the whole hydrologic pattern in complex ways.

In one instance, a lake basin (Myrtle Lake) has existed in the same place for more than 10 000 years. As the surrounding peat built up, the lake margins were maintained, with vertical banks, now about 10 m deep. The lake basin is filled with aquatic peat (Fig. 8.12). Pool systems are common in blanket peatlands, often forming complex patterns (Sjörs 1961).

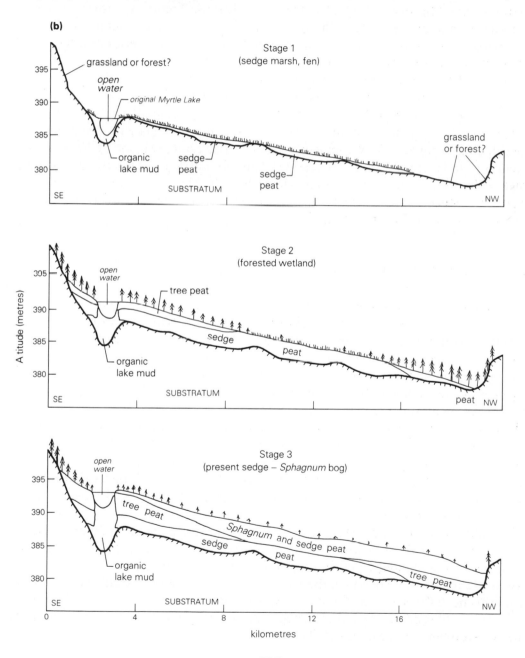

The abundant development of peatlands (muskeg bogs) in Alaska prompted Zach (1950) to suggest that they, rather than forest, seem to represent the most mature (or 'climax') physiography and vegetation on gentler terrain in that area. He saw ample evidence for progressive invasion of forest by bog. Drury (1956) described complex interactions of bog, forest and other vegetation in Alaska. Heinselman (1975) pursued the point that peatland formed the 'climax' vegetation, but noted that, in spite of the long period which has elapsed so far during the development of peatland in the Itasca Park region, the peat accumulation process has not reached the maximum possible in the present climatic conditions. Trends are towards a slow elaboration of landscape types. Change is slow, but ceaseless, predictable only if the full ecological situation is known and depending on a complex series of interactions of vegetation, hydrology, water chemistry, climate, etc.

Sjörs (1980), drawing on experience from Boreal peatlands in Canada and Scandinavia, made similar points and stressed the heterogeneity of vegetation which results from peatland formation. Convergence of vegetation to similar physiognomy and species structure occurs, but there are just as many instances of divergence. In general, long-term mire development results in the establishment of low-statured vegetation and depleted pools of readily available nutrients. This trend is usually irreversible under natural conditions. Heinselman (1975) notes, however, that disturbing influences such as peat decomposition, erosion by stream drainage and climate change can alter the whole situation.

STRATIGRAPHIC STUDIES OF BRITISH MIRES

As a result of observations of mires in North America and Europe, there was a belief by some early workers (e.g. Dachnowski 1912, Clements 1916) that mire development ultimately would always terminate with forest. However, some contemporary ecologists (e.g. Weber 1910) were aware that, in some mires, herbs, heath shrubs and *Sphagnum* moss had formed stable vegetation for many thousands of years. In some instances it was known that forest on mires had been replaced by shrubby or herbaceous vegetation through natural processes.

Tansley (1939) outlined models for the development of British mire vegetation. His view was that, in the suboceanic climate of the lowlands of eastern Britain, there was a progression from reedswamp → fen → carr → forest. Some mire sites there carry oak (*Quercus robur*) woodland so that the belief of Clement and Tansley that the mire vegetation sequence terminates in the 'climax' vegetation for a region seems, on the face of it, to be borne out.

D. Walker (1970) made a study of the stratigraphic sequences in 20 mire sites in Britain, to test Tansley's hypothesis of mire development. These mires could be dated by means of the standard pollen zonation or the radiocarbon method. Sites were chosen to exclude those in which there had been obvious allogenic effects that would have influenced the sequence. Originally each site had been a freshwater basin mire or lake. He established 12 vegetation categories, which could be recognized from the kinds of sediments in the deposits (Table 8.5) and assessed each site according to the

Table 8.5 Vegetation categories recognized from sediments in British mire sites (from D. Walker 1970).

1. biologically unproductive water
2. micro-organisms in open water
3. totally submerged macrophytes (*Myriophyllum*, *Littorella* and *Potamogeton*)
4. floating-leaved macrophytes, with some open water (*Nymphaea* and *Potamogeton*)
5. reedswamp, rooted in substratum, standing in more or less permanent water, with aerial shoots and leaves (*Phragmites*, *Scirpus* and *Carex rostrata*)
6. sedge tussock, rooted in substratum, standing in more or less permanent water (*Carex paniculata* and *C. acutiformis*)
7. fen dominated by grasses (*Phragmites* and *Molinia*) or sedges (*Carex flava* and *C. nigra*), with forbs; rooted in organic deposits, waterlogged most of the year
8. swamp carr with shrubs and trees (*Salix atrocinerea* and *Alnus glutinosa*) growing on unstable sedge tussock, with forbs
9. fen carr with trees (*Alnus*, *Frangula alnus*, *Betula pubescens* and *Fraxinus excelsior*) and undergrowth rich in herbs, ferns; rooted in stable peat
10. aquatic sphagna (e.g. *Sphagnum subsecundum* and *S. cymbifolium*) floating closely below or at water surface
11. bog with sphagna (e.g. *Sphagnum palustre* and *S. imbricatum*) and acid-water tolerant shrubs (*Vaccinium oxycoccus*, *Erica tetralix* and *Myrica gale*) and sedges (*Eriophorum* spp.), in organic substratum
12. marsh with 'fen' forbs and sedges, on waterlogged mineral soil

transitions from one to another of the vegetation categories. The numbers of transitions can be read from Table 8.5, and Figure 8.13 summarizes the preferred transitional sequences.

It is clear from these data that there is considerable variety in the developmental pathways. Not all transitions are 'progressive'. The transitions which are reversed (compared with the expected) are often short-term and possibly result from increases in water levels or other allogenic causes. None of the sites shows a terminal oak forest development. Such sites would be rare now in Britain, anyway, because of human

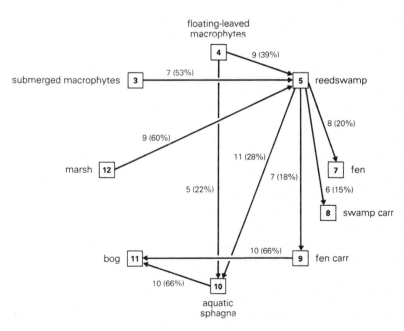

Figure 8.13 Frequencies of transitions between different vegetation types in some British mires (categories 3–12, Table 8.5), based on analysis of sediments (after D. Walker 1970). Only those vegetation categories with five or more transitions are shown, comprising a total of 51% of the 159 transitions. Values alongside arrows are the numbers of transitions with, in parentheses, the percentage that each of these values comprises of all transitions from that category to any other category.

disturbance. Where the few examples of oak forest occur on mires, D. Walker suggests, its persistence on the site depends on the extent to which it can tap and recycle nutrients from the soil underlying the peat. Many of the sites studied by Walker have bog rather than forest as the terminal vegetation. *Sphagnum* mosses, particularly the fully aquatic species, play an important part in the transformation of a site into a bog.

It has been claimed by Drury & Nisbet (1973) and Colinvaux (1973) that the results of D. Walker provide evidence against two of the ideas promoted by Clements (1916). The first is the monoclimax concept, where the terminal vegetation in a region on all kinds of site is the highest physiognomic form permitted by the climate. The second concept is that of predictability of the pathways of sequential change. Walker's data seem to refute the former, but not necessarily the latter except in its strictest sense (that a sequence is predictable because it is inevitable). Mire sediments show that there is a degree of predictability in the development of some mires, but the details vary. Many individual mires do not conform to any 'ideal' pattern.

Tansley (1939), in fact, had postulated that there was considerable variation in wetland vegetation development (cf. Fig. 8.14). His scheme included trends towards bog in higher-rainfall areas in western Britain. It is to be expected, judging by the variety of vegetation patterns and ecological processes occurring in modern mires, that there will have been a corresponding variety in the patterns of vegetation change in mire developmental sequences.

INDICATIONS OF INTERRUPTIONS OF MIRE SEQUENTIAL DEVELOPMENT

Evidence of repeated disturbance of some mires, throughout their history, comes from some well-dated New Zealand stratigraphic sequences (E. O. Campbell *et al.* 1973), where the important peat-forming plants include species of *Sphagnum*, sedges (*Carex* spp. and *Baumea* spp.), members of the monocotyledon family Restionaceae (*Empodisma* and *Sporadanthus*), a fern (*Gleichenia*) and a podocarp tree (*Dacrycarpus dacrydioides*). A repeated sequence: *Baumea* sedge (with or without woody plants) → *Empodisma* and *Sporadanthus* restiad vegetation (with or without shrubs and the fern *Gleichenia*) is recorded in the peat in these mires. The sequence is inferred to result from a change from relatively fertile swamp to infertile, acid, raised bog conditions.

At intervals in the mires studied (Fig. 8.15) the system returned to *Baumea* sedge dominance, probably as a result of flooding of the site. It is possible that climatic variation was involved, but often the change was preceded by deposition of rhyolitic volcanic ash. This probably caused locally impeded drainage and a small nutrient influx, affecting the vegetation temporarily. Frequent fire also disturbed these mires, again probably providing a brief influx of nutrients, released from the burnt vegetation. Before 1000 years ago (the time of first human occupation of New Zealand) the fires were caused naturally.

In raised blanket peatlands in the Netherlands (Casparie 1972) and in Denmark (Aaby & Tauber 1975) detailed stratigraphic analyses of the peat reveal the very

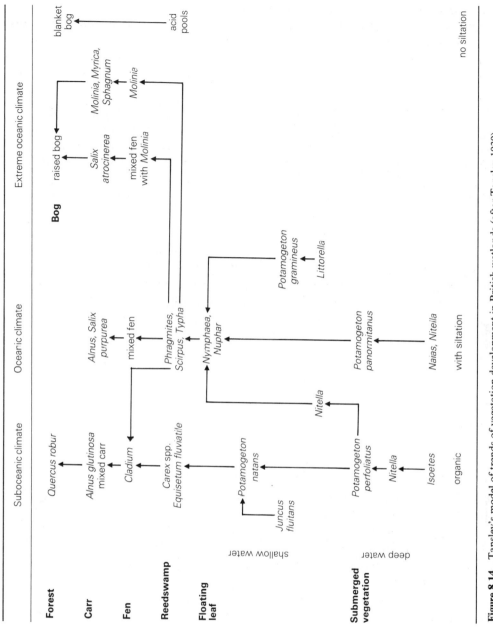

Figure 8.14 Tansley's model of trends of vegetation development in British wetlands (after Tansley 1939).

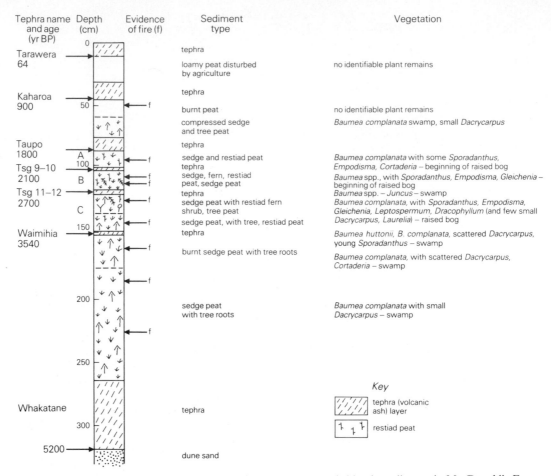

Figure 8.15 Example of recurrent wetland vegetation sequences revealed by the sediments in MacDonald's Farm Mire, Awakeri, Bay of Plenty, New Zealand. The sequences reflect the influence of frequent disturbance by fire and deposition of tephra (volcanic ash showers). At (A), (B) and (C) are well-marked indications of change from fertile swamp dominated by *Baumea* sedges and sometimes the grass *Cortaderia toetoe*, upward into acid raised bog dominated by *Baumea* and the restiads *Sporadanthus traversii* and *Empodisma minus*, with the fern *Gleichenia microphylla* and (in the case of (C)), the shrubs *Leptospermum scoparium* and *Dracophyllum* sp. as well as the trees *Dacrycarpus dacrydioides* and *Laurelia novae-zelandiae*. (See Fig. 8.4 for key to sediment types other than those indicated here.) After E. O. Campbell *et al.* (1973).

complex vegetation history of each mire (Fig. 8.16). Occasional charcoal layers mark fire episodes, and wood layers indicate drying which permitted forest invasion of the mires. Wood layers are also common in the New Zealand peat profiles (E. O. Campbell *et al.* 1973).

'PHASIC REGENERATION CYCLES' IN BOGS

In the British and European literature on mire ecology there has been a long adherence to hypotheses of cyclic regeneration of the small-scale vegetation features, hummocks and pools, on bog surfaces. These were believed (e.g. by Osvald 1930,

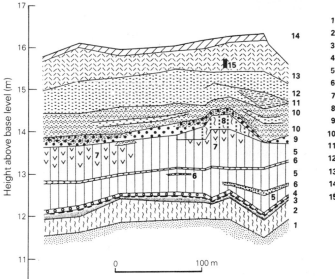

Key
1 substratum (sand)
2 brown moss peat (Hypnaceae and monocotyledonous plants)
3 loess layer (aeolian deposition)
4 twig layer (*Betula*)
5 swamp (fen) peat (Cyperaceae)
6 charcoal layers
7 tree (carr peat (*Alnus, Betula*)
8 siderite (ferrous carbonate) lens
9 tree peat (*Pinus* stumps) (in places *Scheuchzeria* peat)
10 highly humified *Sphagnum* peat with *Eriophorum, Calluna*
11 *Scheuchzeria* peat
12 moderately humified *Sphagnum* peat
13 fresh *Sphagnum* peat
14 surface weathered layer
15 ancient wooden trackway

Figure 8.16 Part of an extensive section in peat of a large raised bog, Bourtanger Moor, Drenthe, The Netherlands. The stratigraphy shows that the wetland began as a moss-rich swamp. Peat deposition was interrupted by loess deposition and shrub growth. Then a sedge swamp persisted for a long time. Several fires occurred during this time. The swamp was invaded in places by trees (*Alnus* and *Betula*) that are tolerant of wet substrates, before undergoing a period of drying, when it was inhabited by pine forest. Then a period of *Sphagnum* dominance converted the mire into a raised bog, initially with *Eriophorum* sedge and *Calluna* shrub components, but later with less of these species. Surface oxidation and peat cutting for fuel have severely modified the bog in recent centuries (from Casparie 1972).

Tansley 1939, A. S. Watt 1947) to be endogenously determined (or autogenic) upgrade or building phases followed by downgrade or degradational phases. The subject is comprehensively reviewed by Barber (1981). He showed, by very careful stratigraphic analysis of peat profiles in Bolton Fell Moss, Cumbria, England, that the hypothesis of cyclic regeneration cannot be sustained. There is no evidence for downgrade phases of the postulated cycles, resulting from endogenous changes in the plant communities concerned. The relative positions and history of hollows and hummocks is determined by climatic variations which affect the growth of either the hummock-forming species (heather (*Calluna vulgaris*), bog cotton (*Eriophorum vaginatum*) and certain *Sphagnum* spp.) or the species of hollows (certain other *Sphagnum* spp.)

During wetter periods hollows with wet *Sphagnum* patches and pools increase in size. The hummocks also tend to extend upwards. During drier periods the hummocks spread, tending slowly to fill hollows, raise the mire surface and create drier substrates. The whole bog surface responds simultaneously to the climatic shifts. In effect, the hummock and hollow vegetation patterns are examples of fluctuation.

Barber's study and other European work on 'recurrence horizons' in bogs (oxidized peat layers caused during dry periods) demonstrate the strong effects of climatic variation (cf. P. D. Moore & Bellamy 1974). Strata indicating periods of flooding, siltation and fire (E. O. Campbell *et al.* 1973, Casparie 1972) show that allogenic interruptions have often affected the course of events. Although particular vegetation sequences may be modified for a time, peat accumulation usually resumes and earlier patterns are re-established. Nevertheless, very strong disturbance, usually effected by human agencies, sometimes causes severe erosion or the total destruction of mires (e.g. Conway 1954).

VEGETATION CHANGE PROCESSES IN WETLANDS

Various interactions of peat-forming plant species with water create the conditions for sequences of change in wetlands such as those from reedswamp to fen and finally bog (with or without an intervening woody vegetation phase). Among the autogenic processes are:

1 Progressive shallowing of water by peat accumulation.
2 Slowing and eventually prevention of free water inflow by the growth of emergent vegetation.
3 Invasion of open water by floating mats of plants.
4 Acidification of the water.
5 Uptake of nutrient ions by plants, and their sequestration in peat.
6 Build-up of peat and maintenance by it of a high water table; in the case of raised bogs this results in the independence of the site from a land-derived water supply.
7 Cutting-off of sites from access to fresh nutrient supplies.

As has been seen, rather unpredictable allogenic events may interrupt or reverse wetland vegetation sequences. As wetland development has often been proceeding for many thousands of years, it is likely that virtually all wetland sites will have been subject to various kinds of strong allogenic influence. Some of these, such as gradual climate change, or the slow downcutting, by stream erosion, of a ridge which is damming a wetland, may have had such subtle effects that they will be very hard to detect from the evidence in the sediments. Other allogenic influences, described above, are more easily detected.

Sphagnum and mire ecology

Sphagnum spp. are so often implicated in vegetation changes in wetlands that it is worthwhile to consider them in some detail (cf. Rigg 1940, 1951, P. D. Moore & Bellamy 1974). These mosses, of which there are some 350 species around the world, more-especially in temperate latitudes, have long, slender, leafy stems (Fig. 8.17). Propagation is largely vegetative, and *Sphagnum* spp. can reproduce from short branches and sometimes from mature leaves. Dispersal over long distances is probably mainly by accidental carriage of portions of leafy stem by aquatic birds – the mosses regenerate vigorously by vegetative means. Spores are formed from time to time, and are dispersed in the air and by water. Different *Sphagnum* spp. have different habitat preferences, with some favouring fully aquatic situations and others able to live on raised hummocks subject to periodic drying out. The different species also differ in their preferences for site fertility.

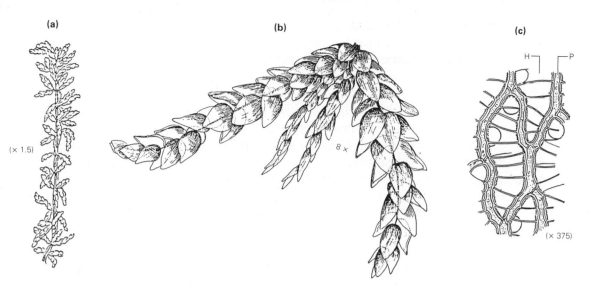

Figure 8.17 *Sphagnum papillosum*, a species from bogs of the northern temperate zone. (a) Habit drawing of a portion of leafy stem. (b) Detail of the end of a leafy stem. (c) Portion of a leaf showing the green photosynthetic cells and the larger, hollow hyaline cells with their pores. The hyaline cells enable *Sphagnum* plants to take up large quantities of water. (a) and (c) after Watson & Richards (1968), (b) after Landwehr & Barkman (1966).

The *Sphagnum* branches grow vigorously at the tip and die at the basal portion, which remains undecayed. Growth rates differ from species to species, and in different conditions. Annual productivity values cited by P. D. Moore & Bellamy (1974) range from 0.3 to 8.0 g dm^{-2} or more. Often the branches grow in crowded masses to the exclusion of other species; sometimes they are supported by the stems of other plants (sedges and shrubs). Underlying the living sward of growing branches, the dead material accumulates. The layers become compressed, and peat forms.

The special properties of *Sphagnum* spp., with respect to peat formation, development of raised peat domes and modifications in the chemical properties of mire water, reside in their leaves (Fig. 8.17). The leaves have large, dead, hollow, perforated hyaline cells which can take up and hold large quantities of water. Almost any of the species can absorb ten times its own weight of water, and some can hold twice that much (Rigg 1940). Living or dead *Sphagnum* causes water with which it is in contact to become acid (pH 3.0–5.0 and sometimes as low as pH 2.0). Acidification is by selective adsorption of the base cations which are exchanged for hydrogen ions (Clymo 1963, P. D. Moore & Bellamy 1974). Other mire plants also have high cation exchange capacity, especially at low pH.

Nutrients in mires

Peat accumulation progressively places nutrients beyond the reach of plants. The high cation exchange capacity of peat favours retention of ions. At the fertile end of the scale, in reedswamp and fens, this phenomenon is least marked because the plants and their litter contain relatively abundant nutrients. The litter decays readily, and further influxes of nutrients occur in the water. At the least-fertile end of the scale, in raised bogs, the lack of mineralization (because only a relatively small amount of oxidation of peat is possible near the surface, and none at deeper levels) permits only minimal recycling of nutrients. In any case the plants and their litter are low in nutrients and, since the only source of water is rainfall (and other airfall materials such as blown dust and plant litter), the influx of nutrients is low.

As the peat sequesters nutrients and there is always a continuous slight loss in drainage water, the fertility of peat-accumulating systems inevitably declines and acidity increases (Malmer 1975). The changed chemistry of the water is an important cause of replacements of plant species. The plant species of acid bogs have generally low nutrient requirements. Some are killed when given access to fertile, basic water; their metabolism is apparently unable to cope with the change in water chemistry (Sjörs 1950, Burrows & Dobson 1972, P. D. Moore & Bellamy 1974). Swamp plants, on the other hand, require neutral to high pH and are killed by acid, infertile water (P. D. Moore & Bellamy 1974).

Nitrogen unavailability is usually an important limiting factor in mires. The more-fertile mires have free-living *Azotobacter* and vascular plants such as *Alnus* and bog-myrtle (*Myrica*) with *Frankia* associations. *Myrica* and some free-living cyano-bacteria can also live in some acid, infertile mires. The *Frankia* in *Myrica* root nodules has been shown to fix nitrogen at pH down to about 3.8 (Bond 1951). *Myrica* is often found in floating sedge mats advancing into lakes. Decomposition of organic matter liberates ammonium ion, and many wetland plants absorb their nitrogen in this form (Etherington 1983). The bog heath plants are well-known for their use of ammonium rather than nitrate.

Some of the most bizarre means of plant nutrition are found among the carnivorous plants of acid bogs (sundews (*Drosera*), bladderworts (*Utricularia*), butterwort (*Pinguicula*), Venus' fly trap (*Dionaea*) and pitcher plants (*Nepenthes*); Lloyd 1942). Their supplies of nitrogen, phosphorus and cations are augmented by the small animals which they trap and digest (possibly with the aid of bacteria which provide digestive enzymes) (Fig. 8.9 & Table 3.12).

Phosphorus deficiency is also an important limitation for plants in acid, infertile mires (Tamm 1954, Gore 1961). Many bog plants, in contrast with fen plants, have low requirements for phosphorus. Some bog plants are known to have very efficient phosphorus economies, obtaining it with the aid of mycorrhizas or resorbing it from dying tissues, or both.

SOME CONCLUSIONS

Mires are relatively stable and permanent landscape features. They can be disturbed but, unless the disturbance is of the grossest kind, they tend to recover. The manifest importance of mires for the study of vegetation change in wetlands over long periods is clear. Careful analysis of mire stratigraphy and modern mire processes reveals good evidence for sequential change proceeding through autogenic processes, as well as allogenically initiated shifts in the direction of vegetation change, including fluctuations. The permanent fossil record in the peat and other sediments is extremely valuable.

In terms of the environmental complex and vegetation composition, each separate wetland (or even different parts of single large wetlands) tends to have a unique history. The combination of environmental factors at each site is distinctive and is reflected in distinctive vegetation patterns. Some western European mires have developed in lake basins through the complex sequence: open water → reedswamp → fen (and sometimes carr) → *Sphagnum*, sedge and heath bog. In eastern North America some mires have originated through the sequence: floating sedge mat → scrub → black spruce or tamarack. There are, however, many variations on themes. The complexity of some large mires makes it difficult to generalize about their history. Of course, some mire systems originate as relatively dry sites and become converted to peat bogs by the autogenic process of paludification.

Although wetland ecology is distinct from the ecology of terrestrial systems in various ways, some of the long-term vegetation change features in wetlands parallel those which occur in leaching environments on land. They result from the decline in nutrient stocks through dissipation in the drainage water, or sequestration in the peat 'sink'.

9

Changes in some temperate forests after disturbance

The rather delicate fabric of vegetation over the Earth's surface is easily damaged by natural forces or by human activities, but usually recovers. The sequences of vegetation change observed after disturbance prompted ideas about plant succession. The recovery phenomenon was termed **secondary succession** in contrast with the **primary succession** which proceeds on newly formed terrain (Chs 4–6). Clements (1904, 1916) defined secondary succession as the redevelopment of plant cover after partial denudation of vegetated sites. Because various propagules or juveniles of plants from the previous vegetation are still present, and because there are supplies of nutrients in the soil organic residues, re-establishment of plants was thought to be relatively rapid compared with colonization, *de novo*, on sterile sites. Vegetation recovery after disturbance, according to Clements, begins at one of the intermediate 'seral' stages and the vegetation ultimately returns to a close facsimile of the original.

It may be acknowledged that there are sequences of change in the plant populations of disturbed sites without necessarily accepting all that Clements said about such change. The early phases of vegetation development, at least, are readily observable within the working lifetime of one observer. Sometimes, also, the course of change over longer periods can be deduced from the population structure of the current vegetation, from the remains of plants on the site or from the composition of stands on a series of similar sites known to have been denuded at different times.

A great deal of observational and empirical research on vegetation change after disturbance has been done on the forests of eastern North America, so it is appropriate to take examples from there. There is some advantage in examining forests in the temperate zone, also, because there the ages of relatively long-lived woody plants reveal facts about the recent history of the populations.

THE MIXED FORESTS OF EASTERN NORTH AMERICA

This chapter considers mainly the vegetation which Braun (1950), in her classification of the deciduous angiosperm forests of eastern North America, placed in the most mature ('climax') associations *oak–pine*, *oak–chestnut*, *beech–maple* and *hemlock–white pine–northern hardwoods* (Figs 9.1–3). (An association in Braun's classification system is a large-scale community characterized by the physiognomic dominance of one or more species.) In these forests are assemblages of large and small tree species (Table 9.1 & Fig. 9.4). Shrubs, climbing vines and herbs, bryophytes, lichens and fungi are also common, but are not listed.

Figure 9.1 View over the canopy of mature mixed deciduous angiosperm forest, Great Smoky Mountains National Park, Tennessee, USA. Among the tree species present are oaks (*Quercus* spp.), beech (*Fagus grandifolia*), sugar maple (*Acer saccharum*) and tulip tree (*Liriodendron tulipifera*) (photo J. L. Vankat).

Table 9.1 Trees of the eastern North American forests and some of their attributes (mainly after Fowells 1965, Forest Service, US Department of Agriculture, 1948).

Name	Usual maximum height (m)	Approximate usual maximum age (years)	Flowering	Heavy seeding	Seedfall	Dispersal agency	Usual maximum dispersal distance	Seed dormancy etc.	Germination time, conditions	Shade tolerance of juveniles	Vegetative reproduction	Browsed by mammals	Other
Acer rubrum red maple	18–27	150	Feb.–May	nearly every year	Apr.–July	wind	100 m+	germinate immediately or may over winter	spring–early summer, low light, mineral soil, leaf litter	moderately tolerant	vigorous sprouter	heavily	
A. saccharum sugar maple	21–36	300	Mar.–May	2–5 year intervals	Oct.–Dec.	wind, squirrels	100 m+	winter dormant	spring, in shade or open, mineral soil, leaf litter	very tolerant	vigorous sprouter, root suckers	heavily, but resistant	
Betula alleghaniensis (= *B. lutea*) yellow birch	27–30	200	mid-Apr.–late May	1–2 year intervals	Nov.–Feb.	wind	hundreds of m to several km	dormant 6–18 months (need after-ripening)	spring, partial shade or open, moist mineral soil, rotten logs	moderately intolerant	stump sprouts	heavily	prefers moist sites
B. lenta sweet (or black) birch	15–20	150	mid-Apr.–May	1–2 year intervals	Sept.–Nov.	wind	hundreds of m to several km	dormant 12 months (if kept cool, moist)	spring, light, mineral soil, rotten logs	moderately intolerant to intolerant	stump sprouts	yes	
B. papyrifera paper birch	21–27	150	mid-Apr.–June	nearly every year	late Aug.–Sept. following Mar.	wind	hundreds of m to several km	dormant 8 months (if kept cool, moist)	spring, open, moist mineral soil, rotten logs	intolerant	vigorous sprouter	yes	
B. populifolia grey birch	10	50	late Apr.-late May	?	Oct.–mid-winter	wind	hundreds of m to several km	dormant 12 months (if kept cool, moist)	spring, open, moist mineral soil	very intolerant	vigorous sprouter	heavily	
Carya cordiformis bitternut hickory	30	250	Apr.–May	3–5 year intervals	Sept.–Dec.	gravity, water*	a few m to 100 m+	dormant over winter	spring, moist leaf litter, humus	moderately intolerant	vigorous sprouter (stump, root)		
C. glabra pignut hickory	24–27	250	early June	1–2 year intervals	Sept.–Dec.	gravity, squirrels*	a few m to 100 m+	dormant over winter	spring, moist humus, litter	intolerant	vigorous sprouter		seedlings frost tender
C. ovata shagbark hickory	39–42	250	Apr.–May	1–3 year intervals	Sept.–Dec.	gravity, squirrels*	a few m to 100 m+	dormant over winter	spring–early summer, moderately moist humus, litter	moderately tolerant	vigorous stump sprouter, root suckers	yes	
C. tomentosa mockernut hickory	30	250	May	2–3 year intervals	Sept.–Dec.	gravity, squirrels*	a few m to 100 m+	dormant over winter	spring, early summer, moist humus	intolerant but can recover from suppression	vigorous sprouter	yes	drought-sensitive; seedlings frost tender
Castanea dentata American chestnut	25–30 (but no living trees now)	350								tolerant	vigorous sprouter		extinct as a mature tree due to chestnut blight

Cornus florida	flowering dogwood	12	150	Apr.–May	2 year intervals	Oct.–Nov.	gravity, birds†	a few m to 100 m+	hard seed coat, dormant over winter (need after-ripening)	spring, moist humus	very tolerant	vigorous sprouter	yes	drought-sensitive
Diospyros virginiana	persimmon	15–22	200	Apr.–June	2 year intervals	Sept.–late winter	birds, mammals, water†	a few m to 100 m+	impermeable seed coat; dormant over winter (dry seeds 2–3 years)	spring, moist humus, litter	very tolerant	stump sprouts, root suckers	yes	
Fagus grandifolia	American beech	18–27	350	late Apr.–early May	2–3 year intervals	after first heavy frost	gravity, rodents	a few m to 30 m+	dormant 1 year, after-ripening	spring, mineral soil, litter	very tolerant	root suckers, layering	slightly	
Fraxinus americana	white ash	21–27	150	Apr.–May	3–5 year intervals	?	wind	150 m+	dormant, need after-ripening 2–3 months, winter chilling, punctured seedcoat	open, full light, mineral soil	intolerant but recover from suppression	stump sprouts	yes, sensitive	
Gleditsia triacanthos	honey locust	22–30	150	May–June	?	Sept.–late winter	wind, birds, mammals	?	dormant, (impervious seed coat)	need scarification or passage through bird gut, open, mineral soil	intolerant	vigorous sprouter	yes	
Juniperus virginiana	eastern red cedar	12–15	300	mid-Mar.–mid-May	2–3 year intervals	Feb.–Mar.	birds, mammals†	a few m to 100 m –	dormant 1–2 years, (need after-ripening) hard seed coat	early spring of second or third year, moist mineral soil, well lit	intolerant	none	no	drought-resistant
Liquidambar styraciflua	sweet gum	20–33	200	Mar.–May	3 year intervals	Sept.–Nov.	wind	180 m	dormant over winter	moist, heavy mineral soil, well lit	intolerant	basal sprouts (young trees)	yes	buds are frost tender
Liriodendron tulipifera	tulip tree	30–35	300	Apr.–June	irregular intervals	Oct.–Jan.	wind	300 m	dormant over winter or longer	light shade, moist mineral soil	intolerant	vigorous stump sprouts	heavily	drought-sensitive
Nyssa sylvatica	black gum (tupelo)	20–30	200	Apr.–June	?	Sept.–Oct.	birds, small mammals, water†	a few m to 100 m+	dormant over winter	moist, mineral soil, light shade	moderately intolerant	weak root suckering		
Picea rubens	red spruce	18–30	400	Apr.–May	3–8 year intervals	Sept.	wind	70 m	some germinate immediately; others over winter	moist mineral soil, humus, thin litter, in shade	tolerant	none	no	affected by spruce budworm
Pinus echinata	shortleaf pine	24–30	80–100	Mar.–Apr.	3–10 year intervals	Nov.–Dec.	wind	400 m	some germinate immediately; others over winter	spring, light shade, then full light, moist mineral soil	intolerant	young trees sprout after damage	no	

Table 9.1 continued

Name		Usual maximum height (m)	Approximate usual maximum age (years)	Flowering	Heavy seeding	Seedfall	Dispersal agency	Usual maximum dispersal distance	Seed dormancy etc.	Germination time, conditions	Shade tolerance of juveniles	Vegetative reproduction	Browsed by mammals	Other
P. resinosa	red pine	22–28	200	Apr.–June	3–7 year intervals	Sept.–summer	wind	200 m	dormant 1–3 years if dry	open light shade, mineral soil, ash	moderately intolerant	none	no	
P. strobus	eastern white pine	30–40	400	Apr.–June	3–5 year intervals	Sept.–Oct.	wind	250 m	dormant	full sun to light shade, moist mineral soil, moss	moderately intolerant	none	no	
P. taeda	loblolly pine	28–35	90–120	Mar.–Apr.	annual to irregular	Oct.–Spring	wind	120 m to 1 km +	short dormant period	full sun, mineral soil	intolerant	young trees sprout after damage	no	
Populus grandidentata	bigtooth aspen	10–12	60–70	Apr.–May	4–5 year intervals	May–June	wind, water	hundreds of m to several km	viable only for a few weeks	moist soil	very tolerant	vigorous stump sprouts, root suckers	yes	
P. tremuloides	quaking aspen	20–27	80–100	Apr.–May	4–5 year intervals	May–June	wind, water	hundreds of m to several km	viable for a few weeks, longer if cool, moist	moist soil	intolerant	vigorous stump sprouts, root suckers	yes	most seeds are inviable
Prunus serotina	black cherry	20–30	150	Mar.–June	3–4 year intervals	June–Oct.	gravity, birds†	a few m to 100 m +	need after-ripening over winter; dormant long periods if buried	full sun, mineral soil or litter	moderately intolerant	vigorous stump sprouts	yes	
Quercus alba	white oak	30	400	Apr.–May	3 year intervals	Sept.–Oct.	gravity, squirrels, jays*	a few m to 100 m +	germinate soon after falling	autumn, full sun, moist, light leaf cover	moderately tolerant	vigorous stump sprouts, root stools		
Q. prinus	chestnut oak	18–25	150	Apr.–May	4–7 year intervals or less	Nov.	gravity, squirrels, jays*	a few m to 100 m +	germinate soon after falling	autumn, light shade, moist, shallow litter	moderately tolerant	vigorous stump sprouts		
Q. rubra	northern red oak	21–28	150	Apr.–May	2–5 year intervals	Nov. of 2nd year	gravity, squirrels, jays*	a few m to 100 m +	dormant over winter	spring, soil with moist leaf cover	moderately tolerant	vigorous stump sprouts		
Q. velutina	black oak	18–25	150	Apr.–May	2–3 year intervals	Nov. of 2nd year	gravity, squirrels, jays	a few m to 100 m +	dormant over winter	spring, mineral soil, light, moist leaf cover	moderately tolerant	vigorous stump sprouts		

Robinia pseudoacacia	black locust	50–75	20–30	May–June	1–2 year intervals	Sept.–Apr.	gravity	a few m	dormant, hard seed coat (if buried retain viability at least 20 years)	spring–early summer; need scarification, open, mineral soil	intolerant	vigorous sprouts, extensive root suckers		
Sassafras albidum	sassafras	75–100	15–30	early spring	1–2 year intervals	Sept.–Oct.	birds, water†	a few m to 100 m +	usually dormant over winter	spring, litter over humus	intolerant	vigorous root and stump sprouts		
Tilia americana	basswood	100	30–35	June–July	nearly every year	late autumn–spring	wind, rodents	180 m	hard pericarp, seed coat, dormant up to 4 years	spring–early summer mineral soil	tolerant	very vigorous stump sprouts	yes	
Tsuga canadensis	eastern hemlock	600	30–35	May–June	2–3 year intervals	Sept.–winter	wind	up to 1.5 km	usually dormant over winter, or longer	spring shaded, moist, cool litter, rotten logs, mineral soil, moss	very tolerant	none	sometimes	suppressed juveniles can survive for long periods
Ulmus americana	American elm	200	15–20	Apr.	most years	June	wind, water	100 m +	may germinate soon after falling or over winter	autumn–spring, moderate shade, mineral soil	moderately intolerant	stump sprouts, root suckers	yes	suffers from Dutch elm disease

* Heavy seeds > 2.5 g.   † fleshy fruit   ? Information lacking.

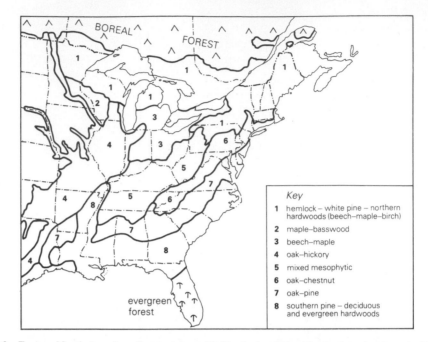

Figure 9.2 Eastern North American forest regions. (1) Hemlock–white pine–northern hardwoods (Beech–Maple–Birch). (2) Maple–basswood. (3) Beech–maple. (4) Oak–hickory. (5) Mixed mesophytic. (6) Oak–chestnut. (7) Oak–pine. (8) Southern pine–deciduous and evergreen hardwoods. After Braun (1950).

Figure 9.3 Summer view of the interior of mature mixed deciduous angiosperm forest, Frederick County, Maryland, USA. Prominent trees include white oak (*Quercus alba*), red oak (*Q. rubra*) white ash (*Fraxinus americana*) and hickory (*Carya* spp.) (photo US Forest Service).

The terrain occupied by this complex of forest types is quite varied. It includes the lower St Lawrence Valley; the ice-smoothed, ancient, metamorphic rocks of the southern rim of the Canadian Shield and the younger, hard sedimentary and metamorphic rocks of the Adirondack and Catskill Mountains; the complex of ridges, valleys and plateaux of the Appalachian Mountain chain, developed on folded, hard, metamorphic, plutonic and sedimentary rocks, and the dissected low plateau of the eastern Piedmont, underlain by hard metamorphic and plutonic rocks. Some of the mountain summits rise above 1800 m, but the mixed forests occur on the lower slopes (up to about 1000 m).

The northern part of the region, south to the Pennsylvania–New York border was, during the last (Wisconsin) glaciation, covered by an ice sheet which left a relatively thin veneer of till. The present pattern of forest distribution originated as tree species migrated northward at the end of the last glaciation (between 14 000 and 10 000 years ago) and some of the forest associations seem to be relatively recent (M. B. Davis 1981). For example, the establishment of beech (*Fagus grandifolia*), maples (*Acer* spp.) and hemlock (*Tsuga canadensis*) in the New York–New England–St Lawrence lowland area was 9000–4000, 10 000–8000 and 11 000–6000 years ago, respectively.

In the northern part of this region, as far south as northern Virginia and Kentucky, the soils are alfisols (grey-brown podsolic types). Further south is an area of ultisols (red-yellow podsolic soils). The grey-brown soils have more organic matter, a pH of 5–6 and are relatively fertile. The red-yellow soils are older, more-acid and weathered. However, the soils show wide differences in qualities such as depth, texture, relative fertility and moisture content.

Several major gradients in climate are evident. Precipitation increases from about 90 cm per annum in the west to about 125 cm in the east. It also increases from north (75 cm) (where more of it falls in the form of snow) to south. Precipitation is relatively evenly spread through the year, but there is occasional summer drought. Low temperatures in the north (120-day growing season, heavy winter frosts, minimum temperatures to $-30°C$, maximum to $35°C$ or more, snowfalls early December to mid-March) grade into the warmer southern conditions (an approximately 200-day growing season; only about 50 per cent of years with snowfalls).

The pattern of forest structure and composition is quite varied. Many of the main dominant tree species are widespread (Fig. 9.2), but their proportions differ in the different associations. In upland areas, relatively undisturbed for several hundred years, there may be stands with a dense overstorey of old trees of species such as beech, sugar maple (*Acer saccharum*) and hemlock, with stems 1 m or more in diameter and the first branches 15 m or more above the ground. On good sites the trees may be 30 m tall or more. In such forests the deep shade inhibits most undergrowth except for suppressed juveniles of the dominant trees and some shrubs and herbs. Otherwise relatively mature forests (which have grown up after various kinds of disturbance in the 19th century or earlier) are usually shorter (25 m or more high and with smaller stem diameters). They have more-varied composition; the canopy species in any stand will include several of the species drawn from those indicated in Table 9.1. Smaller trees such as dogwood (*Cornus florida*), hornbeam (*Carpinus caroliniana*), hophornbeam (*Ostrya virginiana*), redbud (*Cercis canadensis*)

Figure 9.4 Plants of the eastern North American forests. (a) Quaking aspen, *Populus tremuloides*. (b) Eastern red cedar, *Juniperus virginiana*. (c) Black cherry, *Prunus serotina*. (d) Sweet gum, *Liquidambar styraciflua*. (e) Tulip tree, *Liriodendron tulipifera*. (f) Black gum, *Nyssa sylvatica*. (g) Red maple, *Acer rubrum*. (h) Red oak, *Quercus rubra*. (i) White oak, *Q. alba*. (j) Chestnut oak, *Q. prinus*.

(k) Loblolly pine, *Pinus taeda*. (l) White pine, *P. strobus*. (m) Eastern hemlock, *Tsuga canadensis*. (n) Yellow birch, *Betula alleghaniensis*. (o) Sugar maple, *Acer saccharum*. (p) Flowering dogwood, *Cornus florida*. (q) Beech, *Fagus grandiflora*. (r) Chestnut, *Castanea dentata*. (s) Shagbark hickory, *Carya ovata*. (t) Basswood, *Tilia americana*.

and witch hazel (*Hammamelis virginiana*) occur below the canopy. There are shrubs from genera such as *Amelanchier*, *Prunus*, *Rhus*, *Rubus*, *Sambucus*, *Vaccinium* and *Viburnum*; vines such as virginia creeper (*Parthenocissus vitacea*), poison ivy (*Rhus radicans*) and grapes (*Vitis* spp.); many herb species and some ferns. There may be some large stands with relatively uniform composition, but mosaics of smaller, relatively monospecific or mixed stands of differing composition are common.

In winter the ground in the deciduous forests is carpeted with dead leaves; relatively little bare soil is exposed. A few herb species (some orchids, *Viola* spp.) are in leaf through the winter, flower in spring and die back in summer. A few annuals (species of *Floeckea* and *Galium*) germinate in spring and die down in summer. In early spring, after snow melt, many other perennial spring herbs send up flowers and leaves from underground root-stocks. Species from genera such as *Actaea*, *Anemone*, *Arisaema*, *Claytonia*, *Dicentra*, *Erythronium*, *Hepatica*, *Sanguinaria* and *Trillium* are among them. These early-flowering species take advantage of the light allowed through by the leafless trees. As the leaf layer unfolds on the trees in late spring, the light intensity at ground level diminishes and many of the herbs die down. Fern species (*Dryopteris*, *Athyrium* and *Adiantum*) sprout from root stocks then. Other perennial herbs grow to form a summer field layer and flower at intervals, then, until early autumn. They include species of *Aster*, *Eupatorium*, *Maianthemum*, *Trientalis*, *Uvularia* and grasses (*Danthonia* and *Oryzopsis*). All die down in winter. Bright colours develop in the leaves of some of the trees in the autumn as temperature and daylength decline, and finally the leaves are shed before the first of the winter snow.

Braun's (1950) study was done during the 1930s and 1940s when many old-growth forest stands still existed. They have been considerably reduced since then. In fact, as is seen later in the chapter, the forests of eastern North America have been very extensively disturbed by natural and human-induced events. The region has been inhabited by Indians since the ice receded at the end of the Wisconsin Glaciation and by Europeans since the 17th century AD. Fire, selective logging, clear-cutting, agriculture and various other kinds of disturbance, followed by forest regrowth, have contributed to the very complex patterns of forest composition and structure.

Braun and many other ecologists, writing of the eastern North American forests, couched their descriptions of forest development after disturbance in terms of the succession to climax concepts promoted by Clements (1916, 1936) and modified by other workers (Nichols 1923, Lutz 1928, Oosting 1956). Egler (1954) pointed out some field evidence which seems to conflict with the orthodox views on succession. The views of Gleason (1926, 1939), Whittaker (1951) and Curtis (1959) are also in marked contrast to prevailing theory. Other authors (e.g. Raup 1964, 1981, P. S. White 1979, Oliver 1980) have recently questioned the general validity of the succession to climax ideas. They regard as inappropriate the concept of a 'normal' sequence of seral stages, culminating in a stable, self-perpetuating climax, in systems where there is evidence that disturbance is a quite normal phenomenon which may, in fact, be necessary for maintaining certain kinds of vegetation composition and structure, more or less permanently. In later chapters these divergent views are considered at greater length.

In the meantime, the stance will be taken here that, after a major disturbance episode in the eastern forest, provided that there is no further disturbance, there will be a continuous sequence of species replacements culminating, after several centuries, in a relatively stable or only slowly changing community (Bormann & Likens 1979b). Stands of old-growth forest are mainly confined now to a few, upland localities. The relatively non-committal terms **pioneer**, **transitional** and **mature** will be applied to developmental phases of the vegetation. Pioneer species are early colonists on disturbed sites, transitional species are those which replace the pioneers in a simple, or a more-complex sequence, but occupy sites for only one generation. Mature forest species are those which replace the transitional species and form a forest that, in the absence of disturbance, is capable of retaining the same general species composition for more than one generation.

THE DIFFERENT SCALES OF DISTURBANCE

The main disturbing agencies in the eastern North American forests are listed in Table 9.2. Disturbances occur over a wide range of magnitudes. Those considered in most detail are:

1 Weakening and death of single trees or small patches.
2 Widespread elimination of a prominent species.
3 Weakening and death of numerous individuals, or patches, over areas of several to many hectares, but maintenance of the general character of the forest.
4 Virtually complete destruction of large areas of forest caused by major disturbing factors.

Death of single trees or small patches

A normal phenomenon in maturing and mature forest is the opening of gaps in the forest canopy as individual trees are weakened or killed by natural causes. Perennial species seem to have characteristic lifespans and the series pioneer → transitional → mature is one where most of the species in each phase have progressively greater potential longevity. Trees of any age may die of stress brought about by environmental variation (e.g. drought) or competition. Old trees usually die of the accumulated weakening effects of damage by wind, ice, wood-borers, fungal rots and stress. The rots usually enter trees through damaged areas, so that longevity appears to be conferred by a combination of features such as wood strength, thick bark and chemical attributes which make the wood unpalatable to grazers or resistant to fungal attack.

Leaf canopies of ailing trees become less dense, so more light reaches the forest floor beneath them in summer. Canopy gaps are formed by trees which die and remain standing for a time, or by falling live trees, which often carry others with them. The consequent increase in light and reduced root competition allow the establishment and growth of juveniles including those of species other than the canopy dominants.

Table 9.2 Main disturbing agencies in eastern North American forests.

frost
hailstorm
snow breakage
ice glaze storm
lightning strike
drought
wind (including tornado and hurricane)

defoliating insects
wood-boring insects
browsing deer
seed predators (insects and squirrels)
fungal pathogens

fire

logging (including selection logging and clear-cutting)

acid rain

In mature beech–sugar maple forest in New Hampshire, Forcier (1975) showed that gap formation leads to the establishment of individuals or groups of yellow birch (*Betula alleghaniensis*), which are subsequently replaced by sugar maple. In turn, sugar maple is replaced by beech, originating either as seedlings or as sprouts from beech stumps (Fig. 9.5). Forcier found positive associations of sugar maple juveniles with canopy yellow birch and beech juveniles with canopy sugar maple. Conversely, there is a negative association between each of yellow birch and sugar maple canopy trees and saplings of the same species, except where the juveniles are of sprout origin.

Figure 9.5 Patterns of forest regeneration in mature beech–sugar maple forest in New Hampshire, USA, after small-scale and larger-scale disturbances (after Forcier 1975).

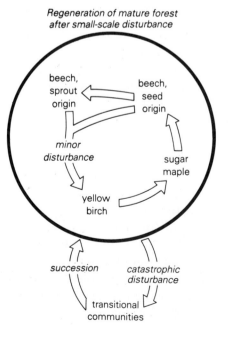

310

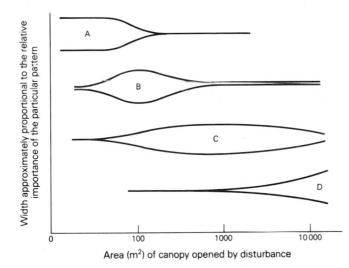

Figure 9.6 Gap-filling patterns in New Hampshire forests according to the size of the canopy gaps. (A) Canopy closure by lateral branch growth of established canopy trees or height growth of subcanopy trees or both. (B) Height growth of hitherto-suppressed juveniles of seed or sprout origin (*Fagus grandifolia*, *Acer saccharum* and *Tsuga canadensis*). (C) Height growth by new juveniles of seed origin (*Acer rubrum*, *Betula alleghaniensis*, *Fraxinus americana*, *Liriodendron tulipifera*, *Quercus rubra* and *Prunus serotina*). (D) Height growth by new juveniles of seed origin (*Betula populifolia*, *B. papyrifera*, *Populus tremuloides*, *P. grandidentata* and shrubs). Modified from Bormann & Likens (1979a).

Seedlings of yellow birch are most common. Its seeds are abundant, produced at 1–2-yearly intervals, very light and widely distributed by wind (Table 9.1). Although the seedlings tolerate some shade, they are relatively light-demanding. Their populations fluctuate considerably from year to year. Yellow birch trees are faster-growing and shorter-lived than sugar maple or beech. The latter have less-frequent heavy seed years, and their seeds are heavier, less abundant and less-widely distributed than those of the birch. Their seedlings are very tolerant of shading, and the seedling populations (especially those of beech) remain numerically stable.

The regeneration properties and relative shade tolerance of the juveniles of the tree species explain the cycle of change after gap-formation. Other similar cycles are evident in forests with different composition in eastern North America and elsewhere (A. S. Watt 1947, Bray 1956, Fox 1977, Lorimer 1977, Vankat *et al.* 1975, Denslow 1980a, Barden 1979, 1980). Bormann & Likens (1979a) outline the general patterns of gap-filling at Hubbard Brook, New Hampshire. There are differences according to the sizes of the gaps (cf. Shugart 1984) (Fig. 9.6). Similar effects could arise after selective logging of individual trees (e.g. white pine (*Pinus strobus*) or hemlock (*Tsuga canadensis*)).

Elimination of a species

Sometimes a species is selectively removed from vegetation by natural causes. Chestnut (*Castanea dentata*) trees have virtually been eliminated from eastern North American forests by chestnut blight and elms (*Ulmus* spp.) have been badly affected by Dutch elm disease. The case of chestnut blight is considered here.

The cause of the blight is a fungous disease, *Endothia parasitica*, accidentally introduced to North America from Asia, in about 1900, on Asiatic chestnut nursery plants and first noticed, in New York city, in 1904. By then it was well-established and apparently uncontrollable. The fungal conidiospores are carried by insects, birds and mammals, and the ascospores are wind-borne. Spores germinate in bark wounds and the fungal hyphae grow through the bark, cambium and outer sapwood. The fungus then spreads concentrically, girdling the stem. The death of a tree is not sudden. It occurs on individual branches and it may take 2–10 years for the whole tree to die (F. W. Woods & Shanks 1959). The dead trees stand for a time until they are weakened by borers and rots. The fungus is a parasite or saprophyte on a number of other tree species (e.g. oaks), but it is never as virulent on them as it is on chestnut.

Sprouts arise from the stumps of killed chestnut stems and these have continued to maintain populations of the species, but as they begin to grow into tall saplings and pole trees they, in turn, are killed by the pathogen. By 1935 mature chestnut had been killed throughout most of the eastern forests (its range had been from Tennessee to Canada, east of the Mississippi River) and its elimination, as a tree, was complete by the 1950s (Braun 1950, F. W. Woods & Shanks 1959).

Chestnut had been one of the most important canopy dominant tree species in several of the forest associations of the region, in particular the oak–chestnut variants of the mixed mesophytic forest type of the Appalachian Mountains. It often formed 20–35 per cent of the total number of trees and 20–30 per cent of the canopy cover. In some stands it was the sole dominant. Trees were up to 35 m tall, often with a stem diameter 1 m or more and a canopy spread 20 m. The trees lived for 250 years or more. Thus, the dead trees left large canopy gaps or extensive spaces where no tall trees remained.

This unfortunate episode of chestnut demise provided unique opportunities to examine the effects that one major forest dominant had had on its associate tree species in the various forest types. Studies were done by Korstian & Stickel (1927), Aughenbaugh (1935), Keever (1953), Nelson (1955), F. W. Woods & Shanks (1959), N. F. Good (1968), McCormick & Platt (1980), Phillips & Swank (1982) and Parker & Swank (1982). The usual method was to examine the composition and dimensions of the tree stands which were growing back into the gaps left by chestnut death. In a few instances statistics were available from permanent study sites established before, or very soon after, the advent of the pathogen (cf. Parker & Swank 1982). The general results from these studies are summarized in Table 9.3. Some quantitative results are included in Tables 9.4 and 9.5.

The most edifying data are those of Keever (1953), Nelson (1955), F. W. Woods & Shanks (1959), McCormick & Platt (1980) and Parker & Swank (1982). A common theme is that adjustments took place slowly and were not yet complete after 30 years or more. The slow death of chestnut trees, so that gaps were only gradually created, meant that some gap-filling was done by increase of the canopies of adjacent trees. Otherwise the gaps were occupied by the growth of young trees arising from populations of suppressed juveniles, or from new seedlings. In some instances there was an increase in the populations of species that are normally present as pioneers or early transitional species in the local forest succession. Later these were usually replaced by other species which normally are later transitional or mature forest dominants or subdominants. A case in point, in Nelson's (1955) study in North Carolina, USA, was black locust (*Robinia pseudoacacia*) which increased from 1934 to 1941 but then declined rapidly to 1953, under attack by insect pests specific to it. Tulip tree (*Liriodendron tulipifera*), normally a transitional forest species, was a

Table 9.3 Patterns of forest development in eastern North America after death of adult chestnut caused by chestnut blight.

Locality	Time since chestnut died/original forest dominants	Nature of study site	Main species increasing	Other important increasers	Author(s)
North Carolina	c. 1930–1953 *Castanea dentata Quercus rubra Q. alba*	6 stands; oak forest with dead standing chestnut (formerly up to 40% of canopy trees)	*Quercus prinus Q. rubra* (predicted *Quercus–Carya* dominance)	*Quercus alba, Acer rubrum, Betula lenta, Pinus strobus, Robinia pseudoacacia, Oxydendrum arboreum, Quercus coccinea, Liriodendron tulipifera, Nyssa sylvatica, Sassafras albidum, Halesia carolina.*	Keever 1953
western North Carolina	c. 1935–1953 *Castanea dentata Quercus prinus Q. rubra Q. velutina Carya* spp.	large forest area, 17 1/5-acre plots; oak forest with hickory and dead chestnut (adult chestnut finally died out c. 1945)	*Liriodendron tulipifera Carya* spp. *Quercus velutina*	*Betula lenta, Oxydendrum arboreum, Magnolia acuminata, Betula alleghaniensis, Tsuga canadensis, Quercus alba*	Nelson 1955
western North Carolina and eastern Tennessee	c. 1925–1958 *Castanea dentata Quercus prinus Q. rubra Q. alba Q. coccinea Carya* spp.	2569 openings caused by chestnut death; oak forest with hickory	*Quercus prinus, Q. rubra, Acer rubrum* (predicted *Quercus* dominance)	*Tsuga canadensis, Halesia carolina, Oxydendrum arboreum, Robinia pseudoacacia, Quercus coccinea, Liriodendron tulipifera, Betula lenta, Quercus alba, Fagus grandifolia, Cornus florida*	F. W. Woods & Shanks 1959
Virginia	c. 1920–1970 *Castanea dentata Quercus prinus Q. rubra, Q. alba*	56 100 m² quadrats; oak forest (chestnut stump sprouts abundant)	*Quercus rubra Q. prinus Carya glabra*	*Quercus alba, Acer rubrum, A. saccharum, Hamamelis virginiana, Fraxinus americana, Betula lenta, Cornus florida*	McCormick & Platt 1980

Table 9.4 Changes in forest composition in relation to chestnut death, Coweeta Basin, North Carolina[*], (after Nelson 1955).

	Stem number per hectare (all stems 1.3 cm diameter or larger)			Basal area per hectare (m^2)		
	1934	1941	1953	1934	1941	1953
Castanea dentata	1146.8	897.5	101.8	12.2	8.8	0.2
Carya spp.	341.5	380.0	450.7	4.3	3.7	4.7
Quercus prinus	187.5	182.4	152.7	2.4	2.3	3.3
Q. rubra ⎤[†]	234.7	240.4	237.0	2.2	2.2	1.2
Q. velutina ⎦				2.2	2.4	4.1
Q. alba	65.5	69.7	82.0	1.0	1.1	0.6[‡]
Q. coccinea	11.6	34.8	26.9	0.3	0.5	1.6
Aesculus octandra	20.3	21.0	21.7	1.0	1.0	1.2
Robinia pseudoacacia	90.9	119.8	40.8[§]	0.9	0.8	0.5
Liriodendron tulipifera	235.5	350.4	375.8	0.7	1.3	3.0
Acer rubrum	140.8	138.6	139.6	0.4	0.6	0.8
Betula lenta	37.0	37.0	62.5	0.4	0.5	0.7
miscellaneous	335.3	379.3	371.6	1.4	1.7	1.8
total	2846.6	2851.0	2063.3	29.6	26.9	23.6

[*] Permanent survey lines established 1934 in a watershed logged 1918–1923. Data are from 17 1/5-acre (0.0809 ha) plots.

[†] Most changes in numbers of red and black oak are for plants less than 12.7 cm diameter. These young plants are difficult to distinguish. Black oak apparently increased.

[‡] Decline in white oak basal area through loss of old trees.

[§] Black locust decline caused by leaf miner moth larvae.

N.B. Death of chestnut was gradual. Crowns of established canopy trees expanded to fill gaps so that relatively few major openings occurred.

Table 9.5 Changes in forest composition in relation to chestnut death, Great Smoky Mountains, North Carolina–Tennessee, USA (after F. W. Woods & Shanks 1959). Relative abundance of major tree species replacing chestnut (5046 individuals in a total of 2569 gaps left by chestnut death).

	No. of individuals	Percentage of total	Percentage of gaps occupied
Quercus prinus	859	17	24
Q. rubra	828	16.4	24
Acer rubrum	678	13.4	22
Tsuga canadensis	287	5.7	9
Halesia carolina	231	4.5	7
Oxydendrum arboreum	228	4.5	8
Robinia pseudoacacia	204	4	7
Quercus coccinea	192	3.8	6
Liriodendron tulipifera	181	3.6	6
Betula lenta	148	2.9	5
Quercus alba	116	2.3	4
Fagus grandifolia	99	2.0	3
Cornus florida	97	1.9	4

Gap filling was: (1) by canopy closure by established canopy trees in sites where chestnut formed up to 30–40 per cent of canopy (Quercus prinus, Q. rubra and Tsuga canadensis); (2) by growth of hitherto-suppressed juveniles (Tsuga canadensis and Acer rubrum); (3) by seedling establishment subsequent to death of chestnut, especially where chestnut was predominant (Acer rubrum, Liriodendron tulipifera, Betula lenta, Tsuga canadensis and Robinia pseudoacacia; the last of these increased, then declined after leaf miner and locust borer attack).

consistent increaser in chestnut gaps, after 30 years, but oaks and hickories were very important in the same sites. Parker & Swank (1982), reporting further work on these sites, show that responses to the elimination of chestnut vary with the site and depend also on the previous species composition of the stands.

Keever (1953) predicted the slow development of oak–hickory mountain forests in North Carolina from the oak–chestnut forests originally studied in 1939. By 1970, on these sites, pignut hickory (*Carya glabra*), formerly a subcanopy tree, had replaced the chestnut to become co-dominant with red oak (*Quercus rubra*) and chestnut oak (*Q. prinus*). Red maple (*Acer rubrum*) has also become important. At lower altitudes white ash (*Fraxinus americana*), hickories and sugar maple had all increased in forests where chestnut was once common (McCormick & Platt 1980). The competitive impact of chestnut on other species can be gauged from four kinds of evidence in the post-blight forest.

1   Expansion of tree species such as oaks and hickories to fill gaps in the canopies of adults.
2   Increase in population sizes of species which were part of the mature immediate pre-blight forest. Red oak, chestnut oak and red maple are consistently among these, but various other species are also involved.
3   Enhanced conditions for growth of trees adjacent to gaps, evident from the substantial increase in annual ring-widths of species such as red oak, chestnut oak and hemlock. In the Woods & Shanks (1959) study this growth increase occurred in the period 1925–1940, after which radial growth returned to normal, as competition from the established trees again imposed limitations.
4   Entry to some of the former chestnut-dominated stands of numbers of species not previously present, e.g. in North Carolina, sourwood (*Oxydendrum arboreum*), *Magnolia acuminata*, sweet birch (*Betula lenta*), yellow birch, hemlock (Nelson 1955); or sugar maple, beech and hemlock (Woods & Shanks 1959).

The means by which chestnut excluded other species seems simply to have been by shading and root competition. A search for possible phytotoxic effects of chestnut leaf and wood, by N. F. Good (1968) proved negative. She suggested that the post-chestnut death forest changes which she observed in New Jersey were complex and superimposed on longer-term trends from oak to maple–beech forest dominance, arising from forest recovery after logging.

Computer simulations by Shugart & West (1977) and Emanuel *et al.* (1978) attempted to predict future changes in forests in Tennessee, USA, in which chestnut previously had formed 20–25 per cent of canopy cover. The present prominence of chestnut oak, black oak (*Q. velutina*), white oak (*Q. alba*), hickories, maples, black gum (*Nyssa sylvatica*) and tulip tree was a likely event indicated by the simulations. The simulations suggest that white oak and chestnut oak will continue to increase.

315

Weakening and death of numerous individuals over extensive areas

If the forest canopy or any of the regeneration phases suffers damage, but most adult trees remain alive, thus preserving much of the physical structure of the forest, various changes in species composition will be set in train. Causes of damage of these kinds include climatic events such as strong winds, severe unseasonal frosts, heavy, wet, clinging snow, hailstorms, ice glaze storms and drought, as well as epidemics of defoliating or cambium-eating insects, or fungal pathogens such as rusts or root-rots, or irruptions of the populations of browsing mammals such as deer, which feed on the forest understorey plants. Epidemics of rodents or birds that remove much of the seed crop of one or more species may also have repercussions for future populations. The forest management literature cites many instances of damage caused by such agencies.

DEFOLIATING INSECTS

From time to time, for reasons that are not well-understood for many of the animals concerned, some insect species whose larvae feed on the leaves of forest trees undergo population increases of epidemic proportions. Moths are especially important in this respect. The larvae occur in such vast numbers that trees may be stripped of most or all of their leaves. This can kill some individuals, particularly juveniles. Survivors respond by producing another crop of leaves. Their vigour is diminished and this, in turn, can affect their competitiveness or their ability to cope with dry spells or fungal pathogens.

One major defoliator in the eastern forests is the tent caterpillar, larva of the moth *Malacosoma disstria*. It feeds on angiosperms, especially quaking aspen (*Populus tremuloides*), but also species of oak, black gum and sweet gum (*Liquidambar styraciflua*), black cherry (*Prunus serotina*) and sugar maple. Several outbreaks have occurred in the eastern forests since 1930. Although tent caterpillars can completely defoliate trees, not many die. However, the weakening effect is severe. An outbreak lasting 3–4 years can cause as much as a 70 per cent reduction in basal area increment of aspen in the first year and 90 per cent in the second year (W. L. Baker 1972). Spurr (1941) noted the near elimination of black cherry in transitional forest stands in Harvard Forest as a result of tent caterpillar defoliation, in conjunction with competitive suppression.

G. R. Stephens (1981) indicates the effects on tree growth and mortality of epidemics of gypsy moth *Porthetria* (*Lymantria*) *dispar* in Connecticut forests in the 1960s and 1970s. These insects escaped from captivity after being introduced from Europe in 1869. They now occur throughout most of New England, southeastern Canada and part of Michigan. Their voracious larvae have a wide host range on angiosperms and some gymnosperms, but prefer the oaks, grey birch and the aspens. Changes in the forest composition are ensuing (Collins 1961), but not solely as a direct result of defoliation. Rather, the loss of leaves renders trees susceptible to other, more-lethal insects (the beetle, two-lined chestnut borer (*Agrilus bilineatus*)) and to the deadly honey fungus (*Armillariella mellea*). In Connecticut, mature oaks (especially white and chestnut oak), but only small maples and birches, died in the year after defoliation. The outcome will be a more-rapid increase of maples and birches in the mature forest.

DAMAGE BY MAMMALS AND BIRDS

Vertebrate animals may have selective effects in forests, by feeding on the juveniles or seeds of particular species. The large seeds of hickories, oaks and beech suffer periodic heavy predation by squirrels (*Sciurus* spp.) and jays (*Cyanocitta cristata*), although they are also dispersed by those animals (Fowells 1975, Stapanian 1986). Hough (1965) found, in a forest in Pennsylvania, USA, that heavy browsing by white-tail deer (*Odocoileus virginianus*) was killing hemlock and *Viburnum* under-storey in a hemlock–beech–sugar maple forest. Beech is resistant to browsing, and it reacts to damage by producing numerous root sprouts. Sugar maple is somewhat resistant, but yellow birch and several other tree species are often severely affected (N. A. Richards & Farnsworth 1971, Tierson *et al.* 1966, Marquis 1974, Fowells 1965, Crouch 1976). If browsing is sufficiently heavy, then virtually all saplings, even of normally unpalatable species, may be killed (Whitney 1984). It is possible that the preponderance of sugar maple seedlings reported by various workers in stands of mature beech–maple forests may be related to reduced deer populations (Vankat *et al.* 1975).

Any of the kinds of damage described here is likely to have prolonged after-effects. This is because of the slow and complex readjustments of the plant populations. In turn these are influenced by vagaries of regeneration and by relatively slow tree .growth, which may also be affected by climatic and soil factors. Long-term effects are, for example, those caused by the entry of rot fungi to damaged trees. They can cause gradual decay and eventual death after many decades.

Widespread forest destruction

The eastern North American forests have a long history of very severe disturbance from human activity and natural causes. Although the Indian people frequently burnt extensive areas and cleared sites around their villages, when the region was first settled by Europeans in the early 17th century, it was predominantly forest clad (Day 1953, D. Q. Thompson & Smith 1970). This situation changed markedly in the 18th and early 19th centuries as the European settlers cut down the forests to clear farmland. The lumber industry in the 19th and early 20th centuries continued this process, and widespread fires occurred then, too. Relatively little of the forest anywhere in the region was untouched by one or other of these activities (Spurr & Barnes 1980, Vankat *et al.* 1975).

Much of the hard-won farmland was abandoned after the mid-19th century, and extensive tracts reverted to forest cover. Damage to the forests has ensued again from the mid-20th century, as increasingly more land is needed for urban settlement, industrial sites, roads, etc., and through logging, but the landscape is still essentially wooded. Acid rain and other pollutants are currently making further inroads in the forests.

Major disturbances in forests are often complex, so generalizations about both the nature of the disturbances and their after-effects are difficult. For example, drought is often accompanied by insect defoliation and then fire (G. R. Stephens & Waggoner 1970). Trees weakened by rots and wood-boring insects (which attack after various forms of damage) are more susceptible to wind-throw. Wind-throw is often followed by fire (Cline & Spurr 1942). As will be seen, different disturbing influences have different effects on the course of subsequent vegetation changes.

HURRICANES

Storms of great ferocity which originate in the Caribbean Sea, strike parts of eastern North America from time to time, causing very extensive damage to the forests. Historical records exist for them in 1635, 1815, 1936, 1938, 1944 and 1954 (Stearns 1949, Ludlum 1963, Horn 1975a, Peet & Christensen 1980a, Spurr & Barnes 1980). In the Harvard forest in Massachusetts the hurricanes of 1938, 1815, one in the early 17th century and another in the late 15th century were identified by analysis of windthrow pits and mounds, formed by the root disc of fallen trees (Raup 1981).

Whole forests are felled by hurricanes, leaving a chaos of broken and fallen trees. Many trees (or their main branch systems) are snapped off by the force of the wind, but most are overturned in the line of the wind direction. A few standing trees survive, although they suffer some damage such as branch breakage and severe foliage attrition. Deciduous species are more-resistant to wind damage than are evergreen gymnosperms. Some juvenile trees are destroyed by burial by forest debris or by breakage, but many of them (and shrubs and herbs) survive relatively unscathed.

Many of the trees that fall lift up a mass of soil on their root disc, leaving a pit on the upwind side (E. P. Stephens 1956). The bared soil of the pit, and among the upturned roots, provides a habitat for colonist species (trees, shrubs and herbs). As the fallen tree trunks rot, they too provide habitats for plant establishment. However, much of the forest regrowth is derived from the juvenile plants that are already established on the site. The resultant forest stand is often patchy, because the ground conditions and the established juvenile populations are heterogeneous.

Spurr (1956) described the regeneration of a hillside forest in New Hampshire, USA, formerly dominated by old-growth beech and hemlock (Cline & Spurr 1942), after it had been completely blown down by the 1938 hurricane. At lower levels on the hill the abundant, established juveniles of beech and hemlock, which formerly had been suppressed in the deep shade of the old forest, gave rise to quite dense, tall sapling stands by four years after the storm. Other species (paper birch (*Betula papyrifera*) in particular, but also sweet and yellow birch, black cherry and red maple) seeded in on bare soil in the open sites. Where suppressed juveniles of the main dominants were less-abundant at the outset, these pioneer and transitional species were much more prominent in the regenerating stands.

Other kinds of forest community in Harvard Forest, Massachusetts, USA, underwent different patterns of regeneration after the storm (Spurr 1956). For example, in storm-devastated, maturing white pine stands, vigorous saplings of the understorey angiosperm trees and shrubs immediately took over the site. Spurr predicted that species such as red maple and red oak would probably dominate the forest as their populations increased in stature. Seedlings and saplings of white pine, grey birch (*Betula populifolia*), yellow birch, aspen, black cherry, pin cherry (*Prunus pensylvanica*) and white ash, which had established from seed, were becoming suppressed by ten years after the storm.

Thus, if the canopy trees are destroyed in the established transitional and mature forests, existing juvenile populations are a very important means of stand replacement. In mature forests, with their intense shade at ground level, the juveniles must be very shade-tolerant to persist. These **suppressed juveniles** are rather like little bonsai trees. They survive with very low productivity and slow growth, often for several decades. If released from shade and root competition by the death of canopy trees, they respond by rapid growth, so that, in ideal circumstances, the composition of a whole mature forest could be reinstated. In reality the forests regenerating after hurricanes are likely to be a patchwork of stands of pioneer, transitional and mature species, each of greatly different size, because of chance factors of site and availability of young populations derived both from new seedlings and suppressed juveniles.

FOREST CLEAR-CUTTING

Responses of forest to clear-cutting have been described in detail from experimental work at Hubbard Brook, New Hampshire, USA (Bormann & Likens 1979a, Bormann *et al.* 1979, Siccama *et al.* 1970) (Fig. 9.7). Prominent in the first two years after cutting are herbaceous species of *Aster*, *Uvularia* and the fern *Dennstaedtia punctilobula* (all usually present in the undisturbed forest), the fireweed *Epilobium angustifolium* (which establishes from immigrant seeds), shrubs, and seedlings, saplings and sprouts of forest trees. The herbs usually dominate. The juvenile trees include species of strictly pioneer as well as transitional and mature phases. After five years dominance passes to the shrubs, which include dense patches of *Rubus* spp. and pin cherry (*Prunus pensylvanica*). Stands of the latter may last for 30 years. Stands of quaking and bigtooth aspen (*Populus grandidentata*) (lasting 40–60 years) and paper birch (lasting 80 years) develop slowly to supplant the shrubs. White ash, red maple and striped maple (*Acer pensylvanicum*) grow up through and suppress the earlier dominant trees, and form a dominant forest for about 100 years. Suppressed juveniles of yellow birch, sugar maple, beech and red spruce (*Picea rubens*) are present in this forest. At first, in the ash–maple forest, gaps caused by deaths through competition are filled by expansion of the canopy trees, but then, as big, old trees die, the larger gaps are filled by growth of the suppressed juveniles. Gradually the sugar maples, beeches and red spruce thin out by competition and increase in size until the remaining big trees come to dominate the forest.

Clear-cutting (and other disturbances), especially if repeated at intervals, causes shifts in species composition towards greater abundance of transitional species (Oliver 1980). In some North Carolina forests clear-cuts in 1939 and 1962 were followed by increases of tulip tree (from seeds), and red maple and chestnut oak (from sprouts). Other species – red, scarlet and black oak, and black locust – changed very little in density, but hickories and other species, including hemlock, beech and sugar maple, decreased (Boring *et al.* 1981, Parker & Swank 1982). Nutrient budget

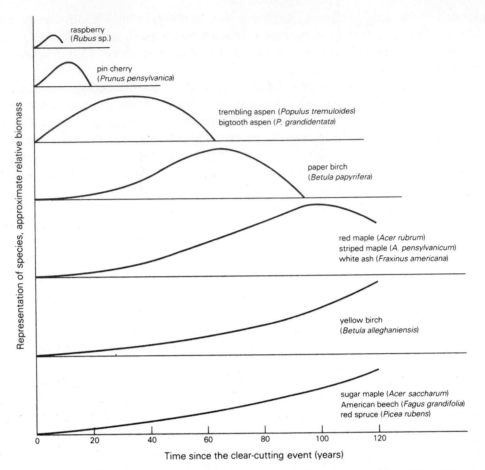

Figure 9.7 Postulated sequence of woody species after clear-cutting of forest in New Hampshire, based on short-term direct measurements and observations of stands of successively greater age (from data in Bormann & Likens 1979a). Note that clear-cutting results in the development of basically even-aged stands of *all* species, which have arisen from seeds, or sprouts, or suppressed juveniles. As the shorter-lived, quickly growing, intolerant species die, their place in the canopy is successively taken by the slower growing, long-lived, shade-tolerant species.

changes during forest recovery after clear-cutting are discussed in Chapters 11 and 12.

FIRE

Fire influences are covered at length in Chapter 7, so this section can be brief. It will serve to illustrate the different impact of burning on communities which are not as flammable as grasslands, Mediterranean-type scrub, Australian eucalypt forests and Boreal gymnosperm forests. Fires of natural and Indian origin have had a very long influence on the eastern forests. Fires are probably less frequent now than before European settlement (Day 1953, D. Q. Thompson & Smith 1971, Kozlowski & Ahlgren 1974, Wein & Moore 1977, Bormann & Likens 1979b). Fire frequency in Indian times, according to D. Q. Thompson & Smith (1971), Niering *et al.* (1970) and Buell *et al.* (1954) was at 1–10-yearly intervals, but covered relatively limited areas.

Pine forests in the region often originated in this way. It is likely also that some of the oak–chestnut, oak–hickory and other oak-dominant forests were induced by frequent Indian fires, because the oaks are relatively fire-resistant. Hemlock and beech are restricted to sites without a recent fire history.

The intensity of individual fires differs, according to circumstances. During hot, dry spells, crown and ground surface fires may destroy many trees, but fires in the mixed deciduous forests are often not as destructive as this (Niering *et al.* 1970, Swan 1970, Ahlgren 1974a, b). Light fires may merely burn the litter and singe shrubs and trees, killing relatively few. The effects are very patchy, with some areas being more comprehensively burnt, sometimes with loss of most of the above-ground living plants and soil litter and humus layers.

Data from the northern part of the region shows the following sequence of regrowth after burning of mature forest (Skutch 1929, Wilm 1936, Martin 1955, Fowells 1965, Swan 1970).

1 Within a few weeks (if a fire is in summer) stands of herbaceous species (*Epilobium*, *Erechtites* from wind-transported seeds, *Carex* and *Polygonum* from dormant seeds), grow up on bared soil. So do the liverwort *Marchantia* and several moss species. Perennial forest herbs, like *Trillium*, *Maianthemum* and the ferns *Dennstaedtia punctilobum* and *Pteridium aquilinum*, sprout from rootstocks.
2 Also within a few weeks there is a rapid growth of sprouts from many transitional and mature forest tree and shrub species. Adventitious buds sprout on the damaged stems of surviving seedlings and saplings of some of the tree species.
3 Within the first two years annual herbs (*Aster* and *Solidago*); some perennial herbs (e.g. grasses such as *Andropogon*); and various shrubs (*Rubus*, *Prunus*, *Viburnum* and *Vaccinium*) increase and may form dense stands.
4 Also, within two years seedlings of light-demanding tree species, such as grey birch or the aspens, germinate and establish, but only on moist mineral soil which has no other plant cover.

Established stands of herbs, shrubs or tree sprouts inhibit the further development of plants from seed. *Pteridium* and angiosperm herb stands may persist for four years or more after a fire and *Rubus*, *Vaccinium* and *Prunus* shrub stands persist much longer. The relatively low vegetation cover forms a patchy mosaic through which the transitional and mature forest species eventually grow.

In any area, different fire intensities have very different effects on the resultant forest composition. For example, in northern Wisconsin, where fire frequency is greater than it is further east, light surface fires favour sugar maple, which regenerates rapidly from seedlings and seedling sprouts. Hotter fires kill the sugar maple seedlings and sprouts, and create openings in the forest which are colonized by patchy stands of shrubs, yellow birch, basswood (*Tilia americana*), elm, white pine and hemlock. Persistent basswood stands result from sprout growth after moderately frequent, hot fires (Maissurow 1941). Similar phenomena in Minnesota were reported by Daubenmire (1936) (Table 9.6).

Table 9.6 Effects of different fire frequency and intensity on the composition of forest regrowth in Minnesota and Wisconsin, USA.

	Prominent species in regenerating forest	
	Minnesota (Daubenmire 1936)	Wisconsin (Maissurow 1941)
infrequent and/or light fire	*Acer saccharum* *Quercus rubra* *Ulmus* spp.	*Acer saccharum* *Tilia americana*
moderately frequent and/or intense fire	*Tilia americana* *Ostrya virginiana*	*Tilia americana* *Betula alleghaniensis* *Ulmus* spp. *Pinus strobus* *Tsuga canadensis*
frequent and/or intense fire	*Quercus macrocarpa* grasses	*Populus tremuloides* *Betula papyrifera* shrubs grasses

SPROUTS AND REGENERATION

Regeneration from sprouts, arising from activated dormant buds in damaged stems, stem bases or roots, is possible for nearly all of the deciduous tree species of eastern North America (Table 9.1). It seems likely that fire and probably also browsing by mammals, have been the strong selective forces which have encouraged this phenomenon. Prior occupation is important in determining the immediate future of sites on which sprout stands develop. Quickly growing but short-lived pioneer trees like grey birch and the aspens may pre-empt sites for a time in this way, preventing colonization by seedlings of other tree species. There will be strong competition between the stems in sprout stands, even when underground connections with mature trees persist. Thinning occurs rapidly. As old trees begin to die, gaps permit species such as sweet birch, yellow birch, black cherry or red maple to become established from seed. Oaks and hickories, or maples and beeches can establish long-lived stands from sprouts. The gymnosperms, except the shrub or small tree species of heathland, pitch pine (*Pinus rigida*) and the juveniles of loblolly (*P. taeda*) and shortleaf (*P. echinata*) pines, do not regenerate vegetatively.

The role of sprouts in the general regeneration of closed forest stands is problematic. Within the depths of a transitional or mature forest stand most sprouts appear to die. They are vulnerable to the transfer of pathogenic fungi from their parents. They are also subject to severe competition, although the nutritional support afforded by the parents may compensate for this.

Clones of tree species can arise from sprouts. While connections remain, stems of connected groups derived from sprouts cannot be regarded as separate, independent individuals. The same applies when root grafts join groups of trees, which then exchange materials through the grafts (Graham & Bormann 1966).

REVEGETATION OF ABANDONED FARMLAND

Forest development after disturbance in eastern North America has often been described by reference to the sequence of events which happens in 'old fields', i.e. fields which were originally forested, then cleared and farmed for a time before being abandoned. Eventually they revert to forest. Some short-term studies have been done by observing the events which follow initial abandonment of fields (Oosting 1942, Keever 1950, Sparkes & Buell 1955, Odum 1960, McCormick 1968). Figure 9.8 illustrates some of the important colonizing plant species. Otherwise, longer-term sequences have been examined by careful selection, in one area, of fields which have been abandoned at different times over a long period (Billings 1938, Oosting 1942, Spurr 1956, Bard 1952, Peet & Christensen 1980a, b, Christensen & Peet 1981).

There are some problems in using abandoned farmland areas as sites for vegetation chronosequence studies. First, few data about the initial condition and subsequent history of a field are ever known. The history of cultivation or other treatments of fields may have differed widely. Some fields may have had fertilizer applications (in various amounts) and others none. Weed plant species which are foreign to the natural flora of the area may have colonized the farmland, building up substantial populations (usually with large numbers of dormant seed). The degree of clearance of adjacent forest and the composition of forest remnants will vary from site to site, so the propagules available to colonize fields abandoned at different times will be of different kinds and will have been dispersed varying distances (Stapanian 1986). Abandoned cropland will differ from abandoned pasture in its capabilities for reafforestation. Pasture, with a dense sward of grass, may be difficult for the establishment of seedlings of shrubs and trees. Bare and upturned soil is usually not as extensive in natural forests as it is in ploughed fields. Burnt forest sites may resemble abandoned cropland most closely in this respect. Some other natural causes of extensive bare areas might include soil erosion or siltation by rivers.

Nevertheless, the abandoned farmland sequences are common, and they illustrate processes of vegetation change which are probably analogous to those occurring in natural circumstances. Perhaps they can be regarded as being no more 'unnatural' than forest areas that are affected by chestnut blight or gypsy moth or logged or burned, or grossly disturbed by humans in other ways. Chapter 11 outlines the events during sequences of change in the herbaceous and forest stands which develop on abandoned fields in the eastern forest zone.

ACTUAL RECORDS OF POPULATION CHANGES

The rarest data on changes in the forests are those consisting of actual censuses of the flux of species in populations over time. They require the marking of the populations and re-measurement at intervals. A long-term experiment of this type was set up in Cockaponsett and Meshomasic Forests, central Connecticut, USA, in 1926 and records have been made, so far, at 10-yearly intervals (except for 1947) up to 1977 (Olson 1965, G. R. Stephens & Waggoner 1970, 1980).

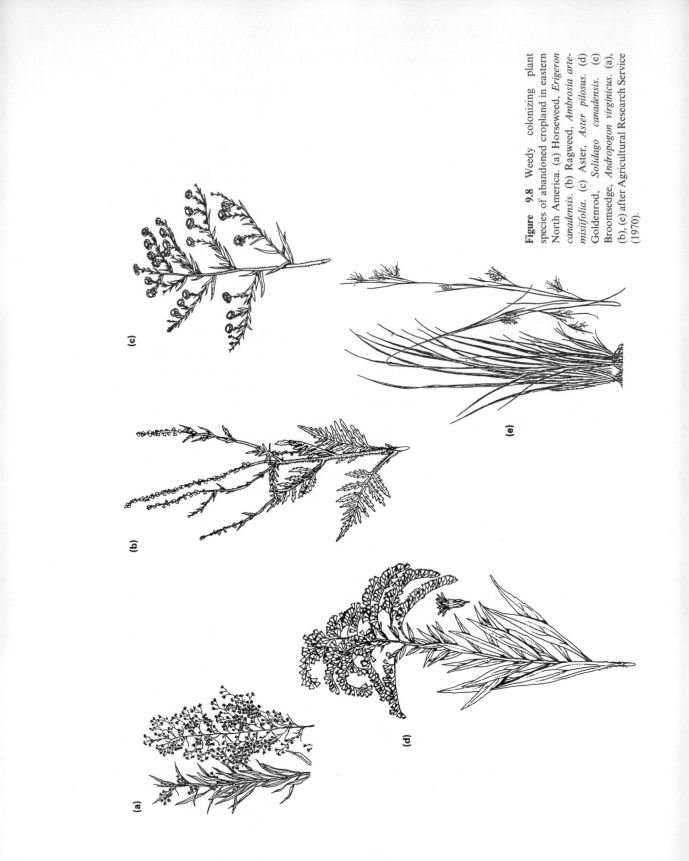

Figure 9.8 Weedy colonizing plant species of abandoned cropland in eastern North America. (a) Horseweed, *Erigeron canadensis*. (b) Ragweed, *Ambrosia artemisiifolia*. (c) Aster, *Aster pilosus*. (d) Goldenrod, *Solidago canadensis*. (e) Broomsedge, *Andropogon virginicus*. (a), (b), (e) after Agricultural Research Service (1970).

The four tracts of forest originated partly as cut-over woodland (some of it burnt) and partly as land cleared for pasture or crops. Chestnut blight had eliminated trees of that species by 1927. Although there are variations in soil moisture and other site characteristics, the detailed data from the different localities are sufficiently similar to be regarded as homogeneous.

When a quantitative survey began in 1926, most of the trees on the tracts were 25–40 years old, with occasional older trees. Sampling was done on strips 5.03 m wide and each stem of 1.52 cm diameter at 1.5 m height, or greater, was plotted on a map, identified and described. The height classes and whether the stem was of sprout origin were noted. A total of 5.56 ha was measured.

The same details were gathered at decadal intervals thereafter (except for 1947), noting deaths and the occurrence of new individuals in each diameter group (ingrowth). In 1957 the minimum d.b.h. was decreased to 1.3 cm and the height of all dominant trees and every tenth tree of subdominant stature were recorded thereafter. By 1967 the careers of 31 000 stems had been followed, many of them for 40 years (Tables 9.7 & 8).

Disturbances in the period 1927–1977 included fire in one tract in 1932 and windthrow of some large trees in the same tract during the 1938 hurricane. (Records from these are excluded from the discussion here.) Severe drought occurred in the decade 1957–1967, with annual average precipitation nearly 18.00 cm less than the normal (111.76 cm). Defoliating moth larvae (gypsy moth, elm spanworm and cankerworm) were very active, particularly in 1962–1964 and 1974–1976, removing 50 per cent or more of the tree foliage in some areas. Dutch elm disease killed *Ulmus americana* in 1937–1957.

Changes in species composition according to the size classes and in contribution to the biomass (expressed as basal area) are given in Tables 9.7 and 9.8.

The main changes in the 50-year period 1927–1977 were the following:

1 *1927–1957.* Decline to low numbers of big tooth aspen, grey birch, eastern red cedar (*Juniperus virginiana*) and black cherry. Much of this occurred in the first decade as dense thicket stands thinned out. Red, scarlet (*Quercus coccinea*), black and chestnut oaks increased proportionately in the canopy but there was little ingrowth.

2 *1957–1967.* Decline to low numbers of hornbeam, hophornbeam, chestnut, scarlet and white oak and pignut hickory (*Carya glabra*). The total basal area declined as many hickories and pole-sized oaks died. The proportion of oaks declined, although they still contributed 47 per cent of the basal area. Many oak seedlings were established following deaths of trees. The proportionate contribution of *Cornus* and *Hamamelis* to total basal area declined markedly, although the numbers remained high. Red and sugar maple and beech all contributed proportionately more. Beech increased steadily, as did sugar maple and yellow birch, especially in moister sites. Red maple became more of a subcanopy tree. Chestnut sprouts were still present.

3 *1967–1977.* The main contributors now to basal area and canopy cover are red, black and scarlet oaks, red maple, sweet and yellow birch and tulip tree. Red and

Table 9.7 Changes in diameter class distribution (stems per hectare) of the main tree species in Cockaponsett and Meshomasic state forests, Connecticut, USA, over the period 1927–1977 (after G. R. Stephens & Waggoner 1970, 1980). Total area sampled 5.56 ha.

	1927				1937				1957				1967				1977			
	Small saplings	Large saplings	Poles	Mature trees	Small saplings	Large saplings	Poles	Mature trees	Small saplings	Large saplings	Poles	Mature trees	Small saplings	Large saplings	Poles	Mature trees	Small saplings	Large saplings	Poles	Mature trees
Acer saccharum	59	35	2	+	47	37	5	+	45	40	5	+	35	40	7	+	35	45	7	+
A. rubrum	348	291	27	2	183	259	42	+	143	168	49	2	148	153	54	2	235	156	59	5
Quercus rubra	82	129	32	5	27	91	52	7	7	30	45	25	5	7	25	30	5	5	15	37
Q. velutina ⎫ *Q. coccinea* ⎬	30	82	30	2	12	45	45	5	2	10	40	20	+	2	17	25	5	+	10	30
Q. alba ⎫ *Q. prinus* ⎬	168	188	40	5	62	136	54	7	17	32	49	15	5	10	20	12	12	2	10	10
Betula alleghaniensis	91	106	17	2	49	94	27	2	54	67	25	5	57	54	25	5	146	69	27	5
B. lenta	131	148	27	5	62	136	40	5	54	74	47	10	69	74	52	10	131	96	52	12
Carya spp.	94	77	7	+	32	64	15	+	7	30	15	+	5	10	12	+	7	7	5	+
Fagus grandifolia	25	10	+	+	20	12	+	−	25	15	+	+	32	32	2	+	54	40	5	+
Liriodendron tulipifera	5	15	7	+	2	7	5	2	+	2	5	5	+	+	5	7	15	2	2	10
Fraxinus americana	74	82	5	+	27	2	12	+	15	20	12	+	7	7	15	+	5	+	7	+

Stem diameter classes: small saplings 1.3–3.9 cm; large saplings 4.0–13.9 cm; poles 14–29 cm; mature trees >29 cm. +, less than 2.

Table 9.8 Changes in basal area (m^2 ha^{-1}) of the main tree species in Cockaponsett and Meshomasic state forests, Connecticut, USA, over the period 1927–1977 (after G. R. Stephens & Waggoner 1980). Total area sampled 5.56 ha.

	1927	1937	1957	1967	1977
Acer saccharum	0.28	0.38	0.44	0.62	0.67
A. rubrum	2.21	2.63	2.67	2.89	3.34
Quercus rubra	1.77	2.80	4.67	4.54	4.90
Q. velutina	0.74	1.13	1.78	1.86	2.00
Q. coccinea	0.74	1.13	1.56	1.45	1.33
Q. alba	1.30	1.69	1.78	0.83	0.89
Q. prinus	1.00	1.32	1.78	0.83	0.67
Betula alleghaniensis	1.18	1.50	1.56	1.45	1.78
B. lenta	1.92	2.26	2.89	3.10	3.56
Carya spp.	0.36	0.47	0.78	0.52	+
Fagus grandifolia	+	+	+	0.21	0.44
Liriodendron tulipifera	0.44	0.56	0.89	1.03	1.33
Fraxinus americana	0.59	0.75	0.67	0.62	0.44
total basal area, for these species	12.85	16.92	21.69	19.95	21.55
total basal area, all major species	14.46	18.82	22.27	20.66	22.27
total basal area for all major and minor species	16.07	20.20	22.73	21.12	22.73

+, less than 0.2 m^2.
Other major species include *Fraxinus nigra, Tilia americana, Ulmus americana, Populus grandidentata, P. tremuloides, Nyssa sylvatica, Robinia pseudoacacia, Juglans cinerea, Prunus serotina, Sassafras albidum, Pinus strobus, Tsuga canadensis* and *Juniperus virginiana*.
Minor species include *Cornus florida, Amelanchier arborea, Ostrya virginiana, Betula populifolia, Hamamelis virginiana* and *Carpinus caroliniana*.

black oaks, red maple, flowering dogwood and witch hazel maintain moderate to high numbers. White ash numbers have declined.

The importance of sprouts in the origins of the stands is clear from the records. Although less than half of major species originated as sprouts, many of the oaks (red, black, scarlet and chestnut) and about two-thirds of stems of minor species (Table 9.8) were of sprout origin.

G. R. Stephens & Waggoner (1980) consider that the forests are now mature. The stem density of major species is about 200 per hectare and their basal area about 3.76 m^2 ha^{-1}. With the aid of transition probability calculations, Stephens & Waggoner predict that the future forests (about 150 years from now) will be dominated by oaks, maples and birches. The large oaks are long-lived and losses are very gradual. The maples and birches increase their basal area only very slowly, so the rate of change in the composition of the stands from now onwards will be correspondingly slow. Other relatively long-term census studies have been done in North Carolina (Christensen & Peet 1981, Parker & Swank 1982).

THE DEVELOPMENT OF OLD-GROWTH FOREST STANDS

After an episodic disturbance it is generally assumed that, provided there is no further disturbance, sequential development will proceed until a relatively stable equilibrium is reached (the 'climax' of Clements 1916, 1936). It is implied that this vegetation is an end-point in vegetation development and perpetuates itself. There are various criticisms of the concept of climax, but we will, for the moment, side-step them. It will be assumed that relatively stable (or only very slowly changing) mature forests can develop where the prominent species perpetuate themselves for at least a few generations. No-one has actually observed the development of such forests, as the timescale is too long. We must use whatever circumstantial evidence is available to gain information about them.

The most usual type of evidence is the presence of stands of old-growth forest with no sign of disturbance (e.g. descriptions by Lutz 1930b, Cain 1935, Hough 1936, Cline & Spurr 1942, Dix 1957, Bormann & Buell 1964, Vankat *et al.* 1975, Wein & Moore 1977, Bormann & Likens 1979b). However, forests with dominant trees 300–400 years old are uncommon now because of human intervention. Acid rain may threaten the survival of forest stands throughout the region.

Among Braun's (1950) 'climax' associations in the old growth eastern forests, which were listed at the beginning of the chapter, canopy dominants are:

Oak–pine. White oak and hickories.
Oak–chestnut. None of this is left in its original state, as indicated above. Adjustment
  to the loss of chestnut is proceeding. Old forests often contain red oak, chestnut
  oak and tulip tree, with increasing frequency of hickories.
Beech–maple. Beech and sugar maple.
Hemlock–white pine–northern hardwoods. Sugar maple, beech, yellow birch, hemlock
  and occasional white pine.

The other main kind of evidence about the development of mature forests is from careful observations on sites which show signs of progression through the transitional stages towards a mature state. Populations of young trees of mature forest species are replacing old populations of transitional species. No other species with the potential to replace the mature forest species occur in the area (Potzger & Friesner 1934, Runkle 1982).

SOME CONCLUSIONS

Briefly, a general conclusion is that vegetation regrowth patterns after disturbance in the forests of eastern North America are extremely diverse, according to the nature of the disturbance and the regeneration capabilities of the species concerned. Many variables may affect the result. The 'orthodox' developmental scheme – a sequence with herbs (and or shrubs) → transitional species (often pines, then angiosperms) → mature forest – is one pattern, but there are several modes, depending on local circumstances. The recovery of damaged forest directly by way of sprout stands, or upsurge of suppressed juvenile populations (often the same species as those originally present) is also quite normal. Often there is no pine phase. Even if there is a sequence from herbs (or shrubs) → transitional forest species there is no clear dependence of the tree species on the previous herb (or shrub) stands. In fact, they often inhibit the trees, at least for a time. There do seem to be some common patterns of vegetation change but, in detail, often they are not very orderly, or very predictable, because of the stochastic nature of disturbing events and forest regeneration processes.

The frequency and magnitude of disturbances in the forests raises questions about the value of adhering to the orthodox 'succession to climax' model. These are considered further in Chapter 12.

10
Changes in some tropical forests

In many respects the patterns of vegetation change in moist tropical lowland forests (Fig 10.1) resemble those in the temperate forests, described in Chapters 9 and 11 (Connell 1978, Whitmore 1975, Foster 1980, Putz 1983, Shugart 1984, Hubbell & Foster 1986a). There are, however, marked differences arising from: the warm, humid conditions, maintained rather evenly through the year; the rapid turnover of nutrients and often infertile, or only moderately fertile soils (but see Jordan & Herrera 1981); the lack of humus, with stocks of nutrients maintained in the living biomass of plants, microbes and animals; the large numbers of seed-dispersing vertebrate animals (bats, birds, monkeys and others); the intense pressure from insect herbivores on leaves, wood and seeds. Distinctive plant and vegetational features are: the often very rich flora of tree species; a wide variety of other life forms; tall forest, with complex vertical stratification; rapidity of plant growth processes; continuation of growth and other plant processes (by at least some of the species) over the whole year (Figs. 10.2–4).

The forests to be considered are those on sites below 1000 m, with more than 10 cm of precipitation each month throughout the year, an aggregate of 200 cm or more, a

Figure 10.1 Exterior of regenerating tropical rainforest, Barro Colorado Island, Panama. *Cecropia* spp. and palms are prominent (photo C. J. Burrows).

330

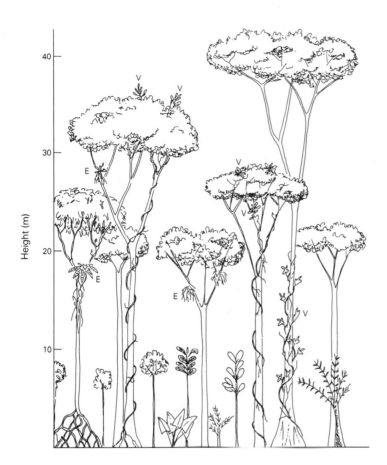

Figure 10.2 Section of a lowland tropical rainforest showing the tree strata and the distribution of other common plant life forms. V, vine; E, epiphyte.

mean annual temperature of more than 24°C and no frost (Fig. 10.5). Emphasis is on processes of vegetation change, reasons for the maintenance of forest diversity, and the impact of the distinctive phenomena in the tropical forests on ideas about vegetation change.

FOREST STRUCTURE AND DIVERSITY

Some tropical forests are the most luxuriant of all plant communities. Projecting through the canopy (30 m tall or more) usually are emergent trees 40 m tall or higher. Many of the canopy and emergent trees may have stem diameters of less that 0.5 m, but some exceed 1.5 m. The canopy is often rich in species. In forests with about 400–700 trees per hectare over 10 cm in stem diameter, total numbers of species may range from about 25 (in some African localities) to 150 or more (in Malaysia, New Guinea, Central and South America) (Paijmans 1976). Whitmore (1975) listed 760 tree species among 30 000 tree stems, with a minimum diameter of 10 cm from about 45 ha in Brunei; 224 species were listed by him among 2607 stems from 5 ha in the Amazon. Gentry (1988) recorded 283 tree species in 1 ha in Upper Amazonia, 63 per cent of them represented by single individuals. Over 90 per cent of the tree flora may

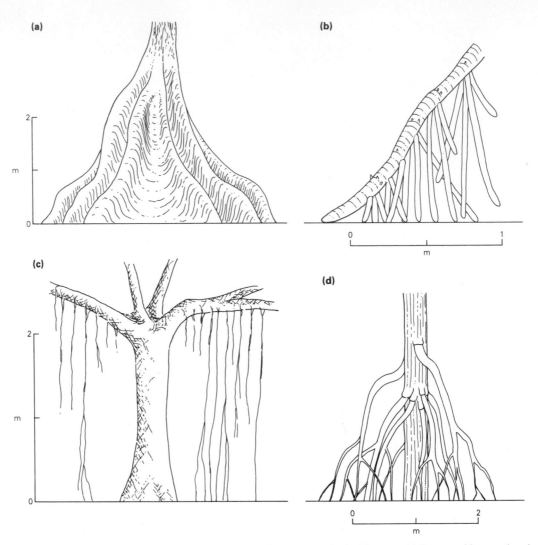

Figure 10.3 Bases of some tropical plants showing specialized morphological features. (a) Buttressed base such as is found in many of the canopy and emergent tree species. (b) Prop roots of a *Pandanus*, a monocotyledonous climber (family Pandanaceae). (c) Aerial roots, some of which will develop into props, on a species of *Ficus* (family Moraceae). (d) Prop roots on the base of a species of *Eugenia* (family Myrtaceae).

change on a short traverse of a kilometre or so in some tropical forests. On Barro Colorado Island, Panama, in a 50 ha plot, only 33, of a total of 186 tree species greater than 20 cm stem diameter, were represented by more than 50 individuals (Hubbell & Foster 1983). More than 300 species of trees, shrubs and climbers occur in this plot (Fig. 10.6).

Some of the tall canopy trees and emergents have buttressed stem bases, and all lack branches below the canopy spread. Beneath them are smaller trees, including palms, of varying height (Fig. 10.2). Vines creep on the tree stems or hang from the upper branches. Herbaceous epiphytes, including ferns and some woody hemi-epiphytes, perch high in the branch systems. There are usually some parasitic plants.

Figure 10.4 A large *Ceiba pentandra* tree in tropical rainforest, Barro Colorado Island. Epiphytes clothe the upper branches (photo C. J. Burrows).

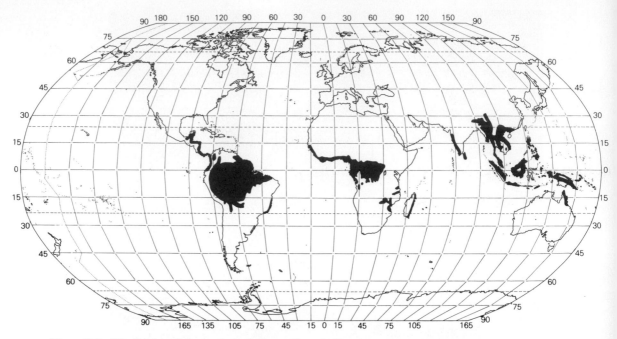

Figure 10.5 Distribution of the tropical rainforests (from Golley 1983).

Bryophytes are scarce and there are few herbs; often the ground is sparsely covered except for juveniles of the taller plants.

Few censuses give counts of the smaller subcanopy species, epiphytes, etc. They may increase total numbers of vascular plants to 300–400 species per hectare. On Barro Colorado Island (15.6 km^2) there are 1196 seed plant species, 452 of which are trees and shrubs. The flora in Malaysia–New Guinea is usually richer (Whitmore 1975).

Attention may be concentrated on the trees. Seasonal rhythms, not necessarily synchronized for different species, are apparent in leaf flush, leaf shed, flowering and fruiting (Table 10.1). Some species have more than one flush of leaves and flowers. Some species flower and seed for most of the year; others flower only at intervals of two years or more. Some species may be deciduous, briefly or for longer periods, while others are evergreen (Mabberley 1983, Shukla & Ramakrishnan 1984). Dry periods or rainy periods are the cues for these rhythms.

Although the tree flora of many moist tropical forests is diverse, relatively species-poor forests, resembling those of temperate localities, also occur in many places. One or a few species may be dominant in localities with relatively extreme soil conditions: e.g. on limestone, or ultrabasic rocks; hard, low fertility and/or well-drained substrates such as quartzite or granite; deeply weathered very low-fertility soils, such as extreme laterite or silica sand podsols; acid peats and other wet sites; steep slopes and sites above 1000–1200 m. The same applies to some relatively fertile sites subject to erosion, siltation and flooding, along rivers; and on sites of recent, extensive disturbance (P. W. Richards 1964, Whitmore 1975, Salo *et al.* 1986).

Most of the ensuing discussion is about species-rich lowland forest. Throughout

334

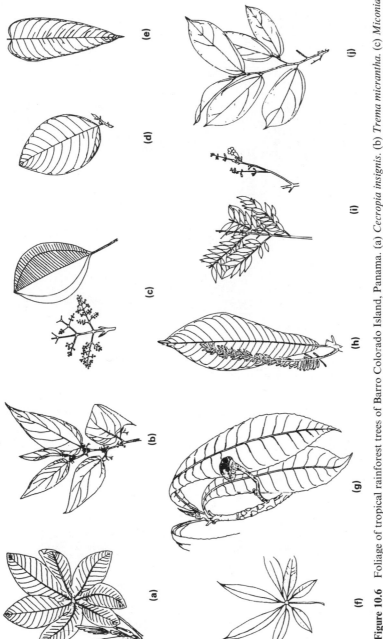

Figure 10.6 Foliage of tropical rainforest trees of Barro Colorado Island, Panama. (a) *Cecropia insignis*. (b) *Trema micrantha*. (c) *Miconia argentea*. (d) *Poulsenia armata*. (e) *Virola sebifera*. (f) *Ceiba pentandra*. (g) *Gustavia superba*. (h) *Alseis blackiana*. (i) *Jacaranda copaia*. (j) *Quararibea asterolepis*. (d), (e), (f), (j) after Dodson & Gentry (1978).

Table 10.1 Diversity in form and function of tropical rainforest tree species (refer to sources such as Longman & Jenik 1974, Whitmore 1975, Mabberley 1983).

Roots

(a) thick, horizontal surface roots, merging into large spurs or buttresses (various shapes); weak vertical sinkers, no tap root	(b) thick, horizontal surface roots, well-developed tap root and vertical sinkers	(c) weak surface roots, prominent tap root, many oblique large roots	(d) weak underground roots, numerous aerial prop roots (various forms)
(often emergents)	(often emergents)	(smaller trees and large climbers)	(monocotyledons, trees on waterlogged sites and some trees on dry sites)

Stems (bases discussed by Whitmore (1975) and see Fig. 10.3)

	(a) emergents	(b) main canopy	(c) subcanopy	(d) small tree	(e) other
heights (m)	45–60	25–45	10–25	6–10	
general form	sympodial, hemispherical, or 'flattened' crowns	sympodial, often irregular or compact, sub-spherical crowns	sympodial, rather irregular, or diffuse, or compact crowns; some have root suckers	some sympodial, many monopodial with more or less compact terminus (palms, other monocotyledons, treeferns, some dicotyledons); some regenerate vegetatively by stem base suckers.	stems compounded from coalesced aerial root systems of epiphytes – 'stranglers' from various families (especially *Ficus* from Moraceae)

various modular forms (cf. Longman & Jenik 1974)

Leaves

	many are simple, entire, oblong-lanceolate, relatively large and leathery, with drip tips		others differ in size, shape, character of margin, thickness, tip. Some are compound (pinnate, palmate); palms and some other monocotyledons have pinnate, pinnatifid, or large, sail-like leaves		
leafing periodicity	evergreen	continuous leaf production or one flush or several flushes (synchronous, asynchronous for a species, asynchronous)	brief period of leaf shed and immediate releafing, various times of year (synchronous, asynchronous)	deciduous for a period (weeks or months): in dry season (beginning or end); or in wet season (synchronous, asynchronous)	

flowers flowers and pollination are discussed by Whitmore (1975), Faegri & Van der Pijl (1971); bat and other mammal and bird pollinators have co-evolved with flowers to produce many elaborate and unusual flower forms; wind pollination is uncommon to rare

flowering periodicity	all year round	regular flowering season, single peak: short period or long period (synchronous for a species, asynchronous)	periods occurring: after dry spells, or after heavy rain, or at beginning or at end of dry season, or in wet season	occasional mass flowering	rare flowering (sometimes at intervals of many decades); polycarpic or monocarpic
flower positions	on branch tips (determinate)	on short shoots (determinate)	in leaf axils, near branch tips or further back on branches (some have very long, pendant inflorescence stalks)	cauliflory: on branches or on stems (some at ground level)	phylloflory (a few species)

Fruit fruit and seed dispersal are discussed by Whitmore (1975), Van der Pijl (1972); particular features (sizes, colours, odours, morphology, seed form, etc.) arise from co-evolution with the dispersing agents, especially bats, birds, monkeys; large sizes and fleshy fruits are common; wind dispersal mechanisms (wings, plumes, etc.) occur but are less common

fruiting periodicity (often related to flowering periodicity, but not always)	small amounts spread over a long time	single peak – fruit ripens rapidly after flowering	single peak, but longer period	delay in fruit ripening for many months to 1 year or more after flowering	mast seeding at long intervals

the world there are few areas of tropical lowland forest which have not experienced shifting cultivation within the past few centuries. However, unless this kind of disturbance is continuous or very frequent, or aggravated by recurrent fire, its influence resembles that of the natural disturbing events in the forest.

Large-scale destruction of some forest areas occurs naturally during storms (Whitmore 1975) as well as through human activity. In mature lowland tropical forest it is usual for changes to take place on a smaller scale, as a result of tree death or treefall and gap-formation, or both. The causes of treefall are manifold. They include wind-gusts, weakening of soil strength and root stability by rain, landslides, lightning strike, weakening by fungal rots or killing by virulent fungal pathogens (Whitmore 1975, Mabberley 1983). In addition there are features peculiar to tropical forests, e.g. collapse of trees which have been eaten by termites, or fall under the weight of epiphyte and vine load (Strong 1977) and death of individuals, or patches of monocarpic trees (e.g. *Tachigalia versicolor* in Central America) (Foster 1977).

Rates of gap formation and gap sizes vary according to various unpredictable events. Trees are more likely to fall if they are alongside a gap than if they are in an undisturbed area. On Barro Colorado Island most treefall occurs in the rainy season (Brokaw 1985). In forest undisturbed for several hundred years, Brokaw (1985) reports that gaps over 150 m^2 are created, on average, at a rate of one per hectare per 5.3 years and Foster & Brokaw (1982) found that the rate for gaps of about 88 m^2 area was one per hectare per year. In the same forest 17 of 328 trees, with more than 60 cm stem diameter, were dead (Putz & Milton 1982). Poore (1968) described a 12.24 ha plot in Malaysia in which 10 per cent of the area was taken up by gaps (75 newly fallen trees, 90 older dead, fallen trunks and more than 40 standing dead trees). Over some 12 years about half of the 12.24 ha area would have consisted of gaps. The largest gap was 600 m^2 and mean gap size 400 m^2. In Costa Rica, Hartshorn (1980) estimated that the return period, on any site, for rejuvenation by gap formation, was about 118 years.

The life expectancy for canopy trees is variable. On Barro Colorado Island, according to Leigh (1982) trees of greater than 20 cm stem diameter live only an average of 60 $\pm$ 11 years (but survivors have a life expectancy of 96.5 years). Ages of more than 100 years were recorded in Costa Rica (Hartshorn 1978) and Whitmore (1975) noted ages of 130–570 years for some Malaysian trees; 200–250 years is an average maximum age.

A falling tree may carry others with it. Usually a mass of branches, vines and other debris, as well as the tree stem, covers much of the ground. The upended tree base usually includes uplifted roots and soil. These create a mound which, with the bare soil in the pit formed alongside, and the rotting tree trunk, are sites for establishment of young plants (Putz 1983). Gap formation initiates a continuous sequence of regrowth.

Small gaps (< 150 m^2) are often filled partly by branch growth from adjacent adult trees, but also by the immediate surge of growth of already-established shade-tolerant juveniles of certain of the canopy species. Shade-intolerant seedlings, which colonize most gaps, even the smallest, are thereby suppressed. The phenomenon is identical with replacement by suppressed juveniles in temperate forests. There is little information about persistence in the Tropics (N. V. L. Brokaw pers. comm. 1988). However, these suppressed individuals probably do not persist beneath the closed canopy for more than a few years. A group of young trees will fill a small canopy break, to subcanopy height, within about one year. Severe thinning soon reduces their number so that one (or a very few) succeed in reaching full canopy height. In larger gaps (>150 m^2) in some African forests, some shade-tolerant understorey plants die when exposed to light, heat and strong evaporation (Vasquez-Yanes & Smith 1982).

The gaps larger than 150 m^2 are colonized mainly by certain common, quickly growing, short-lived and relatively short (25–30 m) tree species (Vasquez-Yanes & Smith 1982, Gomez-Pompa & Vasquez-Yanes 1981, Brokaw 1987). Of about 20 such species on Barro Colorado Island, *Miconia argentea* colonizes gaps of all sizes but is most common in smaller gaps, surviving in those down to 102 m^2; *Trema micrantha* colonizes gaps down to 50 m^2 but survives only in the larger ones (>376 m^2). Some colonists of gaps are species known to require direct light to break seed dormancy, which is induced by the ratio of red : far-red light experienced in deep shade (Gomez-Pompa & Vasquez-Yanes 1981). Other species are known to require strong heat to break seed dormancy. Little is known about seed longevity, but early colonizing trees from genera such as *Miconia*, *Trema* and *Cecropia* appear to have seeds which last in the soil for several years. Seeds of these species are relatively small, produced abundantly for most of the year, and are widely dispersed by birds and bats (Table 10.2). The seed banks in the forests in general are richest in these types (Putz 1983).

Larger gaps (>400 m^2) are colonized by other species. On Barro Colorado and in other Central American forests they include *Didymopanax morototoni*, *Ochroma pyramidale*, *Cecropia insignis* (and other *Cecropia* species) and the palm *Oenocarpus panamanus*. Gap partitioning by the colonist species is a striking feature of these tropical forests (Denslow 1980a). Table 10.2 and Figure 10.7 summarize data for three common and contrasting species on Barro Colorado Island (Brokaw 1987). Some species may have flexible responses to the variable light conditions in the forests (Huber 1978).

Among the colonists of large gaps are vines, some herbs and shrubs, palms and, in some circumstances sprouts from damaged tree bases. The ground, even in the larger gaps, is rapidly covered (Fig. 10.8). Tree seedlings may be inhibited by such cover, either in gaps or beneath the forest canopy (Denslow 1987). Although the general theme is one of rapid growth, relatively slow forest redevelopment has been recorded from some Mexican localities (Gomez-Pompa & Vasquez-Yanes 1981). This seems to result from site pre-emption for a time by the successive waves of species. Grasses may be dominant for a few months, then shrubs for 6–18 months, before low trees such as *Trema* and *Miconia* spp. occupy the ground (3–10 years). Then taller trees (*Cecropia*, *Didymopanax* and *Ochroma* spp.) take over for 10–40 years before mature forest is established.

Very fast growth rates have been recorded for some of the early colonist trees (e.g. 9 m for its first year by *Trema micrantha* in Costa Rica and more than 30 m in 8 years for the same species (Mabberley 1983)). These species either lack mycorrhizae or have

Table 10.2 Relative behaviour of three tree species which colonize treefall gaps in tropical rainforest, Barro Colorado Island, Panama (after Brokaw 1987).

Species	Family	Type of fruit	Seed dispersal agents
Trema micrantha	(Ulmaceae)	drupe	many bird species
Cecropia insignis	(Moraceae)	seeds compressed in peduncles	birds, bats
Miconia argentea	(Melastomataceae)	berry	monkeys, birds

Recruitment	1st year after gap formation	2nd year	Survivors to years 8, 9
Trema	nearly all in first year	none after year 2	only those established in first year
Cecropia	most in first year	many in year 2, none after year 3	from each of these years
Miconia	begin in first year	highest in years 2–5 and continue to years 7–9	from each of these years

Sizes of gap occupied	Years 1–3 (seedlings)	years 8, 9 (saplings, poles)
Trema	50 m^2 and larger	only those >376 m^2
Cecropia	20 m^2 and larger	in largest and also some smaller (>215 m^2)
Miconia	20 m^2 and larger	in largest and some down to 102 m^2

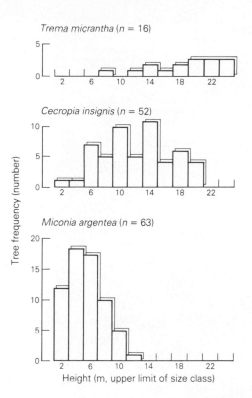

Figure 10.7 Height class distribution of individuals ≥ 1 m tall for three tree species which colonize treefall gaps on Barro Colorado Island, Panama. Most *Trema* are in the largest classes because its recruitment ended early and the survivors grew fast. Most *Miconia* are in small size classes, due to continuous recruitment and comparatively slow growth. *Cecropia* is intermediate in character (after Brokaw 1987).

ectotrophic associations, but they rapidly establish control over the nutrient stocks. Most of the early colonist species are short-lived, have simple branch systems, wood of low specific gravity and large, thin, soft, palatable leaves (Coley 1983). *Cecropia* harbours fierce ants which protect the plant from herbivores as well as from epiphytes and climbers.

One of the very interesting features of tropical lowland forest development is that some shade-intolerant, relatively quickly growing species, which behave as early colonists in large gaps, are very tall (> 30 m) and long-lived. They often persist to become canopy trees or emergents. On Barro Colorado Island about 20 species, including *Anacardium excelsum, Terminalia amazonica, Jacaranda copaia* and *Dipteryx panamensis* are in this category. They are less common than the quickly growing, shorter-lived species. Some emergents are legumes. Perhaps their nitrogen-fixing association boosts their growth. Most of these 'transitional' species have wind-dispersed or terrestrial mammal-dispersed seed. Often the seeds are not dormant. Their wood is usually denser than that of the smaller and shorter-lived early colonists.

The closing of gaps by tree growth sees the establishment of other species, which have seeds that germinate in shaded conditions, and shade-tolerant juveniles. In a Costa Rican forest Hartshorn (1980) found that most tree species of the canopy and immediate subcanopy (108) are shade-intolerant. Only 67 species are shade-tolerant; but the ratio is closer to parity when the lower understorey trees are added to the total (155 intolerant to 142 tolerant). The shade-intolerant mature forest trees contrast with the pioneers and 'transitional' species by being slower-growing and having

341

Figure 10.8 A treefall gap in tropical rainforest, Barro Colorado Island. Note abundant vines and epiphytes and dense ground cover (photo D. Twitchell & N. V. L. Brokaw).

dense wood and smaller (though often relatively large, by temperate standards), thick, leathery leaves that are high in fibre and phenolic compounds. The foliage is not usually so palatable to herbivores (Coley 1983). Mycorrhizal associations are of the endotrophic type, formed by a limited number of fungi. Flowering and fruiting of these trees is at intervals, sometimes annual, sometimes separated by periods of several years, depending on the species. Dispersal of the relatively large seeds is mainly by arboreal mammals and birds. Most seeds fall near the parent plant. The seeds lack dormancy and are only briefly viable. They must remain moist for germination to succeed, but the large food reserves boost seedling growth. Nevertheless, seedling growth is relatively slow, even in canopy gaps, where they may be suppressed to some degree by the earlier colonist species.

The pioneer, building and mature phases resulting from gap-filling occur in a mosaic fashion throughout the forests (and similar events occur in all tropical lowland forests). In any one region, although there are common patterns, the actual species-by-species replacements are somewhat unpredictable because of the different gap sizes, the timing of their occurrence in relation to plant life-cycles, various independent stochastic events and the large numbers of species suited to fill particular roles. One study on Barro Colorado Island (Hubbell & Foster 1986a) indicated, however, that the probability of self-replacement is no lower than the probability of replacement by other species. Yet the saplings of the two most common species avoid their respective adults, presumably because light conditions are unfavourable.

FOREST SPECIES STRUCTURE

Few population analyses covering large areas have been done in tropical forests (Knight 1975, Whitmore 1975, Hubbell & Foster 1983). No time-lapse census studies have been made over more than a few decades (Crow 1980, Whitmore 1985). Examples here are drawn mainly from work on Barro Colorado Island in Gatun Lake, Panama. Gatun Lake was filled and Barro Colorado (about 15.6 km² in area, maximum height above sea level 163 m) created in 1914. The island, at 9° 10'N, 79° 51'W, experiences annual average rainfall of 267 cm, with a dry season from January to April, when rainfall is usually less than 13 cm per month (5 per cent of the total annual fall). Maximum temperature ranges from 21 to 32°C; annual average 27 to 32°C; average daily range 9°C, with little seasonal variation. Relief on the island is gentle to moderate (up to about 18° slope). The soils, on weathered basalt, sandstone and siltstone, are a uniform red, well-drained clay, and in some places a yellowish to reddish-brown and grey-red mottled, gleyed clay (Knight 1975).

Half of the island (Fig. 10.4) is covered by 70–80-year-old forest, regenerating after cutting or cultivation, or both, during the construction of the Panama Canal. The other half is covered by old forest, undisturbed other than by natural events for at least 300 and possibly about 500 years (Foster & Brokaw 1982, Hubbell & Foster 1986a).

Knight's (1975) study showed that the young and old forest canopies were qualitatively similar in composition, although the proportions of species differed. However, larger samples than those that Knight used (0.8–2.0 ha) may be needed to make valid comparisons. Some of the shade-intolerant species, for example (*Miconia argentea* and *Pterocarpus rohrii*), well-represented in the young forest, appear rarely, if ever, in the tall forest. *Terminalia amazonia* is present as an adult in young and old forest, but rarely as a juvenile in old forest. Species of *Cecropia*, *Trema* and *Piper*, usually pioneer or early 'transitional' forest trees, also occur occasionally in the old forest.

Very young forests in Panama, close to Barro Colorado, regenerating after abandonment of cropping, also contain many species found in the old forest, but short-lived species of *Cecropia* are very common. Many species in a 3-year-old forest were reproducing by sprouting, including one, *Gustavia superba*, which is common in the old forest on Barro Colorado (Knight 1975).

In a 50 ha permanent study plot in the apparently mature old forest on Barro Colorado, 33 canopy species, of 20 cm stem diameter or more, are represented by 50 individuals or more, a total of 5697 trees. The canopy is 30–40 m high and there are some emergents (Hubbell & Foster 1983). The most common tree is *Trichilia tuberculata* (1024 stems). Five other species are represented by 200 stems or more, *Alseis blackiana* (430), *Gustavia superba* (208), *Poulsenia armata* (213), *Quararibea asterolepis* (472) and *Virola sebifera* (204) (Fig. 10.6).

Hubbell & Foster (1983) found three patterns of dispersion of the most common 33 species. Some species are distributed randomly. They include *Trichilia tuberculata* as well as other species with abundances ranging from moderately common to rare. Most species, however, are distributed in a clustered fashion. Some of the clusters (e.g. those of *Poulsenia armata*) are associated with topographic or soil features. Often (e.g. for *Beilschmiedia pendula*) no soil or other site conditions could be correlated with the clusters (although they may exist, cryptically, in some instances). Although they are rather diffuse the patch sizes appear to be about 0.5 km^2 in diameter. Species which are mainly grouped in clusters occur, rarely and at random, elsewhere on the plot.

Suggestions by Hubbell & Foster (1983) about the dynamic causes of these patterns of tree dispersion are:

1 For the randomly distributed species, opportunities are equal for them to be established at all sites. Their occurrence will be determined by chance factors connected with disturbance, seed dispersal and perhaps competition.
2 Clusters of species associated with specific topographic or soil conditions may be better competitors there than the randomly distributed species. Possibly they are at a competitive disadvantage elsewhere.
3 Occurrences of species distributed in patches, with no discernible relationships to topography or soils, may result from treefall disturbances or recent spread from colonizing ancestors. At least one of these, *Cecropia insignis*, is well-known as a species which successfully establishes only in large treefall gaps. It is suggested that the sparse, scattered occurrences of otherwise highly aggregated species in other parts of the area are due to continual immigration from the dense population.

MAINTENANCE OF THE DIVERSITY OF TREE SPECIES IN THE VEGETATION

The frequent natural disturbance and consequent temporal site variability in tropical lowland forests, described earlier, must be a potent cause of variable forest composition. Variability will result, simply, from the constant flux of microsites in which different species can become established and grow (Pickett 1983, Connell 1978). Each new gap will have its own distinctive array of microsites (Brokaw 1987). However, disturbance seems to provide only a partial answer to the question of why so many species of forest tree are maintained in the flora. As summarized by Leigh (1982), four other hypotheses have been proposed to account for the maintenance of species richness of these forests:

1 Species are in competitive equipoise: the diversity expresses the balance between speciation and extinction due to chance fluctuations in numbers (Hubbell 1979, 1980).

2 Species coexist because different species occupy different habitats or stations. The different species segregate (i.e. have niches) along the environmental gradients (P. S. Ashton 1969, 1977) (see also Denslow (1980a,b) on gap partitioning and Huston (1980) on nutrient-related variation).

3 Species coexist because they respond differently to environmental fluctuation, with great variation in reproductive success from year to year for different species. Rare species can benefit from being long-lived. Environmental variation favours them, provided that they gain more from a year favourable to reproduction than they lose in an unfavourable year (Chesson & Warner 1981).

4 There are many species of tree because insect predators do not allow any one to become common. Seeds and seedlings under parents are more heavily predated than those dispersed further, because they attract many predators (Janzen 1970, 1975, Hubbell 1980).

The evolutionary balance hypothesis of Hubbell would be difficult to test. We would need to be able to measure evolutionary and extinction rates; few relevant data are available. It has been suggested, plausibly, that the large numbers of species initially arose by allopatric evolution among plants confined to relatively small refuge areas during the last glaciation (Prance 1982). If this is so, then the evolutionary balance hypothesis can be discarded, but we may wonder why so many species remain in existence, rather than becoming extinct through competition.

P. S. Ashton (1969, 1977) suggests other, relatively orthodox, explanations of evolutionary history to account for the rich flora, stressing the highly specialized adaptation of plants to the biotic and physical environment. Relatively narrow fundamental niches are envisaged, and the numbers of tree species present ensure that realized niches will be strictly circumscribed. The 'habitats' or 'stations' may include temporal components.

All of the other hypotheses invoke one aspect or another of niche differentiation in relation to habitat factor variability and chance events, although some authors (e.g.

P. W. Richards 1969, Grubb 1985) have found it hard to believe that there could be so many tree niches. Hubbell & Foster (1986b) propose that there are broad adaptive zones for groups of tree species, rather than highly specialized niches.

The hypothesis of Chesson & Werner (1981) is favoured by Grubb (1985), who discusses variable recruitment as a means for maintaining variability. It would operate through species-specific variations in site availability and year-to-year changes in the establishment of seedlings and survival of juveniles of different species. Of course, it also involves niche differentiation in relation to the environmental complex. In fact, even if there is abundant and frequent seed output or dormant seed populations, or both, there may need to be runs of several years favouring the regeneration of one species, versus its competing neighbours, for this mechanism to operate. Testing of the hypothesis is difficult with long-lived species, but may be possible from censuses of juveniles. Brokaw's (1987) data from Barro Colorado Island are consistent with the hypothesis (Table 10.2).

If tree distributions are clustered (as is true for some species in Malaysia as well as for many of those on Barro Colorado Island and other tropical areas (P. S. Ashton 1969, Whitmore 1975)), Janzen's seed predation model cannot apply. In spite of experimental results and other data in support (Janzen 1975, Clark & Clark 1984), testing by Connell (1979) and others indicates that herbivores are not generally as effective at destroying seeds as the hypothesis predicts. Furthermore, Clark & Clark (1984) suggest that other causes (allelopathy or intracohort competition) could account for the lack of juveniles near adults. Burgess (1969) showed that predation of germinating seed of a Malaysian dipterocarp caused a *clumped* distribution of its population. Further work is needed to test the seed predation idea very thoroughly.

Evidence of niche differentiation among tropical forest species comes mainly from the multiplicity of characteristics of form and function which can be observed (Table 10.1). The reproductive and regeneration properties of species ought, in selective terms, to be more labile than other kinds of property, so we should expect differentiation to be most readily evident in them (Augspurger 1983, 1984a, b, Garwood 1986). Nevertheless, there must also be habitat variability, expressed both spatially and temporally, to permit the niche differences between species to operate, so that each species can establish new populations.

In spite of the modulating effect of the tall forest cover and the relative invariability of climatic conditions, many differences in habitat conditions can be perceived in the tropical forests. Some are intimately related to the structure of the forest and the biology of organisms other than plants. Even the phenomenon of gap formation generates microsite variation according to the ways and times that different species die (by fall, snapping off, or *in situ*).

Among the environmental variations are those in:

1 Soil conditions (e.g. Austin *et al.* 1972, Whitmore 1975, Huston 1980). Soil–plant relationships on a small (individual tree-sized) scale are not well-studied in the Tropics. Many tree species appear to have wide niches for soil factors, but distinctive site properties will be determined by differences in soil morphology, moisture properties and, almost certainly, soil chemistry (cf. Huston 1980,

Hubbell & Foster 1986a). Microtopographic differences affected by tree-fall disturbances and by animal disturbances of the ground, will operate, partly through effects on soil physical and chemical properties, to influence the differential establishment of juveniles.

The soils on which the greatest tree diversity is found are generally not very fertile (Whitmore 1975, Huston 1980). Some differentiation of tree niches is due to mycorrhizal development. Among the trees of closed forest endomycorrhizal infection is almost universal, but there are few fungal types on many different hosts. Janos (1983) believes that this helps to promote diversity; the trees cannot competitively exclude one another. Mycorrhizas have been linked with the very rapid nutrient cycling process in the forests (Jordan & Kline 1972, Stark & Jordan 1978, Golley 1983, Janos 1983). Mycorrhizal roots of trees have even been found high in their canopies (Nadkarni 1981).

Brief mention may be made here of relatively fertile tropical soils on volcanic rocks and steep eroding hill-slopes, where competition for nutrients by trees is not as intense as on poorer or older rocks (Jordan & Herrera 1981). On the poorest rocks there are very infertile tropical podsol soils with thick root mats, relatively low forest canopy and often relatively uniform forest composition. Some of these strongly leached, old soils (e.g. tropical white sands) have experienced such a decline of nutrient stocks over time that they can support only shrubby, heath-like plant communities (Specht 1979, A. B. Anderson 1981, Jordan 1981). The phenomenon parallels that under mor-forming conditions in temperate ecosystems (see Chs 9 & 11).

2   Climate. In spite of the even and mild conditions of precipitation, humidity and temperature prevailing in the Tropics, there are regular differences experienced through the year and, less predictably, from year to year. The modulating effect of the forest on subcanopy microclimate is not total. Subtle microclimatic effects exist (Mabberley 1983). Correlated with the regular changes are the various plant rhythms, many affecting reproductive features. The extension of opportunities for growth and reproduction to the whole year accounts for some aspects of niche differentiation. Gap-forming and more-extreme disturbances, as well as unpredictable events such as drought, or exceptional rainfall, contribute further to the opportunities for diverse plant responses, as indicated earlier.

3   Light. Much the same conditions apply to light intensity and quality as to the climate variability. The deciduous habit of some canopy species creates better-lit zones on the forest floor, for a time. The light climate is quite diverse. Longman & Jenik (1974) point out that daylength differences, although small, operate to control plant processes in the Tropics. Seeds and juvenile trees respond to differences in light regimes in ways that show marked niche-differentiation (Gomez-Pompa & Vasquez-Yanes 1981, Augspurger 1984a).

4   Other biota (especially animals). The huge numbers of animals, both vertebrate and invertebrate, in the Tropics (Tables 10.3–8 & Fig 10.9) must be one of the most potent causes of maintenance of richness of the flora. Tables 10.3 and 10.4 indicate the kinds of influences. Two of the most important are seed dispersal and herbivory. Seed dispersal by wind is relatively uncommon in tropical forests in

Table 10.3 Numbers of vertebrate animal species (except fish) in some tropical rainforest areas (after Bourliere 1983).

	Area (km²)	Mammals	Birds	Reptiles	Amphibia
Barro Colorado Island, Panama	14.8	97 (46 bats)	366 (83 seasonal migrants)	68	32
Kartabo, Guyana, South America	0.6	73 (12 bats)	464 (21)	93	37
Bukit Langan, Malaysia	6.4	>90 (40 bats)	>119	50	23
Makokou, Gabon, Africa	2000	119 (30 bats)	342 (59)	63	38

general, although it is the mode for dispersal of the dipterocarps of Malesia. Wind-dispersed species include some of the canopy or emergent species, and many of the epiphytes and tall vines (Van der Pijl 1972). Wind is considered to be generally ineffective for wide dispersal, but exceptionally there are episodes of distant dispersal (Whitmore 1975).

Arboreal and terrestrial mammals disperse the seeds of many fleshy-fruited species (Leighton & Leighton 1983). Some seeds are even dispersed by fish and a few by insects such as ants (Mabberley 1983). However, bats and birds are among the most-effective seed dispersers (Table 10.9 & 10, Fig. 10.10). Specialized complex co-evolutionary relationships are evident, as well as generalist systems (Howe & Smallwood 1982, Leighton & Leighton 1983, Howe 1983).

Herbivory, on seeds and seedlings as well as other plant parts, creates another dimension to which the plants have responded in many different ways (Table 10.11). By thinning out populations of seedlings, herbivory diminishes competition.

Niche differentiation in seeds, in relation to predation, is evident: e.g. species with seeds produced in large numbers, without apparent defences; species with tiny seeds; species with very thick epicarps; mast seeding; and many species with toxic compounds of various kinds (Janzen 1975). The wide array of flowering period patterns and pollination systems of tropical species is another means by which niche differentiation is achieved.

The enormous variability, giving rise to a staggeringly diverse array of specializations of plant form and function in the Tropics, is summarized well by Whitmore (1975), Bazzaz & Pickett (1980) and Medina et al. (1984). The interactions of the plants with their variable environment, especially the biological components of environment and the many chance events which occur, seem to be all that is needed to account for the coexistence of so many different species.

When large areas of lowland forest are disturbed, relatively uniform stands of trees often arise; uniform forest also occurs on extreme substrates. This suggests that the maintenance of diversity in the tropical forests in general is intimately connected with the biological processes and various forms of feedback taking place, as well as being related to nutrient budgets (Pomeroy 1970, De Angelis 1980, Janos 1983, Huston 1980).

DIVERSITY, COMMUNITIES AND MATURE FOREST

In the most diverse of the tropical forests it is difficult to perceive stands of uniform composition or communities of species, occurring repeatedly over the landscape. It is even difficult to consider individual species populations as a basis for comparison, in the case of some of the rarer species, because they might consist of one adult and a few dormant seeds in areas of many square kilometres. Cases that are even more extreme are listed by Whitmore (1975).

This kind of situation poses some problems for the application of some concepts which are useful in less-diverse vegetation in the Tropics and elsewhere. Mature forest, which replaces itself with similar species composition, is something of a contradiction in terms, for example. On any patch of ground of a few hectares the mean species composition might change drastically between the generations of canopy tree species (but see Hubbell & Foster 1986 a, b). Unless the species become locally extinct, it is likely that representation of adult trees of any of the species will be found somewhere within about a kilometre and some seeds or juveniles may occur on or near the former location of a dead adult. Arbitrary definitions are necessary for the scale of areas to be considered and the plant species population is the best basis for comparison. Of course, this same criterion is applicable to all less-diverse vegetation as well. In effect, if the Barro Colorado Island experience can be applied generally, the diverse tropical forest is probably not as extreme as has just been suggested (also cf. Armesto *et al.* 1986). At least some submature individuals of most of the canopy species can be expected to be found within a few hundred metres of any adult tree which dies, because the tree-fall frequency and seeding frequency of the trees make this possible.

Our view of the reproducing population and stand, in the Tropics, has to be on a broader spatial scale than applies in temperate forests. However, even in temperate forests, or other vegetation outside the Tropics, some members of the *potential* plant population for any site (represented by the parents of widely dispersed seeds) are distant from the site. Most juvenile individuals are probably derived from parents close at hand, but for many species some propagules can originate 1 km or more distant. In some species-rich tropical forests the situation with respect to dispersion of populations is simply an extreme version of that normal in localities outside the Tropics.

Table 10.4 Some vertebrate animal influences on plants in tropical rainforests.

	Eat green leaves	Eat twigs or bark	Eat roots	Eat flowers, seeds*	Eat nectar, pollinate flowers†	Eat ripe fruit and disperse seeds‡
Mammals						
Marsupials						
Opossums, etc. 3	+				+	+
Phalangers, etc. 2	+	+		+	+	+
Carnivores						
Kinkajous, etc. 3	+					+
Coatis	+					+
Civets 1,2					+	+
Mustelids 2,3						+
Bears 2			+			+
Cats 1,2,3						+
Odd-toed Ungulates						
Rhinoceros 2	+					+
Tapirs 2,3	+			+		+
Even-toed Ungulates						
Pigs, Peccaries 1,2,3	+		+	+		+
Chevrotains 1,2	+					+
Deer 2,3	+					
Buffalo 2	+					+
Antelope 1	+					+
Proboscideans						
Elephant 2	+					+
Rodents						
Flying Squirrels 2	+	+				+
Squirrels 1,2,3	+	+	+	+	+	+
Mice, Rats 1,2,3	+		+	+		+
Porcupines 2,3	+	+	+			+
Agoutis 3				+		+
Edentates						
Sloths 3	+					+
Tree Shrews						
Tupaia 2						+
Flying 'Lemurs'						
Cynocephalus 2	+				+	+
Chiropterans						
Bats 1,2,3					+	+
Primates						
New World Monkeys 3	+			+		+
Old World Monkeys 1,2	+			+		+
Gibbons 2	+					+
Pongids (Gorilla, Chimpanzee Orang) 1,2	+			+		+
Birds						
Ratites						
Cassowaries 2						+
Tinamiformes						
Tinamous 3						
Galliformes						
Megapodes 2				+		+
Guans 3	+					+
Curassows 3	+					+
Pheasants 1,2,3	+			+		+

Table 10.4 continued

Galliformes *cont.*				
Jungle fowl 2	+	+		+
Guinea fowl 1	+	+		+
Turkeys 3	+	+		
Hoatzin 3	+			
Gruiformes				
Trumpeters 3				+
Columbiformes				
Pigeons, doves 2,3				+
Psittaciformes				
Parrots, parakeets, etc. 1,2,3	+	+		+
Cuculiformes				
Touracos 1				+
Caprimulgiformes				
Oilbirds 3		+		+
Apodiformes				
Hummingbirds 3			+	
Trogoniformes				
Trogons 1,2,3				+
Coliiformes				
Colies 1	+			+
Coraciiformes				
Motmots 3				+
Hornbills 1,2				+
Piciformes				
Barbets 1,2,3	+			
Toucans 3				+
Passeriformes				
Broadbills 1,2				+
Cotingas 3				+
Manakins 3				+
Orioles 1,2				+
Tanagers, honey creepers, American orioles, caciques 3		+	+	+
Jays 2,3		+		
Titmice 1,2,3		+		+
Bulbuls 1,2				+
Leaf birds 2		+	+	+
Thrushes 2,3				+
Starlings 1,2		+		+
Honeyeaters 2			+	+
Sunbirds 1,2			+	
Flowerpeckers 2			+	+
Whiteyes 1,2			+	+
Vireos 3				+
Woodwarblers 3		+	+	+
Finches 1,2,3		+		

1, Africa; 2, Asia–Malesia–New Guinea–Australia; 3, Central and South America.

* Some mammals and birds collect and store seeds before eating them. Some of these seeds survive and germinate.

† Only the smallest mammals of the particular groups are regular pollinators. Some birds which take nectar crush flowers, or pierce the base of flowers, so may not effect much pollination.

‡ Some seeds are destroyed when mammals or birds eat fruit; monkeys and some birds discard seeds under the parent tree while eating fruit; other birds carry seeds internally for considerable distances before regurgitating or defecating them; bats carry fruit some distance before discarding seeds while eating fruit.

Some reptiles (lizards) eat plants and some transport seeds. Some seeds are also dispersed by fish.

Table 10.5 Some invertebrate animal influences on plants in tropical rainforests[*].

	Eat green leaves	Eat wood or bark of living trees	Eat roots	Eat dead leaves, wood†	Eat flowers, immature fruit, seeds, or young seedlings	Eat nectar, pollinate flowers	Eat ripe fruit, disperse seeds
1. Nematoda							
Roundworms			+				
2. Annelida							
Earthworms				+			
3. Arthropoda							
A. Crustacea							
Hoppers				+			
Slaters				+			
B. Myriopoda							
Millipedes			+	+			
C. Insecta							
Orthoptera							
Cockroaches				+			
Crickets	+			+			
Grasshoppers, etc.	+						
Phasmida							
Leaf, Stick insects	+						
Dermaptera							
Earwigs				+			
Isoptera							
Termites	+			+			
Thysanoptera							
Thrips						+	
Hemiptera							
Cicadas			+				
Cixiids, Jassids, Psyllids	+						
Aphids	+						
Scales, Mealy bugs	+						
Capsids, Lygaeids, Squash bugs, Pentatomids, etc.	+				+		
Lepidoptera							
Swift moths	+L				+L	+	
Pyralid moths	+L			+	+		
Saturnid moths	+L						
Hawk moths	+L					+	
Geometrid moths	+L						
Noctuid moths	+L				+		
Arctiid moths	+L						
Butterflies (several families), etc.							
Diptera						many flies	
Hover flies	+L			+L		+	
Fruit flies						+	
Muscid flies, etc.				+		+	

Table 10.5 continued

Hymenoptera						many wasps	
Saw flies	+L						
Fig wasps						+	
Vespids						+	
Social bees						+	
Ants‡	+					+	+
Coleoptera							
Longhorn beetles		+L	+L		+		
Bruchid beetles		+L		+			
Click beetles		+					
Bark beetles		+L					
Weevils	+	+				+	
Tortoise beetles	+						
Wood borers		+L					
Scavenger beetles			+				
D. Arachnida							
Acarina							
Mites	+			+			
4. Mollusca							
Gastropoda							
Slugs, snails	+			+			

L, Larvae.

* Only a very superficial indication of the great complexity of invertebrate life in tropical forests can be given here. Leaves are eaten in various ways; by chewing, mining, sucking, etc. In some groups (e.g. Hemiptera, Phasmids and beetles) both immature young (or larvae) and adults of particular species feed on leaves. The main defoliators are moth larvae, some beetles (adults and larvae) and leafcutter ants. Leafcutter ants remove large amounts of leaf material, not as food for themselves, but to maintain their fungus gardens. Many Hemiptera and some mites (which mainly feed by sucking plant juices) also have important deleterious effects on plants.

† The main feeders on dead wood and leaves are termites. The detritivores (and the complex of animals which feed on bacteria, fungi and other saprophages) effect rapid mineral nutrient turnover.

‡ Ants are the most numerous animals in tropical rainforests. The various species have many different ways of life, and most are carnivores, but some are important as pollinators, seed dispersers, or guardians of plants or (leaf cutters) in herbivory and nutrient turnover. Faecal and other detritus deposition by any of the animals may benefit some plants. Epiphytic plants often establish on ant or termite nests.

Table 10.6 Numbers of individuals from various insect groups captured by spraying small areas in tropical rainforest with insecticide, Manaus, Brazil (after Erwin 1983).

Isoptera (termites)	129	Orthoptera (cockroaches, grasshoppers, etc.)	1177	Coleoptera (beetles)	4845
Thysanoptera (thrips)	228	Diptera (flies)	1755	Hymenoptera ants	10 311
Psocoptera (book lice)	939	Hemiptera Homoptera (cicadas, jassids, aphids, etc.)	2063	others (including bees and wasps)	1934
		Heteroptera (plant bugs, assassin bugs, etc.)	557		

Table 10.7 Total numbers of individuals of major insect groups captured over 1 month, with light traps, Morawali, Celebes, Malesia (after Sutton 1983).

	Level				
	1	2	3	4	
			Height above ground (m)		
	26	18	9	1	Total
Ephemeroptera	34	179	1543	331	2087
Homoptera	2479	376	164	141	3160
Heteroptera	427	197	131	97	852
Lepidoptera	2253	996	509	288	4046
Diptera	9105	2773	1392	1613	14 883
Hymenoptera	7196	2916	1750	1944	13 806
Coleoptera	5467	3378	2277	1842	12 964

Table 10.8 Analysis of adult beetle fauna from rainforest canopy, Manaus, Brazil (after Erwin 1983).

	Herbivores	Scavengers	Fungivores	Predators	Unknown requirements	Total
no. of families	21	7	10	16	3	
no. of species	795	100	29	145	6	1075
Cerambycidae (longhorns)	42	Tenebrionidae (scavenger beetles) 57		Staphylinidae (rove beetles) 35		
Elateridae (click beetles)	35			Coccinellidae (ladybirds) 50		
Buprestidae (bark beetles)	29					
Curculionidae (weevils)	337					
Chrysomelidae (tortoise beetles)	170					
Helodidae (helodid beetles)	28					
Mordellidae (stem borers)	37					
Anobiidae (wood borers)	36					

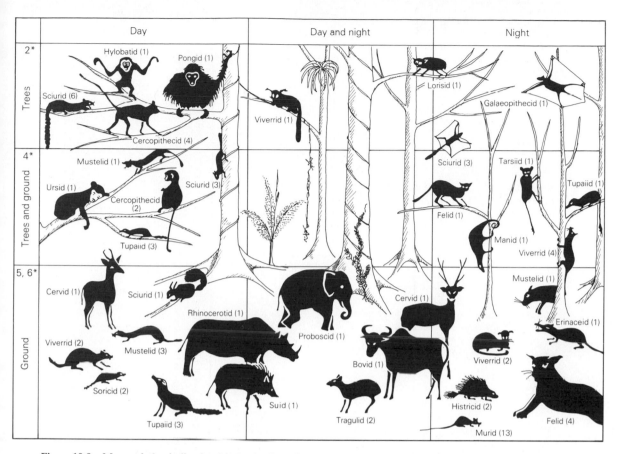

Figure 10.9 Mammals (excluding bats) in lowland tropical rainforest of Sabah, (Borneo), Malesia, according to the forest zones where they are found and the time when they are active. Pongid, orang-utan; Hylobatid, gibbon; Cercopithecid, monkey; Sciurid, squirrel; Viverrid, civet; Lorisid, loris; Galeopithicid, flying 'lemur'; Ursid, bear; Mustelid, weasel-like animal; Tupaiid, tree shrew; Felid, cat; Tarsiid, tarsier; Manid, pangolin; Cervid, deer; Soricid, shrew; Rhinocerotid, rhinoceros; Suid, pig; Proboscid, elephant; Tragalid, chevrotain; Bovid, buffalo; Histricid, porcupine; Erinaceid, insectivore; Murid; mouse, rat. About 40 bat species will be present in the same forests (after MacKinnon 1972, Whitmore 1975).

Table 10.9 Proportions of plant species with particular dispersal types in a Costa Rican wet forest (percentages) (after H. G. Baker *et al.* 1983).

	Fleshy (bats birds, monkeys)	Heavy (fall and are eaten on the ground)	Wind	Projected	Adhesive (spines, burrs, sticky glands)
canopy trees	71	18	10	—	—
shrub	93	—	—	4	—
herb	73	—	—	9	18

Table 10.10 Numbers of plant species with seeds dispersed by particular methods in a forest in Upper Amazonia, Southeastern Peru (after Foster *et al.* 1986, Terborgh 1986).

	Dispersal agent						
	Mainly bats (also birds and other mammals)	Mainly other mammals* (also large birds)	Mainly birds† (also other mammals)	Projected	Wind	Unknown	Total
number of plant species (%)	49 (17.9)	88 (32.1)	100 (36.5)	6 (2.2)	14 (5.1)	17 (6.2)	274
main plant families represented (nos of species)	Moraceae (12) Piperaceae (11)	Leguminosae (21) Rubiaceae (9)	Lauraceae (19) Myrtaceae (10) Meliaceae (10)	Euphorbiaceae (3)	Bombacaceae (4)		

* Mainly howler, spider and other monkeys, and some marsupials and procyonids (kinkajou), agoutis; coatis, tapirs, peccaries and some other mammals probably transport some seeds.
† Mainly pigeons, trumpeters, guans, toucans, trogons and passerines (turdids, cotingas and manakins).

Table 10.11 Particular modes of resistance of tropical plants to herbivory (mainly after Janzen 1975, 1983, Whitmore 1975). Some (M) are mainly effective against mammals, or (I) against invertebrates, others are effective against both invertebrates and mammals.

structurally impenetrable, trapping, repellent or indigestible[*]

thick, tough or corky bark
hard, silica-rich wood
hard, thick seed coats, I
thorns, M
entrapment with gum, sticky hairs, or other types of impenetrable or trapping hairs, I
stinging hairs, M
indigestible fibre-, silica- or calcium oxalate-rich leaves
sensitive plants which collapse when touched, M

chemically unpalatable, repellent, toxic or otherwise interfering with animal physiology[*]
terpenoids, aromatic phenols
gums, resins, latex
mustard oils
phenolic compounds
saponins
alkaloids
cyanogenic glycosides
cardiac glycosides
uncommon toxic amino acids
L-DOPA
strychnine
pyrethrins, derris
rotenone
carcinogens
steroids
chemosterilants
animal hormone substitutes (interfering with maturation)

often the same chemicals occur in leaves, stems, roots, young fruit and seeds (where they may be very abundant)

protection by ants (1 Africa, 2 South-east Asia–Malesia–New Guinea, 3 Central and South America)

ant plant			special features
Myrmecodia	2	Rubiaceae	swollen stems in which ant galleries are formed
Hydnophytum	2		
Clerodendron	2	Verbenaceae	
Lecanopteris	2	fern	swollen, hollow rhizome
Dischidia	2	Asclepiadaceae	adpressed leaves form hollows in which ants live
Tillandsia	3	Bromeliaceae	hollows in stems
Musanga	1	Moraceae	hollow stems, petiole bases or thorns in which ants live, ants attracted by
Macaranga	2	Euphorbiaceae	floral or extra-floral nectaries, or, in, some, by protein and lipid rich food
Cecropia	3	Moraceae	bodies, otherwise ants often harbour scale insects and harvest their sugary
Acacia	3	Mimosaceae	secretion and young

ants keep their host plants clear of insect and mammalian herbivores,
and also of epiphytes and vines

[*] Although similar protection from herbivores is afforded to many temperate climate plants, the numbers of protective modes (especially the chemical types) are much greater and more strongly developed in the Tropics. Multiple protectants (e.g. phenolics and alkaloids) may be present in a single plant.

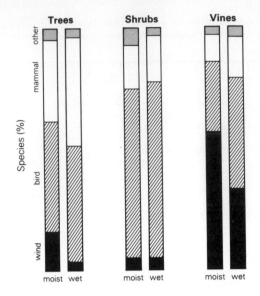

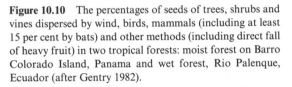

Figure 10.10 The percentages of seeds of trees, shrubs and vines dispersed by wind, birds, mammals (including at least 15 per cent by bats) and other methods (including direct fall of heavy fruit) in two tropical forests: moist forest on Barro Colorado Island, Panama and wet forest, Rio Palenque, Ecuador (after Gentry 1982).

11
Processes of vegetation change

We may now examine in more detail the causes of vegetation changes, chiefly through experiments or observations which ecologists have carried out on the redevelopment of forest vegetation in eastern North America after forest clearance. The examples, augmenting the information in Chapter 9, are used to identify some general principles.

By **process** is meant a set of causally connected events which in this case relate to changes in the plant populations of stands of vegetation. More than shifts in species composition of populations is implied, however. The term process encompasses the interactions of plants with their abiotic and biotic environment; connected processes determine the course of vegetation change.

The main vegetation change patterns to be investigated are sequential replacements: **colonization** of unvegetated areas; **transition** – the sequential changes in plant populations subsequent to first colonization; **maturation** – the development and maintenance of mature vegetation where the stands of dominant plant species replace themselves for at least a few generations. Fluctuating replacement and cyclic replacement are considered only in passing. As far as possible the term succession is avoided because of the confusion about its meaning, noted in Chapter 1 and discussed more fully in Chapter 12.

COLONIZATION OF UNVEGETATED AREAS

A classic site for examining vegetation change in the eastern North American forest regions is the field which, after forest clearance, has been farmed for a time, then abandoned. The problems with 'old field' sequences, as models for what happens after forest is destroyed, are outlined in Chapter 9. However, much of the available empirical and observational data on vegetation change processes come from such sites, so they are considered here. The vegetation sequences on old, arable fields probably resemble the kinds of revegetation processes occurring immediately after an area has been completely stripped of its forest (e.g. by a fire). Fields that have been in pasture are likely to undergo different sequences of vegetation development.

Within a year after abandonment, much of the space in previously cultivated fields in the eastern forest zone is covered by populations of mainly annual and biennial herbaceous species; a few tree and shrub seedlings may also be present (Oosting 1942, Keever 1950, 1979, Bard 1952, Egler 1954). The species comprising the colonizing flora differ from locality to locality, within and between different parts of the region (Table 11.1). Although usually composition will be qualitatively similar, the flora

Table 11.1 Herbaceous plant species prominent in the initial colonizing flora of abandoned fields, eastern North America.

Species	Family	Dispersal mode*	Life history†	Localities‡	After abandonment of field First year	Second year
Agrostis hyemalis	Poaceae	n (? mammal fur)	P	2	+	√
Allium sp.	Liliaceae	bulbils	P	4	+	√
Ambrosia artemisiifolia	Asteraceae	mammal fur, feet	A	1 2 3 4 5	+	√or+
Andropogon scoparius	Poaceae	(? mammal fur)	P	2		√
A. virginicus	Poaceae	(? mammal fur)	P	1 4 5	√	√or+
Antennaria sp.	Asteraceae	wind, stolons	P	5		+
Aristida dichotoma	Poaceae	mammal fur	A	5		+
Aster ericoides	Asteraceae	wind, rhizomes	P	5	√	+
A. pilosus	Asteraceae	wind	B–(P)	1 3 4	√or+	+
Barbarea vulgaris§	Brassicaceae	?projected	B–(P)	3	+	+
Bulbostylis capillaris	Cyperaceae	n	A	5	+	
Chenopodium album§	Chenopodiaceae	n	A	3	+	+
Cynodon dactylon§	Poaceae	(? mammal fur) stolons, rhizomes	P	4	√	+
Cyperus compressus	Cyperaceae	n	A	5	+	
Daucus carota§	Apiaceae	mammal fur	B	2 3 5	+	+
Digitaria sanguinalis§	Poaceae	n	A	1 4 5	+	√or+
Diodia teres	Rubiaceae	(? birds, mammal fur)	A	1 5	√	+
Erigeron annus	Asteraceae	wind	A–(B)	1	√	+
E. canadensis	Asteraceae	wind	A	3 4 5	+	+
E. philadelphicus	Asteraceae	wind	A–B–(P)	3	+	+
E. strigosus	Asteraceae	wind	A–(B)	4	+	+
Gnaphalium purpureum	Asteraceae	wind	A–B	5	+	+
Hypericum gentianoides	Hypericaceae	n	A	5		+
Juncus tenuis	Juncaceae	(? mammals)	P	5	+	
Leucanthemum vulgare§	Asteraceae	n	P	2	+	√
Lespedeza spp.	Papilionaceae	(? mammal fur)	A or P	4	+	+
Oenothera parviflora	Onagraceae	n	B–P	2	+	√
Oxalis stricta	Oxalidaceae	n rhizomes	P	4	√	+
Panicum dichotomum	Poaceae	n		1 5		+
Plantago aristata	Plantaginaceae	n	A	5		+
Potentilla simplex	Rosaceae	n rhizomes, stolons,	P	2	√	+
Rumex acetosella§	Polygonaceae	n, rhizomes	P	2 5	+	√
Setaria viridis§	Poaceae	(? mammal fur)	A	3	+	√
Solidago graminifolia	Asteraceae	wind, rhizomes	P	2	+	+
S. nemoralis	Asteraceae	wind	P	1 2		+
S.sp.	Asteraceae			5		+
Torilis japonica§	Apiaceae	mammal fur	A	4	√	+

* n, no apparent specialization.
† A, annual, B, biennial, P, perennial; (P) rarely perennial; (B) rarely biennial.
‡ 1, Southern Illinois, Bazzaz (1968); 2, New Jersey, Bard (1952); 3, Pennsylvania, Keever (1979); 4, Central Tennessee, Quarterman (1957); 5, Piedmont, North Carolina, Oosting (1942).
§ Not native to North America
+ abundant √, present, not usually abundant

may differ, quantitatively, even between adjacent fields (Swieringa & Wilson 1972, J. Zedler & Zedler 1969).

The propagules which give rise to the adult colonist populations include immigrant seeds and spores, resident (seed bank) seeds and spores and resident vegetative organs. Immigrant seeds are, of course, derived from more or less distant adult populations; any propagules which may reach a site and from which adults eventually arise can be regarded as part of the 'potential population' for that site.

Seed Rain

The seed rain reaching a site over a period depends on the proximity and numbers of parents contributing seed as well as their fruiting period, the abundance of the seed crop, dispersal modes of the species concerned, the operative dispersal agencies and other factors (Table 11.2). Many early colonists of abandoned fields are winter and summer annual herbs. Members of the Asteraceae, with light achenes (single-seeded dry fruit) suited to relatively long-distance dispersal by wind, are well-represented (Bazzaz 1979). Hereafter the term seeds will refer, also, to various kinds of single-seeded fruit which are dispersed like seeds, including the grains of grasses.

Among the immigrant seeds will be those of various perennial herbs and some from the perennial woody species of nearby shrub or forest stands, transported by wind, birds or mammals (Table 11.1). Because of vagaries in the various factors influencing the arrival of a seed at a site, the quantitative composition of the seed rain for any site

Table 11.2 Variable and interacting factors affecting the quantitative seed rain (the numbers of viable seeds reaching an area of ground).

---

1. numerical
   numbers of contributory parents
   abundance of seed crop (and variability year to year, see 3(b))

2. spatial
   proximity to contributory parents
   inherent dispersal distances for the seeds

3. temporal
   (a) annual
       fruit-ripening period
       period over which potentially dispersible seeds remain viable
   (b) year to year
       variability of seed crop (usually connected with variable weather conditions)

4. dispersibility
   general dispersal mode
   particular vectors of dispersal
   availability and effectiveness of vectors at the right time

5. predatory attrition
   before leaving parent
   in transit

6. other chance factors

Table 11.3 Brief list of seedlings which germinated from soil samples on a chronoseries of sites (abandoned fields and developing forest stands), Harvard Forest, Massachusetts, USA (after Livingston & Allessio 1968)

Age of stand (years)	1	2	3	7	25	37	80
Main vegetation	*Erigeron*	*Solidago*	*Andropogon*	*Pinus strobus*	*P. strobus*	*P. strobus*	*P. strobus*
seedling species							
Portulaca oleracea[*]	+	+					
Erigeron canadensis	+	+		√			
Potentilla norvegica	+	+	√	√	√	√	
Panicum capillare	+	+		√			
Rumex acetosella[*]	√	+	+	+	√	√	√
Juncus tenuis	√	+	+	+	+	+	+
J. effusus[*]	√		√		√	+	√
Hypericum perforatum[*]	√	√		+	+	√	
Solidago canadensis		+		√			
Agrostis hyemalis	√			√	+	√	
Panicum lanuginosum	√		√	√	+	√	
Andropogon scoparius			+	√	√	√	
Cyperus spp.			√		√	√	+

√, < 10 seeds germinating per sample; +, 10 or more seeds germinating per sample.
[*] Plants not native to North America.
The seedlings are assumed to have originated from the viable seeds in long-lived seed banks.

may be predicted only within broad limits (Holt 1972). Nevertheless, species such as horseweed (*Erigeron canadensis*) or loblolly pine (*Pinus taeda*) produce such huge quantities of seeds that they tend to saturate available sites, as will be seen.

Seed bank

Studies of the populations of viable, dormant seeds (Livingston & Allessio 1968, Roberts 1981) (Table 11.3) show that a proportion of the population of herbaceous colonists in abandoned fields originates from the seed bank. Not many winter annuals have innately dormant seeds, that is, dormant when they are shed (Keever 1950, Regehr & Bazzaz 1979), so they are poorly represented in seed banks. Seeds of summer annuals have thick seed coats and are innately dormant (Bazzaz 1979). They usually exhibit other dormancy phenomena, also (Fig. 11.1). Many perennial herbs and some shrub and tree species also have dormant seeds (Table 9.1).

Viable seeds of wild plants in agricultural soils may range from a few hundreds to many thousands in each square metre of ground. The qualitative and quantitative representation of seeds in the seed bank depends on the previous history of the site; many dormant seeds originated when their parents grew on or near the site one or more growing seasons earlier; others are immigrant. By cultivation and other means (e.g. falling into crevices or being covered during rodent burrowing) the seeds become buried.

Depending on the species, some dormant seeds reside in the soil for periods ranging from a year to many decades (Kivilaan & Bandurski 1981, Roberts & Feast 1973, Schopmeyer 1974). There is gradual attrition, over time, of the viable seed bank population derived from each year's seed crop, due to predation by animals, fungal attack, loss of viability through ageing and other natural causes (Figs 11.2 & 3).

Soil disturbance, access to light and oxygen, and prior cold treatment are the most common of the various kinds of environmental treatments required to break the dormancy of seeds (Table 3.2). Cultivation brings seeds to the surface, thereby placing them in suitable places both for the breaking of dormancy and for germination (Roberts & Feast 1973). Nevertheless, a proportion of the seeds from any year's fruiting may remain dormant (Bazzaz 1979). Seeds which come out of dormancy, but do not germinate, may return to the dormant state again (see Baskin & Baskin (1985), for an excellent account of the complex nature of seed dormancy phenomena).

Vegetative propagules

The history of the site also determines what vegetative propagules of perennial plants are present after agriculture ceases (Keever 1950, Egler 1954). Among the roots of propagules are rhizomes, bulbs, corms and specialized roots. Often they are the normal method which the species employs to survive unfavourable cold winter or dry summer climatic conditions. New shoots can quickly arise from them. If, before cultivation, the field has only recently been cleared of woody vegetation, there may also be surviving root-stocks of numbers of woody species (Bazzaz 1968).

Establishment

The soil surface environment of open fields is extreme in various ways. Light is unfiltered (high in all wavelengths between 0.4 and 0.7 μm), diurnal and seasonal temperatures are variable, with wide extremes, humidity is often low and water may

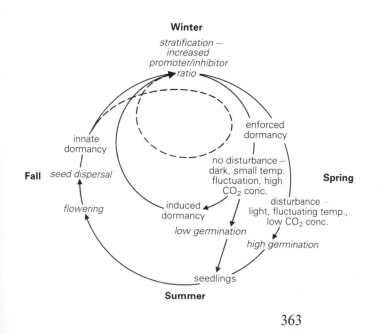

Figure 11.1 Patterns of dormancy and germination in achenes (seed-like fruits) of the annual forb *Ambrosia artemisiifolia*, a pioneer species on abandoned arable land. The broken line represents part of the seed population that requires more than one winter chilling period (after Bazzaz 1979).

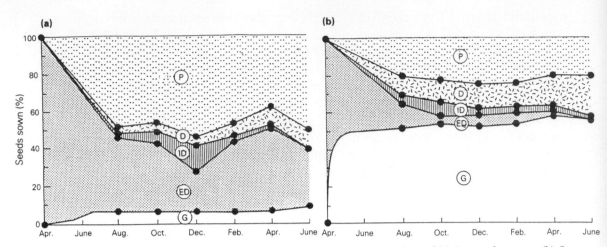

Figure 11.2 Changes in buried seed populations with time. Changing proportions of (a) *Ranunculus repens*, (b) *R. acris* achene populations in soil over seven months. G, achenes observed to germinate; ED, achenes in 'enforced dormancy' (conditions not right for germination); ID, achenes in 'induced' dormancy (physiologically controlled dormancy); D, achenes that decayed; P, achenes that were removed, presumably by predators. After Sarukhan (1974).

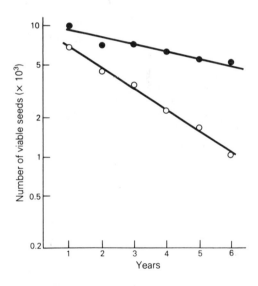

Figure 11.3 Numbers of viable weed seeds remaining after six years in (●) undisturbed and (○) cultivated arable soil (after Roberts & Feast 1973).

be limiting at times. Some perennial herbs may develop from vegetative propagules which survived cultivation. Seeds below the soil surface are buffered against extremes (Fenner 1985). If the field was last cultivated in summer, a population of winter annuals, dispersed in the autumn, usually develops immediately (Keever 1950, 1979). Their seeds need relatively cool, moist conditions and reduced light for germination. If the last cultivation was in autumn, a population of summer annuals and perennials develops from dormant seeds which were brought to, or near, the surface by cultivation and which become preconditioned by the winter cold. Summer annual seeds germinate in spring and early summer. The cotyledons are epigeal and the germinating seeds require light (and a low far-red : red light ratio) (Grime & Jarvis 1975, Wesson & Wareing 1969). Fluctuating temperature is usually beneficial

(Warington 1936) and a flush of nitrate may stimulate germination (Fenner 1985).

Places suitable for germination and establishment of any seed species are termed 'safe sites' by Harper (1977). Above all they must be places where light and moisture conditions can be met without stress to the germinating seed or the young seedling. Differences which individual seeds perceive arise from small-scale heterogeneity of the surface (irregularities, small crevices and clods) and within the soil (structural features, animal tunnels of different sizes and channels left by dead roots) as well as from organic residues, microbial populations, differing concentrations of nutrients and differing amounts of water. Soil heterogeneity, interacting with seeds of different sizes and shapes (Harper *et al.* 1965, 1970) (Figs 11.4–6) ensures that some will be in positions where films of moisture are maintained against the seed micropyle.

Although the soil insulates them from some hazards, seeds that germinate beneath the surface may experience CO_2 narcosis, O_2 deficiency and inadequate light. Their radicles may have problems penetrating a densely packed soil layer. They are vulnerable to animal predators and to fungal and bacterial pathogens.

The combined influences of all of the hazards cause huge attrition of germinating seeds and young seedlings (C. H. Thompson 1981, Fenner 1985). Those that survive grow rapidly; the summer annuals, especially, have high photosynthetic rates; 18–25 mg $CO_2 dm^{-2} h^{-1}$ or more (Bazzaz 1974, Peterson & Bazzaz 1978, Regehr & Bazzaz 1979). Light saturation is at high intensities. Some of the plants are, in any

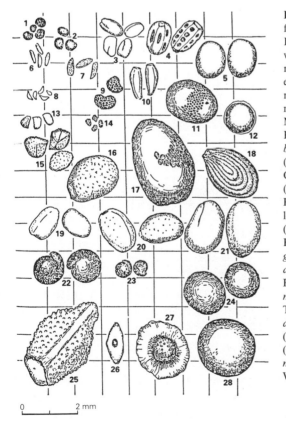

Figure 11.4 A selection of seeds or single-seeded dry fruit of weedy plant species from disturbed sites in Britain (drawn to scale). The seeds or fruits display a wide range of sizes and shapes which will affect their relationship to irregularities in the soil surface or differences in soil structure, with consequent effects on germination. The grid shows millimetre squares at the same magnification. (1) Sandwort (*Arenaria leptoclados*). (2) Mouse-ear chickweed (*Cerastium semidecandrum*). (3) Flixweed (*Descurania sophia*). (4) Petty spurge (*Euphorbia peplus*). (5) Dove's-foot cranesbill (*Geranium molle*). (6) Marsh cudweed (*Gnaphalium uliginosum*). (7) Cudweed (*Filago germanica*). (8) Narrow-leaved rush (*Juncus tenuis*). (9) Corn poppy (*Papaver rhoeas*). (10) Pineapple weed (*Matricaria matricarioides*). (11) Cut-leaved cranesbill (*Geranium dissectum*). (12) Spurrey (*Spergula arvensis*). (13) Hard rush (*Juncus inflexus*). (14) Broomrape (*Orobanche minor*). (15) Pimpernel (*Anagallis arvensis*). (16) Annual mercury (*Mercurialis annua*). (17) Field cranesbill (*Geranium pratense*). (18) Penny cress (*Thlaspi arvense*). (19) Bitter cress (*Cardamine amara*). (20) Milkmaid (*Cardamine pratensis*). (21) Thanet cress (*Cardaria draba*). (22) Fathen (*Chenopodium album*). (23) Red goosefoot (*Chenopodium rubrum*). (24) Charlock (*Sinapis arvensis*). (25) Corn gromwell (*Lithospermum arvense*). (26) Great plantain (*Plantago major*). (27) Yellow toadflax (*Linaria vulgaris*). (28) White mustard (*Sinapis alba*). From Salisbury (1961).

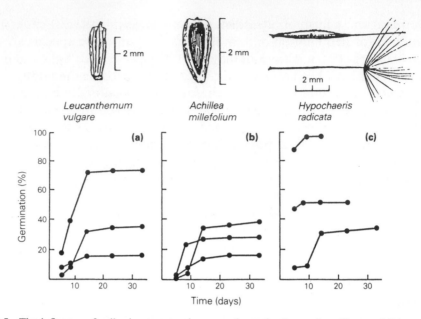

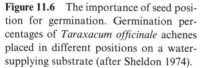

Figure 11.5 The influence of soil microtopography on seed germination and seedling establishment. Percentage germination of achenes and survival of seedlings of *Leucanthemum vulgare*, *Achillea millefolium* and *Hypochaeris radicata* on soil with three grades of microtopography. (1) Flat soil surface, (2) 10 mm grooves, (3) 20 mm grooves. After Oomes & Elberse (1976).

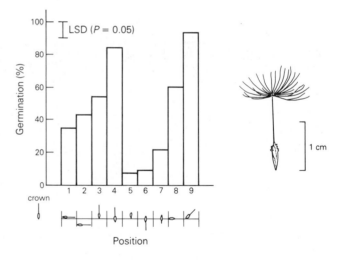

Figure 11.6 The importance of seed position for germination. Germination percentages of *Taraxacum officinale* achenes placed in different positions on a water-supplying substrate (after Sheldon 1974).

case, C4 species which are unaffected by light intensities sufficiently high to cause photorespiration in C3 plants (Fitter & Hay 1981). Metabolism is rapid and dark respiration rates high (Grime 1966, Larcher 1980). Transpiration rates are also high, but rapid photosynthesis rates are maintained at high water potentials (Bazzaz 1974, 1979).

By rapid growth under relatively extreme climatic conditions, annual plants take advantage of freshly disturbed, open sites where, initially at least, competition is minimal. However, ground cover soon becomes denser and the successive appearance of herbaceous species in abandoned fields is very strongly dependent on competitive interactions (Davis & Cantlon 1969). The sequence in North Carolina, USA, fields was explained well by Keever (1950) (Figs 11.7 & 8).

Assuming that a field has been abandoned after the summer harvest, crabgrass (*Digitaria sanguinalis*), a summer annual which had accompanied the crop, is usually abundant. Winter annuals like horseweed germinate in the autumn and overwinter as small rosettes. Some may be lost by winter frost-heaving (Bazzaz 1979). In spring they quickly attain adult size and become dominant, pre-empting resources (light, water and, particularly, nutrients). They flower and die in the summer. Although summer annuals such as *Aster pilosus* germinate in the spring, they remain small. The drought-resistant horseweed easily suppresses them, for, although *A. pilosus* may establish in the shade of other plants, it needs full sunlight to flower. The summer annuals nearly all overwinter and grow to adulthood and dominance in their second year. *Aster pilosus* thus behaves as a biennial, although if it occupies fields lacking winter annuals it flowers in its first year and often dies then. However, it may also produce a second leaf rosette after its first flowering, and flower again in the following year.

Broomsedge (*Andropogon virginicus*) is a tall, perennial grass with dormant seeds requiring cold treatment before they germinate. The drought-resistant seedlings thrive best in full sunlight. A few seedlings appear in the fields in the first year. They

Figure 11.7 A field in the North Carolina Piedmont, USA, one year after cropping ceased, dominated by *Erigeron canadensis* (horse weed), grasses and other weeds. Developing stand of *Pinus taeda* (loblolly pine) in background (photo W. D. Billings).

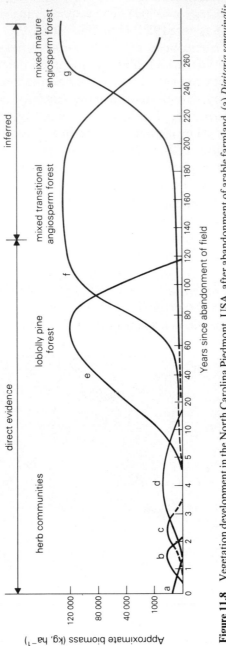

Figure 11.8 Vegetation development in the North Carolina Piedmont, USA, after abandonment of arable farmland. (a) *Digitaria sanguinalis.* (b) *Erigeron canadensis.* (c) *Aster pilosus.* (d) *Andropogon virginicus,* etc. (e) *Pinus taeda.* (f) *Liquidambar styraciflua, Nyssa sylvatica, Liriodendron tulipifera, Acer rubrum, Cornus florida,* etc. (g) *Quercus alba, Q. coccinea, Q. prinus, Carya* spp., etc. (– – –) Sporadic occurrence. After Oosting (1942), Keever (1950), Christensen & Peet (1981).

increase in size but do not flower until the autumn of their second year. Broomsedge is an efficient competitor for water, and it suppresses aster seedings in this way and by shading. Broomsedge does not take over the fields completely until the year following its first seedling. It dominates then until the fields are invaded by pines (about five years after the cessation of farming).

Keever (1950) experimented with the plants of the North Carolina fields to ascertain whether phytotoxicity might be involved with the sequence of changes. Horseweed and aster seedlings were stunted by decaying horseweed roots but Keever (1979) suggested that the inhibition was caused by an unfavourable carbon : nitrogen ratio. Broomsedge also seems to have inhibitory effects on other species, including shrubs and loblolly pine seedlings (Rice 1972, Priester & Pennington 1978), but the suppression may be caused by microbial metabolites. Although it is difficult to demonstrate, unequivocally, that phytotoxicity is involved with abandoned field vegetation sequences, various workers (Rice 1974, 1979) have produced much circumstantial evidence that autotoxicity, and interspecific phytotoxic effects, are implicated in herbaceous vegetation sequences on abandoned fields in Oklahoma, USA. Inhibition by annual needlegrass (*Aristida oligantha*) of nitrogen fixation by *Rhizobium* (in legume root nodules) is an indirect way by which phytotoxic interference may influence the course of events (in this case maintenance of *Aristida* stands). Critical experimental data are needed to verify that the observed effects are actually due to phytotoxicity, and not some other microbial or resource-use influence.

Niches of herbaceous species

Experiments investigating the niche differences of species on abandoned fields in Illinois (Wieland & Bazzaz 1975, Parrish & Bazzaz 1976, 1979, 1982a, b, Pickett & Bazzaz 1978, Bazzaz 1987) help to explain the sequence of replacements in herbaceous communities of old fields (Tables 11.4–6). The mature vegetation in the area is tall grass prairie; weed communities similar to those in the eastern forest zone (but many different species) occur in the recently abandoned fields.

The conclusions, from detailed experiments on their relative behaviour, are that the species of the denser, closed communities are more niche-differentiated, have narrower niches and compete less with their neighbours than do species of open, colonist communities (Figs 11.9 & 10). Root and shoot competition are both important (Pinder 1975, Allen & Forman 1976, Reed 1977, Bakelaar & Odum 1978, Caldwell *et al.* 1987, Grime *et al.* 1987).

Plant productivity, diversity and nutrient use were examined by Mellinger & NcNaughton (1975) in a time series of annuals → herbaceous perennials → shrubs, on abandoned fields in New York, USA. As the system aged they found consistent decreases in the synchrony of resource demand, with later species reaching peak biomass more out of phase with one another. In successively older fields increasing numbers of herbaceous species appear for a brief time and then are replaced. Niches become increasingly narrow.

In this sequence phosphorus, magnesium and calcium available to plants in the soil within reach of root systems are gradually depleted (Fig. 11.11). Potassium increases,

Table 11.4 Niche comparisons between pioneers on abandoned fields, transitional species and mature prairie species in Illinois, USA (after Parrish & Bazzaz 1982a). Niche breadth and proportional similarity of species in pure and mixed stands calculated from total plant growth per species per box on a nutrient gradient.

	Species code	Niche breadth (B)		Proportional similarity (PS)	
		Pure stand	Mixed stand	Pure stand	Mixed stand
pioneer					
Abutilon theophrasti	(At)	0.74	0.62	0.89	0.77
Amaranthus retroflexus	(Ar)	0.80	0.56	0.81	0.75
Ambrosia artemisiifolia	(Aa)	0.70	0.71	0.89	0.77
Chenopodium album	(Ca)	0.88	0.66	0.84	0.66
Polygonum pensylvanicum	(Pp)	0.66	0.72	0.85	0.72
Setaria faberii	(Sf)	0.81	0.67	0.85	0.70
mean		0.77	0.66	0.85	0.74
transitional					
Aster pilosus	(Ap)	0.44	0.61	0.81	0.66
Daucus carota	(Dc)	0.69	0.64	0.78	0.70
Oenothera biennis	(Ob)	0.68	0.60	0.79	0.64
Solidago canadensis	(Sc)	0.52	0.42	0.72	0.61
Vernonia altissima	(Va)	0.38	0.43	0.82	0.62
mean		0.54	0.54	0.78	0.65
mature					
Andropogon gerardii	(Ag)	0.75	0.48	0.87	0.80
Aster laevis	(Al)	0.59	0.50	0.82	0.84
Petalostemum purpureum	(Pp)	0.66	0.45	0.82	0.82
Ratibida pinnata	(Rp)	0.78	0.56	0.84	0.84
Solidago rigida	(Sr)	0.76	0.51	0.87	0.78
Sorghastrum nutans	(Sn)	0.68	0.47	0.79	0.77
mean		0.70	0.50	0.83	0.81

See the original article for the methods of calculating niche breadth and mean similarities. The colonist species have the broadest mean biomass response along the gradient in both pure and mixed stands. The average niche breadth for mature prairie species is slightly narrower than for colonists in pure stands and much narrower in mixed stands. The mean niche breadth of transitional species is the same in pure and mixed stands, but three of five species had narrower niche breadths in mixed stands.

however. When shrubs invade the fields nutrients increase in the topsoil, and this is ascribed to the tapping of supplies from greater soil depth and their return to the upper soil layers by recycling. Fertilizer applications indicate that the most important limiting nutrient controlling community function is phosphorus.

The increase in diversity of the community over time is thought by Mellinger & McNaughton (1975) to be caused by microhabitat diversification, arising from local changes in soil properties around individual plants (cf. Zinke 1962). This could be through nutrient depletion, enrichment or accumulation of phytotoxins.

POPULATION CHANGES OF WOODY SPECIES ON ABANDONED FIELDS

Forests eventually develop and replace the herb communities of abandoned fields. As indicated in Table 11.7, some woody plants appear in the fields virtually as soon as

Table 11.5 Mean percentage concentration of five nutrients per unit dry weight of shoots in relation to nutrient supply, for species from pioneer, transitional and mature prairie vegetation in Illinois (from Parrish & Bazzaz 1982a).

Nutrient element	N				P				K				Ca				Mg			
Concentration	LL	M	HH	r	LL	M	HH	r	LL	M	HH	r	LL	M	HH	r	LL	M	HH	r
pioneer																				
Abutilon theophrasti	1.37	2.75	2.65	0.71	0.13	0.12	0.20	0.08	1.83	2.30	2.77	0.88	1.83	2.13	2.79	0.48	0.34	0.49	0.68	0.87
Ambrosia artemisiifolia	2.68	3.85	4.42	0.88	0.13	0.11	0.27	0.56	2.51	2.75	3.90	0.76	1.90	1.97	2.17	0.45	0.59	0.52	0.65	0.02
Amaranthus retroflexus	1.42	1.47	3.96	0.78	0.29	0.21	0.19	−0.52	2.92	2.63	4.45	0.59	1.04	1.24	2.09	0.76	0.58	0.63	0.89	0.63
Polygonum pensylvanicum	0.88	2.59	—	0.91	0.10	0.16	—	0.16	1.27	2.28	—	0.92	0.52	1.04	—	0.95	0.44	1.16	—	0.85
Chenopodium album	1.89	3.20	4.80	0.94	0.20	0.21	0.32	0.41	4.15	4.68	9.20	0.65	1.07	1.43	1.56	0.51	0.78	0.94	0.83	−0.10
Setaria faberii	1.08	1.80	3.19	0.87	0.06	0.08	0.18	0.61	1.92	2.20	4.81	0.84	0.39	0.49	0.99	0.65	0.26	0.40	0.49	0.86
transitional																				
Aster pilosus	1.50	1.68	—	0.16	0.18	0.18	—	0.03	3.34	3.87	—	0.41	1.17	0.94	—	−0.42	0.26	0.29	—	0.21
Daucus carota	1.32	2.02	—	0.85	0.15	0.10	—	−0.32	3.37	3.38	—	0.21	1.98	1.51	—	−0.30	0.40	0.31	—	−0.63
Oenothera biennis	0.88	2.12	2.37	0.80	0.13	0.15	0.20	0.44	1.46	2.49	2.70	0.65	1.67	1.70	1.85	0.06	0.35	0.61	0.57	0.68
Solidago canadensis	1.13	1.38	1.86	0.78	0.14	0.14	0.20	0.43	1.80	1.86	2.73	0.69	0.78	1.11	0.78	0.23	0.37	0.57	0.29	−0.01
Vernonia altissima	1.73	2.73	—	0.49	0.20	0.15	—	−0.34	2.39	3.18	—	0.61	2.69	1.96	—	−0.67	0.99	0.56	—	−0.73
mature																				
Andropogon gerardii	0.72	0.82	0.76	0.45	0.09	0.08	0.06	−0.29	1.06	1.05	1.30	0.55	0.49	0.56	0.49	0.19	0.26	0.26	0.19	−0.54
Aster laevis	0.88	2.17	5.40	0.73	0.14	0.17	0.26	0.52	2.05	3.56	3.13	0.45	1.32	1.48	1.21	−0.25	0.31	0.35	0.30	−0.15
Ratibida pinnata	0.95	1.18	—	0.84	0.09	0.11	—	0.71	1.89	2.54	—	0.99	2.86	2.60	—	−0.80	0.74	0.77	—	−0.72
Solidago rigida	1.11	1.14	—	0.04	0.13	0.08	—	−0.25	2.22	2.20	—	0.55	1.48	1.28	—	−0.62	0.33	0.28	—	−0.40
Sorghastrum nutans	0.65	0.61	1.42	0.41	0.12	0.06	0.09	−0.13	0.78	0.84	1.13	0.47	0.46	0.56	0.66	0.44	0.27	0.23	0.28	−0.14

LL, no addition; M, medium; HH, large addition.
The r-value is the linear correlation coefficient for the relationship between concentration and all five nutrient levels (i.e. coded LL = 1 to HH = 5)

Table 11.6 Mean niche breadths and proportional similarities of all species in experimental assemblages of herbaceous plants from pioneer and mature grassland vegetation, Midwestern USA (after Bazzaz 1987).

		Early colonist	Mature grassland
(a)	*mean niche breadth*		
	underground space	0.71	0.29
	pollinators	0.20	0.16
	nutrients	0.77	0.70
(b)	*mean proportional similarity*		
	underground space	0.68	0.43
	pollinators	0.31	0.19
	nutrients	0.85	0.83

See the original article for methods and further detail. Species from colonist communities have consistently broader and more-overlapping niche responses on gradients compared with the mature community species, regardless of their stature.

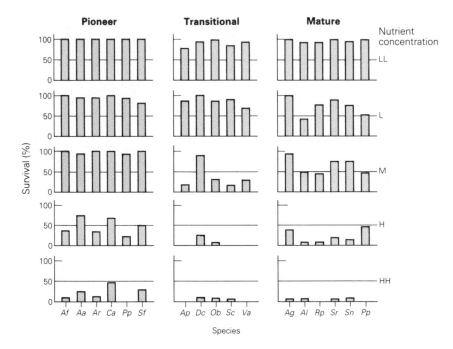

Figure 11.9 Survival to maturity of plants from pioneer, transitional and mature prairie communities in Illinois, USA, on a soil nutrient gradient. Nutrient concentrations: LL, no addition; L, low; M, middle; H, high; HH, very high. Species codes as in Table 11.4. After Parrish & Bazzaz (1982a).

farming ceases. Why do these trees and shrubs not take over the fields immediately? There are several possible explanations.

1   Tree and shrub juvenile populations are slow to build up because their propagules are not dispersed as rapidly or as far as those of herbs.
2   Seed banks and vegetative propagules of woody species are nearly or completely lacking in cultivated fields.

3   Conditions are not very favourable in open, cultivated ground for juveniles of the woody plants.
4   Herbaceous communities exclude the woody plants by rapid growth, pre-empting sites

Chapter 9 showed that disturbed areas within forests often are immediately colonized by shrubs or trees (sometimes including the tree species which form mature forest). Differences between both the 'regeneration potential' and the environments of old fields and small openings in forest account for the different colonizing

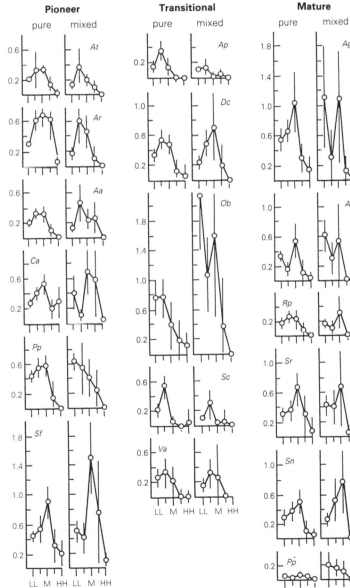

Figure 11.10 Pioneer species contrasted with transitional and mature prairie species, Illinois, USA. Average total weight (g) per individual plant at each soil nutrient concentration (see Fig. 11.9 & Table 11.5). After Parrish & Bazzaz (1982a).

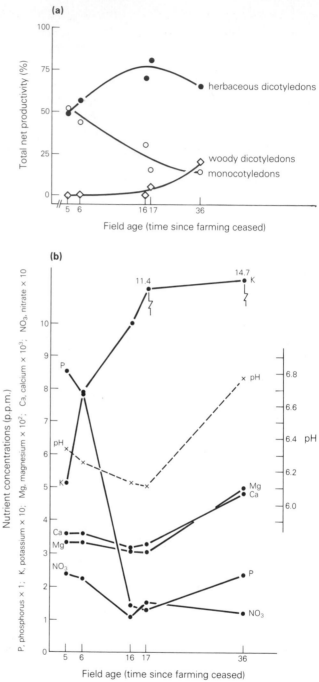

Figure 11.11 Vegetation development on abandoned farmland in New York State, USA. (a) Percentage contributions of different plant groups over time to total above-ground productivity. (b) Depletion over time of nutrients (calcium (Ca), magnesium (Mg), phosphorus (P) and nitrate (NO_3)) and decline in pH, in the upper soil layer (15 cm) beneath herb communities, then increase in all of these except nitrate with increased abundance of shrubs. The increase in potassium (K) is unexplained. After Mellinger & McNaughton (1975).

patterns. First, in disturbed forest areas opportunities for the immediate establishment of woody pioneer stands are very good; they grow up from sprouts, suppressed juvenile populations, seed banks (Oosting & Humphreys 1940, Olmsted & Curtis 1947) and influx of seeds from nearby adults. Sometimes, especially on burnt areas, stands of herbs or shrubs develop by sprouting from surviving rootstocks, by

374

seed banks or by seed immigration. Tree regeneration is then usually inhibited for a time, as the herbs or shrubs pre-empt the sites.

There is less likelihood that tree propagules will be present in, or immigrate so rapidly to, abandoned arable fields; the opportunities for herbaceous populations to take over open fields are excellent, however. Once herb communities have occupied abandoned fields they inhibit the development of forest for a time. Tall and vigorous annual forbs, or sward-forming grasses, are efficient competitors for water and nutrients with the seedlings of woody plants (McCormick 1968, Mellinger & McNaughton 1975). The spaces which arise as old herbs die are available for colonization by woody plant seedlings or other herbs.

Limited dispersal opportunities, as well as the limited seed stocks available, appear to be among the major causes of delay in establishment of woody plants. The study by Livingston & Allessio (1968) showed that woody species are not represented in the seed banks of recently abandoned fields. Although some woody species (sassafras (*Sassafras albidum*) and persimmon (*Diospyros virginiana*)) can establish by sprouting from root fragments (Bazzaz 1968, Egler 1954), this happens only if the trees had been present not long beforehand, or grow on the margins of the fields. Most of the early-established woody plants are early transitional species with wind- or bird-dispersed seeds (Tables 11.7 & 9.1). Although seeds mostly fall or are deposited near the parent plant (Green 1983, Fenner 1985) (Fig. 11.12), if conditions are right some wind- and bird-dispersed shrub or tree seeds may travel 1 km or more (Fowells 1965, Howe & Smallwood 1982).

Although juveniles of some woody plant species establish in recently abandoned fields (Table 11.1), they do not necessarily persist. Bard (1951) observed that the

Table 11.7 Plant species prominent* in time sequences of vegetation on abandoned fields in eastern North America

	Growth form, life history	Time since field abandoned (years) (see Table 11.1 for 1st and 2nd year fields)		
		3	4–10	15
Southern Illinois (Bazzaz 1968)				
Ambrosia artemisiifolia	H,A	√	√	√
Andropogon virginicus	HG,P	√	+	+
Aster pilosus	H,B	√	√	
Diodia teres	H,A	√	√	√
Panicum dichotomum	HG,P	+	√	√
Solidago nemoralis	H,P	+	+	√
Rubus occidentalis	S	√	√	√
Smilax glauca	V	√	√	√
Diospyros virginicus	T	√	√	√
Juniperus virginiana	T		√	
Sassafras albidum	T		√	√
Ulmus alata	T		√	√

+, abundant; √, present.
* Many other herbs occur in each locality. Various other juvenile trees and shrubs occur sporadically, some in 1st and 2nd year fields. Among them are oaks (*Quercus* spp.).

Table 11.7 continued

		Time since field abandoned (years)			
		5	10–15	25–40	60
New Jersey					
(Bard 1952, Buell *et al.* 1971)					
Ambrosia artemisiifolia	H,A	√	√	√	√
Andropogon scoparius	HG,P	+	+	+	+
Aster ericoides	H,P	+	+	+	√
Daucus carota	H,B	+	√	√	√
Hieracium florentinum	H,P	+	+	+	√
Oenothera parviflora	H,B	√	√	√	√
Potentilla simplex	H,P	+	+	+	+
Rumex acetosella	H,P	√	√	+	+
Solidago graminifolia	H,P	+	+	+	+
S. juncea	H,P	+	+	+	+
S. nemoralis	H,P	+	+	+	+
Rhus glabra	S	√	√	√	+
Rubus flagellaris	S	√	+	+	+
Celastrus scandens	V			√	+
Lonicera japonica	V	√	+	+	
Rhus (= Toxicodendron) radicans	V–S	√	√	+	+
Acer rubrum	T	√	√	√	√
Juniperus virginiana	T	√	+	+	+
Myrica pensylvanica	T	√	√	√	+
Sassafras albidum	T	√	√	√	+

		Time since field abandoned (years)			
		3	4–8	15–20	25
Central Tennessee					
(Quarterman 1957)					
Allium sp.	H,P	+		√	
Ambrosia artemisiifolia	H,A	√	√	√	
Andropogon virginicus	HG,P	+	+	+	
Aster pilosus	H,B	+	+	+	
Bromus japonicus	HG,A	√	+	√	
Eragrostis sp.	HG, A or P		+		
Erigeron canadensis	H,A	√	√		
E. strigosus	H,A	√	√	√	
Geum canadense	H,P				+
Lespedeza spp.	H, A or P	+	+	+	
Oxalis stricta	H,P	√	√		
Sanicula marylandica	H,B–P				+
Solidago altissima	H,P	+	√	+	√
Symphoricarpus orbiculatus	S	+		+	+
Campsis radicans	V	√	√	√	√
Rhus (= Toxicodendron) radicans	V–S			√	+
Celtis spp	T	√	√	+	+
Cercis canadensis	T			+	
Fraxinus americana	T			+	
Juglans nigra	T				+
Juniperus virginiana	T			+	
Morus rubra	T				+
Ulmus sp.	T			+	+

H herb, G grass, S shrub, V vine, T tree
A annual, P perennial, B biennial (applying to herbs; all shrubs, vines and trees are perennial).

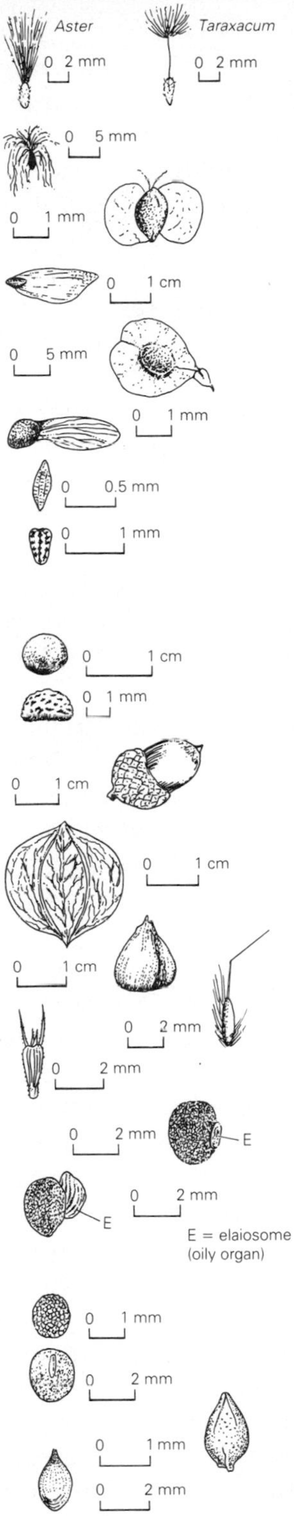

	Characteristic seed weights (g)	Form of seed (with or without fruit)	Distance most seed travels from parent (m)	Usual maximum range achieved by a few seeds (m)
Wind				
...ght, plumed				
...steraceae	0.00001–0.00015			
...opulus	0.00012		< 1 to a few tens	many tens to a few km
...ght, winged				
...etula	0.00033–0.001			
...edium- to heavy-winged				
...inus	0.007–0.025			
...lmus	0.006–0.01		< 1 to a few tens	a few hundred
...cer	0.02–0.3			
...ght, no other specialization				
...uncus	0.0001–0.0003		< 1 to a few tens	a few tens to about 100 (may be much further if accidentally adhering to animals)
...erbascum	0.0001			
Animals				
...wallowed and later voided (mammals, birds)				
...runus (fleshy fruit)	0.09		a few tens to a few hundred	a few km
...ubus (fleshy fruit)	0.001			
...teeth, cheek pouches, ...lls (squirrels, jays)				
...uercus	1.8–4.5		a few tens to 100 or more	? a few hundred
...arya	2.3–15.0			
...agus	0.28			
...xternally, on fur			probably tens to a few hundred	? many hundred
...ndropogon	0.002			
...idens	0.0017–0.0035			
...arried by ants				
...orydalis	0.002		a few to a few tens	?
...helidonium	0.002	E = elaiosome (oily organ)		
Other				
...rojected				
...rassica	0.0017		< 1	?
...cia	0.018		< 1 to 1	?
...ater				
...olygonum	0.001–0.002		a few to a few hundred	many km
...arex	0.0001–0.002			

Some seeds, usually transported by wind, may travel long distances floating on water. Other, usually animal-transported, seeds may also float considerable distances. Water birds transport some aquatic seeds.

Figure 11.12 Dispersal distances of seeds and one-seeded fruit. A few examples of each are given (data from Forest Service 1948, Salisbury 1942).

numbers of seedlings of black cherry (*Prunus serotina*) present in one-year fields in New Jersey, USA, were severely reduced subsequently. In the same area Buell *et al.* (1971) noted that many tree seedlings which appear in recently abandoned fields last for only one season. They identify drought, insects, rabbits, deer and, particularly, frost heave in winter, as the causes of death of these young seedlings. In their locality few trees become established until at least 12 years after agriculture ceases. Elsewhere some woody species establish earlier (Table 11.7).

Bard (1952) and Buell *et al.* (1971) list as early immigrants in New Jersey old fields some late-transitional and a few mature forest species including oaks (*Quercus* spp.). Their heavy seeds usually fall near the parent tree, but they may be transported distances of 100 m or more by squirrels and jays (Howe & Smallwood 1982, Stapanian 1986). Subsequent recruitment of all species with seeds dispersed by birds or mammals is aided by the presence of the first-established woody plants which provide suitable habitat for the animals (McDonnell & Stiles 1983).

Early development of woody plant populations

What determines the order of appearance of woody plant species during vegetation development on abandoned fields? Again the North Carolina example may be cited (Fig. 11.8). The work of Billings (1938) and Oosting (1942) is especially relevant; further insights arise from the experimental programme of Christensen (1977), Peet & Christensen (1980a, b) and Christensen & Peet (1981).

Angiosperm tree species, sweet gum (*Liquidambar styraciflua*), black gum (*Nyssa sylvatica*) and tulip tree (*Liriodendron tulipifera*) occur occasionally in fields in the first few years after farming ceases, but the angiosperms never become as prominent as do pines; loblolly (*Pinus taeda*) on moister and more-fertile sites and short leaf (*P. echinata*) on less-favourable sites.

Kramer *et al.* (1952) and Bormann (1953, 1956) showed that effective seed dissemination and rapid establishment by pines (beginning in the first five years, with the young pine stand being well-developed by ten years) (Christensen & Peet 1981) (Fig. 11.8) accounted for their rapid rise to dominance. Loblolly pine seedlings have shade-tolerant primary leaves. They can establish in shady but bare sites among weeds. The herbs provide protection from climatic and other extremes. The secondary leaves of seedling pines are not at all shade-tolerant but, by the time they develop, height growth has lifted the foliage to a well-lit position (Bormann 1956).

Pines compete more effectively for water than do angiosperms among the herbaceous stands (Bormann 1953, 1956). Dense, closed pine stands grow up. There are virtually no other canopy species, but many of the angiosperm tree species of the region soon establish beneath the adolescent and adult pines (Figs 11.13 & 14). The species with wind- and bird-dispersed seeds are especially abundant (Tables 11.7 & 9.1). The young angiosperm trees form a more or less suppressed understorey. Self-thinning of the pine stands ensues as the trees grow taller and extend their branch systems laterally to form a dense, closed canopy. The shape of the pine thinning curve (Fig. 11.5) approaches the ideal inverse size–density relationship which is achieved rapidly in dense populations (Silvertown 1982, p. 120). The stands

Figure 11.13 A young *Pinus taeda* stand on a field 8–10 years after cropping ceased, North Carolina Piedmont, USA (photo W. D. Billings).

appear to reach the carrying capacity of the site, presumably determined by nutrient availability (Peet & Christensen 1980a, b). Virtually no pine seedlings ever establish in these stands unless there are major canopy breaks (Kramer *et al* 1952). Even then pine recruitment is usually poor (Christensen & Peet 1981).

The angiosperms which colonize the young pine stands are themselves subject to considerable attrition over time, but there is continuous recruitment of their populations. Even the large-seeded species, oaks (*Quercus* spp.) and hickories (*Carya* spp.) are consistently present (Fig. 11.8 & Table 9.1), carried in by squirrels. Loblolly pines hold the ground for about 100 years, but from about 80 years (when the period of their peak productivity is over) the canopy of the pine stand opens up and dead pines are replaced by transitional angiosperms such as tulip tree, sweet gum and red maple (*Acer rubrum*).

Experiments comparing the relative performance of pine and angiosperm seedlings show that the latter have lower light compensation points (Kramer & Decker 1944). Their photosynthesis is better at low light intensity than that of pines, and under these circumstances they compete more effectively for water and nutrients (Oosting & Kramer 1946, Kozlowski 1949, Ferrell 1953).

A thinning experiment in young pine stands (Christensen & Peet 1981) showed that, when the pine canopy is removed, established juvenile angiosperms quickly grow to canopy height. Recruitment is not affected much (as shown by measurements of 53-year-old pine stands thinned many years previously), although there is a slightly increased influx of the more shade-tolerant species.

Figure 11.14 A 60-year-old mature *Pinus taeda* stand, Duke Forest, North Carolina Piedmont, USA. The understorey deciduous angiosperm trees include sweet gum (*Liquidambar styraciflua*), red maple (*Acer rubrum*) and tulip tree (*Liriodendron tulipifera*) (photo N. Christensen & R. R. Peet).

The first angiosperm-dominated forests which naturally replace the pine stands are themselves gradually replaced, over a period of about 200 years, by species such as white oak (*Quercus alba*), scarlet oak (*Q. coccinea*) and species of hickory (*Carya*) which dominate the mature forest of the region (Figs 11.16 & 17). However, the mature oak–hickory forest usually has a scattering of large trees (but few juveniles except in large canopy gaps) of the less shade-tolerant tulip tree, black gum and sweet gum. Red maple often occurs in the understorey. During the chain of species' replacements in the sequence from open field to mature forest there are consequential changes in productivity and biomass similar to those depicted in Figure 11.18.

SEQUENCES IN OTHER LOCALITIES

Differences in the timing of major changes during early forest development and differences in the roles of the participating species are apparent in localities throughout eastern North America where vegetation change sequences on abandoned fields have been examined (e.g. A. G. Chapman 1942, Buell *et al.* 1971, Bazzaz 1968, Horn 1971) (Table 11.7). Bard (1952) found, in the New Jersey Piedmont, that the most important woody species early in the sequence were the red cedar tree (*Juniperus virginiana*) and the shrub *Rubus flagellaris*. They form a scattered, park-like community with other woody species, among forbs and grasses. The woody plants gradually increase in density until, by about 60 years after farming has ceased, an almost closed stand develops. Red cedar is the main tree, but vines and shrubs contribute much of the cover. Young trees of the submature to mature forest of the area – flowering dogwood (*Cornus florida*), black cherry, red maple, sassafras, white oak, and red oak (*Quercus rubra*) are well-represented.

Presumably the slow development of forest in New Jersey is the result of the persistence of a dense herbaceous cover, absence of pines and climatic conditions which inhibit tree seedling establishment (Sparkes & Buell 1955, Buell *et al.* 1971). It is likely that, if pines were absent in North Carolina, the fields there would undergo a similar, slow development of angiosperm-dominant forest, with sweet gum, tulip tree and species of shrubs and vines playing more prominent roles. A sequence like this, in

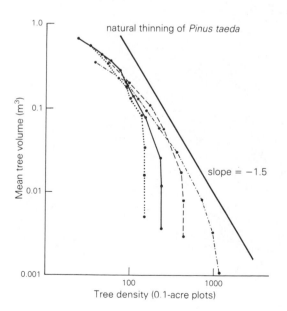

Figure 11.15 Self-thinning curves. Logarithmic plots of changing tree volume versus tree density in four *Pinus taeda* stands of different initial densities (149, 234, 431 and 1172 per 405 m²). A trend towards a −3/2 slope for thinning applies at first, but then the slope shifts towards −1 as the trees mature and reach limiting dimensions (after Peet & Christensen 1980a).

Figure 11.16 A developing mixed deciduous angiosperm stand in winter, Duke Forest, North Carolina Piedmont, USA. Prominent tree species are red oak (*Quercus rubra*), *Fagus grandifolia* and *Acer rubrum* (photo N. Christensen & R. R. Peet).

Figure 11.17 A mature mixed deciduous angiosperm stand, North Carolina Piedmont, USA. Prominent tree species are white oak (*Quercus alba*) and mockernut hickory (*Carya tomentosa*). Subcanopy trees include flowering dogwood (*Cornus florida*), sourwood (*Oxydendron arboreum*) and *Acer rubrum* (photo N. Christensen & R. R. Peet).

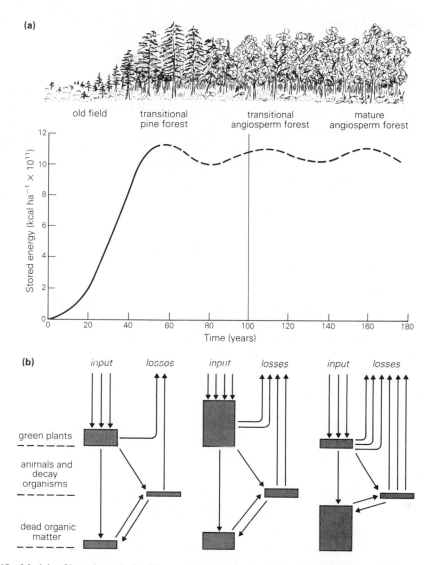

Figure 11.18 Models of broad trends for biomass accumulation and energy flow in a developmental sequence from abandoned fields to mature, mixed angiosperm forest in eastern North America. (a) Rapid rise of biomass after invasion of weedy communities by pines (usually peaking later than is shown here), then a fall and oscillations about an equilibrium level. (b) Relative differences in the compartment sizes for the main ecosystem components in a developing system during pioneer, transitional and mature phases. Modified from Woodwell (1963).

Tennessee, where *J. virginiana* is present but not prominent and a different array of tree species occurs, was outlined by Quarterman (1957) (Table 11.7).

Observation of occurrences of late-transitional or mature forest species in recently abandoned fields in Connecticut led Egler (1954) to suggest that the apparent sequence of different species populations during forest development is really an artefact. He proposed that, in reality *all* woody species were established at the outset and, because they grow at different rates, this creates an impression of successive replacement. This hypothesis can be tested only if individuals are marked and followed through the entire sequence. Unfortunately, no such experiment has been performed (Christensen & Peet 1981), but the circumstantial evidence is that many of the individual early-established seedlings of late-transitional or mature forest species do not persist (Buell *et al.* 1971).

Sassafras often establishes in abandoned fields from root sprouts (Bard 1952, Bazzaz 1968). It then expands, clonally, by continued root sprouting, and may maintain itself into mature forest. The sparse undergrowth and limited flora beneath sassafras canopies in abandoned fields in Tennessee led Gant & Clebsch (1975) to undertake studies of its possible phytotoxic effects on other tree species. Aqueous leachates from leaves, litter and roots reduced the growth of seedlings of other species. Sassafras and possibly some other species including oaks (McPherson & Thompson 1972) appear to be able to occupy and hold ground in this way, as pines do, by possessing a dense shady canopy which suppresses competitors.

Aspens (*Populus tremuloides* and *P. grandidentata*) and grey birch (*Betula populifolia*) are important early transitional species in the northern part of the eastern forest region (G. R. Stephens & Waggoner 1970, Horn 1971). They are also propagated vegetatively and (especially the aspens) may form extensive and sometimes long-lived stands in this way (A. W. Cooper 1981).

It is worth briefly mentioning other patterns of behaviour among tree species which begin their life on bare, open sites. Some of the species cited occur in western or northern North America. Balsam fir (*Abies balsamea*) dominates some forest areas in the north-east of North America. It is a lesser component in some other northern and northeastern forests. If pure balsam fir forests are destroyed, they may regenerate immediately. Balsam fir can behave both as a colonist and a member of mature, mixed forest (Bormann & Likens 1979a). Similar behaviour is found in Boreal and Rocky Mountain gymnosperm forest species (see Ch. 7).

White pine (*Pinus strobus*), red pine (*P. resinosa*) and pitch pine (*P. rigida*), like loblolly and short-leaf pines of the south-east, are common early transitional species in sequences of forest development after forest disturbance (Braun 1950, Whittaker 1975a, Spurr & Barnes 1980). However, scattered individual white pines often persist in mature forest in the north-east. They are able to do so not only because the species is an opportunist, which sometimes establishes in canopy gaps in the mature forest, but also because it is long-lived (450 years or more) and a very tall tree (30 m or more). These properties allow it to coexist as an adult with mature forest species like beech (*Fagus grandifolia*) and sugar maple (*Acer saccharum*), although it is not able to regenerate beneath their canopies.

Some potentially very long-lived tree species have shade-intolerant seedlings. One such species is giant redwood (*Sequoia sempervirens*), of California and Oregon, which regenerates only on disturbed sites, e.g. following fire. Species that are more shade-tolerant, with shorter life-histories (e.g. western hemlock (*Tsuga heterophylla*) and tan oak (*Lithocarpus densiflorus*)) grow under its canopy and, having replaced themselves over several generations, may eventually replace veteran redwoods when they die (Stone *et al.* 1972). Douglas fir (*Pseudotsuga menziesii*), of Washington and Oregon, also behaves like the *Sequoia* under some circumstances (Franklin & Hemstrom 1981).

THE CAUSES OF CONTINUED CHANGES IN WOODY PLANT POPULATIONS

So many variables influence species regeneration and survival that the processes affecting replacement sequences of trees may be expected to be complex. From what has been written above, it is clear that the individual properties of the tree species which occur in any region are a major determinant of the patterns of sequential change.

The environment under tree canopies during most of the growing season differs from that of open sites in ways expressed in Table 11.8. In the following sections contrasts are drawn between the tree species (mostly angiosperms) which customarily establish in open sites (colonists or early-transitional species) and those which establish under tree canopies (late-transitional or mature-forest species).

Temperature, light and humidity effects

Above the soil surface within forest the modified light conditions in summer, as well as the modulated temperature and atmospheric humidity conditions, are the main features contrasting with open sites. Shading by leaf canopies reduces the incoming solar radiation by amounts ranging up to 95 per cent or more. However, the pattern

Table 11.8 Climate and other contrasts between open, unforested sites and sites beneath forest canopies, in fine weather, during the growing season.

	Open sites	Forest sites
light intensity	high	low (down to a small percentage of full sunlight); variable in places – sunflecks, gaps
light spectral quality	unfiltered, complete range	attenuated blue, red, high far-red
air temperature	high daytime, relatively low night-time	relatively even, lower daytime and higher night-time
vapour pressure deficit	relatively high	relatively low
evaporation rate at soil surface	high; transpiration losses high if plant cover present, but drawn from a shallow soil depth	relatively low; transpiration losses high, but drawn from a considerable soil depth
soil organic layers	thin mulch of litter, if any, and humus incorporated into upper mineral soil horizon	thicker mulch of litter and (mor) accumulation of humus above mineral soil horizon or (mull) incorporation of abundant humus into upper mineral soil horizon
soil temperature	very high surface temperature and warm upper soil layer	relatively low surface temperature and upper soil layer (insulation by canopy and litter)
soil moisture	variable, depending on rainfall; depleted rapidly due to strong radiation and high evaporation rate	more even as a result of shading, litter mulch and maintenance by humus

of light distribution within the forest is highly variable. The light quality is also altered (Fig. 2.3). This creates an extreme and often limiting environment for the growth of subcanopy plants, including the juveniles of trees. Those species which can cope with the conditions are the specialized subcanopy shrubs, the juveniles of late-transitional and mature forest trees, and the specialized field layer herbs. Seasonal phenological changes in the deciduous forest are matched by the behaviour of the many subcanopy herbs (Chs 2, 3 & 9) (Hicks & Chabot 1985), but they are not considered further here.

'Tolerance'

Explanations of the replacements of the earliest tree colonists on a disturbed site by transitional species, then their replacement by mature-forest species, are usually in terms of their relative shade-tolerance. 'Tolerance' is a rather vaguely defined concept based on the field experience of foresters (Table 11.9). At the one extreme are the 'very intolerant' species which require open, well-lit sites for the germination of their seeds and growth of young plants. The sites are relatively well-supplied with nutrients, but some environmental extremes (e.g. high temperature or low soil moisture) may be experienced at times. At the other extreme are 'very tolerant' species which have the capacity to germinate and grow well in crowded, deeply shaded sites. Nutrients may be in limited supply and root competition fierce, but temperatures are damped down and moisture supply is more even and reliable. The greatest 'tolerance' is correlated with the capacity of the species concerned to produce the densest canopy (Horn 1971). Relative shade-tolerance, clearly, is involved, but there are wider

Table 11.9 'Tolerance' ratings for eastern North American forest tree species (after Baker 1949).

Very intolerant	Intolerant	Intermediate	Tolerant	Very tolerant
Juniperus virginiana	Juniperus virginiana	Pinus strobus	Picea rubens	Picea rubens
Pinus rigida	Pinus rigida	P. taeda	Abies balsamea	Abies balsamea
P. echinata	P. echinata			Tsuga canadensis
Betula populifolia	P. taeda			
	Carya cordiformis	Carya ovata	Acer rubrum	Acer saccharum
			Tilia americana	
Populus tremuloides	C. tomentosa	Castanea dentata	Nyssa sylvatica	Fagus grandifolia
P. grand-identata	C. glabra	Betula alleghaniensis	Diospyros virginiana	Carpinus caroliniana
	Betula papyrifera	B. lenta		Ostrya virginiana
	Prunus serotina	Prunus serotina		Cornus florida
	Quercus coccinea	Ulmus americana		
	Q. velutina	Quercus velutina		
	Q. rubra	Q. rubra		
	Q. alba	Q. alba		
	Liriodendron tulipifera	Q. prinus		
	Sassafras albidum	Tilia americana		
	Liquidambar styraciflua	Nyssa sylvatica		
	Nyssa sylvatica	Diospyros virginiana		
	Gleditsia triacathos	Fraxinus americana		
	Diospyros virginiana	Magnolia spp.		
	Fraxinus americana			

implications; the term 'tolerance' refers also to other aspects of environmental relationships and the competitive ability of the species concerned.

Temperature and tree growth

The first colonist (early transitional) angiosperm tree species in abandoned fields have a long shoot-growth period which begins after frost finishes in spring and ends when the first frosts of autumn initiate leaf fall (Marks 1975). These species also have indeterminate branching systems and wood of relatively low specific gravity and strength, which is not resistant to wood borers and fungal attack. The morphological pattern allows the trees to utilize favourable weather to the maximum by rapid growth responses. The late-transitional and mature-forest species have a shorter period of shoot growth (in sugar maple 15 weeks) and are determinate. Their wood has high specific gravity and strength, and it resists attack by pests and diseases. Their growth response to favourable temperatures is relatively slow.

Tree architecture

Horn (1971, 1975a) proposed a general mechanism for the replacement of the light-demanding early-transitional tree species in New Jersey forests by more shade-

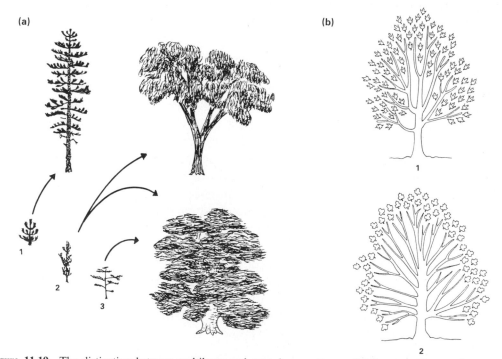

Figure 11.19 The distinction between multilayer and monolayer patterns of foliage arrangement in trees. (a) General shape of tree crowns: (1) multilayer conifer, (2) multilayer angiosperm, (3) monolayer angiosperm. (b) Distribution of leaves in (1) a multilayer tree (*Acer saccharinum*, silver maple) and (2) a monolayer tree (*Acer saccharum*, sugar maple). After Horn (1971, 1975a).

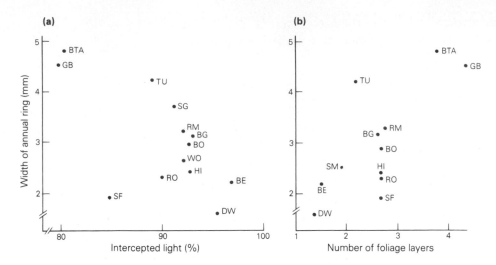

Figure 11.20 The correlation of annual growth layer width in stems of eastern North American trees with each of: (a) light interception by the leafy canopy and (b) number of foliage layers. BE, beech (*Fagus grandifolia*); BG, black gum (*Nyssa sylvatica*); BO, black oak (*Quercus velutina*); BTA, big tooth aspen (*Populus grandidentata*); DW, flowering dogwood (*Cornus florida*); GB, grey birch (*Betula populifolia*); HI, hickory (*Carya* sp.); RM, red maple (*Acer rubrum*); RO, red oak (*Quercus rubra*); SF, sassafras (*Sassafras albidum*); SG, sweetgum (*Liquidambar styraciflua*); SM, sugar maple (*Acer saccharum*); TU, tulip tree (*Liriodendron tulipifera*); WO, white oak (*Quercus alba*). After Horn (1971).

tolerant later-transitional and mature-forest species. The early transitional species (e.g. grey birch) tend to have their foliage arranged in 'multilayers' and to be relatively highly productive and fast-growing. The late-transitional and mature-forest species (e.g. sugar maple) tend to have 'monolayer' foliage arrangements and to be less productive and slower-growing (Figs 11.19 & 20).

Multilayered trees, by having a greater leaf area index (total leaf area per unit area of ground), are able to grow more quickly in the open conditions than monolayer species, and thus out-compete them. However, the multilayer species create so much shade that their own offspring are excluded. The shade-tolerant seedlings of monolayer species are at an advantage. Multilayer species can only establish within the forest if a major gap occurs (cf. Forcier 1975, Shugart 1984).

The 'multilayer–monolayer' contrast does not seem to have been examined more widely, although Horn believes that it applies to some gymnosperm trees. He knows of no species which have both 'multilayer' and 'monolayer' configurations (Horn 1974). Perhaps certain gymnosperms, such as balsam fir, would fit this description.

It may be that the 'monolayer–multilayer' contrast is not quite as clear-cut as Horn has indicated. Most open-grown angiosperm trees, whatever the species, assume a 'multilayer' form. At least some early-transitional tree species (e.g. black locust (*Robinia pseudoacacia*)) can assume a 'monolayer' form. New Zealand beeches (*Nothofagus* spp.) also assume both forms.

PHYSIOLOGICAL ECOLOGY OF JUVENILES

Relative shade-tolerance

Species replacements are decided at the level of juveniles, starting with seeds and continuing with the establishment of seedlings and the growth of young plants to adulthood. The site contrasts which influence seedling and juvenile growth are most pronounced during the growing season (spring to autumn) because of the prevailing climatic conditions and the presence of canopies of leaves then.

Seed properties for trees of the eastern North American forests are shown in Table 9.1 and Figure 11.12. The seeds of the early-transitional species require the unfiltered red light and fluctuating temperatures of open sites for germination. The shade and high far-red : red light ratio under leaf canopies suppress their germination (H. Smith 1981, Forcier 1975). The germination of seeds of the dominants of the mature northeastern forests – beech and sugar maple – is unaffected by light.

A trend of correlation of the seed reserve weight with the maximum height gained by young trees in dense shade conditions was noted by Grime & Jeffrey (1965) (Fig. 11.21). However, the seed reserve weight does not perfectly match the appearance of the species in the sequence early-transitional → mature-forest. For example, relatively heavy-seeded honey locust (*Gleditsia triacanthos*) and white pine are early-transitional species. Lighter-seeded red maple occurs in a wide range of forest conditions. Oaks and hickories have heavy seeds, but, in localities where relatively lighter-seeded beech, sugar maple and hemlock thrive, they supplant oaks and hickories.

Tolerance of shading by seedlings, affecting establishment and survival under the

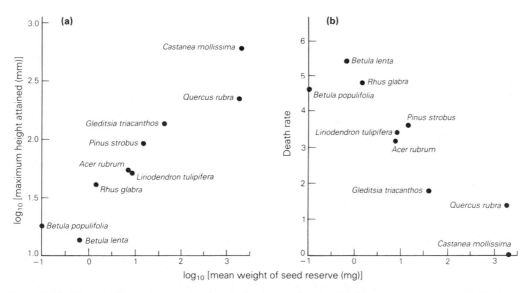

Figure 11.21 The relationship between seed reserve weight and each of seedling growth and mortality for trees of eastern North American forests. (a) Maximum height attained by seedlings after 12 weeks in a 55 cm shade stratum versus seed reserve weight. (b) Number of fatalities per container, over 12 weeks, of seedlings grown in a 55 cm shade stratum versus seed reserve weight. After Grime & Jeffrey (1965).

Table 11.10 Photosynthetic properties of some deciduous tree species from eastern North America (after Hicks & Chabot 1985)*.

Status in forest development	Light-saturated photosynthesis $(\mu mol\ CO_2\ m^{-2}\ s^{-1})$	Respiration	Light saturation level (Percentage of full sun†)	Compensation point
pioneer – early transitional				
Liriodendron tulipifera	6.7–10.7	0.25–1.4	25–30	—
Diospyros virginiana	9.5	1.1	53	5.5
Populus tremuloides		2.0	32	—
Sassafras albidum	6.2	0.7	30	~1
late transitional – mature				
Acer rubrum	3.8	0.4	22	—
Quercus rubra	4.5	0.5	14	—
Acer rubrum	2.8	1.1	8	—
Betula alleghaniensis	4	0.2	7	~0.5
Quercus alba	11.2	—	29	0.8
mature				
Cornus florida	2.6	0.2	25	—
Acer saccharum	4.4	0.3	30	—
Fagus grandifolia	4.4	0.9	17	~0.5

* Bazzaz (1979) cited the following photosynthetic rates (mg $CO_2\ dm^{-2}\ h^{-1}$) *Juniperus virginiana* 10, *Liriodendron tulipifera* 18, *Quercus velutina* 12, *Q. rubra* 7, *Q. alba* 4, *Fraxinus americana* 9, *Fagus grandifolia* 7, *Acer saccharum* 6.
† Full sun assumed to be 2000 $\mu E\ m^{-2}\ s^{-1}$, or 50 000 lux. Rates at 25–30°C.

summer leaf canopy, is one of the main determinants of sequential replacements. A syndrome of morphological and physiological effects is evident (Hicks & Chabot 1985) (Tables 11.10–12). The leaves of shade-tolerant species are relatively broad but thin, usually having one layer of palisade cells. Leaves of intolerant species are smaller but thick, with more than one layer of palisade cells. Intolerant species change their leaf attitude readily, tolerant species do not (McMillan & McClennon 1979). Intolerant trees are light-saturated at high light intensity (25–50 per cent of full sunlight) (Bacone *et al.* 1976). Saturation in tolerant species is mostly less than 20–25 per cent of full sunlight (Loach 1967, Wuenscher & Kozlowski 1970). The tolerant species respond rapidly to increased light (Bourdeau & Laverick 1958, Loach 1967). The intolerant species respond less rapidly. This means that shade-tolerant species utilize sunflecks efficiently.

Photosynthetic rates are relatively low in shade plants (Boardman 1977) and there is a decline in rate with increasing shade-tolerance (Table 11.10). Concentrations of the primary carboxylase enzyme are low. Chlorophyll concentrations are high. In tolerant plants, also, the photosynthetic rate declines in light intensities above saturation (Kozlowski 1949). Presumably this results from photorespiration (W. A. Jackson & Volk 1970). The photosynthetic rate remains unchanged in tolerant tree species that are exposed to high light intensities (Hicks & Chabot 1985).

The rate of dark respiration is low in beech and other shade-tolerant species (R. J. Taylor & Pearcy 1976). The rate is higher in intolerant species. The compensation point (the light intensity level at which photosynthesis balances respiratory losses) is low for tolerant species and higher for intolerant species (Wallace & Dunn 1980).

The photosynthesis to respiration ratio (P/R) reflects the ability of species to cope with shaded conditions. The P/R of sugar maple is highest in the shade and lowest in openings. The opposite is true for white oak. Sugar maple has a more favourable carbon balance in the shade, whereas white oak's carbon balance is more favourable in gaps. Thus, relatively intolerant species are competitively superior in gaps, because their growth is enhanced there, compared with tolerant species (Bazzaz 1979, Hicks & Chabot 1985).

Moisture relations

The mature forest species, with high stomatal and mesophyll resistance, have relatively low transpiration rates compared with early-transitional species (Federer 1976, Holmgren *et al.* 1965, Wuenscher & Kozlowski 1970, 1971a, b). The water use

Table 11.11 Contrasts between the properties of seeds and seedlings of eastern North American tree species which establish in open, unshaded habitats and those which establish in the shade of forest canopies.

Open	Shaded
seeds	
usually small	most are large
large numbers produced per plant	few seeds per plant
many are widely dispersed	less widely dispersed, although squirrels and jays are effective for some trees
often viable for relatively long period (not universal); some have relatively long-term dormancy and form large seed banks	mainly shorter viability, long-term dormancy less common
seedlings – germination requirements	
prefer mineral soil seed bed	prefer (or tolerate) litter, humus seed bed
require unfiltered light (high in red)	unaffected by light quality (which is high in the far-red)
require considerable fluctuation of day and night temperature	require a lower amplitude of temperature fluctuation
seedlings – growing conditions	
Change leaf attitude readily	Leaf attitude does not change readily
Have relatively high compensation points; light saturated at 25–50% full sun; photosynthetic rate high and undiminished in high intensities, but respond slowly to increased light	Have lower compensation points; light saturated at <20–25% full sun; photosynthetic rate lower and declines in high light intensities (photorespiration) but respond rapidly to increased light
Photosynthetic to respiration ratio low in shade, high in openings	Photosynthetic to respiration ratio high in shade, low in openings
Extension growth poor in shade, where far-red wavelengths are high in proportion to red wavelengths	Extension growth proportionately greater where far-red wavelengths are high
Low stomatal and mesophyll resistance; stomata open slowly	High stomatal and mesophyll resistance, stomata open rapidly
Efficient water use. Drought and high temperature tolerant	Less efficient water use. Intolerant of drought and high temperature

Table 11.12 Characters of eastern North American tree species correlated with pioneer – early-transitional or late-transitional – mature-forest status (after Wells 1976).

Character	Pioneer–early-transitional	Late-transitional–mature forest
A. light absorption factors		
apical dominance	pronounced	weaker
branch orders	2–3	5 or more
phyllotaxis	spiral, whorled	two-ranked
leaf division	pronounced	less so
area of leaf unit	small	large
leaf orientation	mainly vertical	mainly horizontal
leaf flushes	usually continuous growth	rhythmic leaf flushing
cuticle, lustre	thick, lustrous	thin, dull
leaf pigmentation	light yellow-green	dark blue-green
B. reproduction factors		
age at first fruiting	< 15 years	> 30 years
sex structure of population	some are dioecious	monoecious or hermaphrodite
periodicity of seed years	annual	3- or 4-yearly intervals
seed weight, number	light, large number	heavy, relatively few
dispersal mechanisms	wind, birds	gravity, mammals, water
shade-tolerance of juveniles	low	high
C. growth maintenance and longevity factors		
growth rate of young trees	rapid	slow
longevity of mature trees	< 100 years	> 250 years
maximum height	variable	variable but dominants are tall
wood density	low (specific gravity 0.2–0.4)	high (specific gravity 0.5–0.7)
modulus of wood rupture	weak (400–800 kg cm^{-2})	strong (1150–1250 kg cm^{-2})
decay resistance	slight	considerable
tyloses	absent	present
wood ray volume	low	high
leaf palatability to herbivores	some are palatable	most are unpalatable

efficiency (the amount of water expended in transpiration per unit carbon gain) is low for sugar maple.

Part of the reason for shade-tolerant species being able to take rapid advantage of sunflecks is that stomata open in a few minutes with increased light, whereas intolerant species such as white ash (*Fraxinus americana*) need 30 min or more (by which time the sunfleck will have moved) (D. B. Woods & Turner 1971, Hicks & Chabot 1985). However, a reliable water supply is needed by the tolerant species – water may become limiting at times. Photosynthesis in sugar maple, for example, begins to decline at − 200 kPa and is very low at − 1000 kPa. The water demands of beech are similar. Red cedar has its maximum photosynthesis rate at low levels of available water (Bazzaz 1979). The same is true for pines (Fowells 1965).

The seedlings of shade-tolerant species such as beech and hemlock (*Tsuga canadensis*) cannot withstand high temperatures. In full sun conditions they die from the effects of direct heat injury (Diller 1935, Olson 1958).

The many morphological and physiological contrasts between species that are able to germinate, and grow to adulthood in shaded conditions, compared with those which require well-lit conditions, account for replacement series in situations where leaf canopies become progressively denser. There are, of course, degrees of tolerance and intolerance of shading. Some of the later-transitional species have seedlings which (e.g. in yellow birch (*Betula alleghaniensis*) and white oak) are intermediate in characteristics between the species which establish only in sunlit sites and those which establish only in very shaded sites. Others (e.g. tulip tree and white pine) have a requirement for forest gaps for establishment, but the adults can persist in crowded conditions.

Seed bed and soil

A further array of seed bed and soil differences is superimposed on the light, atmospheric moisture and temperature patterns which differentially affect germinating seeds and establishing seedlings of trees, in the sequence from open fields to mature forest.

The surface soil layers of forest interiors (Fig. 11.22) have: abundant litter, and the humus derived from it; reliable moisture supplies; and extensive development of permanent root systems of adult trees, permeating the soil from its surface to several metres depth. These features contrast with the properties of the soils of open sites, and they strongly influence young plants in various ways, particularly in relation to nutrient supply.

The North Carolina sequence of loblolly or short-leaf pines → mixed angiosperm transitional forest → mature oak–hickory forests may again be used as a main example. As the pine forests develop on old fields (Billings 1938, Coile 1940) an organic A mor–humus soil profile forms after about 30 years. Under angiosperms undecayed or only partially decayed litter layers are not as deep as under pines. Transitional forest trees such as tulip tree create mull humus profiles (Fig. 11.22), but under the mature angiosperm forest the soils tend to be moroid; the dominant mature soils in the region are classed as red-yellow podsolic (or ultisols). Further north, in the northern hardwoods region the transitional tree species (and sugar maple) are also mull formers, whereas other mature-forest species like beech and hemlock are mor-formers and the mature soils are grey-brown podsolic (or alfisols). Forest A soil horizons form within the first 100 years of the colonization of North Carolina old fields by trees, but it would probably take several hundred years of oak–hickory occupation to reconstitute a mature soil profile, showing no signs of the agricultural disturbance.

Soil changes during forest development (cf. Ch. 2) are a *result* of the characteristic surface litter accumulation and decay, the mineralization of nutrients and the development of root mats. However, these changes, interacting with canopy effects on light and moisture conditions (Phares 1971, Bourdeau 1954), determine the specific conditions for seedling establishment, and thus *cause* subsequent events (i.e. generate autogenic change in the plant populations). The supply of seeds is of obvious importance (Marquis 1975a, b, Stapanian 1986).

The organic matter is a major stored stock and source of nutrients (Bormann & Likens 1979a). It also provides the appropriate conditions for ion-exchange between

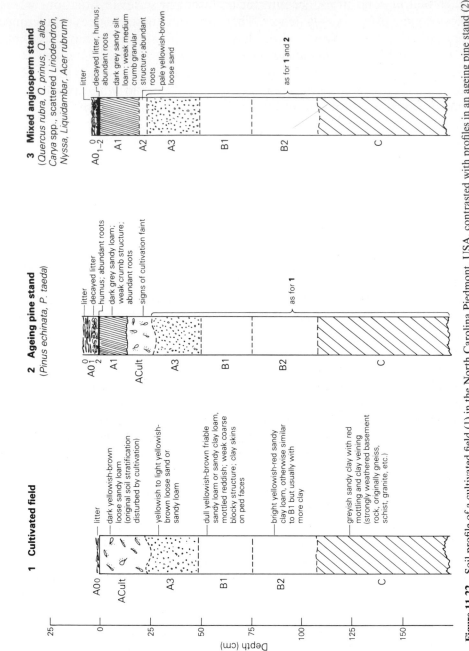

1 Cultivated field

A00 — litter

ACult — dark yellowish-brown loose sandy loam (original soil stratification disturbed by cultivation)

A3 — yellowish to light yellowish-brown loose sand or sandy loam

B1 — dull yellowish-brown friable sandy loam or sandy clay loam, mottled reddish; weak coarse blocky structure; clay skins on ped faces

B2 — bright yellowish-red sandy clay loam, otherwise similar to B1 but usually with more clay

C — greyish sandy clay with red mottling and clay veining (strongly weathered basement rock, originally gneiss, schist, granite, etc.)

Depth (cm): 25, 0, 25, 50, 75, 100, 125, 150

2 Ageing pine stand
(*Pinus echinata, P. taeda*)

A0 1 2 — litter; decayed litter; humus; abundant roots

A1 — dark grey sandy loam; weak crumb structure; abundant roots

ACult — signs of cultivation faint

A3

B1

B2

C — as for **1**

3 Mixed angiosperm stand
(*Quercus rubra, Q. prinus, Q. alba, Carya* spp., scattered *Liriodendron, Nyssa, Liquidambar, Acer rubrum*)

A0 1–2 — litter; decayed litter, humus; abundant roots

A1 — dark grey sandy silt loam; weak medium crumb granular structure; abundant roots

A2

A3 — pale yellowish-brown loose sand

B1

B2 — as for **1** and **2**

C

Figure 11.22 Soil profile of a cultivated field (1) in the North Carolina Piedmont, USA, contrasted with profiles in an ageing pine stand (2) and an old, mixed angiosperm stand (3). Based on Billings (1938), McCaleb & Lee (1956).

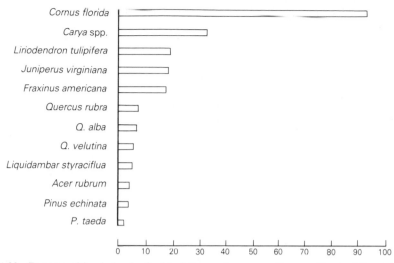

Figure 11.23 Decomposition index for freshly fallen, undecomposed leaves of trees of the eastern North American forests (after Coile 1940). Decomposition index derived from $\{(N \times Ca)/[H^+]\} \times 10$, where N is the total nitrogen content of the leaves, Ca is the total calcium content of the leaves, and $[H^+]$ is the buffering effect of 5 g of leaf material in 100 ml of 0.05M HCl. Leaves of *Cornus*, *Carya*, *Juniperus*, *Liriodendron* and *Fraxinus* are known to decompose rapidly under field conditions. *Liquidambar* and *Acer rubrum* leaves decompose relatively rapidly, more so than those of oaks and pines, which decompose only slowly. The last two of these contain more tannins or resins, or both, than those of *Liquidambar* and *Acer*.

the soil and plant root hairs. Mycorrhizas extend the capacity of roots to obtain nutrients and water.

Pine litter was shown by Coile (1940) to decay slowly, because of its resinous nature and the low content of calcium and nitrogen in pine leaves. The litter is poorly buffered. By contrast the relatively high decomposition rate of the better-buffered mull litter of flowering dogwood, tulip tree and white ash is allied to the higher nutrient content of their leaves. Beech has relatively low leaf nitrogen and its leaves decay slowly (Fig. 11.23). The litter of oaks, sweetgum and red maple is intermediate in these respects between that of pines and most of the transitional species.

Nutrient contrasts in different forest stands are complicated by the nutrient content differences between above- and below-ground parts. Root : shoot ratios are shown in Table 11.13 (Harris *et al.* 1977). Ralston & Prince (1965) found that loblolly pine leaves and stems contain much lower amounts of nutrients than those of certain angiosperm trees (Fig. 11.23 & Table 11.14), but Reichle (1971) points out that oak–hickory root systems contain only 40 per cent of the nutrients maintained in each plant, whereas white pine roots contain 65 per cent of its nutrients. As nutrients are made available underground by root sloughing, root exudations, and root consumption and recycling by herbivores, measurement of all the individual nutrient inputs contributing to plant responses is difficult.

Cromack (1981) notes that below-ground root and ectomycorrhizal death account for substantially more of the organic matter turnover in young douglas fir forests than do above-ground parts. Mycorrhizal sheath tissue alone contributes 40 per cent of the total below-ground organic matter input. Of total tree nitrogen input to the

Table 11.13 Biomass of some eastern North American forests and shoot : root (S/R) ratios of tree species of pure stands* (after Cole & Rapp 1981).

Forest type and locality	Above-ground biomass		Biomass values (kg ha^{-1}) Root biomass	S/R ratio
Pinus echinata Oak Ridge, Tennessee	leaves branches stems	4600 27 000 89 000	34 000	3.55
		120 600		
Liriodendron tulipifera Oak Ridge, Tennessee	leaves branches stems	3200 27 100 94 400	36 000	3.50
		124 700		
Liriodendron – Quercus spp. Oak Ridge, Tennessee	leaves branches stems	5100 21 000 83 000	31 000	
		109 100		
Quercus prinus Oak Ridge, Tennessee	leaves branches stems	5300 30 000 102 000	36 000	3.80
		137 300		
Quercus spp.–*Carya* spp. Coweeta, North Carolina	leaves branches stems	5580 25 600 106 310	52 525	
		137 490		
Acer saccharum– Betula alleghaniensis– Fagus grandifolia Hubbard Brook, New Hampshire	leaves branches stems	3180 38 980 91 840 130 820	28 560	

* It is likely that the total root mass is underestimated as it is very difficult to isolate fine roots.

soil, 86 per cent is returned by below-ground components. Few comparable data seem to be available for eastern North American forests, although Hermann (1977) and Harris *et al.* (1977) indicate that about half of the annual production of trees there is contributed by the root systems. Cox *et al.* (1977) found that the annual input of nitrogen from roots to the soil in tulip tree forest in Tennessee is at least 1.5 times the above-ground contribution (leached from the leaf canopy, or from leaf decay).

Seed bed requirements

The seeds of many transitional angiosperm tree species (e.g. pines, tulip tree and sweet gum) germinate best on bare mineral soil (Fowells 1965) (Table 9.1). The risk of failure is high if conditions are too dry, but at least some of the transitional species have dormant seeds, ensuring that seed populations are available when conditions are right. Others produce large amounts of well-dispersed seeds annually.

The relatively drought-resistant pine seedlings can survive drying conditions of open fields, which would be fatal to late transitional and mature forest species. Summer drought is a main cause of mortality of white, red and scarlet oak seedlings, for example (Bordeau 1954).

The large seeds of most late-transitional to mature forest species do not form long-lasting seed banks. They are shed in autumn and germinate in spring. If they dry out, oak and hickory seeds die. Thus, the well-developed litter layers of pine, or mixed angiosperm forest, form a suitable seed bed in which they can germinate (Bourdeau 1954). Shallow burial by rodents ensures that some, at least, will survive. Pine seeds lie on top of the litter, and most are eaten by birds and rodents.

Seed bed conditions encourage the replacement of pines by angiosperms. The annual accumulation of pine needles forms a thick, hygroscopic litter mat which the short roots of pine seedlings find it hard to penetrate. They die through summer

Table 11.14 Nutrients in above-ground parts and roots (and ratios) for some eastern North American tree species growing in pure stands (after Cole & Rapp 1981).

Forest type and locality		Nutrient values (kg ha^{-1})				
		N	P	K	Ca	Mg
Pinus echinata	leaves	51	12	36	43	12
Oak Ridge,	branches	64	11	29	100	11
Tennessee	stems	100	17	60	164	17
	total above ground	215	40	125	307	40
	roots	117	21	128	187	17
	litter	290	23	21	256	23
	ratio above ground/root	1.8	1.9	0.97	1.6	2.3
Liriodendron	leaves	54	6	52	35	—
tulipifera	branches	79	31	46	144	—
Oak Ridge	stems	172	10	75	277	—
Tennessee						
	total above ground	305	47	173	456	—
	roots	123	19	112	151	—
	litter	78	5	9	100	—
	ratio above ground/root	2.5	2.5	1.5	3	—
Quercus prinus	leaves	75	5	52	58	12
Oak Ridge,	branches	139	10	61	294	13
Tennessee	stems	183	11	129	500	20
	total above ground	397	26	242	852	45
	roots	132	22	140	249	18
	litter	298	18	26	318	22
	ratio above ground/root	3	1.2	1.7	3.4	2.5

desiccation (as well as lack of light). Roots of angiosperms like oaks are long and faster growing. They penetrate the litter mat to reach the more-reliable moisture supply in the mineral soil (Billings 1938, Coile 1940). In the northern hardwood forest type establishment of beech and hemlock seedlings is favoured by the development of mature mor humus (Fowells 1965).

Soil nutrients and seedling growth

Coile (1940) suggested that nutrient conditions in abandoned fields in North Carolina are about as favourable for angiosperm trees as they are in young pine stands. However, in old pine stands the carbon : nitrogen ratio is more favourable and, indeed, the old pine stand soils are better for nutrient supply than the soils of mature oak–hickory forests. Calcium and nitrate are readily mobilized. Flowering dogwood is a mull-forming understorey species in the old pine stands, which apparently helps to confer a favourable nutrient status, through recycling from litter. The ageing pine stands also probably contribute to high nutrient availability. Their nutrient demands are low, and part of the nutrient stock is released by tree death.

Water-holding and percolation properties are also good in the old pine soils. In combination these conditions temporarily favour the transitional angiosperms in contest with: (a) pine seedlings which are inhibited by shading; and (b) mature forest species with relatively slow growth rates. Inexorably, however, the oaks and hickories establish increasing control over the nutrient stocks. Through mor formation and leaching of the upper mineral soil horizons, as well as through uptake and immobilization in the trees, the total available nutrient supplies are diminished. Seed supplies of oaks and hickories and seed dispersers are increasingly abundant. The shade and seed-bed conditions favour establishment of the seedlings of oaks and hickories. Root competition becomes intense, and mycorrhizal systems are well developed. The early-transitional angiosperms are gradually excluded through competition for light and nutrients. The mature-forest species, shade-tolerant and with lower nutrient demands than the transitional species, eventually hold all of the ground except where gaps caused by tree-fall bring about release from both root and shoot competition.

Little seems to be known of the spatial heterogeneity of soil nutrient availability in the forests under consideration. The results of Tilman (1982) for herb communities suggest that supplies of nutrients are also likely to be very variable over short distances in the forests. Such differences probably influence the specific establishment of different tree species.

Trenching experiments in dense forest stands, where tree root systems are cut around small areas of ground, and the trenches refilled with soil, result in the growth of many more tree juveniles (including early transitional species) than in untrenched adjacent sites (Korstian & Coile 1938). In North Carolina such experiments were done in pine stands and in oak forests, with similar results. The results are usually interpreted as reflecting the release of seedlings from competition by adult root systems for water, and this is probably true to some degree. However, the stimulatory effects of trenching on seedling growth seem to be complex. Rice (1979) concludes that maturation of forest is accompanied by decreased nitrification. This is probably due to ammonium immobilization and plant uptake (Woldendorp 1978, Sharpe *et al.* 1980). Cutting of roots will release supplies of nutrients from various compartments (e.g. the stocks weathering from rock, in the soil solution, and from decaying roots and mycorrhizas). Moleski (1976) found that trenching in oak forest eliminated the inhibitory effect of tree root exudates on nitrification; nitrate became available in the trenched area.

NICHES

The outcome of experimental investigations having a bearing on 'tolerance' is that the concept, as used by foresters and presented in Table 11.9, is a little simplistic. For tolerance of light conditions by establishing seedlings, for example, the array of species on the gradient is likely to be more like Fig. 11.24a than Fig. 11.24b, which is how tolerance tables express the relationships. Many other factors than light alone are involved with the regeneration niche of plant species (e.g. mineral nutrient requirements, resistance to herbivory and fungal attack, and water relations) (Vartaaja 1952, Grubb 1977). Some way is needed to indicate the response of species to each factor gradient which influences them, with and without competitors (i.e. a means of expressing both their fundamental niche dimensions and realized niches in particular circumstances).

It is much more difficult to experiment with trees than with herbs; most experimental study which defines tree niche dimensions has been done with seeds, seedlings

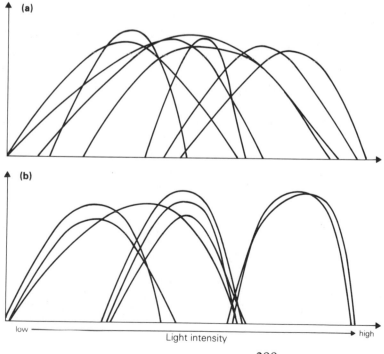

Figure 11.24 Array of eight hypothetical tree species with respect to the productivity of their seedlings in particular light intensities. (a) Pattern of response of the eight species to a light intensity gradient. (b) The pattern of response as expressed by tolerance tables.

and saplings (Table 11.15). Although the establishment and later juvenile phases are critical times in the life of a tree, information is also required on poles (near-adults) and adults. The niche dimensions of adults may differ in various ways from those of juveniles (Bazzaz 1987).

Much of the species-specific experimental information outlined earlier in this chapter is relevant to the definition of plant niches. However, it is useful to be able to examine whole suites of the species which occur together in stands, rather than to compare one or two at a time. The 'tolerance' tables do this in a rather imprecise way. Considerably more detailed is the 'climax index' method of Wells (1976) (Tables 11.12 & 16). From this ecomorphological survey it is evident that most species conform to orthodox views on the order of their appearance in a sequence from early-transitional to most-mature (and cf. Whitney 1976).

Shugart (1984) classes species into four role-types in relation to their performance in forest dynamics (Fig. 11.25) It can be expected, however, that where there is a rich tree flora, there may well be an almost continuous, or at least a very complex, array of properties of trees fitting them for particular roles. Classifications are arbitrary. Detailed examination of the data which Wells (1976) uses to derive his 'climax index' shows that this is so. Clusters of species appear here and there on the continuum, but each species has some distinctive features.

Direct gradient analyses by Whittaker (1956) of forests in the Great Smoky Mountains of Tennessee, and by Curtis (1959) in Wisconsin, reveal interesting facts about the differences in distributions of species, relative to one another, on environmental 'gradients' (Figs 11.26–28). The two 'gradients' in Whittaker's study are really gradient complexes, since various controlling factors are likely to be involved. Syndromes occur, however, where certain factors vary in parallel. For example, the altitudinal 'gradient' integrates temperature, increased soil moisture, wind exposure and possibly decline in pH. However, it is also likely that gradients of soil moisture, soil fertility (for many different nutrients) and temperature vary independently, so the moisture 'gradient' is unlikely to be smooth and uncomplicated (cf. Loucks 1962).

Table 11.15 Niche breadth (B) and proportional similarity (PS) of seedlings of tree species from early-transitional and late-transitional to mature-forest phases on three resource gradients (after Parrish & Bazzaz 1982c).

Gradient	Moisture		Nutrients		Light	
	B	PS	B	PS	B	PS
Early-transitional						
Gleditsia triacanthos	0.77	0.80	0.91	0.94	0.95	0.85
Pinus taeda	0.97	0.81	0.91	0.94	0.97	0.85
Crataegus mollis	0.92	0.86				
mean	0.89	0.82	0.91	0.94	0.96	0.85
Late-transitional–mature						
Tilia americana	0.86	0.74	0.78	0.79	0.95	0.92
Quercus rubra	0.54	0.75	0.82	0.86	0.97	0.93
Acer saccharum	0.56	0.61	0.82	0.84	0.99	0.92
mean	0.65	0.70	0.81	0.83	0.97	0.92

See original article for methods and further details.

Table 11.16 'Climax index' for 30 angiosperm tree species of eastern North America (after Wells 1976).

	'Climax index'		'Climax index'
Salix nigra	15.1	Diospyros virginiana	60.5
Populus deltoides	17.0	Carya cordiformis	60.5
Oxydendrum arboreum	30.2	Castanea dentata	62.3
Betula populifolia	32.1	Ulmus americana	62.3
Liriodendron tulipifera	35.8	Tilia americana	64.2
Liquidambar styraciflua	41.5	Carya ovata	68.0
Gleditsia triacanthos	49.1	Quercus alba	68.0
Nyssa sylvatica	49.1	Ostrya virginiana	70.0
Sassafras albidum	49.1	Quercus rubra	71.7
Quercus coccinea	51.0	Betula lenta	71.7
Acer rubrum	52.9	Cornus florida	71.7
Prunus serotina	52.9	Carpinus caroliniana	75.5
Robinia pseudoacacia	52.9	Quercus prinus	77.4
Fraxinus americana	54.8	Acer saccharum	88.8
Quercus velutina	58.6	Fagus grandifolia	96.2

Distribution of species on a scale from 15 ('extreme pioneer', intolerant of competition, requiring open sites) to 96 ('extreme climax', very tolerant of competition, inhabiting deeply shaded sites). The numerical index is derived by scoring each species against 26 'ecomorphological' characters (see Table 11.12 for a summary of them). The results are expressed as a percentage of the maximum possible value for all characters.

Whittaker's analysis (1956, 1965) shows that each species tends to behave individualistically in relation to every other species, as is to be expected from niche theory (Ch. 3).

Indirect gradient analysis (e.g. Bray & Curtis 1957, Curtis 1959, Loucks 1962, Austin 1977) is another, more-abstract way of surveying the niche differentiation among the tree species. Figure 11.29 shows 11 species on two gradients. (See also Ch. 12, Figs 12.4 & 5, where the 'population centres' for numbers of tree species on three axes are extracted from a simple ordination in Wisconsin forests.) The abstract axes may represent real environmental gradients or gradient complexes, or they may relate to the vegetation properties.

Stochastic or orderly species replacements?

Horn (1974, 1975b) proposed that the perceived changes in forest development are really the result of probabilistic events, involving plant-by-plant replacements and analogous to, if not the same as, statistical processes known as 'regular Markov chains'. Markov chains are discussed in a little more detail in Chapter 12.

Botkin et al. (1972) applied transition probability methods in a simulation of the changes in a New Hampshire forest. The results closely resembled the species replacements in the forest. The data required for the computer simulation were the birth and death rates of each species, as well as growth patterns, responses to environmental factors and plant–plant and plant–environment–plant interactions. Many other simulation methods have been developed for the study of forest dynamics (cf. Shugart 1984).

Horn (1975b), in a study of the dynamics of a New Jersey forest, assumed that each

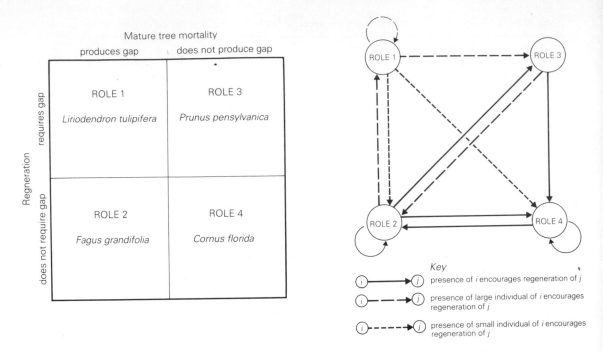

Figure 11.25 Role-types postulated by Shugart for forest trees, with respect to site formation and site requirements for regeneration. (a) The four essential role types, with examples from eastern North American forests. (b) The directions of positive or strongly positive interactions among the four role types, with respect to their influence on regeneration, if species filling the roles occur in the same neighbourhood. Modified from Shugart (1984).

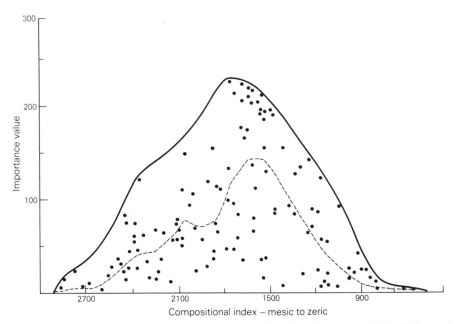

Figure 11.26 Behaviour of red oak (*Quercus rubra*) on a compositional gradient from mesic (moist) to xeric (dry) in upland forest, Southern Wisconsin, USA. Each dot indicates the importance value of this species in a separate stand sample, with the stands located along the abscissa according to their compositional index. The solid line includes all of the stands containing red oak. The dotted line is the average importance value of this species as determined for sequential groups of stands of 100 compositional units each (after Curtis 1959).

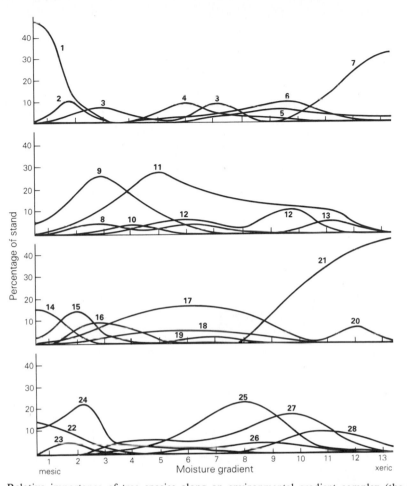

Figure 11.27 Relative importance of tree species along an environmental gradient complex (the topographic moisture gradient from mesic (moist) ravines (at left) to xeric (dry) southwesterly facing slopes (at right), between elevations of 460 and 760 m in the Great Smoky Mountains, Tennessee, USA). Smoothed curves for major tree species are plotted by percentage of the total numbers of tree stems over 1 cm in diameter, 1.4 m above the ground, in sample plots; all of the species illustrated are part of the same vegetation gradient, but they are separated into four panels for the sake of clarity. Although, with 28 species and 13 steps of the gradient, some species must have their importance modes in the same step, the modes of the species appear to be scattered along the gradient. Pairs of species having their modes in the same step of the gradient may be shown to be differently distributed in relation to an elevation gradient. Some species are bimodal, with two ecotypes having different population centres (see Fig. 11.28). Plant communities intergrade continuously from cove forests (transect steps 1–4), through oak forests (steps 6–8), to pine forests (steps 10–13). The species are (1) *Halesia monticola*; (2) *Acer saccharum*; (3) *Hamamelis virginiana*; (4) *Carya tomentosa*; (5) *Nyssa sylvatica*; (6) *Pinus strobus*; (7) *P. rigida*; (8) *Quercus rubra*; (9) *Tsuga canadensis*; (10) *Fagus grandifolia*; (11) *Acer rubrum*; (12) *Q. alba*; (13) *P. echinata*; (14) *Aesculus octandra*; (15) *Betula alleghaniensis*; (16) *B. lenta*; (17) *Cornus florida*; (18) *Carya glabra*; (19) *C. ovalis*; (20) *Q. marilandica*; (21) *P. virginiana*; (22) *Tilia heterophylla*; (23) *Cladrastis lutea*; (24) *Liriodendron tulipifera*; (25) *Q. prinus*; (26) *Q. velutina*; (27) *Oxydendrum arboreum*; (28) *Q. coccinea*. after Whittaker (1965).

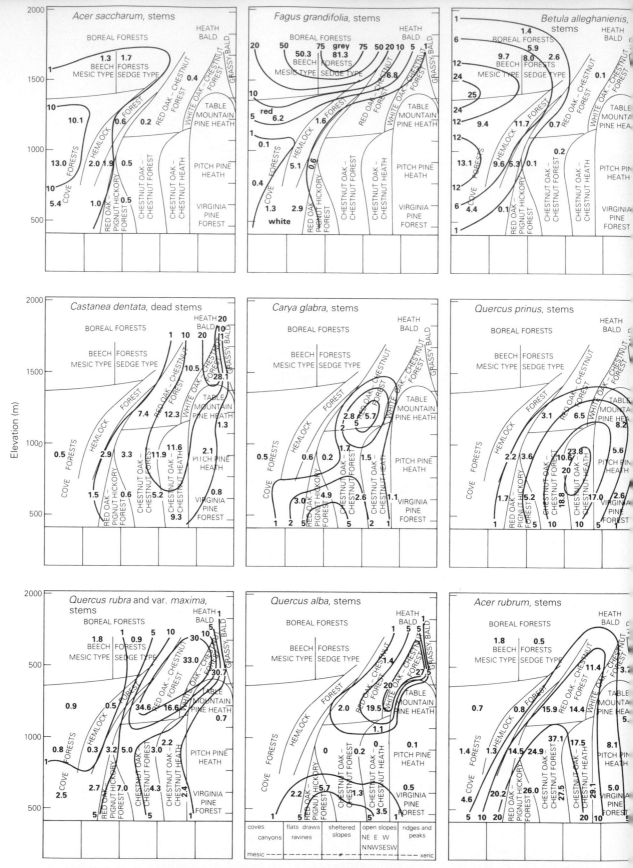

Figure 11.28 Relative density of distribution of each of nine tree species over the entire range of major habitat–vegetation types, Great Smoky Mountains, Tennessee, USA. Plotted values are percentages of tree stems over 1 cm diameter, 1.5 m above ground level, in composite samples of approximately 100 stems (after Whittaker 1956).

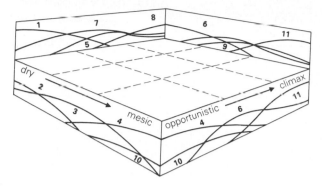

Figure 11.29 Inferred species occurrence on a continuous, two-dimensional gradient in forests of northeastern USA. The horizontal axes represent (left) a sequence from dry to mesic (moist) and (right) a sequence from opportunistic (pioneer and early transitional species) to 'climax' (mature forest) species. (1) *Quercus velutina*. (2) *Q. macrocarpa*. (3) *Q. alba*. (4) *Q. rubra*. (5) *Ulmus americana*. (6) *U. rubra*. (7) *Carya ovata*. (8) *Prunus serotina*. (9) *Tilia americana*. (10) *Acer rubrum*. (11) *A. saccharum*. After Loucks *et al*. (1981).

adult tree was likely to be replaced by a new generation drawn from the saplings beneath it. The 'successful' individuals were chosen on the basis of the proportions of each species initially present. The transition probabilities, expressed as percentages, are summarized in Table 11.17. The predicted abundances in the next generation are combined with the transition matrix to calculate the composition of the forest two generations later, and so on. Table 11.18 is the matrix in Table 11.17 applied for five generations.

In applying these methods, various assumptions must be made which are probably much too simplistic. Various events may confuse the issue. The different ecological behaviour of species is a complicating factor (e.g. differences in shade-tolerance, growth rates, resistance of juveniles to diseases and grazing, sprout or seed origin of juveniles, etc.). The stochasticity of nature (e.g. random deaths of adults and sap-

Table 11.17 Transition matrix for Institute Woods, Princeton, New Jersey: percentages of saplings of the various species under particular canopy tree species (after Horn 1975b).

| Canopy species | Sapling species (%) | | | | | | | | | | | Total |
	BTA	GB	SF	BG	SG	WO	OK	HI	TU	RM	BE	
Big-toothed aspen	**3**	5	9	6	6	—	2	4	2	60	3	104
Grey birch	—	—	47	12	8	2	8	0	3	17	3	837
Sassafras	3	1	**10**	3	6	3	10	12	—	37	15	68
Black gum	1	1	3	**20**	9	1	7	6	10	25	17	80
Sweet gum	—	—	16	0	**31**	0	7	7	5	27	7	662
White oak	—	—	6	7	4	**10**	7	3	14	32	17	71
Red oak	—	—	2	11	7	6	**8**	8	8	33	17	266
Hickories	—	—	1	3	1	3	13	**4**	9	49	17	223
Tulip tree	—	—	2	4	4	—	11	7	**9**	29	34	81
Red maple	—	—	13	10	9	2	8	19	3	**13**	23	489
Beech	—	—	—	2	1	1	1	1	8	6	**80**	405

The number of saplings of each species listed in the row at the top is expressed as a percentage of the total number of saplings (last column) found under individuals of the species listed in the first column. The entries are interpreted as the percentages of individuals of species listed on the left that will be replaced one generation hence by species listed at the top. The percentage of 'self-replacements' is shown in boldface.

The total number of recorded saplings is 3286. A dash implies that no saplings of that species were found beneath that canopy; a zero, that the percentage was less than 0.5 per cent.

The species are: BTA, *Populus grandidentata*; GB, *Betula populifolia*; SF, *Sassafras albidum*; BG, *Nyssa sylvatica*; SG, *Liquidambar styraciflua*; WO, *Quercus alba*; OK, *Quercus* sp; HI, *Carya* spp; TU, *Liriodendron tulipifera*; RM, *Acer rubrum*; BE, *Fagus grandifolia*.

Table 11.18 Transition matrix of Table 11.17 applied for five generations (calculated from successive predicted abundances obtained from the original and subsequent transition matrices) (after Horn 1975b).

	Species (%)										
	BTA	GB	SF	BG	SG	WO	OK	HI	TU	RM	BE
theoretical											
generation:0	0	49	2	7	18	0	3	0	0	20	1
1	0	0	29	10	12	2	8	6	4	19	10
2	1	1	8	6	9	2	8	10	5	27	23
3	0	0	7	6	8	2	7	9	6	22	33
4	0	0	6	6	7	2	6	8	6	20	39
5	0	0	5	5	6	2	6	7	7	19	43
⋮	⋮	⋮	⋮	⋮	⋮	⋮	⋮	⋮	⋮	⋮	⋮
stationary distribution (%)	0	0	4	5	5	2	5	6	7	16	50
longevity (years)	80	50	100	150	200	300	200	250	200	150	300
age-corrected stationary distribution (%)	0	0	2	3	4	2	4	6	6	10	63
empirical											
years since 25	0	49	2	7	18	0	3	0	0	20	1
cultivation 65	26	6	0	45	0	0	12	1	4	6	0
ceased 150	—	—	0	1	5	0	22	0	0	70	2
350	—	—	—	6	—	3	—	0	14	1	76

Abbreviations are as in Table 11.17. The theoretical approach results from taking a row vector of species proportions in a 25-year-old stand, expressed above as percentages, and multiplying it once for each generation by the transition matrix of Table 11.17.

The stationary distribution is the solution of five linear equations (see Horn 1975b). Note that initial changes are rapid, but that a close approximation to the stationary distribution is reached in only five generations. Several species (e.g. RM = red maple) show damped oscillations. The age-corrected stationary distribution is simply the stationary distribution, weighted and normalized by the rough approximations to the specific longevities listed above it.

The empirical approach results from independent measurements of 639 trees in stands that have been abandoned from cultivation for at least the number of years indicated. The percentages are of total basal area, calculated from diameters measured at 1.5 m. Note the similarity between the age-corrected stationary distribution and the composition of the oldest forest.

lings) might well exceed that of the mathematical analysis. Also there is no guarantee that saplings of any species, observed at an instant of time, will survive to replace any tree which may occupy a place on the ground for many decades or 100 years or more (McIntosh 1980b).

In spite of the randomness of so many events in regeneration of populations of plants, there is also regularity imposed by the interaction of individual (and uniquely held) plant properties of different species with the conditions created by the developing vegetation. Especially this concerns the development of a forest canopy, forest litter, humus and other soil conditions, and the control by the vegetation of microclimate, light conditions and the specialized forest fauna and microbiota.

MAINTENANCE OF MATURE FOREST

What is meant by mature forest? So far the term has been used in a general, not very precisely defined way, as referring to vegetation where the dominant plants maintain themselves for a few generations, at least. A good discussion of the problems of definition of community stability is found in K. D. Woods & Whittaker (1981). The topic is considered further in Chapter 12.

K. D. Woods & Whittaker (1981) present a 'simple, functional' definition of self-maintaining, stable forests. These are forests in which several species can coexist, with relative importance and population structure remaining constant for at least three or four generations of the longest-lived species. The spatial scale is on the level of the stand, of relatively uniform composition, over areas of 1 ha or more. Small-scale disturbances, e.g. when one or two trees die and are replaced, can be envisaged as part of the pattern, because replacement then will usually be by growth of suppressed juveniles of the canopy species, which have been awaiting their opportunity. The characteristic species of the more-important eastern North American forest types, which have been regarded as capable of maintaining themselves, are noted in Chapter 9.

Acceptance of the definition of K. D. Woods & Whittaker (1981) as a general model for mature forest and other kinds of mature vegetation creates some difficulties. The matters which have to be reconciled are: constancy of composition, longevity of the plants, persistence in space and time; spatial scale; and type, degree and frequency of disturbance. The definition used hereafter for mature vegetation is that it consists of any stand of vegetation, comprising one or more dominant species, which maintain themselves on an area for more than one generation, with approximately the same stand composition. If the whole stand (or any portion of it) is disturbed, there is immediate replacement by the same species in approximately the same proportions. This definition could be applied to almost all kinds of vegetation, including stands of annuals which replace themselves, monospecific forests which are disturbed frequently and regenerate straight away and chronically disturbed vegetation which maintains a constant composition. The definition might be difficult to apply in very diverse tropical forests (Ch. 10).

Although most of the forest in the region is in a state of flux through disturbance (Ch. 9), some tracts of mature, old-growth forest in eastern North America are dominated by combinations (and sometimes monospecific stands) of beech, sugar maple, basswood (*Tilia americana*), eastern hemlock (*Tsuga canadensis*), red spruce (*Picea rubens*), balsam fir, oaks (red, white and chestnut (*Quercus prinus*)) and hickories (Hough 1936, Williamson 1975, Bormann & Likens 1979a, K. D. Woods & Whittaker 1981, Whitney 1984). Held & Winstead (1975) found that the basal area of trees with more than 10 cm stem diameter in several kinds of mature forest was constant at about 30 m^2 ha^{-1}. Provided that human or large-scale natural disturbances do not intervene, these old-growth forests are likely to maintain themselves more or less in steady state.

The concept of the mature stand is sufficiently flexible for these forests to incorporate relatively small-scale disturbances like those caused by the death of a few canopy trees; occasional episodes of insect damage which partially defoliate the canopy trees; or occasional episodes of deer browsing which destroy some juvenile plants. Larger-scale disturbance would permit transitional tree species to enter the stands, and some of these may persist for long periods. The presence of some patches of immature forest or single individuals of species such as yellow birch, black cherry, red maple, white pine or tulip tree, which are unlikely to replace themselves on that particular spot because their juveniles are not sufficiently shade-tolerant, should not exclude a tract of forest from consideration as being, essentially, mature. The species forming most mature forest in some parts of the region (e.g. beech, sugar maple, hemlock, basswood) are missing from other parts of the region for climatic or historic reasons. Other species (e.g. oaks and hickories) then assume the local role as dominants of mature forest. In places, on relatively severe sites, pine species are favoured by local conditions and form the mature vegetation.

Dynamic interactions between canopy and juvenile plants, in relation to microsite conditions for seedling establishment, are invoked by K. D. Woods (1979) & Whittaker (1981) to explain the regeneration patterns which maintain beech–sugar maple stands and other, more-complex, forest composition. Dominance by shade-tolerant species indicates that canopy gaps which form are small and replacements are drawn from the suppressed juvenile stock of the dominants. Different characteristics of the juvenile plants, in relation to the mosaic of microsites available to them, are believed to be important for the coexistence of the species. One way for the forest to maintain itself would be for each species to replace itself; e.g. beech root sprouts may replace adults which die. However, K. D. Woods & Whittaker (1981) present much evidence for reciprocal replacement of beech and sugar maple, although the latter is relatively shorter-lived. K. D. Woods & Whittaker (1981) suggest that, although maple forms the denser canopy, beech seedlings and saplings are more shade-tolerant than maple juveniles. Maple juveniles just survive the deep shade under maple canopy. The beech juveniles, although they grow slowly, surpass maple under these conditions, and suppressed beech sapling populations develop. Maple juveniles can grow more rapidly than beech juveniles in the higher light levels under the beech canopy. When canopy gaps arise, the sapling populations of each of beech and maple, under the other's canopy, begin more-rapid growth and maintain their respective lead.

K. D. Woods & Whittaker (1981) (Fig. 11.30) present a model for patterns of canopy tree replacement in mature forest containing five species, drawn from experience in the northern part of the eastern North American forest region. Basswood (by basal sprouts) and hemlock (from suppressed shade-tolerant juveniles) customarily replace themselves; beech and sugar maple alternate as indicated above. However,

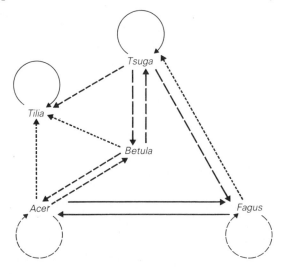

Figure 11.30 Patterns of behaviour of five tree species (*Acer saccharum, Betula alleghaniensis, Fagus grandifolia, Tilia americana* and *Tsuga canadensis*) in forests in the Lake States, Pennsylvania and New York, USA, which may lead to their coexistence in a stable dynamic equilibrium. The patterns of canopy tree replacement are suggested by distributions of sapling species with respect to canopy species. The arrows are directed toward species being replaced. The solid lines indicate transitions that occur most frequently, broken lines less-common transitions and dotted lines those that probably occur occasionally. Other possible transitions probably occur but more rarely (after Woods & Whittaker 1981).

many other variations occur commonly or occasionally, including the maintenance in gaps of yellow birch which has small, light seeds and shade-intolerant seedlings.

There are innumerable vagaries, both spatial and temporal in the way that microhabitat conditions for juveniles may be determined by canopy and other influences by adult trees, or by the forest field layer and the animal and microbial populations that are present. Canopies are not often perfectly even and dense; light-flecks or other irregularities in the light climate are experienced. Tree bases or large roots, fallen logs and differential litter accumulation, animal trails and burrows affect surface or subsurface microtopography. Variable microsite factors include water and nutrient availability, competing roots, mycorrhizas, pathogens, herbivorous or other influential animal populations. The field layer plants create spatially variable (and competitive) conditions. Seeds, seedlings and larger juveniles perceive and respond to these small-scale differences which form a veritable multidimensional environmental mosaic. Only broad generalization is possible, given such variability of likely influences on health, productivity and survival of the young plants.

Many of the eastern North American forests are subject to frequent and severe disturbance (cf. Loucks 1970, Bormann & Likens 1979b, P. S. White 1979) (Ch. 9); it is difficult to sustain concepts of stability in these forests on anything but a very large-scale (the *forests* are probably maintained over periods of about a millennium, or more, although compositional changes on each patch of ground might occur at intervals of one generation length of the potentially oldest tree species, or less). Nevertheless, some forests in the region appear to have maintained themselves with much the same composition for more than one generation. As K. D. Woods & Whittaker (1981) indicate, however, even in the absence of major disturbance, maintenance of constant stand structure for more than a few generations on any patch of ground is unlikely. Changes in at least some of the species populations are inevitable because the factors promoting change are ubiquitous and powerful. Of course, wherever one component of a plant community changes, readjustments of the other species take place. Mature vegetation must be regarded as being in only approximate steady state. Perception of change in such a forest may be difficult because of the shortness of human lifetimes relative to the lifetimes of trees or because the change is so slow, or for both reasons. Changes in the field layer or of understorey species might occur without any perceptible change in the canopy. They cannot be ignored, however, because they may determine conditions affecting the future performance of juveniles of the canopy species.

INFLUENCES OF OTHER BIOTA

Several references are made in Chapter 9 and in this chapter to influences of biota other than plants on the course of events (e.g. seed dispersal, herbivory, reduction of detritus, plant diseases, mycorrhizal return of nutrients and nitrogen fixation and micro-organisms). Figure 11.31 summarizes some important effects of biota other than plants on the course of vegetation development from abandoned fields to mature forest (cf., for example, Golley 1960, Shanks & Olson 1961, Reichle 1971, Cates &

Orians 1975, Stapanian 1986). Refer to Chapters 7, 9 and 10 for more information on animal influences.

SOME COMMUNITY PROPERTIES

During the sequence of vegetation development from abandoned farmland to mature forest, the dominant plant species which replace one another tend to increase in size. In the early forest phases productivity increases markedly (Fig. 11.18), but, as the forest approaches maturity the productivity declines (Loucks 1970, Whittaker 1975a, Bormann & Likens 1979a).

Using the evidence from studies of community changes after forest clear-cutting, both peak standing crop biomass and productivity are reached during the late transitional phase. The change in production rates is thought by Bormann & Likens (1979a) to be because the developing forests tend to be even-aged. Later the forests comprise asynchronous, all-aged patches which sustain lower biomass and have lower production rates overall. The old trees grow slowly and contain large amounts of non-photosynthetic structural biomass and the forests contain much dead wood (Bormann & Likens 1979a, Franklin & Hemstrom 1981) (Fig. 11.32).

The model of Botkin *et al.* (1972) predicted that standing crop in mixed deciduous forest would reach a peak by about 200 years, then decline by 30–40 per cent in the following 200 years. Peet (1981) outlines some variant models for production (Fig. 11.33), including the 'time-lag' one where, after a period of dominance by one species, production declines, then rises again as one or more other species takes over. Root-biomass is usually about 30 per cent of the shoot biomass, or less, and the ratio declines as stands age (Table 11.13). Mature forest trees are slow-growing, with dense wood, and the mineral ratio increases (Whittaker 1975).

Whittaker (1975) emphasized increases in species diversity (in its simplest form, the number of plant species per unit area) during old field to forest sequences, but there is not necessarily a regular change in this direction. In the Hubbard Brook area of New Hampshire, USA, Bormann & Likens (1979a) note that the recently disturbed site is most diverse, sites with transitional forest (they call it the 'aggradation' phase) are lowest in diversity and mature forest (their 'steady-state' phase) is probably inter-mediate. In a sequence from herbs through pines to angiosperm forest it might be expected that changes in diversity would follow a similar pattern, although the pine forest could be as diverse as the herb vegetation, and mature forest might be much more diverse. Some old forests are probably much less diverse than the transitional forests which preceded them. Diversity is not necessarily correlated in any regular way with vegetation development. In some instances it appears to be correlated with site quality (fertility) (Tilman 1982). In Chapter 10 it is shown that in tropical forest areas the most diverse forests are on rather infertile soils. Very fertile and very infertile soils both have reduced diversity, by comparison.

Fields under agriculture and recently disturbed forest sites tend to lose nutrients relatively easily in runoff and by leaching to the ground water (Ballard & Woodwell 1977). In the absence of plants the nutrient stocks in such sites are maintained in

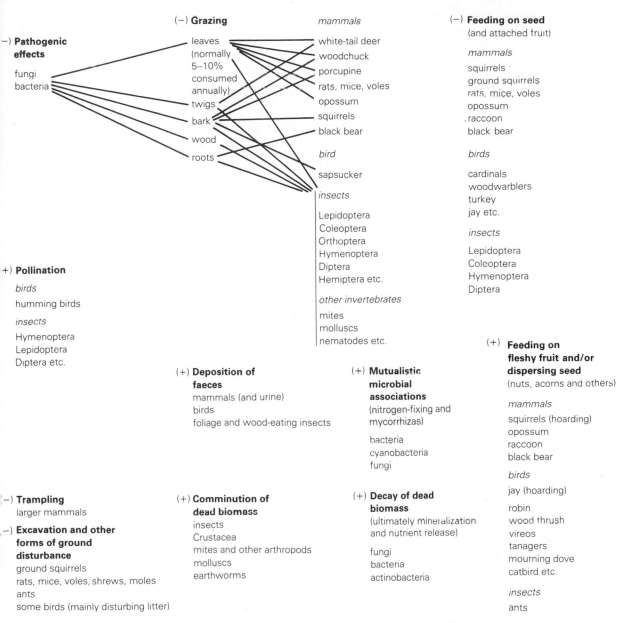

Figure 11.31 Important influences of biota other than plants in Eastern North American forests. Influences are classed according to whether they have positive (+) or negative (−) effects on the immediate recipients. Various negative influences may have indirect positive influences on either the immediate recipients or on other plants.

unweathered minerals, the colloidal complex, the soil solution and organic matter. Annuals or other short-lived plants (crop or weeds) are not as efficient as established woody plants are at abstracting nutrients. Because of their small size they do not take up much of the nutrient supply.

As woody vegetation develops, this situation changes. Pines and other early-transitional trees, having well-developed mycorrhizal systems, pre-empt nutrient

411

Figure 11.32 Winter view of the interior of mature mixed deciduous angiosperm forest, Hueston Woods State Nature Preserve, Ohio, USA. Windthrown trees in foreground. The main tree species include *Fagus grandifolia*, *Acer saccharum*, ash (*Fraxinus americana*), black cherry (*Prunus serotina*) and *Liriodendron tulipifera* (photo J. L. Vankat).

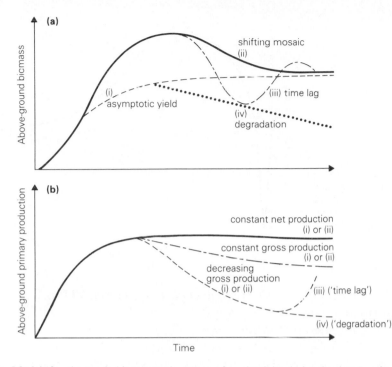

Figure 11.33 Models for changes in biomass and patterns of productivity during development from bare ground to mature forest. (a) Curves for shifts in biomass. (i) 'Asymptotic yield'. Smooth logistic curve with the upper limit fixed by site conditions. (ii) 'Shifting mosaic'. Biomass increases to a peak, then falls in the mature forest. (ii) 'Time lag', with initial biomass increase, then fall, then recovery, with damped oscillations around a lower average level. (iv) 'Degradation'. Increase to a peak, then a steady fall, experienced on sites deficient in nutrients. (b) Curves for productivity, explained mainly in terms of time lags in regeneration and synchroneity of mortality in relation to inhibition of regeneration and resource pre-emption by dense developing stands, then occurrence of regeneration after canopy thinning releases resources. The different patterns shown with roman numerals relate to the models in (a). Modified from Peet (1981).

supplies from increasingly greater soil depths. The early-transitional shrub, pin cherry (*Prunus pensylvanica*), was shown by Marks (1974) to do this very efficiently in New Hampshire forests. After forest disturbance pin cherry arises *en masse* from long-lived seed banks. As these plants shed leaves, twigs and roots and exude materials from roots and leaves, nutrient cycling begins, and when the early transitional species begin to die, other species can take over the released nutrient stocks (Bormann & Likens 1979a, Christensen & Peet 1981). In the developed forest, because of the well-established root–mycorrhiza network, nutrient losses are minimal. The same applies to water retention; experimental studies show that forests are very efficient in this respect (Ballard & Woodwell 1977, Bormann & Likens 1979a).

Some ecologists emphasize particular components of ecosystems as being important controls of processes, and their rates (including the development of vegetation after disturbance). For example, Odum (1971) stressed the importance of energy flow (see Ch. 12 & Fig. 11.18) in ecosystem development. Other ecosystem ecologists consider the nutrient component of ecosystems to be the most important overall

413

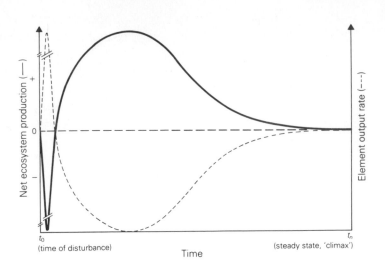

Figure 11.34 Model for change in net ecosystem production and element output rate (assuming constant input rates) on a vegetation development sequence, after disturbance (after Miles 1987 *ex* Gorham *et al*. 1979).

control (e.g. Gutierrez & Fey 1980, Reichle *et al*. 1975, Tamm 1975). The generalized changes in nutrient budgets during vegetation development in relation to plant productivity were outlined by Gorham *et al*. 1979) (Fig. 11.34).

The slow recycling of nutrients from bound organic forms can be a bottleneck in the maintenance of nutrient supplies for plant growth. Detritus-feeding organisms are emphasized as playing a very significant role in nutrient mobilization by some ecologists (e.g. O'Neill 1975, 1976). Mattson & Addy (1975) believe that herbivorous insects regulate forest primary production, on the whole in beneficial ways, through the removal of photosynthetic tissue (partly detrimental and partly beneficial) and increase in nutrient turnover by consumption and defecation of plant material which is rapidly recycled.

Mycorrhizas have been considered by some ecologists to play a critical role in community development. Their part in making phosphorus available is important in the model of T. W. Walker & Syers (1976) for control of soil–vegetation development. Stevens & Walker (1970) have also considered nitrogen-fixing associations to be connected with the maintenance of rapid rates of productivity early in vegetation change sequences. Tilman (1982) suggests that vegetation diversity and structure are maintained by competition for nutrients.

Although each component has a part to play in ecosystem functioning, and some may be more important than others in numbers of ways, it seems unwise to try to single out any one segment of a complex interactive system as playing a sole (or even a most important) regulatory role. In doing so, it is possible to lose sight of the fact that each species which participates in the interactions is exploiting available opportunities to the best of its individual ability.

The plants themselves are essential regulators, for by their very presence, as well as their modifying effect on climatic and other conditions, they establish conditions that are appropriate for all other interacting organisms (many of which they could not do without). Figure 11.35 illustrates important interactive features of the forest ecosystem.

The maintenance of order in complex ecosystems seems to be enhanced by 'niche-fit'. Plant species which are potential competitors because they utilize the same array of resources, are selected in such ways as to reduce the intensity of contest between one another (Ch. 3, Whittaker 1969b). Rather than overlapping absolutely, their niches tend to mesh with one another to some degree. A degree of order also seems to be maintained by the reciprocities and synergisms between the many interacting species of all kinds of organism (Knapp 1974b). Change in ecosystems implies that there is lack of order, apparently arising from imbalance in niche relationships both between plants and between them and their moneran, fungal and animal neighbours, as well as from the influences of abiotic components.

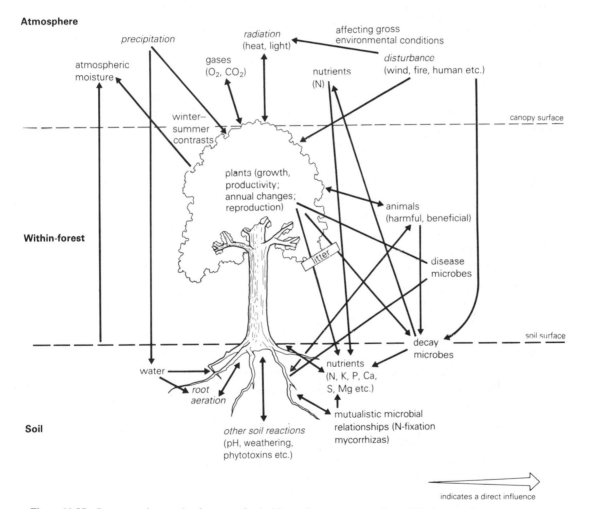

Figure 11.35 Important interactive features of a deciduous forest ecosystem (especially from a plant point of view).

SOME CONCLUSIONS

From the mass of facts about causes of sequences of change in the species populations of the vegetation during forest development, is it possible to establish any general principles? Do the models of vegetation change which various authors have proposed (Figs 11.25, 30 & 36 and others in Ch. 12) throw light on the subject?

A very important point is that the individual ecological behaviour of prominent plant species determines the actual course of events. Species have evolved to fill roles which are circumscribed by the abiotic and the biotic environmental conditions prevailing. The timing and spatial distribution of events in relation to the lifestyles of particular species is important. Horn's (1981) model reminds us of these and other important matters. The other schemes describe events during changes in plant and other components of ecosystems from early to late forest development, but are not necessarily very edifying about the *causes* of change.

The alternative hypotheses, proposed by Connell & Slatyer (1977) to explain the course of events – *inhibition, facilitation* and *tolerance* – can be used to guide our thoughts. Inhibitory effects are common. Herbs inhibit other herbs, and also woody species, in old fields for a time. Pines inhibit angiosperms for a time. Inhibition operates through the pre-emption of sites by species which have been favoured by the conditions prevailing at the establishment phase. They may effectively prevent other potentially dominant species from getting established for a time (sometimes a very long time), by assuming temporary control over resources such as light or nutrients. Often, however (as is true, for example, in sequences where pines are early dominants), the inhibitors also, in due course, provide nursery conditions for other species which ultimately will replace them and which find it difficult to establish in the absence of the nursery. Other species invade or grow vigorously only when the dominant resident species are damaged or die, providing space and microsites for seedling establishment and releasing nutrients. This is facilitation operating at a simple level. It appears to operate, also, in more-complex ways where microhabitats with specific litter, nutrient, microbial and other requirements for seedling establishment gradually develop under a long-lasting forest cover. Facilitation (otherwise 'reaction' of Clements (1916), autogenic influence of most subsequent authors) is implicit also in the modification of environment by species such that it becomes unsuitable for recruitment of those species, but suitable for others. The failure of some late-transitional or mature forest species to establish early in a sequence supports this view. Another simple example of facilitation is the enhancement of the spread of animal-dispersed species by the development of a woody plant cover of, for example, pines. Facilitation applies in most (if not all) cases of the establishment of juveniles, beyond the first colonists of bare ground. However, it is probably not as specific as Clements and other authors thought. Alternative patterns are often evident, due to vagaries in the prevailing conditions and in seedling establishment.

Tolerance is the model where the modifications wrought on the environment neither increase nor reduce the rates of recruitment of later colonists, and their growth to adulthood. Species that appear as adults later are simply those that arrive at the beginning, or later, and then grow very slowly. There is little evidence for and

416

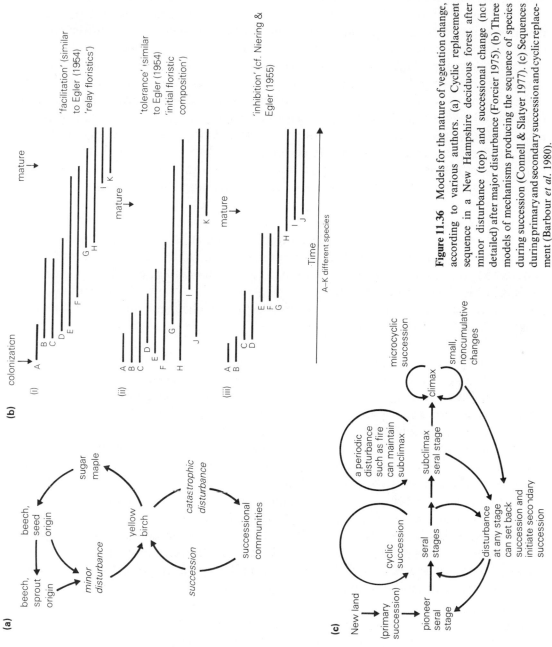

Figure 11.36 Models for the nature of vegetation change, according to various authors. (a) Cyclic replacement sequence in a New Hampshire deciduous forest after minor disturbance (top) and successional change (not detailed) after major disturbance (Forcier 1975). (b) Three models of mechanisms producing the sequence of species during succession (Connell & Slatyer 1977). (c) Sequences during primary and secondary succession and cyclic replacement (Barbour *et al.* 1980).

some evidence against this scheme, operating in the form of the 'initial floristic composition' notion of Egler (1954). However, the idea needs rigorous testing by marking young plants and monitoring them throughout their life. If all tree species could establish on fresh, arable soil, then the model might be sustained; but this is not the case. Neither is there evidence that growth rates of the mature forest species are slower under favourable conditions than, say, some late-transitional species. Nevertheless, although there are some exceptions (e.g. *Sequoia sempervirens*), late-transitional and mature forest tree species are usually slower-growing and longer-lived than the species that preceded them, and this contributes to the impression of sequential replacement. Experiments to try to establish beech or oak juveniles in eastern North American pine stands, in competition with an understorey of species such as flowering dogwood, sweet gum or red maple, might be edifying.

Species-by-species replacements begin at the level of seedling establishment, so competition between seedlings, and older juveniles, and interaction between them and adults must be understood well before we can be sure about the causes of any local patterns of vegetation change. The information available on the eastern North American forests so far provides various inklings about what is actually happening at the level of leaf, stem and root physiology. Gaps remain to be filled, by comparative experiments on the requirements of species for particular 'safe site' conditions for seedling establishment, and other aspects of the underground systems and nutrient supplies.

The notion that any one of inhibition, tolerance or facilitation is the sole mechanism capable of explaining a whole sequence of vegetation change can be discarded. Each may operate simultaneously in the same sequence, depending on which species, or which part of a sequence, attention is focused. It may be more fruitful to think of sequential forest development simply in terms of the properties of individual prominent species. For sets of species in a sequence the properties of the earlier residents determine conditions for establishment, rates of growth and development to maturity of other species which eventually replace them. The properties of juveniles of the replacing series determine their capabilities for competing with juveniles of earlier residents and taking over the sites as the resident adults die. Table 11.19 summarizes some important mechanisms promoting changes. Horn (1981) quite aptly expresses the whole process of vegetation change, simply as 'the differential expression of life histories'. Pickett *et al.* (1987) (and others before them) also emphasize site availability, differential species availability and a differential species performance.

After episodes of forest disturbance, in each locality there are limited assemblages of species which can play particular roles in forest redevelopment. This causes a semblance of orderliness, although at the scale of the individual plant on the ground and its regeneration potential, there is considerable variability in the course of events.

Maturation of forests, or the attainment of 'stability', is in part the manifestation of slowing of rates of change. Many of the tree species of mature forests (at least those in eastern North America) are among the longest-lived. They can maintain their ground for long periods, aided by their strong wood, which is resistant to snow and wind damage, herbivory and fungal attack. Their canopies cast deep shade, but their

Table 11.19 Conditions promoting changes in plant species populations during sequential forest development (affecting the distribution, abundance and survival in time of the main species which occupy space in a locality).

1. modification of site conditions by external events such as climatic shift, change in herbivore populations, fire (*chance*); continued modification of local site conditions in cumulative ways by resident populations (e.g. modification of light climate, nutrient depletion, accumulation of litter and humus, changing pH, soil moisture retention etc.)

2. site provision (i.e. microsites for seed germination, seedling establishment; (a) disturbance by outside forces (particular intensity, duration, size of area affected, timing are important) (*chance*); (b) natural death of residents (*chance*)

3. availability of seeds and other propagules of particular species which can take advantage of prevailing conditions; influences on this of relative reproductive success of the species concerned, timing, appropriate dispersal vectors, seed predators (*chance*)

4. establishment conditions at ground level: conditions created by disturbance; conditions created by prior occupants (litter, etc.) and present surrounding plant, animal and microbial populations; local factors of soil fertility, microclimate, etc. – physicochemical stress (*chance*)

5. differential species performance with respect to germination and establishment, continued growth, productivity, competition

6. interactions of both juveniles and adults with organisms critical for influencing competition, survival; mutualisms such as nitrogen-fixation, mycorrhizas; pathogens; herbivores (*chance*); phytotoxic effects

7. differential ability of adults to continue to maintain integrity and reproduce on particular sites in the face of habitat changes

(*chance*) means that various stochastic events may affect the particular situation.

juveniles are shade-tolerant. They resist browsing and defoliation better than other species do. Nevertheless, they are not totally invulnerable, as was evident from the sad demise of chestnut (*Castanea dentata*) as well as from slow changes which have occurred in other mature forests. Exceptions to the general rule are found where long-lived colonist species persist in the forests and are replaced by shorter-lived species when they eventually die. Local conditions, with respect to the whole milieu of environment in a region, will determine which species have the competitive versatility to form the 'most-mature' vegetation. This ideal concept must not be taken too seriously in localities where disturbing influences occur at intervals that are more frequent than the generation period of the species potentially capable of forming 'most-mature' vegetation (P. S. White 1979).

12
Community phenomena in vegetation change

It is appropriate now to look at the broader issues of community properties linked with vegetation change. At this *macro-ecological* scale, although the role of individual species must not be neglected, it is the functioning of entire vegetation patches, plant communities or ecosystems that is being considered.

An historical perspective is justified because the theory relating to vegetation change is in a state of flux. It is necessary to know how and why this came about. Older ideas (up to about 1975) are being re-examined in the light of advances in ecological science. Newer ideas are still emerging and being tested. Fundamental conflicts have arisen between ecologists who favour either holistic or reductionist approaches to the subject. These have generated fierce philosophical arguments. Can the differences be resolved? Discussion of the theoretical problems which have arisen is found in Drury & Nisbet (1973), McIntosh (1980a, b, 1981) and Finegan (1984).

We can begin by examining the evolution of ideas on the theory of developing communities and ecosystems. Several main, interrelated themes are apparent. They are:

1   The fundamental nature of community processes and organization.
2   The concept of the plant community or ecosystem.
3   The way in which changes in plant species populations are conceived to take place.
4   The concept of stability.
5   Means of expressing or describing the phenomena of vegetation change.

OLDER THEORY ON SUCCESSION TO CLIMAX

Some ideas about vegetation change, which began to be promulgated about 100 years ago, were maintained up to the 1930s. In modified form they are still adhered to by many plant ecologists and land managers.

The term **succession** came into use in the 19th century to describe sequences of vegetation change. Early writers, influential in developing the theory of what may be termed **succession to climax**, were O. R. Hult and H. C. Cowles. Hult (1885, 1887) described vegetation changes beginning in water, or on bare ground in heath vegetation and forest in Sweden, leading to stable types. He postulated convergence of the developing vegetation towards a few types. Cowles (1899a, b, c, d, 1901) wrote of the genetic relationships of plant communities on Lake Michigan sand dunes. He defined the **climax** as the end-point in vegetation development and the most 'meso-phytic' vegetation which the climate will support.

Cowles (1910, 1911) also noted the occurrence of vegetation cycles and three important phenomena relevant to vegetation change:

(a) The modifying effects of plants on moisture conditions and soil fertility.
(b) The likelihood that nitrogen fixation, mycorrhizas and the production of toxic compounds by plants affected soil–plant interactions.
(c) Differences in the ecological amplitude of different plant species.

Clements (1904, 1905) was influential in the development of theory. In 1916 Clements published his major book *Plant succession: an analysis of the development of vegetation*. Like some other ecologists of his time, Clements had rather rigid views on the way in which species are organized into communities.

Clements' book, republished as *Plant succession and indicators* in 1928, was a statement of the theory of plant succession in the form of an elaborate descriptive account, with much new terminology (most of it omitted from the present discussion).

The essentials of Clements' ideas are the following.

1 Succession is initiated by **nudation** [provision of new sites]; it proceeds by **ecesis** [establishment and development of plants until they reproduce]; **action** [the effects of environment on plants]; **reaction** [the effects of plants on the physical environment and feed-back effects on the plants]; **competition**; and, finally, **stabilization** [control of habitat conditions by the vegetation]. These processes may be successive in the initial stages or interact in a most complex fashion in all later stages. Much emphasis is placed on competition for water; water needs are regarded as the best basis for indicating the direction of successional development.

2 A new term **sere** was applied to a unit succession, from appearance of the first pioneers, to the final or **climax** state [hence the adjective seral, often used for developmental phases].

3 **Primary succession**, beginning on sterile new surfaces, is distinguished from **secondary succession**, which starts part-way along a sere, following disturbance.

4 **Cliseres** are those set in motion by a major climate shift.

5 The rate of succession, fast at the beginning, slows down towards climax.

6 Species are replaced during succession, because their seedlings cannot establish under the conditions which they themselves have created, but the seedlings of their successors can [i.e. 'reaction', prepares sites for establishment of later phases of the sere].

7 Community reactions, affecting physical environmental factors, are more than the sum of the reactions of the component species and individuals.

8 Reaction explains the orderly progression of succession, by stages, and the increasing stabilization, which produces the final climax. The reactions of species of the climax vegetation exclude all other species.

9 The overall effect of stabilization is a trend towards mesophytic conditions. The climax dominants are those which make the greatest demands on the resources of the site. The climax is the highest possible life-form for the particular climatic region.

10 As succession advances, vegetation complexity (e.g. as reflected by vertical stratification) increases. The species in such communities are of complementary nature, as reflected by differences in stratification of shoot and root layers and seasonal periodicity.

11 Given sufficient time, convergence of seres towards a common climax occurs.

12 Succession is inherently and inevitably progressive. Regression is impossible, as are cyclic successions.

13 Climax **formations** recognized by the life-form types of the dominant plants, and determined by the regional climate, are the basic vegetation units. Subunits are **associations**, characterized by their species' dominants. **Subclimaxes** are communities that are not quite mature, or those which are prevented, for the time being, from succeeding to full climax status. [N.B. The later use (Clements 1936) of **proclimax**, for several types of 'arrested' successional communities; **serclimax** – held in early developmental stages by special environmental conditions; **subclimax** – where maturation to full climax is delayed in a late successional phase; **disclimax** – held in stable conditions different from climatic climax by animal or human disturbance, etc].

14 After disturbance of climax vegetation succession is reinitiated (beginning part-way along a sere) and the vegetation will return, eventually, to climax.

15 Climax formations are organic entities. As an organism, the formation arises, grows, matures and dies.

This last statement appears at the beginning of Clements' book and, with statements 7 and 12, helps us to understand his stance on various other matters. His was a holistic viewpoint in a rather extreme form. The organismic view of the nature of biotic communities was held by numbers of his contemporaries (cf. Allee *et al.* 1949).

The book made an immediate and profound impact on the subject of vegetation ecology (and other branches of ecology, cf. Shelford & Olson 1935). The views expressed in it were accepted without demur by many (but not all) ecologists (W. S. Cooper 1926, J. Phillips 1934, 1935). The impact of the book was reinforced by its republication, in slightly modified form, in 1928 and by Clements' collaboration with J. E. Weaver (Weaver & Clements 1929), in writing an ecology textbook. Modifications to some of the terminology were made in Clements (1936) and in the second edition of Weaver & Clements (1938).

There are numerous problems with Clements' succession theory. Some arise from his lack of scientific rigour, the rigidity of his scheme, into which some phenomena were unnaturally squeezed, his rigid adherence to some doubtful concepts and his tendency to ignore evidence which conflicted with his own views. The most unfortunate effect, however, was the uncritical acceptance of his ideas by many ecologists and the incorporation of these ideas into derivative textbooks on plant and animal ecology and geography. In some respects Clements' forceful presentation of his ideas, the compliance of many contemporaries and the pedagogical consequences, held back the development of plant ecological research and teaching for 25 years or more. Among Clements' most-important contributions to ecological science and the study of vegetation change were his development of quadrat studies and census methods to follow the history of changing populations (e.g. Pound & Clements 1898, Clements 1910).

THE ORTHODOX SUCCESSION TO CLIMAX THEORY

Very little of Clements' special terminology is still in use in the modern ecological literature. The conceptual parts of his scheme, which survived in somewhat modified form at least to the 1970s, (e.g. in Odum 1971, Collier *et al.* 1973, Ricklefs 1973, Whittaker 1975b) are outlined below.

1   Succession begins on newly formed substrates (primary) or after disturbance (secondary).
2   Sequences occur: pioneer → seral phases → climax.
3   Secondary succession culminates in the re-establishment of climax.
4   Frequent disturbance or difficult habitats maintain some vegetation in an almost perpetual state of non-maturity (pro-climax).
5   Climax vegetation is stable and self-perpetuating.
6   The sizes of plants, species diversity and complexity of structure all increase as succession progresses.
7   Autogenic processes (i.e. 'reaction') provide the impetus for successional changes. Among these are soil modifications. Soils mature as succession proceeds.
8   Replacement of species populations, as succession proceeds, is by competition.
9   Succession can be driven, also, by allogenic events.

Directly contradicting Clements, most ecologists also recognize that:

10   Different climax species-assemblages may occur in any climatic region; forming a mosaic pattern according to different substrate or other persistent local site conditions.
11   Degradation of ecosystems (regression) occurs, often as a result of changes in soil characteristics, induced by vegetation.
12   Cyclic vegetation change processes are quite common.

Recent schemes summarizing different types of vegetation change are shown in Figures 12.1 and 12.2. Ideas on increased diversity during succession were dropped by many ecologists by the mid-1970s.

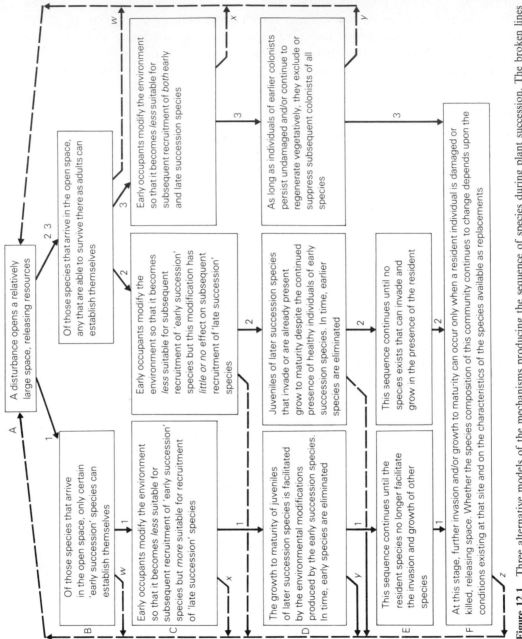

Figure 12.1 Three alternative models of the mechanisms producing the sequence of species during plant succession. The broken lines represent interruptions of the process, in decreasing frequency in the order *w*, *x*, *y* and *z* (after Connell & Slatyer 1977).

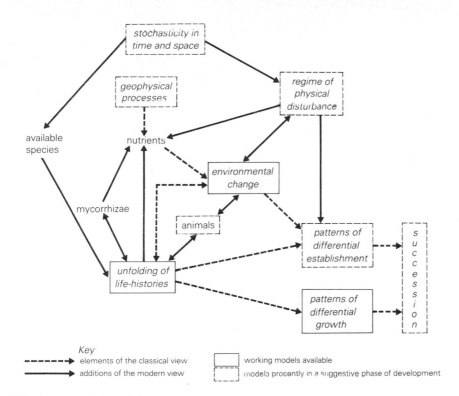

Figure 12.2 A representation of the succession concept. Although not shown explicitly, all factors affect life-histories when evolutionary time is considered. However, the resulting adaptations may have originated in a different environmental context from that of the present, especially in a different array of available species. (Compiled and arranged by H. S. Horn & J. F. Franklin). After D. C. West *et al.* (1981).

HOLISM AND DETERMINISM IN SUCCESSION THEORY

Although it was greeted with derision by some contemporary ecologists (e.g. Tansley 1916, Gleason 1917, Gams 1918, Du Rietz 1919, W. S. Cooper 1926), Clements' promotion of the idea of vegetation having properties like organisms has influenced subsequent thought on the bioenergetics of coenoses and developing ecosystems (Lindeman 1942, Margalef 1968, Odum 1969, 1971). Lindeman (1942) originated the trophic–dynamic approach to ecosystem succession. In it he emphasized the control of processes by the ecosystem. Odum (1971) tabulated 'trends to be expected' in the development of ecosystems (Table 12.1, and see modified versions in Whittaker 1975b, Gutierrez & Fey 1980).

According to Odum, succession culminates in a stabilized 'mature' ecosystem in which maximum biomass (high 'information' content) and symbiotic function between organisms are maintained per unit of available energy flow. The 'strategy' of succession is increased control of the physical environment or **homeostasis** (self-regulating stability) in the sense of achieving maximum protection from environmental disturbance. Reichle *et al.* (1975b) and O'Neill (O'Neill 1976a, b, O'Neill *et al.* 1975) pursue these ideas further. Ecosystems, according to them, are energy-processing units, regulated by the availability of water and essential nutrient elements and

Table 12.1 Contrasts in various community attributes during ecological succession (after Odum 1969).

Ecosystem attributes	Developmental stages	Mature stages
community energetics		
1. gross production/community respiration (P/R ratio)	greater or less than 1	approaches 1
2. gross production/standing crop biomass (P/B ratio)	high	low
3. biomass supported/unit energy flow (B/E ratio)	low	high
4. net community production (yield)	high	low
5. food chains	linear, predominantly grazing	weblike, predominantly detritus
community structure		
6. total organic matter	small	large
7. inorganic nutrients	extrabiotic	intrabiotic
8. species diversity – variety component	low	high
9. species diversity – equitability component	low	high
10. biochemical diversity	low	high
11. stratification and spatial heterogeneity (pattern diversity)	poorly organized	well organized
life-history		
12. niche specialization	broad	narrow
13. size of organism	small	large
14. life-cycles	short, simple	long, complex
nutrient cycling		
15. mineral cycles	open	closed
16. nutrient exchange rate, between organisms and environment	rapid	slow
17. role of detritus in nutrient regeneration	unimportant	important
selection pressure		
18. growth form	for rapid growth ('r-selection')	for feedback control ('K-selection')
19. production	quantity	quality
overall homeostasis		
20. internal symbiosis	undeveloped	developed
21. nutrient conservation	poor	good
22. stability (resistance to external perturbations)	poor	good
23. entropy	high	low
24. information	low	high

constrained by climate. Energy is the fuel through which ecological processes operate, but the rates at which the processes proceed are controlled, in natural systems, by the nutrient availability. The deterministic hypothesis of these authors is that ecosystems function to expend readily available energy so as to minimize the constraints imposed by limiting nutrients and water. O'Neill (1976b) even took the extreme view that ecosystems can retain their identity although their species composition changes.

Reichle *et al.* (1975b) stressed the importance of the reservoir energy base, the living and dead biomass of the system, in providing inertia in case of disturbance. Regulation and persistence of a system are mediated by the remobilization of the nutrient resource. We return to consider the validity of these points, when discussing the role of nutrients in community development.

REDUCTIONISM VERSUS DETERMINISM

Drury & Nisbet (1973), in detailed dissent about the currently held succession to climax theory, stated that some of the points made by Odum, Whittaker and others about changing productivity and associated energetic phenomena in developing ecosystems, were not supported, even by their own evidence. Undoubtedly community productivity increases after the initiation of succession, but some empirical data show that, although biomass in 'climax' communities is greater than in seral phases, the productivity of 'climax' communities is lower than during some of the seral phases (Odum 1960, Kira & Shidei 1967). The rate of gross (and net) production in the early phases appears to increase in a series of steps associated with the particular plant species of different stature present (Holt & Woodwell *ex* Whittaker 1975b, (Fig. 12.3). Odum (1960, 1969) had already suggested this and that, with transition from one phase to another, there may be a temporary peak of production as the plants exploit limiting materials not available to their predecessors, then a levelling off to 'steady state' before the next phase is initiated. Eventually the overall production rate levels off, and then declines as the system approaches maturity.

The hypothesis of Margalef (1963, 1969) and Odum (1969) that succession is directed towards maximum ecological efficiency and the efficient management of 'information' (e.g. the accumulation of biomass and efficient cycling of nutrients) was attacked strongly by Colinvaux (1973) and Drury & Nisbet (1973). Drury & Nisbet maintained that some of the trends in maturing ecosystems, which Odum (1971) described, are simply an outcome of increased biomass and vegetation density. Increased 'information', rather than causing increasing complexity, results from it.

The evidence showing that observed structural or functional characteristics of developing communities do not (or do not always) conform to the generalizations about successional systems made by Margalef and Odum, called into question the general validity of the bioenergetic succession model. Drury & Nisbet (1973) maintain that the view that succession is connected with the 'emergent' properties of communities is unacceptable as a generalization, even though there may be some empirical data which are consistent with it. (By 'emergent' properties are meant features of a system which are more than the sum of effects created by its constituent species.) A strict reductionist view of ecology or biological systems does not permit an organizing, deterministic principle to operate during succession. Neither can there be direction by negative feedback in ecosystems. Ecosystems, according to Colinvaux, are merely concepts useful for empirical study and have no biological identity.

However, Reichle *et al.* (1975b) emphasize the viewpoint that the conception of behaviour of ecosystems as units is legitimate and does not contradict the principle

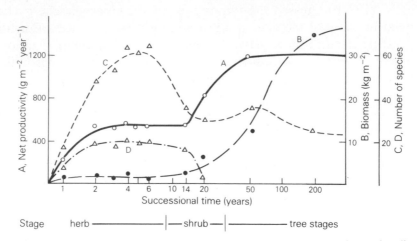

Stage   herb ——————|— shrub —|—————— tree stages

Figure 12.3 Changes in productivity, biomass and diversity during succession after major disturbance in the Brookhaven oak–pine forest, New York, USA. (A) Net primary productivity increases to a stable level in the herb stages (2–6 years), then increases as woody plants enter the community (14–50 years), to a stable level that may persist into the mature phase. (B) Biomass is low through the herb stages and then increases steeply with the accumulation of woody tissues of shrubs and trees; a stable biomass is probably not reached until after 200 years. (C) Species diversities – number of species in 0.3 ha samples – increase into the late herb stages, decrease into shrub stages, (14–20 years), increase again into a young forest, (50 years) and then decrease into the mature phase. (D) Numbers of exotic species (present only in the herb and early shrub stages). After Whittaker (1975b), Whittaker & Woodwell (1968).

that individual populations react to environmental gradients in a relatively independent manner. Populations must evolve within their own biological environment, the ecosystems in which they live (Goodall 1963). Whatever individual population responses occur, the populations establish homeostatic feedback mechanisms to maintain the system. Similar views were put forward by Dunbar (1972). These statements relate to the philosophical basis for disagreements between reductionist evolutionary population biologists and holist ecosystem ecologists (cf. Levandowsky & White 1977). Conflicts have arisen because of divergent views on the 'units' on which evolutionary processes operate (individuals, populations or species, on the one hand, or aggregations of organisms living together, on the other). It seems reasonable to assume that the two approaches are not mutually exclusive. Species living in communities can behave individualistically only within the constraining bounds placed by their neighbours, with which they may have positive or negative relationships (cf. Chs 3 & 11).

These questions have been addressed by other authors, who describe functional groups at levels of biological organization greater than populations, and ranging up to large-scale ecosystems, e.g. Valentine (1968), Wilson (1980), Whittaker (1969b, 1977), Van Valen (1980), J. N. Thompson (1982). Information on **co-evolution** (Gilbert & Raven 1975, Janzen 1969), **taxocenes** (Levandowsky & White 1977) **ecological guilds** (Root 1967, 1973) and **group selection** (Alexander & Borgia 1978) is relevant to the discussion. So too are data on **emergent properties** of biotic communities (Salt 1979) and evidence for parallel ecotypic responses of different wide-ranging plant species occurring together in communities (McMillan 1959, Vaartaja 1959).

ECOSYSTEM NUTRIENT BUDGETS

Although the discrepancies in Odum's model of developing ecosystems cannot be denied, support for the theme that nutrients are an important control of vegetation development is strong (see e.g. Chs 5, 6, 8 & 11). In any system the immediate nutrient balance is important for the maintenance of any particular array of plant species (Reichle *et al.* 1975b, Tamm 1975, Gorman *et al.* 1979, Tilman 1982). The stock of nutrients in organic matter is especially important. Elements are stored in biomass in forms which can readily be made available to plants. Regulation and ultimately persistence of the system can be mediated by the remobilization of this nutrient resource (Reichle *et al.* 1975).

Climatic conditions are an overall control of nutrient availability to plants. In moist climates (i.e. where monthly evapotranspiration does not exceed monthly precipitation, and leaching losses from the upper soil layers are possible) nutrients are among the most important limiting factors affecting plant growth and productivity. In turn, nutrient budgets govern long-term vegetation changes and the relative stability of mature systems. Differences between ecosystems in a range of climatic conditions, with respect to nutrient budgets, are summarized in Table 12.2.

Contrasts in nutrient budgets occur between immature and mature systems. Although there is some dissent (Vitousek & Reiners 1975) many ecologists (e.g. Ballard & Woodwell 1977) hold that young, developing systems tend to leak nutrients, whereas mature systems are more efficient at retrieval and storage of nutrients. Conservation of nutrient capital can be achieved by durability of tissues (evergreenness and decay-resistant wood), withdrawal of nutrients from organs which are purely structural or about to be shed, and mycorrhizal associations. In mature vegetation, both in tropical and temperate systems, mycorrhizas are very prominent. Through the maintenance of tight cycling systems they can be vital in permitting plants to retain phosphorus and other nutrients when the supply is scarce. Marks & Bormann (1972) and Bormann & Likens (1979a) explain the steady-state nutrient cycling that occurs in mature systems. 'Steady-state', as opposed to slow change, may, however, be a condition that is difficult to detect by means of the measurement techniques that are available at present.

The significance of animal and microbial components of ecosystems for nutrient mobilization may be stressed here. The faunal herbivores and detritivores are very important in processing organic matter, before its mineralization, because they comminute coarse material, providing readily available substrate for microbial decomposer populations (Olson 1963, Edwards *et al.* 1970, Mattson & Addy 1975, Swift *et al.* 1979) and also because of the important effects of their own enzyme systems (Greenfield 1981). Damage to the consumer and decomposer ecosystem components (or, of course, to the organic or inorganic nutrient stocks themselves) will be a source of instability (Harte & Levy 1975).

Resilience of ecosystems was theoretically linked to nutrient-cycling by De Angelis (1980). Recovery is slow after disturbances of mature systems, which interfere with the tightly cycled essential elements. Destruction of all or part of the biomass of mature ecosystems (e.g. by fire or clearing of natural vegetation for agriculture), and

Table 12.2 Nutrient regimes for ecosystems in different climatic regions.

Landscape or vegetation type	Physical limiting factors affecting plant growth and nutrient budgets[*]	Litter and humus content of the soil	Nutrient availability of mature soil	Nutrient recycling rate
arctic tundra	very low temperature, short summer, frozen ground, wet soil	high	low	very slow
boreal forest	low temperature, short summer, wet soil, fire	high	low, leached	slow
temperate deciduous forest	seasonal low temperature	high in mor soils, lower in mull	moderately high in mull-forming vegetation; low, leached in mor-forming vegetation	slow to moderately fast
temperate grassland	seasonal low temperature, seasonal dry conditions, fire	high	high during the moist season, low during the dry period; accumulation of $CaCO_3$, etc.	moderately slow
mid-latitude savannah	seasonal dry conditions, fire	moderately low	high during the moist season, low during the dry period, but generally low unless released by fire or animals	moderately slow to moderately fast
mid-latitude desert	high temperature, very dry soil most of the time	very low	low through lack of water, though abundant salts are usually present	very little cycling is possible
tropical rainforest		very low	low, leached in most-mature soils	very fast

[*] Locally there may be various other limiting factors, e.g. extreme waterlogging, salinity or other toxic soil conditions. Grazing may have important influences in all regions.

with it the loss of much of the pool of nutrients, leads to irreversible or only slowly reversible changes in mineral nutrient status and soil structure. Although factors promoting the conservation of nutrients in ecosystems are stabilizing influences (cf. Reichle *et al.* 1975b, O'Neill 1976a, Harte & Levy 1975), they also make the systems vulnerable to gross disturbance. Grasslands and temperate forests should be able to recover more readily after major disturbance (inertia conferred by the relatively large store of detrital organic matter) than tropical forests can. In the latter the nutrient stocks are very vulnerable because there is little dead biomass (Pomeroy 1970, Webster *et al.* 1974, Swift *et al.* 1979). However, other constraints, especially climate, must affect the rates of recovery.

Notwithstanding the often detrimental effects on nutrient stocks of ecosystems by destruction of biomass, some forms of disturbance may rejuvenate the nutrient supplies of any locality. Deposition of fresh, new mineral material (e.g. by river siltation, loess accumulation or fallout of volcanic ash) is one means of doing this.

Another is by erosion of the surface soil down to fresh, unweathered rock. Nitrogen supplies cannot be replenished in this way, unless organic matter from elsewhere is deposited but all other nutrients could be made available. In hilly or mountainous terrain constant soil rejuvenation is possible, but in places with gentle terrain, or where the rocks are very deeply weathered (as in tropical regions), inherently infertile or hard, it may not be a very effective way of replenishing nutrient stocks. In forest areas tree-fall can lift up unweathered rock fragments caught among the tree roots.

Gradual decline in nutrient stocks of ageing ecosystems in humid environments is a general theme outlined in Chapters 5, 6 and 10. As just noted, rejuvenation of soils may occur through disturbances. There are also a few reports of rejuvenation of the nutrient stocks of ecosystems through plant activity. Miles (1981) showed that birches (*Betula* spp.) grown on poor, podsolized, mor heath soils in Britain transformed them to mull condition. *Agrostis–Festuca* grasslands also improved heath soils. Greenland and Nye (1959), Nye (1961) and Aweto (1981) describe similar improvements in the upper soil horizons of some poor African soils after invasion by pioneer tree species. Improved nutrient status is achieved through nitrogen fixation by legume symbionts or free-living organisms, by concentration of precipitation-borne nutrients and by tapping and concentration of nutrients from weathering minerals deep in the solum.

PLANT COMMUNITY AND CLIMAX CONCEPTS

In Clements' **monoclimax** concept all community development is supposed to converge on and culminate in one kind of vegetation in each major climatic region – the climax, of **formation** status. The formation is a major physiognomic unit of wide geographic extent (Table 12.3).

Table 12.3 Climax vegetation formations of North America (after Clements 1928).

grassland climaxes
    Stipa–Bouteloua climax – prairie
    Carex–Poa climax – tundra

scrub climaxes
    Atriplex–Artemisia climax – sagebrush
    Larrea–Franseria climax – desert scrub
    Quercus–Ceanothus climax – chaparral

forest climaxes
    Pinus–Juniperus climax – woodland
    Pinus–Pseudotsuga climax – montane forest
    Thuja–Tsuga climax – coast forest
    Picea–Abies climax – subalpine forest
    Picea–Larix climax – boreal forest
    Pinus–Tsuga climax – lake forest
    Quercus–Fagus climax – deciduous forest
    Isthmian forest
    Insular forest

It is possible to accept the climatic climax idea (without necessarily accepting the dynamic or organismic connotations of Clements' succession theory) to the extent that grand scale ecosystems (biomes), such as the Arctic tundra, Boreal conifer forests, semi-arid grasslands, or tropical evergreen rainforests appear to fit Clements' definition. They maintain their integrity in space and time, and are controlled, essentially, by climatic conditions (Holdridge 1967).

At the regional or local scales, however, monoclimax is an unrealistic concept because the habitat conditions are usually non-homogeneous and corresponding vegetation patterns are diverse. Plant ecologists of the 1910s to 1930s (e.g. Nichols 1917, 1923, Tansley 1920, 1929, 1935, 1939) developed the **polyclimax** concept to cope with this problem. According to this concept it is possible for many persistent climax associations to be found in any region, corresponding to soil, topographic and biotic, as well as climatic controls. Convergence of community composition, over time, does not necessarily occur. Associations, in this scheme, are characterized by one or a few physiognomic dominants and have constant structure. Many American ecologists have followed this sort of community–climax concept (e.g. Braun 1950, Oosting 1956, Daubenmire 1968b).

Whittaker (1953) suggested that even the polyclimax concept was too static and typologic. He proposed the alternative **prevailing climax** concept. In this, climax communities are relatively stable and self-maintaining portions of the vegetation continuum, associated with the pattern of environmental gradients.

Some American contemporaries of Clements – Shreve (1915), W. S. Cooper (1926) and, particularly, Gleason (1917, 1926, 1927, 1939), had concepts of plant communities and climax very different from the typological and hierarchical systems used by most ecologists at the time and since (cf. De Vries 1953, M. C. Anderson 1966, Shimwell 1971). Gleason, whose **individualistic concept** of the behaviour of plant species in communities has influenced many ecologists recently, rejected the idea that communities are like organisms, with the plant species that comprise them being held together by close bonds of interdependence. His view was that any vegetation units which can be recognized are temporary and fluctuating phenomena. Their fortuitous composition is determined by the availability of species in the surrounding vegetation. They have come together because their environmental requirements overlap. Species behave as individuals, responding each in their own way to environmental factors, including interactions with other species. No fixed and definite community composition exists. Thus, vegetation varies continuously in space and time, and communities are simply arbitrary segregants from the continuum.

Succession, according to Gleason, is not an orderly process. It does not lead to the establishment of any definite climax, for change is universal and constant. Differences in the lifespan of plant species affect our perception of the relative stability of stands of vegetation. Similar views to Gleason's were advanced independently by a Russian, Ramensky (1924), with much more supporting data (cf. McIntosh 1983).

Gleason's individualistic concept was taken up by other American ecologists, e.g. Whittaker (1953, 1974), Curtis (1959) and Glenn-Lewin (1980). According to Whittaker (1953, 1974) no two stands are quite alike. Each stand is visualized as a system of superimposed species population lattices where the populations differ in density,

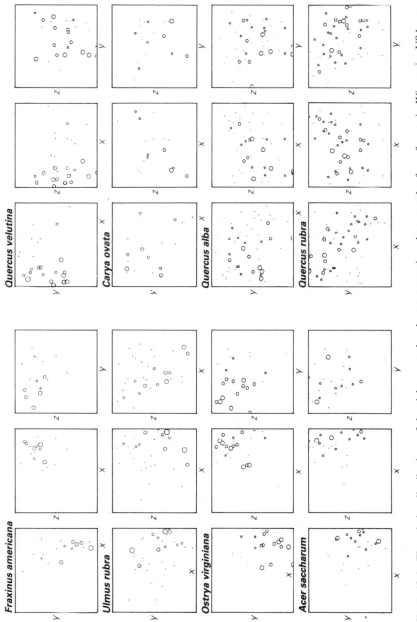

Figure 12.4 The relative distributions of the eight most abundant tree species when samples from forests in Wisconsin, USA, are subjected to a simple ordination. The distribution of each species is shown in relation to three abstract ordination axes x–y, x–z, and y–z. The relative importance of each species, in each sample, is shown by the sizes of circles (after Bray & Curtis 1957).

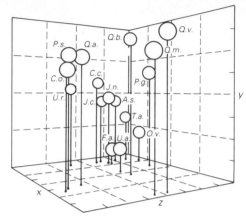

Figure 12.5 Summary of the kind of information shown in Figure 12.4. A three-dimensional plot of the centres of species distribution for 16 species, derived from the simple ordination. Tree species indicated by genus and species initials: *Acer saccharum, Carya cordiformis, C. ovata, Fraxinus americana, Juglans cinerea, J. nigra, Ostrya virginiana, Populus grandidentata, Prunus serotina, Quercus alba, Q. borealis, Q. macrocarpa, Q. velutina, Tilia americana, Ulmus americana, U. rubra.* After Bray & Curtis (1957).

spacing and degree of association with one another in space and time: 'The vegetation of a landscape consists ... of a complex and subtle pattern of communities interrelated by continuous intergradation and by sharing of species'. Whittaker (1975b) used the term **coenocline** for the plant community gradient corresponding to the spatial environmental factor gradient. The two gradients, together, comprise an **ecocline**. The idea can also be adapted to encompass gradients of species composition in time (e.g. during succession).

Another, abstract, approach to the explanation of the organization of vegetation, by objective comparisons of many field samples, **indirect gradient analysis** (or **ordination**), has been developed (Bray & Curtis 1957, Williams *et al.* 1966, Whittaker 1973, Orloci 1978, Hill 1979a, b). The results of ordinations, using the wide variety of methods available, tend to confirm that the species composition of vegetation varies continuously (Figs 12.4 & 5).

Some ecologists have disagreed with Gleason's view that plant communities lack bonds of interdependence between the constituent species (e.g. Poore 1964, Langford & Buell 1969), but Gleason's actual position was not quite as extreme as they indicate. This point is addressed later, when examining the evidence for the stability and integrity of plant communities. In the meantime, it may be acknowledged that field observational and experimental evidence accord with Gleason's individualistic concept, to the extent that species composition of vegetation varies continuously in space (the exceptions are discontinuities caused by differences of substrate-type, disturbance or strong competitive interactions). Nevertheless, as is to be expected of organisms which have evolved together, there is evidence for a degree of integration between species in communities (Curtis 1959).

A recent development promoted by O'Neill *et al.* (1986) and Urban *et al.* (1987) (Table 12.4) is that an hierarchical approach is an appropriate means for viewing ecosystems and ecosystem dynamics. The components of the hierarchy are systems organized at increasingly large spatial scales and increasingly long temporal scales, with relationships such that lower-level units interact to generate the higher-level patterns. The hierarchical approach is useful to provide guidelines of scale, but sight must not be lost of the essential continuity between vegetation systems at any level of scale, beginning with individuals.

Table 12.4 Levels of a landscape-vegetation (and ecosystem) hierarchical classification (modified from Urban *et al.* 1987).

Level	Scales of definition	Spatial scale	Temporal scale for persistence
regional landscape	physiographic provinces; biomes	10 000s of ha	many 1000s of years
local landscape	watersheds and other similar sized landform units; vegetation cover types	100s to 1000s of ha	100s to a few 1000 years
stand	local topographic positions; patches, stands	1s to 10s of ha	1 to 10s to a few 100 years
individuals (and gaps)	local sites; individual plants and gaps	<0.01 to 0.1 ha depending on plant size	1 to 10s to a few 100 years

PREDICTABILITY AND CONVERGENCE IN SUCCESSION

Ideas of the predictability of successional change were generally accepted by most ecologists up to the 1960s. They depended on observations that similar temporal vegetation patterns occur in similar circumstances in a region (Daubenmire 1968b). However, Drury & Nisbet (1973) criticized this view on the grounds that there was conflicting evidence for predictability of successional pathways (e.g. from D. Walker's (1970) study of British mire succession (Ch. 8)). Field ecologists are generally aware that successional pathways, in localities where both site and flora are diverse, may be quite varied (e.g. Clements 1916, Olson 1958, P. Wardle 1980). However, common patterns of succession, particularly in various kinds of temperate zone forest, have been described by so many authors that the general predictability of successional pathways (within broad and approximate limits) is not questioned. In detail, there will often be variations on themes. To some degree this will depend on the opportunities for seedling establishment and the diversity of the participating flora, as well as the lifestyles of species that are able to participate during the successional phases. These were all points raised by early writers on succession (Cowles, Tansley and Cooper).

One reason for there being a semblance of orderliness and predictability in succession in many regions, even if species-by-species replacements result from random events (as Drury & Nisbet, Colinvaux and others maintain), is the occurrence in each region of a limited number of species with niche or lifestyle appropriate to fill different roles, as the vegetation develops. If there is a rich flora (for example, in tropical rainforest), then the detailed sequences may be more difficult to predict than if the flora is limited in composition.

The old idea that there is inevitable convergence of seres in any region on a single climax type went out with the monoclimax concept. Sequences in a region may remain distinct, diverge, or, if the dominants of the mature vegetation have wide amplitude, sequences may converge later in succession.

PROBLEMS WITH THE CLIMAX CONCEPT

Some ecologists have suggested dispensing with the term climax altogether, as its use and meaning are fraught with so many problems. For example Selleck (1960) objected to the connotation of finality which is attached to the concept of climax. He proposed alternative terminology to 'seral' and 'climax' – **intermediate** and **most advanced phases** of vegetation development. Because absolute stability (i.e. lack of any change in species composition) is unlikely, and because stability would be difficult to perceive or verify, as human lifetimes are shorter than those of many plants, Bray (1958) and Park (1970) recommend replacing the term climax with **steady state**. Steady state denotes 'a temporary state of dynamic equilibrium in an open system'. Odum (1971) and other ecologists have preferred the term **mature** instead of climax.

The phenomenon of degradation of nutrient pools (and of the vegetation structure) of some types of ecosystem is considered in Chapters 5, 6 and 8. Over long periods vegetation in many regions might deteriorate in stature and change in species composition through similar causes. What, then, is to be regarded as the climax vegetation? The concept of some ideal, optimal type of vegetation, against which all other vegetation is to be measured, is not very realistic in these circumstances. Another problem of perception of stable, persistent communities arises from the occurrence in some regions of very long-lived dominant plant species which nevertheless do not regenerate, and are ultimately replaced, possibly by other, shorter-lived species.

THE KINETIC CONCEPT

The frequent disturbance of vegetation and the importance of disturbance for the regeneration and maintenance of natural and semi-natural vegetation (C. H. Muller 1940, 1952, Raup 1957, 1981, D. Walker 1982, Vogl 1974, 1980, Connell 1979, P. S. White 1979, J. C. Noble 1982, Whitmore 1982) leads other authors to suggest that the climax concept should be abandoned. The logical extension of the polyclimax concept is that *any* factor which controls the composition of persistent communities, including continuous, severe disturbance, can be regarded as a determinant, limiting factor. Such vegetation would have to be regarded as climax. This, says P. S. White (1979) hardly seems to accord with the idea of stable, mature vegetation, which the climax concept requires.

In the *kinetic* scheme (P. S. White 1979, Veblen 1979, Veblen *et al.* 1977, 1980, Whitmore 1982) a radically different conceptual framework is advanced to replace the succession to climax concept, with its emphasis on progression towards a 'normal' stable condition. The kinetic concept does not require a stable end-point;

disturbance and continually changing vegetation are regarded as normal and stable, self-maintaining vegetation is rather unusual. These ideas draw strength, also, from the pattern and process concepts of vegetation dynamics enunciated by A. S. Watt (1947), where cycles, with upgrade and downgrade phases are evident, but no particular phase can be regarded as a stable end-point. Once these kinetic ideas are adopted, the concept of the *sere* – the complete sequence from colonization to development of climax (and the associated term *seral* for the intermediate developmental phases) – comes into question. Spurr (1964) has already discussed the various problems inherent in the climax concept. He preferred (Spurr 1956) the term *transitional* for the intermediate phases.

The kinetic concept can be appropriately applied, also, in mountainous or broken terrain and in places such as the Arctic tundra, or savannah grassland, where disturbance and fluctuations in vegetation composition are incessant. The point that a stable, self-perpetuating climax model was not very meaningful, in circumstances such as these, was made long ago by W. S. Cooper (1913) and Shreve (1914, 1915) and iterated by Griggs (1934) and Vogl (1974). Rather than climax, P. S. White (1979) uses the term **potential natural vegetation** to indicate the arbitrarily defined vegetation which would develop if the general environment was stable for sufficiently long to allow a relatively stable equilibrium of the plant populations to be reached.

In several respects the kinetic scheme is philosophically more satisfying than succession to climax. It takes into account the perceived vegetation variability which results from disturbing pressures. It does not require an elaborate terminology to explain exceptions to general rules, which was necessary in the succession to climax scheme. It can accommodate, without difficulty, all kinds of vegetation change phenomena including fluctuations, allogenically initiated changes, cycles, degradation of ecosystems through nutrient loss and maintenance of persistent vegetation by severe disturbance.

INDIVIDUALISTIC PLANT LIFESTYLES ('STRATEGIES')

The kinetic scheme has much in its favour as a means for considering various natural phenomena in vegetation. However, the variety of lifestyles among plant species suiting them for specific conditions is evidence for the view that a sequence pioneer → transitional → mature (and relatively long-lasting) vegetation is a common phenomenon over much of the Earth's surface, especially in the climatically less-extreme environments. The lifestyle range includes a gamut from those well-suited to life in open, uncrowded habitats, to others which can live and regenerate their populations in crowded conditions. In recent times the conceptual continuum from r- to K-selected species (Pianka 1970) has been applied to this situation. The r-species are those with attributes suiting them for reproductive success, including the capacity for wide dispersal and rapid population growth in uncrowded environments. The K-species are tolerant of competition in conditions of crowding. They are relatively slow-growing, often large or woody, or both, and proliferate vegetatively, or regenerate by means of relatively few, not widely dispersed seed.

Figure 12.6 Summary of the ten vital attributes associated with the method of persistence of plants in vegetation subject to disturbance (particularly by fire). An open bar indicates that a method of persistence is available at a particular lifestage of a species population. The method will usually result in only juvenile material being present immediately after a disturbance, but in some cases (solid bar) mature tissue will persist. D, Well-dispersed propagules (sole method); Δ, both well-dispersed propagules and adults able to resist the disturbance; S, long-lived propagule store in the soil; Σ, long-lived propagule store in the soil and adults able to resist the disturbance; G, whole propagule store germinates or is otherwise lost at a first disturbance – until plants reach reproductive maturity there will be no propagule pool; Γ, whole propagule pool dissipated at first disturbance but adults resist disturbance and survive, V, plants resprout from underground when top killed; U, species remains reproductively mature after disturbance; C, only adults survive disturbance but they lose all of their reproductively mature tissue; W, reproductively mature tissue of adults resists disturbance (sole method). After I. R. Noble & Slatyer (1980).

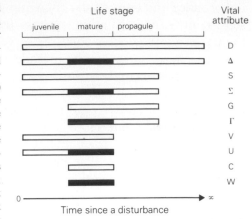

However, neither the r–K 'strategy' system, nor the more-elaborate 'ruderal/competitor/stress-tolerator' model of Grime (1979) is adequate to explain the full range of plant attributes which would help us to understand the role of individual species in vegetation change (cf. Grubb 1985). For example, among very versatile species with broad fundamental niches, there are some which are weak and others which are strong competitors. The 'strategy' schemes do not accommodate these variations easily. In very rich floras it might be expected that there would be an almost infinite range of permutations and combinations of vital attributes. Usually the number of alternatives is limited (cf. I. R. Noble & Slatyer 1980) (Fig. 12.6).

CAUSES FOR SEQUENTIAL REPLACEMENTS

Causes for the order of the appearance of species in sequential replacement series are considered in Chapter 11. The orthodox theory on the reasons for sequential replacements is that:

1 The species that are suited to the conditions on a site colonize it.
2 By modifying the conditions (changing the light, moisture, nutrient and other site parameters) they make conditions favourable for other species, which hitherto have been unable to occupy the site, but can now do so.
3 These species compete with the resident species populations and replace them.
4 This process continues until no more species are capable of invading the established vegetation (which is in equilibrium with its environment). The rate of the process is fast near the beginning of succession, but slows as the vegetation becomes more mature and the lifespan of dominant species longer (Braun 1950, Keever 1950, Daubenmire 1968b).

Autogenic influences, thus, are seen as the main driving force for succession. Allogenic forces are thought to be a different class of causative factor which also may drive some successions; often they cause such damage to vegetation that they initiate secondary succession.

The argument by Drury & Nisbet (1973) and others against universal autogenic control of species replacements during succession arises from observations that in some primary successions (W. S. Cooper 1916, Sigafoos & Hendricks 1969) and some secondary successions (Cowles 1911, Tansley 1929, Egler 1954) species which occur in the mature vegetation enter the sequence at its inception. An important inference from this is that it is not *necessary* for there to be a sequence, beginning with pioneers, followed by seral species and culminating in climax, neither does there necessarily need to be a sequential process of 'site preparation' by autogenic means (Drury & Nisbet 1973).

Species participating in direct replacement (Ch. 7) in desert, tundra, boreal forest and on Hawaiian lava flows, etc., also do not require site preparation. Wide ecophysiological amplitude of the species concerned seems to apply in these cases.

Egler (1954) proposed that there are, in abandoned field successions in the eastern USA, two possible models of change (Fig. 12.7). One is the conventional successive replacement of one group of species by another as a result of site modification, which he called **relay floristics**. The second model, which he called **initial floristic composition** (i.f.c.), entails the development of groups of species all of which were initially

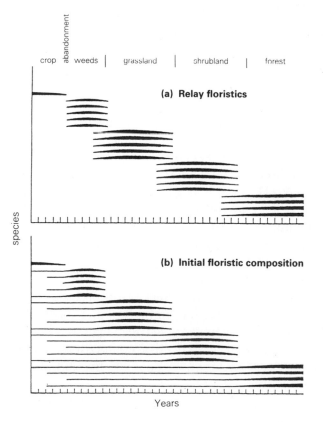

Figure 12.7 Two possible models for the behaviour of species during succession on abandoned cropland (after Egler 1954).

present in the form of seeds or other propagules. Different species mature and come into prominence at different rates. Thus, replacement of the species of earlier phases by later ones is not related to autogenic site modifications. Egler suggested that the initial floristic composition model may predominate in secondary successions.

The initial floristic composition model is a quite different conceptual framework from the relay floristics model and, if it were to be generally applicable, as Drury & Nisbet (1973) point out, the apparent vegetation sequences beginning with small plants, which give way to successively taller species, could be nothing more than relative changes in conspicuousness of the plants involved, caused by differences in potential stature and correlated differences in their growth rates.

In Chapters 9 and 11 it is seen that the initial floristic composition pattern applies, up to a point, to some secondary forest successions but not to others. The particular pattern depends on local circumstances. It is less likely that i.f.c. would occur during primary succession because of dispersal constraints. Initial colonists usually have light, easily dispersed seeds. Late-successional species often have heavy seeds, dispersed over short distances. Nevertheless, young spruce trees (*Picea sitchensis*) establish on raw till on the Glacier Bay moraines during the early herb phase of the primary succession there (W. S. Cooper 1931, 1939) (Ch. 6). However, they are not common and their growth is poor compared with those establishing during the *Alnus* phase, when the soil is enriched with nitrogen. Livingston & Allessio (1968) found that viable seed of transitional and mature forest species do not appear in the seed bank until late in abandoned field succession in eastern North America.

There is evidence, too, that species of mature vegetation in various localities could not serve as initial pioneers. Their seedlings would die in the conditions existing at the beginning of succession (e.g. rimu (*Dacrydium cupressinum*) on volcanic scoria at Mt Tarawera in New Zealand (Ch. 4), red and white oaks (*Quercus rubra* and *Q. alba*) or beech (*Fagus grandifolia*) on ploughed fields in the eastern USA (Buell *et al.* 1971, Held 1983)). For such species, prior site modification by other plants is essential.

Niering & Egler (1955) and Drury & Nisbet (1973) also cited instances where, rather than facilitating succession, some dominant pioneer or early seral species form dense stands which *suppress* seedlings of the next, potentially taller species in the sequence, at least for a time. This inhibition phenomenon may be rather general in sequential vegetation change (cf. Chs 6 & 11). As well as efficient competition by the prevailing species for resources such as light or nutrients, a possible cause of the phenomenon is the inhibition of competing species (which potentially could take over the site) by release of phytotoxic compounds from established plant species (Keever 1950, Rice 1979).

Odum (1969) had noted the suppression of potential successors by the established population, but regarded it as a temporary phenomenon. Succession proceeds by a series of steps and plateaus, the latter representing the lifetime of the particular temporary dominant. As such 'seral' species age, their canopies open up and possibly their root systems also die back. They are then less-effective competitors. Other species, which eventually will succeed them, grow up under their canopy and finally overtop it, suppressing the resident population.

The alternative explanation, proposed by Drury & Nisbet (1973) as a general

440

theory for the successional phenomena, is based on the observed congruence of temporal sequences of vegetation and spatial sequences along environmental gradients. They suggest that most of the phenomena of succession can be understood as consequences of differential growth, differential survival (and possibly differential colonizing ability) of species adapted to growing on different parts of environmental gradients. Each species is specialized and competitively superior over a limited range of conditions (i.e. has an individualistic niche). The stepwise successive replacement of one group of species by another results, in part, from interspecific competition which permits one group of plants temporarily to suppress slower-growing successors. The structural and functional changes associated with successional change result, primarily, from correlations between size, longevity and growth rates. 'Species whose seeds travel far and grow fast in harsh conditions cannot also grow large and live long.' These ideas, explaining successive change in terms of the properties of individual species, were a shift away from emphasis on vegetation-controlled processes, although they are not necessarily incompatible with them. The explanations specially apply to reproductive 'strategies' (lifestyles) of plants, in relation to prevailing habitat conditions, and to competitive mechanisms for successive replacements. Such ideas were favoured by many population biologists in the 1970s (e.g. Horn 1971, 1974, 1975a, Ricklefs 1973, Colinvaux 1973, Pickett 1976, Diamond 1975, Grime 1979, Miles 1979). In their view the deterministic concept 'reaction', or autogenic change, must be replaced by models based on stochastic (probabilistic) processes of distribution of seed and establishment of seedlings, and the lifestyles of the species concerned.

A weak point in Drury & Nisbet's (1973) argument is their failure to discuss just how the environmental gradients in the temporal vegetation sequences arise. Few vegetation ecologists would argue against the importance of competition in successional replacement. Indeed, its importance was stressed by Clements and all of the other early writers on the subject, and they also emphasized the roles played by individual plant species. However, they and many later ecologists would take issue with the view that competition and the different lifestyles of the plants are the only determining factors. The temporal environmental gradients arise from differential, reciprocal and modifying effects of different plant species on their environment, especially light, moisture and soil characteristics (i.e. autogenic influences). The environmental changes engender long-term as well as short-term vegetation changes. The outlook of Drury & Nisbet (1973) and other population biologists was a little myopic, focused as it was on short-term secondary succession.

The alternative models for successional change in vegetation proposed by Connell & Slatyer (1977), **facilitation, tolerance** and **inhibition** are useful for focusing our attention on the relative behaviour of species as vegetation develops. As shown in Chapters 6 and 11, there is evidence that each model may apply during different phases of the same vegetation sequence. They may even operate simultaneously for different species. None can be accepted as the sole explanation for the course of events during vegetation development.

COMMUNITY STABILITY

Change in vegetation has been dealt with at great length in the previous pages. The field evidence for the dynamic nature of the species composition of vegetation is abundant. Nevertheless, some attention must be given to states of relative equilibrium. It has been seen that the succession to climax theory sets great store by the ultimate 'stabilized' condition, the climax community, which is regarded as the norm for vegetation development. As an ecosystem matures, it is supposed to become more stable, so that it remains near an equilibrium state and returns to it after disturbance.

In fact the concept of stability in biotic systems is quite complex (cf. Orians 1975). The most important meanings are:

1 **Constancy** – a lack of change in some parameter of a system such as number of species, taxonomic composition, life-form structure of a community, or feature of the physical environment. There is an implication that the system is self-sustaining.
2 **Persistence** – the survival time of a system or some component of it. The timescale must be specified.
3 **Inertia** – the ability of a system to resist and recover from external disturbances, with respect to the parameters noted under constancy. To these may be added the ability to resist invasion by other taxa that are not normally part of the community. It is usually implied that inertia increases during succession. (*Disturbance* is defined in Ch. 2.)
4 **Resilience** – the speed with which a system returns to its former state after a disturbance. (Orians used the term elasticity for this phenomenon, but resilience, used by Grime (1979) seems more appropriate.) Resilience is supposed to decrease as succession proceeds.

Constancy and persistence are supposed to be characteristics of mature vegetation. What evidence do we have for this?

A general conceptual stance needs to be outlined here on spatial and temporal *scale*. To a considerable degree our perception of stability depends on the spatial scale which is being contemplated. Species populations of stands may not be very constant or persistent, whereas the species populations of whole land systems or regions, although exhibiting much local temporal variability (fluctuations, cycles and successions), in a mosaic spatial pattern, may be quite constant and persistent (A. S. Watt 1947, Loucks 1970, Whitmore 1982). Perhaps this could be the justification for the retention of a climax concept. Unless otherwise indicated, constancy will refer, in the discussion below, to the species composition of stands.

The *duration* of a relative state of stability (constancy or persistence) also must be measured on an arbitrary scale. This is because of the different lifespans of the plants which comprise the vegetation (and because of independent behaviour of different species in communities). Stable (persistent and constant) stands are defined here as those in which populations of the main dominant species are maintained for more than one generation.

442

The main evidence for maintenance of stability is that there are old-growth stands (observed in forests, some grasslands, wetlands, scrub, etc.) which seem to be unchanging (or only very slowly changing). The problem is to actually demonstrate that this is so; very few long-term permanent study sites are established (cf. J. White 1985, Falinski 1986) and in most (all?) of these, change of at least some components is evident. Nevertheless, the circumstantial evidence is that there are stable stands:

1 In some forests, based on ages of trees, there are stands that are hundreds of years old. The same composition has probably been maintained for at least a few generations, judging by pollen analytical data. Changes in the subordinate, shorter-lived species may have occurred.
2 In some tussock grasslands, based on estimation of ages of the spot-bound, evergreen plants, there are stands 100 years old or more, which probably have been maintained with much the same composition.
3 In some wetlands, based on stratigraphic studies, a similar stand composition has been maintained for many generations. Note that in points 2 and 3 vegetative reproduction of plants (but not vegetative spread) contributes to the maintenance of stands. It may do so also in some scrub and forest communities. Determining how long a plant or branch system has lived (and thus the number of generations in a given time) may be difficult.
4 Some annual plant communities seem to have been stable through replacement of their populations for at least a few years (Grubb *et al.* 1982).

How are stable stands, in which different species coexist, maintained? This has been the subject of much speculation, some observation and a limited amount of experimentation. Several hypotheses, not necessarily all mutually exclusive, have been proposed:

1 Physical structure and environmental modification by dominants exerts control.
2 Communities are well-integrated by evolution together – competition – niche-fit of the component organisms – mutualistic interactions.
3 Diversity of communities tends to enhance stability.
4 Nutrient balance (and relationships between decomposer and producer organisms) promotes equilibrium (noted earlier in this chapter).

Structural control

Large plants certainly modify the micro-environment, as seen in earlier chapters. Not only light intensity, light quality, temperature and moisture regimes on a site, but also litter effects, nutrient budgets and soil conditions can be determined by the presence of long-lived, large dominants (Grime 1979). If they can regenerate *in situ*, then such conditions could be maintained. The modulation of environmental extremes by dominants of wide ecological amplitude is seen by Langford & Buell (1969) as the main determinant of stability and integration of communities.

Kershaw (1963, 1973) described non-random patterns in communities (equally spaced individuals and clumps) caused by: the combination of plant–environment interactions; the vegetative proliferation of many species, forming intricate groupings related to the morphology of the species concerned; and the presence of plants which create microhabitat conditions and enhance the growing conditions for other species. The importance of the patterns developing around individual colonists in successional sequences, which they call **nucleation**, was stressed by Yarranton & Morrison (1974). Scott (1961) showed that dominant plants in grasslands create an 'aura' of distinct environmental conditions around themselves.

Evolution–competition–niche-fit–mutualism

Goodall (1963) suggested that species become adapted to the more common types of site. If a number of species grow together habitually, the conditions of the sites where this happens will tend to become uniform under their mutual influences. The creation of an integrated community is, he thought, a process which is likely to increase the pressure on other species to become adapted to it so that the plant community will evolve as a whole.

According to Poore (1964) the key to the integrity of plant communities is the competitive balance between the components, which selects out unadapted biotypes. The behaviour of each species is modified to some extent by the presence of others, and the whole assemblage functions as a self-perpetuating system. The resultant plant communities are organized spatially – in the horizontal scale by pattern, in the vertical scale by stratification of root and shoot systems. Both kinds of organization arise from the relative differences in size, general form and modes of growth of the component species. The community is also organized temporally, arising from differences and similarities in the active growth periods of component species. A. S. Watt's (1947) work suggests that the fit of requirements in time and space in stable communities can be very exact. Curtis (1959) stressed the same point with respect to the integration of field layer herbs with the phenologic phases of canopy trees in deciduous forest.

In the North American Great Plains grasslands, McMillan (1959) showed that there is a harmony of vegetation and habitat over a wide geographic range (in which there are marked environmental gradients). This is achieved by close ecotypic adaptation of plant populations (drawn from the same suite of species) to the specific conditions of each locality.

According to Harper (1977), order and organization appear in plant populations because of feedback from existing populations to new recruits and because these recruits themselves interact. These were the features considered by Forcier (1975) and K. D. Woods & Whittaker (1981) to permit mature forests to replace themselves (Ch. 11).

The importance of co-evolution of species in communities was stressed by Whittaker (1962, 1967, 1969b, 1972, 1977), and Ehrlich & Raven (1969). By competition their niches diverge, enabling them to become adjusted to one another. Rather than their niches overlapping absolutely (thus generating strong competition), they mesh with one another, at least to some extent. Species population centres are scattered in

444

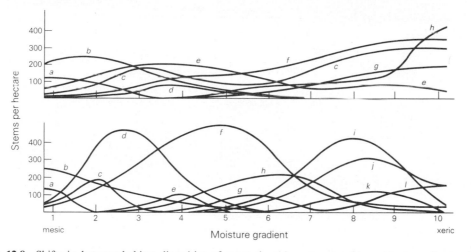

Figure 12.8 Shifts in between-habitat diversities of vegetation along topographic moisture gradients. Above, moderately low diversity (less change in composition along the gradient, with widely dispersed population curves) at 460–470 m elevation, Siskiyou Mountains, Oregon, USA. (From Whittaker 1960). Below, high between-habitat diversity (narrower population curves, greater change in composition along the gradient) at 1830–2140 m, Santa Catalina Mountains, Arizona, USA. (From Whittaker & Niering, 1965). Half-change values expressing relative change in composition in the ten-step transects were 1.1 for the tree stratum of the Siskiyou transect, 3.4 for that of the Santa Catalina transect. Species: above (a) *Taxus brevifolia*; (b) *Chamaecyparis lawsoniana*; (c) *Castanopsis chrysophylla*; (d) *Abies concolor*; (e) *Pseudotsuga menziesii*; (f) *Lithocarpus densiflora* (× 0.5); (g) *Quercus chrysolepis*; (h) *Arbutus menziesii*. Below (a) *Abies concolor*; (b) *Quercus rugosa*; (c) *Pseudotsuga menziesii*; (d) *Pinus ponderosa*; (e) *Arbutus arizonica*; (f) *Quercus hypoleucoides* (× 0.5); (g) *Pinus chihuahuana*; (h) *Quercus arizonica*; (i) *Arctostaphylos pringlei*; (j) *Pinus cembroides*; (k) *Garrya wrightii*; (l) *Quercus emoryi*. After Whittaker (1977).

relation to environmental gradients (Fig. 12.8). Niche specialization thus permits the integration of communities in which species have different, but complementary, roles (cf. Dobzhansky 1950, McIntosh 1970, Pickett 1976, Fox 1977). Nevertheless, in natural vegetation, Grubb (1977) pointed out that groups of species occur with strongly similar habitat niches. Poore's (1955a, b, c, 1956) work on noda also supports this view. So does that of Ellenberg (1978), on the overlap of distribution ranges of European forest tree species in relation to moisture and soil acidity gradients (Fig. 12.9). Whittaker (1969b, 1977) explained the paradox in terms of opposing evolutionary trends towards relatedness (in the sense of interactions with other species) and unrelatedness (in the sense of distributional individuality).

As stabilizing influences Langford & Buell (1969) and Knapp (1974b) cite the positive interactions such as mycorrhizal relationships, interdependences and synergisms which increase as communities mature (and also negative interactions which exclude species). Component species of temperate grasslands, for example, show complex patterns of association between species, which are stable through time (De Vries 1953, Turkington *et al*. 1977, Aarsen *et al*. 1979). Root exudates of gregarious species of *Anthoxanthum, Cynosurus* and *Holcus* stimulate their own growth. Kershaw (1973) described patches where shoots of particular species emerge from the soil together even when they are derived from different rhizomes.

Third-party interactions, including those with animals, may also be important (Pianka 1981, Pickett 1976, J. N. Thompson 1982). In competitive mutualism two species which compete moderately with one another may mutually benefit each other

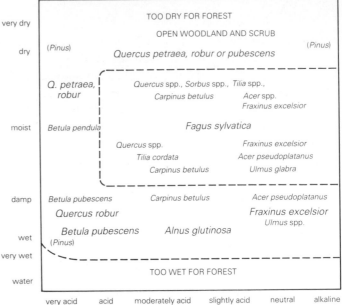

very dry

TOO DRY FOR FOREST

OPEN WOODLAND AND SCRUB

dry — *(Pinus)* — *Quercus petraea, robur or pubescens* — *(Pinus)*

Q. petraea,
robur

Quercus spp., *Sorbus* spp., *Tilia* spp.,

Carpinus betulus — *Acer* spp.

Fraxinus excelsior

moist — *Betula pendula* — *Fagus sylvatica*

Quercus spp. — *Fraxinus excelsior*

Tilia cordata — *Acer pseudoplatanus*

Carpinus betulus — *Ulmus glabra*

damp — *Betula pubescens* — *Carpinus betulus* — *Acer pseudoplatanus*

Quercus robur — *Fraxinus excelsior*

Ulmus spp.

wet — *Betula pubescens* — *Alnus glutinosa*

Figure 12.9 'Ecogram' of Middle European montane forests showing the ecological range and relative prominence of tree species in relation to soil moisture and soil acidity gradients (after Ellenberg 1978).

(Pinus)

very wet

water — TOO WET FOR FOREST

very acid — acid — moderately acid — slightly acid — neutral — alkaline

by having strong negative effects on a third competitor (Lawlor 1980). Although they have been arrived at from different directions, views about mutualistic interactions which bind species in communities converge to some extent with views of Odum (1971) and Reichle *et al.* (1975b) about community homeostasis.

Grubb (1977) maintains that three features are necessary to explain the long-term coexistence of plant species in communities: competition, complementarity and dependence. Complementary life-forms and life styles (especially with respect to active periods) allow the main species in communities to coexist. Dependent species (epiphytes, parasites and saprophytes) have specific habitats provided for them by the main species, of course. Grubb sees differentiation of the regeneration niche as being a very important determinant of the long-term coexistence of different species.

The inertia of mature communities is judged by their resistance to invasion. They also retain their species longer (i.e. are constant and persistent) (Shugart & Hett 1973). Examples, where foreign species are excluded, are seen in dense forest and alpine vegetation in New Zealand (Cockayne 1928, Norton & Burrows 1979) and in some tropical forests (van Steenis 1969). Such communities may be considered to be 'saturated' with species that are well-adapted by long-term evolutionary adjustment of species to one another and to the local environment. Unadapted foreign species find no vacant niche space in undisturbed vegetation, but adjacent disturbed areas are readily invaded. (See also Pimm (1982) for a computer simulation of this point.)

Another facet of the integrated community-coexistence problem is that some authors think that species mixtures occur because of the microhabitat heterogeneity of sites (Harper 1961, Hickman 1977, Werner & Platt 1976). Coexistence is only 'apparent'. However, in many communities microhabitat heterogeneity which might favour differential establishment of seedlings is autogenically created by the adult plants, living or dead.

446

Diversity

In the 1960s it was held to be a fundamental ecological truth that the stability of ecosystems was engendered by the *diversity* of organisms which lived in them (e.g. Odum 1971, Margalef 1968). There are several different meanings to the term diversity (Whittaker 1965, 1975b, Huston 1979), but here is meant the simple concept *richness*, the numbers of species per unit area. A corollary of the supposed increase of species richness during succession (Odum 1971) is reduced dominance (Margalef 1968). Homeostasis was thought to be increased by the nexus of interspecific interactions as a system matures. 'Climax' communities are not necessarily more diverse than 'seral' communities (Shafi & Yarranton 1977, Horn 1974). Many mature and apparently stable communities in temperate regions have a single or a few dominants. Diversity, on a regional scale, can be generated by the constant provision of new sites by disturbance (Loucks 1970, P. S. White 1979). Diversity does not necessarily confer stability on communities (May 1975, Pimm 1982). There may not be any very important connection of diversity with community maturation. It is behaviour of individual species in response to environment, including neighbours, which generates stability or instability.

Stable communities may be species-rich or quite simple in species composition. May (1975) suggests that ecosystems evolve to be as rich and complex as is compatible with the persistence of most of the component populations in the face of the magnitude of environmental hazards experienced. In a stable environment a system need only cope with relatively small disturbances, and a 'fragile' and persistent complexity can be achieved. Another way of putting this is that plant species of stable environments have to allocate fewer energy resources to survive the climatic and other vicissitudes of their habitats than do species of unstable and fluctuating, or more-severe, environments. Connell & Orias (1964) wrote of this in relation to the difference between the rich species composition of tropical communities and the poorer species composition of communities nearer the Poles. Species of deserts and polar regions *are* highly specialized to deal with the climate conditions that they experience. The vegetation in such places may be species-poor and lack resilience, but it seems to possess considerable inertia. In the climatically more-equable Tropics there are other environmental hazards, arising from the presence of neighbours. Competing plants and predator animals create severe environmental pressures with which the plant species present cope by other kinds of specialization (Janzen 1970).

Diversity seems to be related to the opportunity and time for groups of plant species to have evolved before they came together. To a considerable degree chance factors and phylogenetic history must determine whether species in a region have evolved to have wide or narrow niches or to be dominant or subordinate.

In closing this section reference must also be made to Chapter 11, where it is pointed out that the persistence of constant stand structure in forest for more than a few generations is unlikely. Stability is a relatively ephemeral property of any kind of vegetation.

MATHEMATICAL AND MODELLING APPROACHES TO COMMUNITY DEVELOPMENT

Attempts to describe vegetation change by means of mathematical equations have not proved very successful or enlightening. Widely applicable generalizations have not so far been possible.

Olson (1958) set out an equation for succession:

$$E_d = f(t)E_i + \sum \frac{\partial E_d}{\partial E_i} \Delta E_i + E$$

where E_d denotes dependent variables of a local ecosystem (organisms, soil, microclimate, moisture status and microrelief), E_i independent variables (time, initial substrate, topographic relief, hydrographic factors, regional climate, biotic factors and human influence), t is time and E is the chance of random variability.

This was developed from the state factor equations which Jenny (1941) had used to explain the process of soil formation. These ideas were extended to ecosystem development by Jenny (1958, 1961, 1980) and Major (1951). Major (1974) applied them to an analysis of rates of succession. Jenik (1986) has also presented equations for vegetation change according to the patterns of change, the initiating factors and the timescales (Fig. 12.10). Although such equations direct attention to the main interacting components of ecosystems, they have not yet shed much light on causal relationships. One problem has been the inexactitude or variability of the separate factors which are the terms that must be entered in the equations. Terms for plant variables, including niche dimensions, are missing. Also, there remains the problem of trying to deal with the complex and chaotic effects of many kinds of dynamic events occurring simultaneously, or over short periods. It remains to be seen whether they can ever be expressed mathematically.

The Lotka–Volterra equations (Table 12.5) (see Whittaker 1975b for a fuller account), derived originally from microcosm experiments between competing protozoan animals (Gause 1934), have been applied by animal population ecologists to questions of competitive replacement of species and predator–prey relationships. Harper (1977) suggests that the population theory derived from the application of such quantitative methods should be the basis for understanding ecosystems and is applicable to plants. However, Schaffer & Leigh (1976) hold that theoretical animal population ecology methodology is not appropriate for plants, because their populations are too heterogeneously distributed.

Another approach, the use of the mathematical expressions known as Markov chains was attempted by Horn (1975b) to explain the probabilistic events as species populations replace one another during succession. He examined the tree-by-tree replacements and estimated, for each tree in a forest in the northeastern USA, the probability that it would be replaced by its kind or by another species (judged from the identity of the surrounding juveniles). From a matrix of these probabilities he calculated how many trees of each species should be found at any 'stage' of succession and considered that the model made accurate predictions for this forest. G. R. Stephens & Waggoner (1970), in an account of secondary forest succession over 40

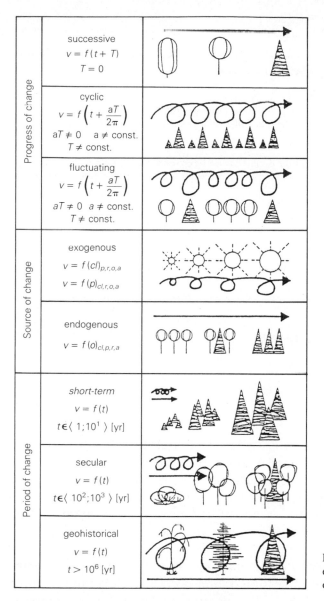

Figure 12.10 Formal and pictorial symbolic expression of some vegetation change phenomena (after Jenik 1986).

years in the same region of the USA (Fig 12.11), also suggested that change in the forest conformed to a simple Markov chain.

Van Hulst (1979a, b) is more cautious, however, and points out that it first must be demonstrated that successional species replacement *is* a Markovian process, and that the Markov chain is regular before the routine statistical properties of a plant-by-plant replacement follow, as asserted by Horn (1975b). Van Hulst suggests that a first-order Markov chain is the appropriate model to investigate succession, if knowledge of only the present vegetation is required in order to predict future vegetation (i.e. if there is no dependence on past interactions). As was seen earlier present patterns probably are always dependent on past interactions, e.g. in the nutrient budgets of the systems (see Chs 6, 8 & 11).

449

Table 12.5 Application of the Lotka–Volterra equations for assessing competitive relationships between species (after Barbour *et al.* 1980).

The logistic equation of population growth for a species in isolation is:

$$\frac{dN}{dt} = rN\left(\frac{K - N}{K}\right)$$

where r is the natural, intrinsic rate of population growth, N is population size and K is population size at saturation (the environment's carrying capacity). Multiplied out, the equation contains an *intraspecific competition* component:

$$\frac{dN}{dt} = rN - \frac{rN^2}{K}$$

where rN is the unlimited rate of growth and rn^2/K is the decline caused by crowding (competition).

If a second species is introduced, a second negative component is required, to account for *interspecific competition*. The equation for species 1, viz.

$$\frac{dN_1}{dt} = r_1N_1\left(\frac{K_1 - N_1 - \alpha N_2}{K_1}\right)$$

where α is the inhibiting (competitive) effect on N_1 for every individual of N_2. Similarly, the equation for species 2 is

$$\frac{dN_2}{dt} = r_2N_2\left(\frac{K_2 - N_2 - \beta N_1}{K_2}\right)$$

where β is the inhibiting effect on N_2 for every individual of N_1.

If $\alpha > K_1/K_2$ and $\beta > K_2/K_1$ then only one of the pair of species will persist in the mixture. That species will depend on the starting proportions.

If $\alpha > K_1/K_2$ and $\beta < K_2/K_1$, species 1 will be eliminated.

If $\alpha < K_1/K_2$ and $\beta > K_2/K_1$, species 2 will be eliminated.

If $\alpha < K_1/K_2$ and $\beta < K_2/K_1$, both species persist together in equilibrium.

Various other writers on the subject (e.g. Jeffers 1978, McIntosh 1980b, Gibson *et al.* 1983) believe that Markov chains are inapplicable as a model for succession. McIntosh (1980b) pointed out the fallacy of expecting that a particular juvenile beneath an adult tree at a particular time would necessarily replace it. Most juveniles do not survive to adulthood. Auclair & Cottam (1971) showed that, for many forests, there is an inverse relationship between the importance of trees and saplings of a species.

Modelling, with the aid of computer simulations, has been used to attempt comprehensive descriptions of plant succession. Gutierrez & Fey (1980) use a simple feedback loop model (Fig. 12.12) to express the functional relationships of compartments of an ecosystem. This model is then extended to the processes occurring during succession. The dynamic hypothesis of Gutierrez & Fey (1980) takes into account the interactions of five variables: energy, matter, diversity, time and space (indicated as soil carrying capacity).

When the soil carrying capacity of the site ('space') becomes saturated, according to Gutierrez & Fey, surplus energy is allocated to quality functions such as building up community diversity. By permitting niche specialization, this can bring about increased carrying capacity and, in turn, further build-up of plant biomass. Eventually, as a succession approaches 'climax', increased diversification does not result in further increased carrying capacity because: (a) niche specialization cannot proceed any further, (b) trophic equilibrium becomes limiting, or (c) some environmental

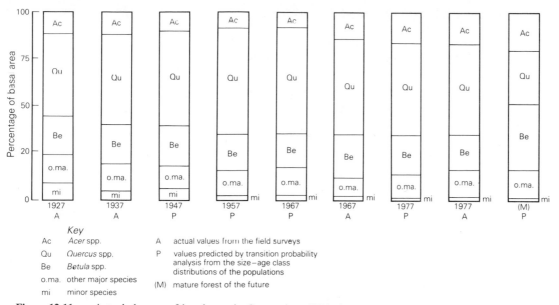

Figure 12.11 Actual changes of basal area in Connecticut, USA, forests, and changes predicted by transition probability analysis (after G. R. Stephens & Waggoner 1970, 1980).

factor such as temperature or moisture becomes limiting, so that succession can proceed no further.

Considering only the plant components, one problem with the analysis described here arises from some of the assumptions made about increased diversification and carrying capacity. It was seen earlier that succession does not necessarily entail increase in diversity. Species are lost as well as being gained as succession proceeds and some mature communities are simple in species structure. It is uncertain that increased carrying capacity arises from increased diversity (through niche specialization). In fact, it may work in the opposite direction (increased carrying capacity could permit the development of greater diversity). Increased specialization (the narrowing of niches) may occur while the carrying capacity remained constant. Increased diversity could be achieved by an increase in efficiency in the cycling of nutrients, so that functions occur at different times, although the stock of nutrients remains the same. Increased carrying capacity probably arises, as vegetation develops, by the extension of root systems to tap nutrients at deeper levels in the soil. This accounts for an increase in plant size through the course of plant succession.

The second problem with the analysis of Gutierrez & Fey (1980) is that it ignores the point that the carrying capacity (in the sense of the nutrient stock) is not the only control of successional change. Ecological constraints of other kinds, arising from the properties of plant species, are important. Successional sequences are completed when no more species which can replace the established species populations are available in a region. Loop analysis is outlined more formally by May (1981).

Systems analysis by simulation modelling is a method of trying to deal with the multiplicity of interactions in ecosystems with the help of computers. The variable components that are thought to be significant are entered into computer programs

451

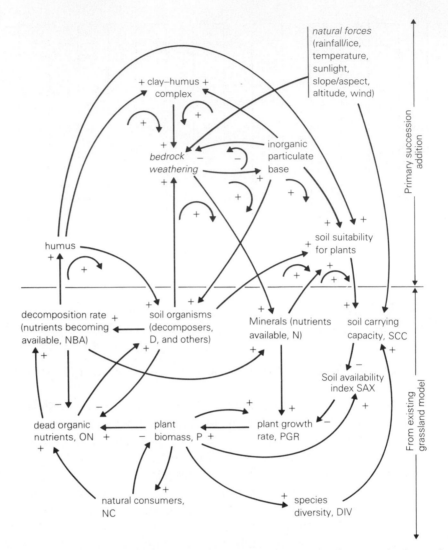

Figure 12.12 Feedback loop model for interactions of ecosystem components during succession in grassland (after Gutierrez & Fey 1980). Links with arrows represent cause-and-effect relationships between variables: the variable at the tail affects the variable at the head. Positive signs mean that, when the variable at the tail increases, so does the variable at the head. Negative signs mean that when the tail variable increases the head variable decreases.

which are then run to see what will be the outcome of varying specific items. The aim is to generate models that can be used to predict what will happen in the real world. An adequate database must be available to prepare the computer programs. This means that many parameters need to have been measured for a long period before the analysis. Figure 12.13 shows the results of Gutierrez & Fey (1980) for manipulations of grasslands on the Great Plains in Colorado, USA.

In somewhat more-sophisticated analyses Shugart & Noble (1981) and Shugart (1984) describe compositional dynamics of Australian eucalypt forest subject to periodic disturbance by fire (Fig. 12.14). Individual species parameters and environmental parameters are used in the simulations. Shugart (1984) describes various other

452

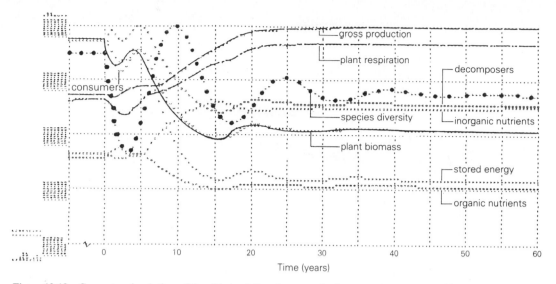

Figure 12.13 Computer simulation of the effects of disturbance and subsequent recovery on various components of a grassland ecosystem (after Gutierrez & Fey 1980).

simulations and uses the results to develop a theory of forest dynamics after gap formation (Fig. 12.15). Other systems analysis studies which consider vegetation change phenomena have been done by Bledsoe & van Dyne (1971) and Loucks *et al.* (1981).

Critiques of modelling and description of the practical difficulties of carrying it out are given by K. E. F. Watt (1975) and Pielou (1981). The main problem has been, in Watt's words, 'the difficult path between unrealistic simplicity and unwieldy and untestable complexity'. Enough critical factors, measured over a sufficiently long period, must be available to be fed into the computer program for there to be confidence in the results. The number of elements requiring measurement in a changing system is very large and the number of potential interactions increases with the square of the number of elements. Large-scale simulations are very expensive to run. Some of the models so far constructed have been fairly simple descriptions of the course of events, not providing profound answers to unanswered questions. Nevertheless, work such as that of Loucks *et al.* (1981) and Shugart (1984) provides some useful insights.

FLUCTUATIONS AND CYCLES

Much less attention has been given to the vegetation change phenomena known as fluctuations and cycles than to 'linear' sequential change phenomena (succession) after disturbance. How do fluctuations and cycles relate to concepts of the plant community and its development?

453

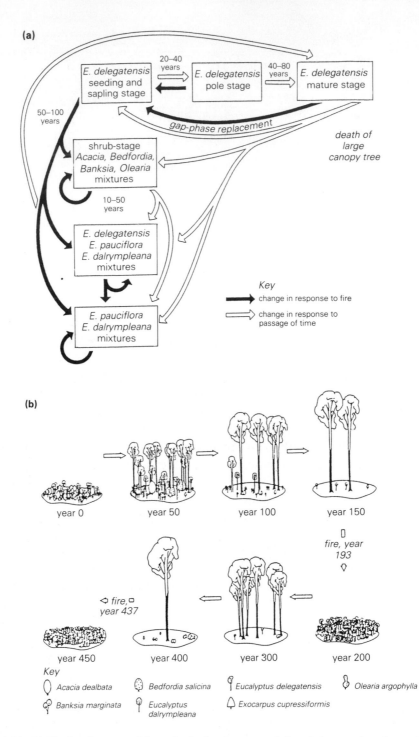

Figure 12.14 (a) Simulated pattern of dynamics for forest types on sheltered slopes and southeasterly aspects, about 1300 m, Brindabella Range, Australian Capital Territory, Australia. Boxes indicate forest types, light arrows indicate changes expected over time, without fire. Dark arrows indicate changes expected over time, in response to fire. (b) 450 years of change in a single simulated plot for alpine ash (*Eucalyptus delegatensis*) forest on the Brindabella Range (after Shugart 1984).

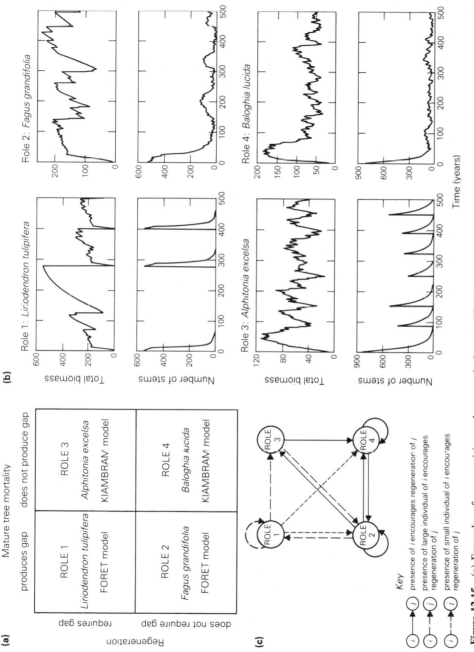

Figure 12.15 (a) Examples of gap-requiring and gap-producing tree life-history traits (roles 1–4). *Liriodendron* and *Fagus* from eastern North American deciduous forests; *Alphitonia* and *Baloghia* from eastern Australian evergreen rainforests. (b) Number and biomass dynamics of hypothetical patches in forests of single species that conform to the roles shown in (a). *Liriodendron* and *Fagus* each on 1/12 ha plots simulated by the FORET model. *Alphitonia* on a 1/30 ha plot and *Baloghia* on a 1/20 ha plot simulated by the KIAMBRAM model. (c) The directions of positive or strongly positive interactions among four roles of tree species with respect to their influence on regeneration (after Shugart 1984).

Fluctuations

Fluctuation is said to be the continual state of flux in vegetation composition around a notional mean (Coupland 1974, Rabotnov 1974, Miles 1979). Fluctuations, caused by year-to-year or longer-period environmental variation (for example, in precipitation, temperature, groundwater conditions, herbivores or fungal pathogens) (Coupland 1974), have mainly been recognized in herbaceous vegetation, e.g. in deserts and semi deserts, semi-arid and temperate zone grasslands (Bykov 1974, Weaver & Albertson 1956, Albertson & Tomanek 1965, Rabotnov 1966, Miles 1979, Rabotnov 1985) (Ch. 7). Examples are also known from woody vegetation composed of long-lived species (e.g. Korchagin & Karpov 1974, Mattson & Addy 1975, Falinski 1986).

Plant features affected by short-period environmental variation range from productivity of foliage, fine roots, flowers or seeds, to survival of seedlings or of adult plants. The alterations to relative vitality of plants, or their number, will cause readjustments of the populations of the species in the local flora in response to changed competitive power of some species and the creation of gaps. Some species will gain at the expense of others, and the proportions of the species in the populations will change. Some species may experience fluctuations, whereas other, hardier species are little affected by a particular stressful environmental episode.

The environmental changes causing fluctuations seem to be virtually random, but, even if they were of regular occurrence, differences in their intensity would introduce a stochastic element to the vegetation responses. Rabotnov (1966) also attributed fluctuations to peculiarities of the life-cycles of some plants, but these are considered below, among cyclic phenomena.

In forests the same kinds of causative factors may cause fluctuations, but overall effects may differ in degree because of the longer response time of trees; forests may be better buffered against short-term fluctuation than herbaceous vegetation. Korchagin & Karpov (1974) cite critically low temperature in spring, causing the death of trees, shrubs and herbs in mixed Boreal forests. Similar effects are caused by frost in savannah woodland in East Africa (B. H. Walker 1981) and by drought and insect attack in New Zealand beech (*Nothofagus*) forests (J. A. Wardle 1984). Falinski (1986) notes long-period fluctuation due to large mammal browsing in a Polish forest.

The most readily observed fluctuations are short-term and reversible changes in the relative abundance of plant species in stands from year to year (or periods of a few years). The detection of fluctuations will be more difficult in longer-lived vegetation experiencing long-term change.

It is most likely that climatic variations influence plants not only through direct effects on plant productivity and survival, but also through changes in consumer and decomposer populations and the nutrient flux. This will have subsequent feedback influences on the plant populations.

The actual degree of fluctuating change which is perceived depends on the spatial scale being considered, and possibly on the richness of the flora. On a scale of a square metre to a few tens of square metres, during an oscillation moist–dry–moist in semi-arid grassland, the species composition on any patch of ground will vary considerably from decade to decade. It is unlikely that exactly similar composition would be reconstituted either qualitatively or quantitatively, during successive 'moist' peaks of the climatic oscillation. On a larger scale, say a few tens or hundreds of square kilometres, the concept of a 'normal' or mean, around which the vegetation composition varies, is more realistic. If the flora is poor (for example, in desert), then it is possible that, even on a small patch of ground, species composition would shift around a notional mean.

Is vegetation which undergoes fluctuation 'stable'? The points just discussed are apposite. Stability, in this context , must be related to spatial scale.

How is fluctuation related to other kinds of vegetation change? It is evident from the writing about fluctuations that some ecologists have difficulty in deciding where to draw lines between what they would see as fluctuations and cycles (Rabotnov 1974, Kershaw 1973), or fluctuations and successions (Rabotnov 1974, Knapp 1974a). When does a fluctuation become an allogenically driven succession, or a cyclic process? Miles (1979) points out that distinctions are arbitrary. However, fluctuational shifts in composition, or trends over one year or a run of years, may be reversed in the next year or run of years. Successional changes are supposed to be directional and to have end-points that are different from the starting points. Cycles are supposed to occur as regular, repeatable sequences, returning to the same point at which they started. It will be seen shortly that these distinctions are not necessarily clear. Changes in vegetation initiated, for example, by a long-term climatic oscillation (over many decades) might well be recognized as an allogenic succession because only part of the sequence of change could be observed. It could be accompanied by fundamental shifts in the regional flora, or soil conditions. Nevertheless, this sort of change is not essentially different in kind from shorter-period fluctuation. Neither are fluctuations in grasslands, for example, essentially different from changes in forest due to death of trees through drought, which are regarded usually as successional or cyclic.

Cycles

Vegetation cycles are differentiated as sequences of change where there is a regular progression of species, or suites of species, on the same patch of ground, beginning and ending with the same (or very similar) composition. Some apparently rather different kinds of situation have been classed as being cyclic (Table 12.6).

In those examples at the top of the table (group A) a strong, periodically occurring exogenous environmental factor causes the death of an individual or a patch or a stand and replacement is through a simple or a complex sequence. Autogenic processes may operate throughout the replacement series. Different ecologists classify such changes differently. To some (e.g. Grubb 1985) sequences like this are simply regeneration; to others they are community regeneration (Miles 1979) or regeneration complexes (Knapp 1974a). Others regard such sequences as secondary succession (Shugart 1984) or cycles (Forcier 1975).

Regeneration was defined in Chapter 1 as the immediate replacement of the populations with populations having the same species composition as before. Direct

Table 12.6 Cyclic patterns of vegetation change.

A. change forced by periodic strong exogenous environmental pressure (with important autogenic components also)

Southern California (Sweeney 1956, 1968)

moribund *Adenostoma* chaparral —fire→ forbs, grasses → chaparral species (seedlings and sprouts from ligno-tubers) → increasingly dense, maturing chaparral, *Adenostoma* dominant → moribund chaparral

New Hampshire (Forcier 1975)

mature *Fagus grandifolia* (beech) forest —treefall→ *Betula alleghaniensis* (yellow birch) (from seeds) → *Acer saccharum* (sugar maple) (from seeds) → *Fagus grandifolia* (from seed and sprouts)

Iowa (Platt 1975)

mature grassland with forbs
Andropogon gerardi
Poa pratensis
Panicum oligosanthes
Solidago speciosa
Melilotus alba
Rosa suffulta

—badger excavation→ forb community
Rosa suffulta
Physalis heterophylla
Viola pedatifida
Amorpha canescens
Helianthus laetiflorus
Liatris punctata
(from rhizomes, seeds) + some annuals, biennials (dispersed as seeds)

→ redevelopment of grassland; mainly invasion by rhizomes

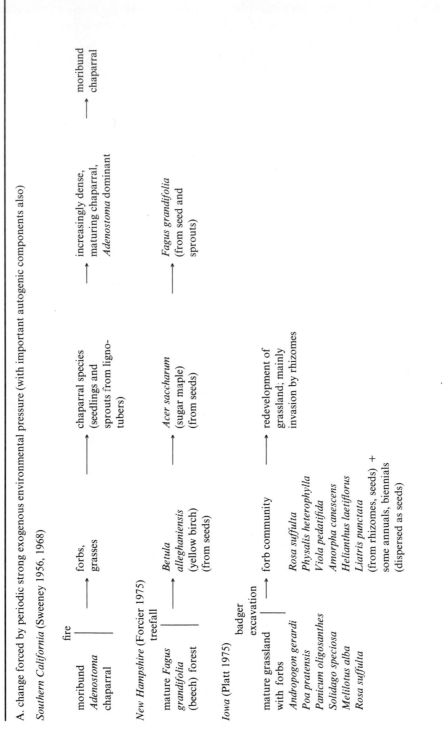

B. change forced by interaction of strong, persistent (but not necessarily continuous) exogenous environmental pressure with autogenic effects of plant growth and development

Cairngorm Mountains, Scotland (A. S. Watt 1947)

bare ground → vegetatively advancing band of *Arctostaphylos* invading; *Arctostaphylos* (bear-berry) in turn invaded vegetatively from behind by stunted *Calluna* (heather) which provides wind protection for *Arctostaphylos* → rise in height of *Calluna*, blasting by wind-driven sand, snow, erosion of rear of *Calluna* band → bare ground with old, dead *Arctostaphylos* and *Calluna* stems remaining

Manawatu, New Zealand (Esler 1970)

colonization of level sand area in dunes by marram grass (*Ammophila*) (from seeds) → rise of dune height caused by sand trapping and continued upward growth of *Ammophila* → blowout of dune crest, erosion and lowering of sand level

Alaska (Benninghof 1952)

bare soil, retreating permafrost table → establishment (from seeds) of *Salix* (willow) shrub cover, rising permafrost → invasion by *Picea mariana* (black spruce) forest (from seeds); thick litter, insulated soil, high permafrost table → disturbance by ice-lift, differential subsidence, falling permafrost table → bare soil

C. change induced apparently by ageing and senescence of dominant plants, causing their decline and death

Britain (A. S. Watt 1955, Barclay-Estrup & Gimingham 1969)

mature *Calluna* (heather) beginning to degenerate → death of *Calluna* → dwarf shrubs, grasses, sedges, rushes, (from seeds and vegetative propagules), bryophytes and bare ground → colonization by young *Calluna* (from seeds) → growth to maturity

Breckland, England (A. S. Watt 1955)

mature *Pteridium* (bracken) begining to degenerate → death of *Pteridium* → grassland → colonization by young *Pteridium* rhizomes → growth to maturity

D. direct replacement after vegetation disturbance

Maine (Sprugel 1976)

band of mature *Abies balsamea* balsam fir affected by damage on windward side of stand — ice glaze damage → death of *Abies* adults → replacement by young *Abies* plants originating from juvenile population → growth to maturity

Rocky Mountains (Clements 1910)

dense mature *Pinus contorta* (lodgepole pine) stands — fire → rapid establishment and growth of juveniles from seeds (released from serotinous cones by fire) → growth to maturity

South Island, New Zealand (J. A. Wardle 1984)

mature *Nothofagus solandri* (mountain beech) stands — drought, insect damage → death of *Nothofagus* adults → replacement by young *Nothofagus* plants originating from large suppressed juvenile population → (in some sites other species may achieve prominence for a time after the death of the adult mountain beech)

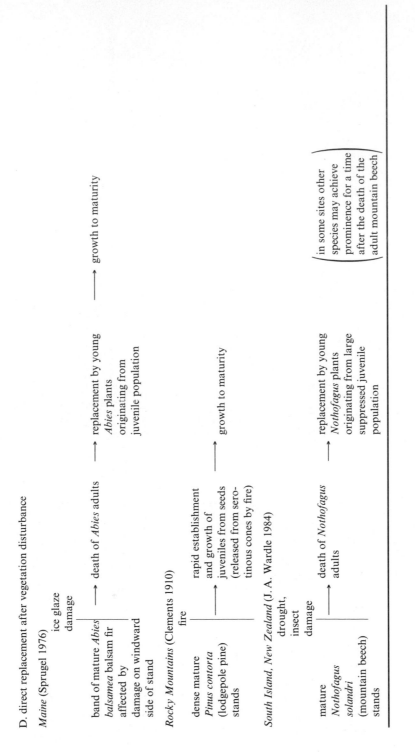

replacement (Group D in Table 12.6) is effectively the same. However, no strong case can be made for classifying the sequences in group A as being either examples of succession or cycles. Qualitative differences do occur in sequences such as the forest gap-filling, or badger disturbance colonization, according to the size of the disturbed patch (Shugart 1984, Platt 1975). Again the magnitude of area being considered determines our outlook to some degree. Most ecologists would regard the sequential development of moderate to large patches of vegetation (larger than about 2500 m²), after disturbance, as secondary succession. Therefore, by these criteria the well-known chaparral fire cycle must also be regarded as succession. It is a matter of taste whether the filling of smaller gaps, after disturbance, is called succession.

The examples in Group B of Table 12.6 are all initiated by strong environmental pressures. Endogenous factors are important, too (mainly the growth and regenerative properties of the plants), as are autogenic processes which, in modifying the environment, create feedback effects that tend to influence the plants adversely. Again, in their essentials, they do not seem to differ much from examples in other groups in the table, or from succession, although the plant responses to change may be relatively slow.

The Group C examples are those which A. S. Watt (1947) described from several British localities and which seem to result primarily from endogenous ageing and senescence factors which cause the decline and finally the death of dominant plants on a site, thus initiating rejuvenation of the cycle (see also Grime 1979, Grubb *et al*. 1982). The heather (*Calluna*) cycle is the best-known of these (Ch. 7), but A. S. Watt's work on bracken (*Pteridium*) and *Festuca ovina* also contribute a great deal to our understanding of phasic development in plant communities (Figs 12.16a, b). The interaction of heather and bracken (Fig. 12.16a) is especially interesting. However, from evidence in Southern California, USA, the bracken pattern is thought, by Gliessman & Muller (1978) to be due not to senescence *per se*, but to autotoxicity, caused by phenolic compounds leached from bracken fronds. It may eventually turn out that microbially produced toxins are involved.

Kershaw (1973) describes other sequences that are apparently related to plant senescence. Others, which might also be thought to resemble fluctuations occur through the death of monocarpic species, which may be non-synchronized for different individuals (*Tachigalia* in Central American tropical forests) or synchronized (bamboos in Assam or China). Periodic decline of *Trifolium* spp. in Russian meadows (Rabotnov 1974) seems to result, also, from senescence.

The gap-causing process in forests, (e.g. the death of old trees which have declined because they can no longer withstand disease, pests or drought), also has points in common with the Group C examples. So too do the chaparral and eucalypt forest fire cycles where the characteristics of the plants promote their 'suicidal' removal by fire.

The Group D examples are heterogeneous. They are inserted to demonstrate that 'cyclic' processes seem to be extremely varied. Although there is no change in species composition, 'regeneration waves' in Appalachian forests, and Rocky Mountain or boreal forest regeneration after fire resemble both Group A and B examples. So does gap-filling after 'murderous' treefall under the weight of a heavy epiphyte or vine load in tropical forests.

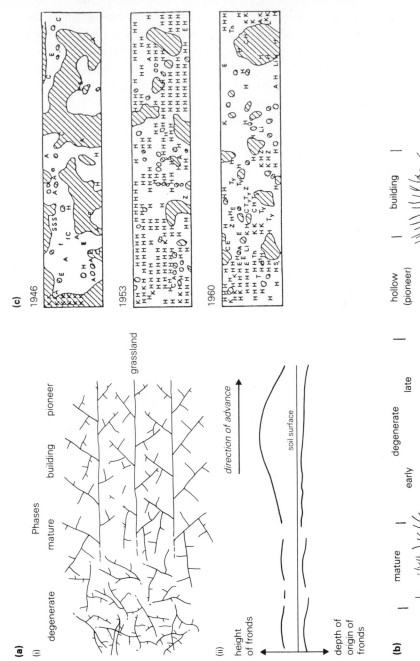

(a)

(i)

degenerate   mature   building   pioneer

Phases

grassland

(ii)

height
of fronds

direction of advance

soil surface

depth of origin of fronds

(b)

mature   early   degenerate   late   hollow   building
                                         (pioneer)

cm
3
2
1
0

3
2
1
0
cm

(c)

1946

1953

1960

When all is considered, there are quite close similarities between fluctuations, cycles and successions. So much is this so, that it seems appropriate to regard them all as forming a continuum of styles of vegetation change, differing in manifestation according to whatever plant properties, environmental influences, autogenic processes or scale of area are to be emphasized. There is less difficulty in incorporating the dynamic situations, represented by fluctuation and cycling, into kinetic theory (where no stable end-points are required), than into a succession to climax theory. If a climax concept is to be maintained, it must be sufficiently flexible to accommodate the fluctuation phenomena and cycles. It may be difficult in these circumstances to discern 'stable' vegetation in an absolute sense of lack of change in the species composition of stands of vegetation. Only at the coarse scale of the vegetation formation or the biome is the integrity of vegetation composition of an area likely to be maintained for centuries or millennia.

PROBLEMS WITH THE SUCCESSION CONCEPT

Confusion reigns about what should be incorporated within the term succession. Usually succession has meant development from pioneer to mature vegetation; conceptual problems with succession to climax have been considered earlier in this chapter. There are other vegetation change phenomena (direct replacement; allogenically driven changes, including fluctuations; cycles; and long-term changes resulting from nutrient losses). Are any or all of these to be regarded as succession?

The search for pattern and order in the vegetation change phenomena has led ecologists a merry dance. So far a unifying theory to link components of the apparent kaleidoscope of interacting variables has proven elusive. A useful analogy may be with a theme which has attracted attention in physical science recently, namely the theory of **chaos** (McDonald 1978). The chaotic nature of some natural and artificial systems (e.g. turbulent fluid motion, computer models of stellar oscillations and enzyme dynamics), with innumerable random variables, places limits on prediction. Reductionism has its limitations under such circumstances. However, even chaos has an underlying, orderly form (Mayer-Kress 1986, Crutchfield *et al.* 1986, Degn *et al.* 1987).

Figure 12.16 (a) Changes in vigour and orientation of bracken fern (*Pteridium aquilinum*) rhizomes on and behind a front which is advancing into grassland, Breckland, England. (i) Plan of the bracken rhizomes. The young, vigorous main axes are parallel, with primary branches emerging at an angle but directed forwards. In the 'hinterland', where the old fronds are much less vigorous, rhizomes form an irregular network. (ii) Section showing the relative heights of fronds in the various phases. From A. S. Watt (1955). (b) The changing spatial (and temporal) relationships of patches dominated by sheep's fescue (*Festuca ovina*) in a matrix of bare ground and other herbaceous plant cover in Breckland, England. The phasic pattern of change is induced by colonization of open areas by *Festuca*, then development of vigorous clumps of the grass, which trap soil particles and build up the soil level. Then the old *Festuca* patches degenerate and the soil level erodes down to a basement lag of stones and sparsely vegetated soil (after A. S. Watt 1962). (c) Changes over the period 1946–1960 in a 10×50 cm quadrat, fenced from rabbits in Breckland, England (*Festuca ovina* areas hatched; H = *Hieracium pilosella*). The areas of *Festuca* which occurred in 1953 and 1960 may have been influenced in part by competition of the mat-forming *Hieracium* with *Festuca* seedlings (after A. S. Watt 1962).

From another direction some authors (e.g. Horn 1981, P. H. Zedler 1981) have begun to have doubts about the succession concept. Perhaps, Horn notes, the concepts of succession (and facilitation, inhibition and tolerance) 'are nothing more and nothing less than statements about the comparative life histories of species that are found in association with one another'. The questions raised by this point are considered in Chapter 13.

One of the most useful recent statements about succession is that by Pickett *et al.* (1987). In their paper these authors outline a three-level hierarchical approach to the explanation of the causes and mechanisms of succession. The highest level in the hierarchy defines the general and universal conditions under which succession occurs: (a) the availability of open sites; (b) the differential availability of species; and (c) the differential performance of species at a site. The second level is outlined in terms of the contributing ecological processes or conditions. The third level is that of the defining factors that are required to elucidate the mechanisms of succession at particular sites and to make detailed predictions.

13

On the theory of vegetation change

Examples of changes in vegetation in many kinds of land system and some aquatic systems were described in earlier chapters. Chapter 11, especially, emphasized empirical information on vegetation change processes, and Chapter 12 summarized the theoretical and empirical bases from which ecologists have hitherto tried to understand vegetation change.

It is now time to sum up the reasons for belief that the existing 'succession to climax' theoretical structure is inadequate as an overall description of vegetation change and to present a new, comprehensive theoretical framework for the study of vegetation change phenomena. The means for revision of vegetation change theory are sketched out by considering the faults in existing theory, the problems which must be tackled and essential concepts through which vegetation dynamics may be understood. A statement then outlines the new theory. Some consequences of the adoption of this scheme and further research requirements are noted. The final section of both the chapter and the book briefly describes the past, present and future of natural vegetation, and suggests desirable broad aims for human relationships with it.

IMPORTANT RECENT VEGETATION CHANGE LITERATURE

The summing-up of diversely held theoretical views on vegetation change is not easy. Many hundreds of articles on field and laboratory research, computer modelling and philosophical speculation about plant succession have been written by ecologists in the past few decades. Far fewer articles, but still a considerable number, have been written about other vegetation change phenomena.

Aspects of the development of vegetation change theory are conveniently covered in reviews, symposia or books: McCormick (1968), Drury & Nisbet (1973), Pianka (1974), Knapp (1974), Horn (1976), Golley (1977), Miles (1979), Bazzaz (1979), Gorham et al. (1979), P. S. White (1979), Grubb (1977, 1985), McIntosh (1980a, b, 1981), Van der Maarel (1980), Beeftink (1980), Poissonet et al. (1981), D. C. West et al. (1981), Shugart (1984), Finegan (1984), J. White (1985), Diamond & Case (1986), Jenik (1986), Falinski (1986), O'Neill et al. (1986), Huston & Smith (1987) and Gray et al. (1987). Reference to other critical literature is made in earlier chapters.

Modern textbooks of plant ecology are beginning to incorporate the more-varied viewpoints of recent decades about the concepts of vegetation change (e.g. Barbour et al. 1980, Spurr & Barnes 1980 and, most comprehensive of all, McNaughton & Wolf 1979). However, they do not yet go far enough. A unifying theory about vegetation change has proved to be elusive. Whittaker & Levin (1975) even implied that it may never be achieved, because of the heterogeneity of plants and of vegetation.

HOLISM VERSUS REDUCTIONISM

On the whole, the controversy, holism versus reductionism, of the 1970s was an unfruitful exercise. The extravagant language used by some ecosystem ecologists in writing about the systems that they studied (e.g. O'Neill 1976b, Margalef 1968) (whatever their actual philosophical stance on the matter) led to accusations of ultra-determinism. We are considering the science of behaviour of living systems, not phenomena driven by some metaphysical lifeforce. Ecosystems are not really homeostatic, because the plants and animals of which they are composed are not capable of the degree of organization and cognition which would be required to achieve this. The properties of apparent homeostasis of maturing ecosystems are drawn from the properties of individual species, particularly those that contribute a lot to the biomass or to nutrient cycling. Among them are prominent species which control environmental conditions for other species.

On the other hand, some of the supporters of individualistic or reductionist viewpoints (Lewontin 1969, Ricklefs 1973, Harper 1977) prosecute the case too strongly. Certainly the behaviour of individual species must be very well known if we are ever going to make progress in the understanding of ecosystem dynamics. Much more such work must be done. However, no organism is an island. Many communities are well-integrated, with close functional relationships between pairs or groups of species. The dependent systems involving nitrogen fixation, mycorrhizal associations, obligate epiphytism or parasitism are examples, and there seem to be other, looser, positive and negative relationships between higher plants of which we are only dimly aware (cf. Ch. 3). It is a matter of semantics (and of personal choice) whether some of the collective features of whole ecosystems can be regarded as emergent (*sensu* Salt 1979) or not. Barbour *et al.* (1980) list some ecosystem characteristics which they regard as emergent. Furthermore, an holistic standpoint for scrutinizing some aspects of the biology of ecosystems is quite legitimate. Reductionism and holism are simply different ways of looking at the same phenomena. Also, holism is not synonymous, necessarily, with determinism, or vitalism.

Progress in the understanding of ecosystem dynamics will also be made by studying whole, functioning systems, however difficult that may be. Successful work of this kind was done by the systems analysts of the International Biological Programme. Nevertheless, it seems doubtful whether all of the possible interactions in some of the more-complex ecosystems can be fully understood unless ampler real data sets, acquired over long periods, are available, and unless more-powerful and more-cheaply operated computers are in use. K. E. F. Watt (1975) recommends modelling particular, limited compartment sets, rather than trying to take on whole ecosystems; that makes sense.

PROBLEMS WITH IDEAS ON 'SUCCESSION TO CLIMAX'

McIntosh's (1980b) review showed that much of the criticism of the orthodox succession theory of the 1960s, presented as new ideas by population ecologists (Drury & Nisbet 1973, Horn 1974, Diamond 1975), was by no means new and in some instances was misguided or erroneous. Most of the criticisms of Clements' ideas had already been made by earlier workers, some of them soon after his 1916 book was published. In any case the errors of Clements' model were understood sufficiently well by most active vegetation ecologists, although they were perpetuated in many textbooks. In their paper critical of orthodox succession theory, however, Drury & Nisbet (1973) made some timely points. Many ecologists had been quite careless and unscientific in their application of the modified succession to climax theory to the situations that they were studying. This particularly applied to ideas of climax and assumptions about successional stages based on observations made at one time. Inferences often had been on the basis of slim evidence, or preconceived ideas. Teleological thinking was rife. Drury & Nisbet quite rightly pointed to the frequent lack of empirical data, and to the contradictory evidence pertaining to some aspects of succession theory.

The restriction of the scope of study to secondary forest succession by Drury & Nisbet (1973) and some other workers in the 1970s (on the grounds that long-term succession cannot legitimately be studied, and that there are not adequate observations in unforested areas) was defeatist. By assiduous search, long forest chronosequences can be found, and there are some relatively long-continuing observations in other kinds of vegetation (McIntosh 1980a, b, J. White 1985). In fact, the failure of many ecologists to comprehend the full complexity of vegetation change phenomena arises from a myopic preoccupation with short-term change sequences after disturbance in forests of the temperate (and tropical) zones. The Boreal forests, the Mediterranean-type vegetation, the tundra of alpine and sub-polar zones, the deserts and the natural savannahs and grasslands also can tell us much about the kaleidoscope nature of changes in vegetation. So can well-chosen chronosequences, and the sediments and fossil assemblages preserved in mires.

Thus, the idea of succession (orderly, predictable sequential change) to climax (a stable, self-restoring community), believed to take place, for example, in temperate deciduous forests over periods of a few centuries, has been overemphasized. Direct replacement, a common phenomenon, is seldom mentioned in the vegetation change literature; it can hardly be considered to be succession. Other patterns of change, also common in nature – fluctuation, cyclic replacement and gradual, long-term change, including that which results from decline in available nutrients – have been neglected.

In Chapters 11 and 12 it was shown that there are various problems in sustaining commonly held views on the regularity of successional sequences, convergence of successional trends and causes of replacements. Only broad generalizations are possible, given the variability of factors affecting species regeneration. Locally, vegetation change patterns are variable; they depend on the properties of the individual participating species. Also, the 'regularity' of change sequences in most places, other than the tropical rainforests, arises from the circumscription which is enforced by a limited array of available species in each locality.

Realization that vegetation change must be explained in terms of the individual niche properties of the participating species has been consolidated by the contributions of I. R. Noble & Slatyer (1980), Austin (1981), Grubb (1977, 1985) and P. H. Zedler (1981). Gleason's (1917, 1926, 1939) original contributions to this aspect of theory have been well and truly vindicated, although they neglect the evidence for positive relationships of species in stands of vegetation.

Until recently (cf. Connell & Slatyer 1977 and Chs 4, 5, 6 & 11) relatively little critical attention had been given to the kinds of phenomena occurring during sequential change. In a rather general way autogenic processes were thought to be a major driving force, although allogenic events and competition were also invoked as causes of change. The early contributions of W. S. Cooper (1916), A. S. Watt (1925) and Tansley (1920, 1935, 1939) and later, those of A. S. Watt (1947), Keever (1950), Egler (1954), Niering & Egler (1955), Sigafoos & Hendricks (1961), Connell & Slatyer (1977), Bazzaz (1979), Tilman (1982) and Huston & Smith (1987) are important.

Each of facilitation, inhibition and tolerance may occur, depending on the behaviour of the plant species that are present. No one of these models is a complete description of the events occurring during a sequence of species replacements.

PROBLEMS WITH IDEAS ON STABILITY AND 'CLIMAX'

Drawn-out arguments about 'climax' and stability have contributed little to our understanding. Some ecologists prefer to abandon the term climax and replace it with a less-committal one such as 'approximate steady state' or 'most mature', which would denote vegetation in a temporary and relative state of equilibrium only. It is, in any case, necessary to use arbitrary definitions of the limits of relative stability, in terms of the attributes of the plant species in the vegetation being considered (cf. Ch. 11). The definitions must take into account the spatial scale and relative timescale in relation to disturbance regimes, the dimensions of plants or stands, their regeneration patterns and the turnover time of their generations. As outlined in Chapter 11, a mature stand can, realistically, be defined as one which maintains itself with approximately the same composition for more than one generation. Provided that disturbance is minimal (spatially and temporally), mature herbaceous vegetation stands might persist for a few years to a decade or more, whereas mature forest might persist for a couple of centuries or more. In due course, however, it is likely that shifts in the composition of any mature stand will occur, through various causes.

The concept of climax, if it is to be retained as an ecologically useful idea, must accommodate the spatial heterogeneity, as well as the relatively short-term vegetation variation in the regional and local contexts (e.g. in vegetation such as semi-arid grassland, or tundra, as well as forest). In the sense of the overall stability and persistence of vegetation of a given composition over periods of many centuries, it can only be applied to grand-scale ecosystems. In this sense the meaning of climax seems to be close to that which Clements intended. Climax is an inappropriate term, at the local scale, for the mature stands of vegetation which develop a century or so after an episode of disturbance. It is also inappropriate at the local scale for vegetation that is in a constant state of flux through fluctuating or cyclic change. It is inappropriate, too, for the persistent vegetation which results from incessant disturbance. The recently enunciated hierarchical perspective (O'Neill *et al.* 1986, Urban *et al.* 1987) is an appropriate alternative means for summarizing the complexity of landscapes and ecosystems.

'KINETIC' IDEAS ON VEGETATION DYNAMICS

The perspective of the 'kinetic' model ecologists (cf. P. S. White 1979), who point out the importance of strong environmental pressures in maintaining persistent vegetation, or particular vegetation patterns, must be seriously considered. The 'kinetic' viewpoint, which requires no stable and preconceived end-points, cuts right across a theory which emphasizes stability as the norm. The 'kinetic' model enables all kinds of vegetation change to be considered without recourse to elaborate and restrictive terminology. It is usefully applied on local and regional scales, where direct replacement, cyclic or fluctuational patterns of change may be apparent. It is also useful in systems where 'degradation' is occurring, e.g. from forest to heath. This seems to be important in a branch of ecology which claims to be dynamic. Nevertheless, the 'kineticists' have tended to neglect the indications that mature vegetation, lasting at least a few generations of the dominant species, is commonly to be found (although under human assault it is rapidly becoming less so).

FORMULATING A THEORY OF VEGETATION CHANGE: THE ESSENTIAL PROBLEMS

The main problem is that some vegetation change phenomena, observed in nature, contradict or are not encompassed by the orthodox succession to climax theory. As a result, ecologists lack a comprehensive theoretical framework which readily accommodates and explains all kinds of vegetation change (Table 13.1).

In seeking to explain vegetation change in any of the patterns outlined in Table 13.1, ecologists must also explain the proximate causes for replacements of species populations. Relative steady-state conditions of vegetation also require explanation. Appropriate means are needed for expressing the various manifestations of vegetation change phenomena.

Appropriate concepts

Some of the older concepts confine us within an inadequate theoretical straitjacket and stultify progress towards adequate theory. That being so, they should be abandoned, as indicated below. The fresh outlook that is needed must be based on criteria which apply to all vegetation and which can be used to describe and explain the fundamental features of all kinds of vegetation change. The particular criteria chosen will govern how we think about the subject. Quite a simple, straightforward conceptual approach is desirable.

SCALE

Of the four scales of magnitude of vegetation which might be used to develop ideas about vegetation change (Table 13.2) the most appropriate for most purposes is judged to be the patch or stand. Stands are monospecific or mixed species populations (adults, juveniles and propagules) living together. Stands occupy areas of

Table 13.1 Phenomena to be encompassed by a theory of vegetation change.

(a) colonization and sequential replacements following the formation of new sites or following the disturbance of established vegetation
(b) direct replacement following the disturbance of established vegetation
(c) cyclic replacements in response to endogenous or exogenous influences
(d) fluctuating replacements in response to exogenous influences
(e) vegetation maintained by frequent or continual disturbance
(f) vegetation in relative equilibrium for a time (composition unchanging for more than one generation of the dominant species)
(g) long-term, gradual change in response to autogenic or allogenic influences

Table 13.2 Scales of magnitude for considering vegetation change phenomena.

1. *individual plants*

2. *stands*, comprising populations of one or more species living together on an area of ground in a patch of homogeneous composition (less than a square metre to a few hectares)

3. *plant communities*, each consisting of several to many stands of similar composition, in some instances forming almost continuous, uniform vegetation cover over extensive areas (a few to many hectares), but often forming more or less complex mosaics with stands of different composition

4. *landscape – vegetation types* with stand mosaics, or plant communities covering extensive or very extensive areas; the dominant plant species are usually of similar stature and general physiognomy, but the species composition often differs widely from place to place

ground of a size which is dependent on the stature of the resident species. They tend to function as units. Individual stand sizes differ within and between different kinds of vegetation, and their boundaries in space are not necessarily clearly delineated; vegetation varies continuously in space. The essential requirement is that the patch of vegetation be relatively homogeneous in composition. Individual plants in the stand will influence one another in various ways. Around the margins of any stand there will be mutual influences of the individuals with those of adjoining stands.

It is in stands of plants that the various vegetation change processes are readily discernible. Nevertheless, there is no clear demarcation between stand-sized vegetation patches and larger-scale vegetation units; or between stands and the populations of individual adult species of which they are composed. Adults (those individuals which can reproduce) receive most emphasis at a stand-based scale of magnitude. However, propagules (spores, seeds and vegetative offshoots) and juveniles (non-reproducing young seedling and adolescent plants) are very important. Propagules (especially seeds and spores) permit the plant species to move or to be transported from site to site and to survive adverse circumstances; they also give rise to new populations of juveniles. Juveniles provide the immediate potential for the replacement of adults when they die.

The other unit of scale, which is used here as a temporal yardstick, is the mean generation length of each species (the time from the establishment of seedlings to the death of the adults which arise from them). It is realistic to use the generation length of the prominent species in a stand as a gauge of its persistence. This means that stands of, for example, herbs, shrubs or trees may persist for equivalent periods in terms of numbers of generations, but very different periods, in terms of years.

DEFINITION OF WHAT CONSTITUTES VEGETATION CHANGE

Vegetation change is defined as being *a net shift, over a given period, in the numerical composition of the adult (reproducing) plant species populations of a stand occupying an area of ground*. Change in open sites, where there are untapped resources, may occur simply by the increase in numbers of individuals of one or more species. In these fully open or partially open sites, not saturated with adult individuals, immigrant species that were not previously present may also establish and grow to adulthood. Also, in well-occupied sites that are, nevertheless, not saturated by adults, change may occur through the immigration of species with niches such that they can occupy a stand without displacing resident adults.

Assuming that stands of established vegetation are saturated with adult individuals (i.e. no living space is immediately available for further adults of the locally available species assemblage to become established), change may occur through the addition or subtraction of individuals of one or more species under two sets of circumstances.

1   When resident individual adults die or emigrate from the stand.
2   When one or more species new to the stand immigrate. This will lead to the narrowing of realized niches of the species present, or the competitive elimination of one or more of them by the immigrant species.

Suppose, now, that we are observing a vegetation stand consisting of a given number of plant species, each different to some degree in form, life-history, reproductive potential, longevity, etc., and each with different numbers of adult individuals (and juveniles and propagules). Birth and death phenomena will be occurring continually or periodically in the plant populations. By the definition at the beginning of this section, a change in the numbers of adults of any one species will cause a change in the stand. There will be proportionate shifts among the different species, as well as probable readjustments of stand composition as other species increase or decrease in response to the initial changes. The profundity of change will depend on how many individuals and how many species are involved in the numerical shifts. It is possible that the numbers of two or more species may change reciprocally while the numbers of others remain constant.

MANIFESTATION OF NUMERICAL CHANGE IN ADULT POPULATIONS IN STANDS

In sites that are saturated with adults, numerical changes in stands can be envisaged as occurring through one or more of:

1   The death of adults without replacement by equal numbers of individuals of the same species throughout the stand.
2   An increase in the numbers of adults of one or more of the species that are already established in the stand. This must be accompanied by a decrease in the numbers of adults of one or more of the other species.
3   The immigration by seed dispersal or vegetative means of one or more species that

471

is not already present, then development to adulthood. Again this must be accompanied by a decrease in the numbers of adults of one or more of the resident species.

4   The emigration of individuals of one or more species by vegetative means and a consequent increase in one or more of the other species that is resident in the stand.

NON-CHANGE IN VEGETATION

Although change of greater or lesser magnitude is probably continually occurring at various rates in most vegetation, some stands seem to exist in a state of dynamic equilibrium. If this is so, as adults of different species in the stand die, they are replaced by similar numbers of the same species, so that the adult population remains constant over a given period. In such a stand replacements of any individuals that die are not necessarily by the same species on the same spot, but replacement of their numbers is accomplished, on average, throughout the stand, so that the proportional composition of the adult population does not change.

Other stands may be out of equilibrium, as outlined in the previous section. Relative rates of change or equilibrium states in stand adult species composition can best be measured against the lifespans of the species concerned. Change in the adult population of any species after only one generation of occurrence in a stand indicates that the species is not in equilibrium with the prevailing conditions. The maintenance of populations of adults in the same proportions for more than one generation signifies a relatively steady state for the species concerned.

The generation may be difficult to define for species which habitually reproduce by vegetative means. The age of the modular shoot system can then be used as a basis for comparison. The generation length may also be difficult to define for any species which has no clear indication of an annual growth period.

CAUSES OF DEATH OF ADULT PLANTS

Adults die through one or more of: (a) inherent ageing (senescence); (b) damage; (c) stress; and (d) competition.

The endogenously determined phenomenon, ageing, seems rarely to be the cause of death of plants, other than monocarpic species. Death is most commonly caused by the environmental pressures, damage and stress. Their effects are usually imposed allogenically, but stress (and rarely damage) may also originate autogenically. Stress effects are usually more or less gradually applied; damage may affect plants gradually or suddenly.

Resource use or direct interference competition affect plants stressfully. Although adults may be directly killed by competition, it more frequently regulates population numbers by causing regeneration failure (preventing reproduction by adults, inhibiting seed germination or seedling establishment or inducing the death of juveniles).

Variation in exogenous influences may tip the competitive balance in favour of one or more species at the expense of others. Gradual, cumulative influences, such as direct interference or autogenic habitat modification, may have similar effects.

In real situations it may be very difficult to determine what the actual cause of death of adult or juvenile plants is. Stands must be monitored for generations in order to know what is happening to the population structure. Vegetative proliferation complicates the picture. Random environmental events, as well as influences of historical factors that affected site conditions long ago also complicate the observation of causes of death.

The self-impelled emigration of plants from stands by vegetative means is almost the only way in which removal of individuals can occur, other than through death. Rare instances of physical removal exist (e.g. some moss stands, where portions of plants are removed by wind and establish elsewhere, or similar transportation and re-establishment of herbaceous or sometimes woody vascular plants by agencies such as soil creep, slump or fall on hillslopes, snow avalanche or rivers). Emigration is usually restricted to species with an efficient extension of their shoot or root systems, through which the whole plant can move about. In reality, however, since adult portions of the plants will remain fixed in position and ultimately die, as the youthful parts extend away from them, the phenomenon is a variation of the ageing process.

SOURCES OF REPLACEMENTS

Dead individuals in a stand are replaced: (a) from the population of juveniles already present in the stand (which may have originated from seeds or as vegetative off-shoots); and (b) directly by germination of seeds which were resident in the seed bank, or that are newly immigrant.

Many random factors will influence the availability of specific propagules to replace any individual adults which die. Probably most of the replacing individuals will have been derived from the resident adult population. Some (seed-bank) seeds may be derived from long-dead adults, possibly species that are no longer present in the stand. Some of the replacing individuals will have arisen from newly immigrant seeds, or seed-bank seeds which may have originated considerable distances from the site of the stand. The original seed parents of any replacing individual may, thus, be difficult or impossible to identify. This applies also, of course, to pollen parents. Plant populations can be very far-flung!

CONDITIONS FOR ESTABLISHMENT

Establishment from seeds (and spores) requires rather specific sets of conditions. The particular light, climate, temperature and moisture conditions are critical for seed germination (cf. Chs 2 & 11). Their combined influences will determine which species are available for establishment in any locality.

The maintenance of favourable moisture, nutrients, light and temperature conditions will decide which species finally do become established in the range of microsites available in any locality. The amounts of seed reserves and the rates of growth of both root and shoot systems of seedlings will be critical for the continuity of life of young plants.

The vulnerability of the germinating seeds and juvenile plants does not stop with the climatic, light, soil moisture and nutrient hazards, because they are also subject to attacks by pathogens and predators. The survivors, thus, will have endured a severe environmental sieve.

The establishment of vegetative offsets from adults is often thought to be more reliable than establishment from seeds. In general this may be so, although more plant species seem to have specialized on seeds alone as a means of continuing their kind than on vegetative modes alone. Also, there are problems for many vegetatively proliferating species of not being able to disperse their propagules as easily as seeds. They can be, also, subject to disease spread directly from the adults, as well as to parental competition. Vegetative proliferation is advantageous in some severe conditions (e.g. tundra, desert and sites that are frequently burnt, such as grassland or chaparral). At least in some circumstances (e.g. aquatic sites or windy locations), vegetative offshoots may be spread rapidly and widely.

NICHES AND LIFESTYLES

The niche concept is so important for the theory which is to be enunciated here that a brief iteration of the points made in Chapter 3 and subsequently is desirable. Fundamental niche dimensions are those defined by the ability of each species to respond to the complex of exogenous environmental factors (i.e. its ecological amplitude). Its realized niche is that which is actually expressed when a species is in competition with neighbour species.

The n-dimensional niche concept (cf. Hutchinson 1958) expresses in a very general way the variety of attributes which fit plants for life in their habitats and communities. The importance of a multidimensional outlook towards the attributes of plants in vegetation is stressed in Chapter 3 and discussed, also, by Wuenscher (1969, 1974), Van den Bergh & Braakhekke (1978), Tilman (1982), Giller (1984) and Grubb (1985). Unfortunately, it is difficult to measure niche attributes very precisely, and niche differences are usually gauged from the observed characters of the plant species concerned, from their relative behaviour in field situations, or from both. Such observations provide only a rough measure of the relative niches. The relative contributions of fundamental and realized niche dimensions are not easily distinguished through them. Whittaker's (1956) or Beals' (1969) direct gradient analyses; Wells' (1976) 'climax-index' method (Table 11.16) and I. R. Noble & Slatyer's (1980) 'vital attributes' method (Fig. 12.6) are examples of approaches to niche definition.

Local data sets of these types, preferably amplified by further critical experimental data (cf. Parrish & Bazzaz 1982a, b, Newman 1982, Garbutt & Zangerl 1983, Grime et al. 1987, Caldwell et al. 1987), are required for the potentially interacting plant species in the flora of each locality under study. The potential for great complexity in niche dimensions can be gauged from the data outlined in Chapters 2 and 3. Matters are complicated further by the differences in niche dimensions that are evident at different stages of the life-history of each plant species.

In any region with a relatively rich flora there is a more or less continuous range of the expression of both niche breadth and competitive power among the species. Although niche overlap is evident among plant species that live together, each species appears to have a distinct 'population centre' where it is competitively superior. The way in which a species performs in its habitat is referred to here as its **lifestyle**. The term 'strategy', often used by population biologists, is avoided, as it has an unfortunate connotation of cognition, which is not a plant attribute.

COMPETITION

Sequential replacements are usually considered to result from interspecific competition. Species are in equilibrium in specific sets of environmental conditions. They replace one another competitively if the environment changes so that the original species are no longer able to reproduce or to replace themselves. In some circumstances competitive replacement may occur simply through the increase in size and spread of one or more individual, allowing pre-emption of the resources of a site. Otherwise shifts in allogenic effects such as the increase of a pathogen or a decline in rainfall, or autogenic shifts such as intensified shading or modified organic soil layers, or exudation of a toxin, may be sufficient to tip the balance in favour of the replacing species. Competition of both the resource use and direct interference kinds is described more fully in Chapter 3.

CRITERIA FOR CLASSIFYING KINDS OF VEGETATION CHANGE

Among possible alternative criteria for classifying kinds of vegetation change are:

(a) *Lifestyles.* Although the niches of plants that live together overlap to some degree and it is possible to group species with roughly similar lifestyles, the wide range of permutations and combinations of factors makes it difficult to employ this as the sole criterion.

(b) *Replacement patterns.* The patterns of replacement by individual species or groups of species could be classified (Figs 11.25, 30 & 36). More-complex patterns, including some that are random, may occur. The models outlined in Figure 11.36 etc., may be useful descriptors of the actual replacement details, but they do not necessarily explain the associated processes. Replacement patterns are also not suitable as the sole criterion.

(c) *Individual phenomena reflecting the ultimate causes of change.* Single classes of phenomena, such as: the causes of death of adults; types of propagules giving rise to replacements; physiological properties of seeds or other propagules, and of juveniles, in relation to sets of environmental conditions, especially those which influence establishment; and competitive processes, are not themselves adequate descriptors of kinds of vegetation change. Combinations of them seem to offer an appropriate range of features which fall into several recognizable classes. One approach could be to use the hierarchical table of 'successional causes' of Pickett *et al.* (1987) (extendable to cover vegetation change causes) (Table 13.3). The table summarizes and organizes in an appropriate order many of the criteria which must be considered. In the statement below a somewhat simpler approach is taken.

The processes of vegetation change result from interactions among complex sets of factors which create particular kinds (sometimes syndromes) of effects. At the outset only the ways in which habitat changes (or certain other situations) cause responses in plant species populations will be invoked. These will form the fundamental basis for the explanation of vegetation change which follows. Superimposed on this framework is a brief consideration of how species with

Table 13.3 An hierarchical table of successional (or vegetation change) causes. The highest level of the hierarchy represents the broadest, minimal defining phenomena. The intermediate level contains the mechanisms of change or causation of the highest level. The lowest level consists of the particular factors that determine the outcome of the intermediate-level processes, and are discernible or quantifiable at specific sites. For simplicity, interactions among factors at each level are not shown (modified from Pickett *et al.* 1987).

General causes of succession	Contributing processes or conditions	Defining factors
site availability	coarse-scale disturbance	size, severity, time, dispersion
differential species availability	dispersal	landscape configuration dispersal agents,
	propagule pool	time since disturbance, land use
differential species performance	resource availability	soil conditions, topography, microclimate, site history
	ecophysiology	germination requirements, assimilation rates, growth rates, population differentiation
	life-history strategy	allocation pattern, reproductive timing, reproductive mode
	stochastic environmental stress	climate cycles, site history, prior occupants
	competition	presence of competitors, identity of competitors, within-community disturbance, predators and herbivores, resource base
	allelopathy	soil characteristics, microbes, neighbouring plants
	herbivory, disease and predation	climate cycles, plant vigour, plant defence, community composition, patchiness

different lifestyles fit into the complex kaleidoscope of spatially and temporally variable conditions, and a brief account of how particular modes of change are manifested in field situations.

A THEORY OF VEGETATION CHANGE

The following statement of theory is intended to encompass all kinds of vegetation change phenomena in all kinds of vegetation (Fig. 13.1).

Proximate causes of changes in vegetation

Among the sets of circumstances through which vegetation change may occur on areas of ground are the following.

OPEN UNVEGETATED AREAS

1   An increase in the numbers of individuals on otherwise unoccupied ground. This type of change will be referred to as Mode 1A.

AREAS ALREADY FULLY COVERED BY VEGETATION

2   The destruction, more or less instantaneously and more or less completely, of individuals, patches or larger areas of existing adult populations by allogenic factors. Although in some circumstances there may be direct replacement of adults by juveniles of the same species, often the altered habitat conditions permit other opportunist species to become established and to persist for a time. This type of change will be referred to as Mode 1B. Of course, the removal of vegetation cover, disturbance of the soil and certain other effects, immediately change habitat conditions at a site, and this will influence the subsequent pattern of regrowth of vegetation.

3   The shift of one or more habitat conditions so that some or all of the resident species populations are killed or are no longer able to regenerate, or both. They are replaced by other species populations that are better suited to the new conditions.

   (a)   Changes in habitat are brought about by allogenic factors causing disturbance or stress, acting relatively slowly on individuals, patches or larger areas. The adult plants may be killed as they stand, or prevented from regenerating so that when they die no replacements are available to fill gaps. Very often the altered habitat conditions simply make some species less competitive than others, which replace them. This type of change will be referred to as Mode 2A.

   (b)   Changes in habitat conditions are induced by the resident plants themselves (i.e. are autogenic), with a similar outcome to the Mode 2A type. Some kinds of autogenic habitat change (such as soil modifications) are often slow to develop. Others (such as changes in the light conditions) are more rapid. Their effects on the inhibition of regeneration of species populations in a stand are similar to those induced by allogenic factors. This type of change is referred to as Mode 2B.

   In the circumstances of the Modes 2A and 2B type of change, where habitat changes other than rapid destruction of vegetation are an inherent underlying reason for shifts in species composition in stands, differential propagule survival or dispersal rates of the invading or replacing species may be a component in the immediate changes in composition, independently of the changed habitat conditions, just as they are in the changes of Mode 1 type.

4   Some species are programmed to die (annual, biennials and longer-lived

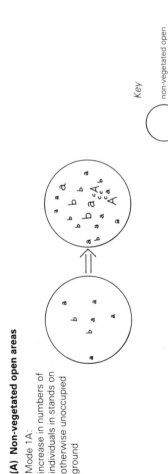

(A) Non-vegetated open areas

Mode 1A:
increase in numbers of individuals in stands on otherwise unoccupied ground

(B) Areas already fully covered by vegetation

Mode 1B:
increase in numbers of individuals in stands on unoccupied ground after sudden destruction of individuals, patches or larger areas of existing vegetation

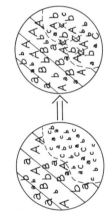

Mode 2A:
replacement of species in stands after more or less gradual habitat changes, due to allogenic influences, that cause death or prevent regeneration of individuals, patches or larger areas of vegetation

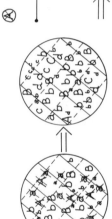

Key

○ non-vegetated open area

⊞ allogenically caused habitat change

⌀ stand with complete vegetation cover

⌀ autogenically caused habitat change

A B C D adults
a b c d adolescents
a b c d juveniles } plant species

✗ killed by stress, damage, competition

⊗ killed through senescence

↕ migration
(it is implicit that there will be a regular influx of propagules to the stand from the array of species in nearby stands. Similarly there will be efflux of propagules from the mature plants of the stand to other stands)

⇒ change, over a period

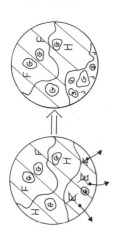

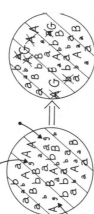

Mode 2B:
replacement
of species in stands
after more or less
gradual habitat changes,
due to autogenic
influences by the
resident plants, that
cause death or prevent
regeneration of
individuals, patches or
larger areas of
vegetation

Mode 4:
replacement of species
which emigrate from
stands by vegetative
means

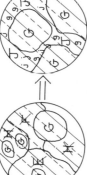

Mode 3:
replacement of species
in stands after death of
individuals through
senescence

Mode 5:
immigration to stands
of species not
previously present, with
or without displacement
of resident species

Figure 13.1 Outline of a theory of vegetation change on areas of ground according to proximate causes: circumstances through which vegetation change may occur. Each of the circular diagrams represents a patch of vegetation, or a stand at an instant of time. Change is often continuous or recurrent, but at different rates for different species.

monocarpic species, which die after flowering, as well as at least some other perennial species which undergo senescence). Vegetation change occurs if the individuals or populations which die are replaced by other species. This type of change is referred to as Mode 3.

5 Plant species which habitually proliferate vegetatively may be able to extend themselves (rapidly or more slowly) so that they emigrate from a stand. This may initiate change in the species populations of the stand. This type of change is referred to as Mode 4. In effect, Mode 4 change occurs when parts of vegetatively dividing clones or the old, established adult portions of the emigrating plants die, so that it closely resembles Mode 2 or Mode 3 change.

6 The natural introduction of plant species by the spread of propagules to localities where previously they were absent, may bring about change by invasion of existing stands, with or without displacement of species of the resident population. This type of change is referred to as Mode 5. Circumstances include the invasion of a region by one or more species that are locally absent for historical reasons, the spread of newly evolved species or the spread of species adventive to the region.

In practice, in any area of vegetation over a period of a few generations of the plant species present, Modes 1B, 2A and 2B, and sometimes Modes 3, 4 and 5, types of change are all likely to be occurring practically simultaneously, so that the actual details of population changes are likely to be most intricate. Some characteristic patterns expressed in the vegetation are noted later.

The array of plant lifestyles

The spatially and temporally variable environmental complex, within which plants must operate, comprises sets of conditions that are dependent on purely abiotic as well as biotic factors, including those created by the presence, properties and activities of the plants themselves.

In each region the flora is composed of an array of species with different lifestyles. To some degree species which live together overlap in their habitat requirements, but, overall, each individual species is suited by its lifestyle for a particular set of habitat conditions (its fundamental niche) and is uniquely distinct in one or more respects, in relation to other species. Each species also has the ability to withstand a particular degree of competitive pressure from its neighbours in the vegetation (determining its realized niche). The competitive power of each species is probably different and is independent of (or a distinct component of) its fundamental niche breadth. It is these different properties of plant species which allow them to find living places in the almost infinitely variable habitat complex (incorporating many factor gradients). Chance, in relation to allogenic circumstances, and the reproductive and establishment success of different species will often be important.

The niche dimensions of propagules and seedlings are of critical importance in permitting them to become established under particular sets of circumstances, but the niche dimensions of adolescent and adult plants (often differing in various ways or degrees from those of seedlings) are also important, for example in determining competitive interactions and longevity.

Common, broad lifestyle types of places in the temperate zone include those of:

1   The set of plant species which produce many, light, well-dispersed seeds and seedlings that must establish quickly in open sites. The short-lived adults cannot persist in crowded conditions.
2   The set of plant species which produce relatively few, heavy seeds, dispersed over short distances, and seedlings which can establish in deep shade. The long-lived adults grow slowly and persist in densely crowded conditions.
3   The set of plant species which are specialized for growing in water.
4   The set of plant species which are specialized for growing in alpine tundra. etc.

A wide variety of subsets within each of these sets (and any other sets which could be specified) can be recognized according to: lifespan; growth form and stature; growth periodicity; the occurrence or not of dormant seeds; requirements for nutrients; fit to specific habitat conditions, etc.

In reality the array of lifestyles might form an almost continuous variation pattern in any extensive region with a relatively rich flora. For example, some colonist species have attributes which allow them to persist, also, in crowded, resource-limited conditions, whether other potential competitor species are present or not. Some species customarily found in densely crowded conditions, where resources are limited, have light, well-dispersed seeds. The heavy seeds of some species are widely dispersed, and they tend to establish in open, uncrowded sites. Some species have very wide ranges of tolerance for resources such as nutrients or water. Others are extremely limited in their tolerance ranges, and are restricted to very limited sets of habitat conditions. Some of the species with the widest fundamental niche dimensions are very weak competitors. Other species, also with wide fundamental niches, are such strong competitors that they exclude all other species. The range of lifestyle types forms a virtually infinite array – a multidimensional continuum. Correspondingly, the manifestations of plant species in dynamic vegetation patterns will be multiform. In effect, each species must be taken on its own merits.

Common replacement patterns in vegetation undergoing change

The immediate manifestations of vegetation patterns (Table 13.1) by which ecologists usually recognize vegetation change were called *styles of change* in Chapter 1. It is important to realize that these styles are segments of a multidimensional continuum. Changes in vegetation will differ in style, degree and rate according to the local expression of environmental and plant factors and the frequency, extent and stochasticity of events which affect plant lifespans, deaths and replacements. Among the most important factors are site history, relative site quality, environmental forcing, differences in plant lifestyle characteristics (e.g. life-history mode, fecundity, propagule dispersal, size, competitive power, resistance to hazards, and longevity). Combinations of attributes are individualistic for each species.

The individual behaviour of dominant species during particular phases of vegetation change sequences contributes a lot to the character of vegetation (cf. Ch. 11). The dominants control conditions for other species, but their prominence may lead to the behaviour of the lesser components of the vegetation being overlooked.

In terms of the lifespans of the species concerned, rates of change in species populations may be rapid, intermediate or slow. The numbers of plants involved may be limited to single individuals, be intermediate, or include multitudes of individuals and vast stretches of vegetation. The areas of ground affected are correspondingly diverse. Changes may be almost continuous, with no, or only short, periods of relative stability; intermediate; or sporadic, with long periods of virtual steady state, broken by short periods of change. Specific details will depend on the interactions between autogenic and allogenic factors, and on the modes of change which are operating.

Mode 1A type changes occur whenever totally new sites are created. All of the plants on them are immigrants. Mode 1B changes, consequent on disturbance of existing vegetation, are very common and easily recognized (cf. Chs 7, 9 & 10, in particular). Immediately after the establishment of the first colonists on a disturbed site, however, the mode of change shifts (in cases other than those of direct replacement). Mode 2B, induced by autogenic effects, is probably the most common type subsequently.

Mode 2A changes resulting from stress effects or damage, such as might be caused by periodic oscillations in the amounts of precipitation or in grazing intensity are common causes of fluctuating replacement in vegetation. As the intensity of the peaks of stress or damage differs over time and the stressful or damaging periods usually occur somewhat randomly, the resultant vegetation adjustments are irregular in degree and direction. Different plant species take advantage of the different sets of conditions.

Fluctuations have been recognized mainly in vegetation composed of herbaceous and often relatively short-lived species, but long-term environmental oscillations could also affect the populations of long-lived species. The oscillatory nature of such phenomena could be very hard to detect. Vegetation changes caused by long-term environmental oscillations could appear to be directional, sequential replacements (i.e. 'successional') trends.

Changes initiated by some long-term and profound environmental shifts may be regular and directional rather than fluctuational. A secular climatic change or the introduction of a new pathogen might bring about such changes. Forcing factors such as wind may also initiate some Mode 2A type changes recognized as cyclic replacement series (cf. Chs 7 & 12).

Mode 2B changes occur wherever there is a continuing shift in environmental conditions induced by the presence and activities of plants. In the relatively short term, changes in light, temperature or moisture conditions, or immediate nutrient availability, are commonly their cause. Such causes are impermanent – if the plants are removed, so is the effect. Relatively long-term changes in forest, and possibly in grassland and scrub, often appear to be related to the progressive intensification of shading. Slower, but more-permanent plant-induced changes occur when the soil properties are modified. Among such modifications, operating mainly through litter deposition above and within the soil, are increased acidity, altered ion-exchange complex, redistribution of nutrients, leaching and nutrient depletion.

Soil change may be the most usual cause of long-term sequential replacement. Vegetation of diminished stature and variety associated with waterlogging or nutrient loss, or both, results, often, from long-continued operation of Mode 2B type changes. Disturbance (e.g. fire) may also cause a sharp reduction in nutrient stocks and initiate Mode 1B change.

Cyclic replacement series may occur through Mode 2B changes; examples are the sequences resulting from the accumulation of toxins in the soil or the build-up of alpine cushion plants until they are exposed to frost effects or abrasion by wind-driven snow. The auto-inhibition of mature plants or juveniles by their own toxic exudates, or the inhibition of any species in the stand in this way by another of the species present, are means by which Mode 2B change may occur.

Mode 3 type change, resulting from the senescence of resident individuals, occurs in some situations which have been recognized as cyclic replacement series (e.g. the *Calluna* cycle). Mode 3 changes appear, usually, to take the form of simple replacements with no sequential change. If senescence causes the death of individuals, the failure of the individuals of the senescent species to occupy the site immediately is likely to be because it is taken over rapidly by opportunists (thus, the situation resembles Mode 1B type change) or because autogenic modification of the site has taken place under the influence of the resident adults which became senescent. In some circumstances the senescence of adults may simply result in direct replacement.

Mode 4 type changes may be less common than the other types recognized here. Nevertheless, examples are known in nature, especially in herbaceous vegetation, e.g. migratory white clover (*Trifolium repens*) in pastures, or bulrush (*Typha* spp.) in aquatic habitats. Trees like *Robinia pseudoacacia* or *Maytenus boaria* can also migrate, slowly, in this way.

The potential for Mode 5 type change arises in any region wherever species which could occupy a site are slow to disperse to it through accidents of history rather than climatic or soil limitations. When they finally do reach the site they must be sufficiently strong competitors to enter existing vegetation and may or may not displace existing individuals. Mode 5 type changes are probably common in natural vegetation, but are not well-documented. In many instances they would be difficult to distinguish from Mode 2B type changes. The normal operation of the Mode 5 type change may involve a simple replacement of one or more species by an invader. There may or may not be sequences of change.

The ability to predict the occurrence and outcome of vegetation change depends on having very good knowledge of the nature of the plants and the habitat conditions of a region, as well as the modes of change which operate in each situation. The lifestyles of the particular plant species present will determine the phenomena actually being expressed in the vegetation, but so many variables affect the magnitude and direction of changes that prediction is possible only within broad limits.

Lack of change in vegetation is explained simply as being:

1 The lack of habitat change, or other changed conditions which would advantage species that are not part of the current resident population.
2 The replacement (direct or reciprocal) of dead adults by juveniles of other resident species, such that the composition remains unaltered.

Although some vegetation may appear to be unchanging over periods of a few generations of the resident dominants, it seems likely that most stands of vegetation will be undergoing at least some change. Because of the longevity of many plant species, this may be difficult to detect. The need for long-term study is obvious.

NEW THEORY IN RELATION TO ORTHODOX THEORY

The theory just outlined readily accommodates direct replacement, fluctuating replacement, cyclic replacement and 'linear' sequential replacement (succession), as well as changes in vegetation which are the outcome of decline in site quality. Relative steady state in dominant populations of plants over periods of a few generations is also accommodated within this theoretical framework. However, stable end-points are not a necessary part of the theory, so it can also encompass the ideas about the influences of disturbance promoted by the 'kinetic' school of vegetation dynamics. The theory is compatible with the continuum concept of species composition in space; indeed, the continuity of vegetation and vegetation change processes in space and time are integral parts of the theory. The theory is also compatible with vegetation patterns arising from 'tolerance' and 'inhibition' (Connell & Slatyer 1977). The specific patterns depend on the local circumstances.

There are some clear points of difference between the new theory and the received 'succession to climax' theory. Several of the older concepts are inadequate and can be discarded.

According to the new theory, for example, repeated similar patterns of sequential vegetation change in a region occur simply because, in that region, there are limited numbers of species with lifestyles that suit them for inhabiting particular broad sets of habitat and competitive circumstances. The exception is in very species-rich tropical forests where floristic diversity permits replacement by species drawn from a wide array (Ch. 10). Nevertheless, even in tropical forests species can be recognized with distinctive lifestyles, fitting them to play particular roles. In temperate regions there may be species which can play (a) similar roles, or (b) particular, distinctive roles, which determine the courses of change patterns of approximately similar or distinctively different kind. It is shown in Chapters 9 and 11 that the detailed sequences of change in some temperate forests follow a variety of pathways according to the prevailing circumstances. Unpredictability is contributed by increased species richness of floras. Some of the most predictable vegetation change patterns are those in regions with poor floras and severe environments. Versatility of the species of such regions, in their ability to occupy a wide range of habitat conditions, compensates to some degree for the dearth of lifestyle types.

The convergence of different vegetation change sequences towards a particular type of species composition might occur where there are relatively few strongly dominant species that are able to take over most of the habitat. However, there is not much evidence for convergence. Divergent sequences may be just as common.

Mature is the preferred term for vegetation in which the dominant species are maintained on a site for more than one generation. By the criteria outlined here, some change in subordinate species populations, and in the spatial arrangement of individuals of different species in a stand from generation to generation, is permissible in mature stands. In a region there may be many stands of mature vegetation of different composition. Some may show no sign of change for several to many generations. Maturity does not imply that further change is unlikely; the habitat conditions may change, or new species may invade the site, thus creating disequilibrium. In the long term (tens of generations of dominants or more) it is highly unlikely that sites will remain in stable equilibrium, so the concept of steady state, for habitat conditions and vegetation composition, is perhaps only ever applicable over a few generations. Exceptions include certain kinds of monotypic vegetation, such as some *Pinus contorta* or *Abies balsamea* forests in North America and *Nothofagus solandri* forests in New Zealand, which replace themselves immediately after disturbance (Ch. 7, Ch. 12).

If a developmental sequence of vegetation is evident, leading up to a mature stand, the preferred terms for phases earlier than mature are **pioneer** for the earliest colonists and **transitional** for all intervening phases before the mature one. Seral is considered inappropriate because it harks back to the 'sere' – the whole sequence in Clements' deterministic succession to climax theory. There is no such thing as a sere, because vegetation change goes on forever. Vegetation will always be variable in space and temporally in a state of flux, now or at some future time. The composition of stands can never reach a stable end-point. Therefore, it is probably best to abandon the term climax, because of its chequered history and the impossibility of defining a preconceived ideal form of vegetation for a region.

The new vegetation change theory de-emphasizes 'linear', sequential replacement series (succession) as the main organizing principle of vegetation dynamics. It allows due emphasis to be placed on all of the various styles of change which are apparent in vegetation, as well as some other phenomena which do not rest easily in the 'succession to climax' theory.

In the new theory, also, there is no difficulty in considering any particular type of vegetation to be the most mature under the prevailing circumstances. There is no implication of a requirement that sequences should end up with the tallest life-forms (or the most productive, most diverse communities) for a region. Each stand is taken on its merits, according to the relationships of the array of species with different lifestyles to prevailing habitat conditions. For example, raised peat bogs or heathland may be the mature vegetation on some sites in an otherwise forested region. Sparse lichen cover may be the mature vegetation on the rock outcrops of an otherwise shrub-covered region. Where change is incessant, e.g. in some grasslands, tundra or salt marshes, no stand of vegetation, particularly, may be differentiated as being more mature than any other stand.

A further suggestion is that the terms primary and secondary succession do not have much explanatory value, and may be abandoned. The problem with the concept of primary succession is to know what is to be taken into it. The problem goes back to Clements' original definition, and it is inherent in the deterministic approach. Succession beginning on a new site may proceed towards a relatively mature steady-state condition. However, over a given period, continual autogenically driven soil changes may eventually create disequilibrium, initiating further change until one or more new steady-state conditions is reached. Does this still represent a primary succession? Of course a 'primary' sequence may be interrupted at any time by allogenic forces which cause slow change. Alternatively there may be total or partial destruction of much of the vegetation, followed by recolonization and sequential development, leading to mature vegetation which may or may not resemble that which was present before the disturbance.

When does a primary succession become a secondary succession? Presumably as soon as a disturbance deflects the original primary sequence. There has always been difficulty with the original criteria for what constitutes the 'primariness', or 'secondariness' of some sites. For example, when a landslide occurs, much new, sterile substrate may be present in the debris, but some propagules or detritus from the previous vegetation of the area may remain. Mixed 'primary' and 'secondary' conditions then pertain.

Some ecologists have had difficulty in deciding what actually constitutes succession. Does the term apply to single plant replacements, or gap-filling, which seem to be cyclic in nature? Does it apply to allogenically caused shifts in composition? There is so much confusion in the use of the term that it seems best to abandon succession in favour of the non-committal **vegetation change**.

Vegetation change phenomena are, then, seen as complex continuous variation patterns arising from the operation, in nature, of the causal Modes 1–5, interacting with plant lifestyles and allogenically or autogenically determined and multidimensional site factors. They are explained by reference to the different properties of individual plant species, in competition with one another, interacting with variable environmental conditions. Complex descriptive terminology for vegetation change phenomena seems to be unnecessary.

REQUIREMENTS FOR FURTHER RESEARCH

A lack of empirical data on dynamic processes in vegetation taken from actual censuses over time and long-term experiments in stands of vegetation has been a major hindrance to progress. Population census data, preferably in long runs of a century or more, are required for forests. Some other kinds of vegetation may furnish useful data from shorter census (e.g. the British heath grassland studies of A. S. Watt (1960) and chalk grassland studies of Grubb (1976) and Grubb *et al.* (1982)), but usually the need is for long runs. The important need to accumulate such information has been emphasized recently, along with the need to base theory on experience from a wide range of vegetation types (Austin 1981, Usher 1981 and other papers in the symposia edited by Knapp (1974a), Beeftink (1980), Van der Maarel (1980), Poissonet *et al.* (1981) and J. White (1985)). Some valuable data from long-term forest studies are accumulating (e.g. Falinski 1986, Christensen & Peet 1981, G. R. Stephens & Waggoner 1980, Parker & Swank 1982).

Ideally the gathering of long-term plant population census data should be coupled with very careful assessment of the complex environmental interactions of all kinds which affect species-by-species replacements. The ultimate causes of population changes must be a prime concern of ecologists. Studies of this nature were initiated during the International Biological Programme, e.g. the German Solling project (Ellenberg 1971), the Belgian (Duvigneaud 1971a) and American (Reichle 1981) forest projects and many others. The emphasis in the IBP work was on ecosystem productivity, energy flow and nutrient cycles. Greater attention needs to be focused on plant–plant interactions and feedback processes associated with temporal change in plant population composition. Very valuable work of this kind has been done in the Hubbard Brook study (Bormann & Likens 1979a), in Alaskan sequences (Van Cleve & Viereck 1981) and in forest studies in Poland (Falinski 1986).

Among the plant–environment interactions requiring further study, in relation to vegetation change, are the roles of microbial and animal components (in ecosystem nutrient relationships and otherwise). The significance (if any) of phytotoxicity in plant species interactions must be very carefully unravelled.

Of extreme importance is the requirement to obtain details of the relative lifestyles, or niche dimensions, of the species participating in vegetation change sequences. It is only by understanding thoroughly the niche differences of each species in the local array, with respect to reproductive potential, seed dispersal, seed longevity, requirements for seedling establishment, productivity rates, nutrient requirements, resistance to herbivory or disease, tolerance ranges for temperature, light and water conditions, adult longevity, causes of death, etc., that sense can ultimately be made of vegetation change phenomena. The great diversity of many floras, as well as habitat diversity, and the inherent difficulty of experimenting with long-lived plants, makes systematization of these variables a formidable task (cf. Grubb 1985). It is evident that studies on permanent sites in a range of vegetation types and concurrent detailed environmental recording are needed for long periods – a century or more, if possible.

THE PAST, PRESENT AND FUTURE OF NATURAL VEGETATION AND HUMAN RELATIONSHIPS WITH IT

The human species has had an enormous direct and indirect modifying influence on the vegetation of the world, so it is reasonable to contrast vegetation that is natural (i.e. unmanaged by humans in any way) or semi-natural (only lightly managed, so that natural processes prevail) with vegetation which has been profoundly altered in various ways by human activities, or actually takes the form of human artefacts (plantations, crops, pastures and gardens).

As human population and economic expansion place ever-increasing demands on land and other natural resources, the natural and semi-natural vegetation of the world shrinks. Ten thousand years ago, except for the effects of fire in places of

487

human settlement, most of the world's vegetation was pristine, affected only by natural processes. Now very little natural vegetation remains. We may examine the evidence from New Zealand, one of the best-documented examples of human impact on natural vegetation, as well as one of the most recently occupied regions. Until 1000 years ago the islands were uninhabited by people. They were covered mainly by dense, evergreen, temperate forest, dominated by angiosperm and gymnosperm tree species. On the drier east the forest merged into tall scrub. The vertebrate fauna consisted mainly of birds; the only mammals were species of bats and marine species such as seals, whales and dolphins. Natural fires were rare, but sporadic volcanic outbursts affected the central part of North Island and tropical cyclones occasionally disturbed the forests.

The islands were discovered by Polynesian seafarers with a stone-tool culture, who brought some food plants, rats and dogs. They soon modified the vegetation, especially with the help of fire. By the early 19th century, when European settlement and a great influx of foreign plants and animals began, some 50 per cent of the original forest vegetation had already been destroyed and most other kinds of vegetation had been burnt at least once.

The European settlers mounted a second assault on the vegetation. By now only about 25 per cent remains of the area covered by native forest a millennium ago. In the most intensively farmed and settled areas only small pockets of native vegetation (if any) survive. Natural vegetation is extensive now only in the remote, rugged south-west of South Island, on Stewart Island and in the most mountainous areas. Even there, introduced herbivorous mammals of many kinds make inroads. New Zealand is truly a paradise disturbed.

Much the same course of events, with different timing, has affected most other parts of the world. In East Africa the changes began, perhaps, several hundred thousand years ago; in Australia at least 40 000 years ago; in Northern Europe (where the vegetation was recovering from the last ice age) 14 000 years ago; in the Mediterranean region, Middle East, India and China considerably earlier; possibly 20 000 years ago in North America and, in some tropical forest areas, in the past few hundred years. Bulldozers, chainsaws and herbicides hasten the conversion of natural vegetation, but slash-and-burn agriculture, wood-gathering, fire and goats are also powerful modifying forces in less-developed regions. Little remains of the original natural vegetation of Africa, Europe, Asia or North America, except in areas that are essentially hostile to human occupation – the deserts, high mountains and tundra. Nevertheless, if damaged vegetation is left alone, it recovers some of its original qualities.

The decline of natural or otherwise unmanaged vegetation seems likely to continue unless political decisions are taken to retain areas of land with few or no permanent human inhabitants. The rationale for a deliberate policy of preservation of wild lands, in competition with economic pressures of all kinds, is that there are many important values inherent in untamed landscape and vegetation. They include:

- Aesthetic and spiritual benefits to people deriving from natural beauty (part of our cultural heritage).
- Natural museums provide habitat and havens for wildlife and for plants,

microbes and smaller animals. They also contain examples of interesting landscape and soil types.

- Recreational areas.
- Benefits to the tourist industry.
- Sources of stock for hitherto unknown practical purposes: e.g. plants for crossbreeding or genetic engineering with cultivars, or for amenity planting, new drugs, antibiotics or other useful products.
- Areas for the preservation of particular types of natural system for scientific study, and to provide baselines against which changes in managed systems can be measured.

The future for the natural or otherwise unmanaged vegetation of many regions seems bleak unless strong voices from ecologists and environmentalists and others can foster a conservation ethic. Absolute necessities are the development of a world population policy and the rationalization of economic systems, because human population growth, together with untrammelled economic growth, are the ultimate causes of the problem of the decline of our nature heritage and the extinction of species.

If the decline of natural systems is not halted, the world will be left with little but human artefact, and, in less-intensively managed areas, monotonously uniform vegetation. The uplifting wildland values which have helped to mould human culture will be confined to a few relatively small parks and reserves, and the world will be the poorer. If that happens, then this book will have been written to no avail.

References

Aaby, B. & H. Tauber 1975. Rates of peat formation in relation to degree of humification and local environment as shown by studies of a raised bog in Denmark. *Boreas* **4**, 1–17.

Aarsen, L. W., R. Turkington & P. B. Carver 1979. Neighbour relationships in grass-legume communities II. Temporal stability and community evolution. *Canadian Journal of Botany* **57**, 2695–703.

Adams, M. S., B. R. Strain & M. S. Adams 1970. Water-repellent soil, site and annual plant cover in a desert shrub community of southeastern California. *Ecology* **51**, 696–700.

Agricultural Research Service 1970. *Selected weeds of the United States.* Agriculture Handbook No. 366. US Department of Agriculture.

Agrios, G. N. 1969. *Plant pathology.* New York: Academic Press.

Ahlgren, I. F. 1974a. The effect of fire on soil organisms. In *Fire and ecosystems*, T. T. Kozlowski & C. E. Ahlgren (eds) 47–72. New York: Academic Press.

Ahlgren, C. E. 1974b. Effects of fires on temperate forests: North Central United States. In *Fire and ecosystems*, T. T. Kozlowski & C. E. Ahlgren (eds), 195–225. New York: Academic Press.

Ahlgren, I. F. & C. E. Ahlgren 1960. Ecological effects of forest fires. *Botanical Review* **26**, 483–533.

Albertson, F. W. & G. W. Tomanek 1965. Vegetation changes during a 30-year period in grassland communities near Hays, Kansas. *Ecology* **46**, 714–20.

Albertson, F. W., G.W. Tomanek & A. Riegel 1957. Ecology of drought cycles and grazing intensity of grasslands of Central Great Plains. *Ecological Monographs* **27**, 27–44.

Alcock, M. B. 1964. The physiological significance of defoliation on subsequent regrowth of grass-clover mixtures and cereals. In *Grazing in terrestrial and marine environments*, D. J. Crisp (ed.), 25–41. 4th Symposium of the British Ecological Society. Oxford: Blackwell.

Alexander, R. D. & G. Borgia 1978. Group selection, altruism and the levels of organization of life. *Annual Review of Ecology and Systematics* **9**, 449–74.

Allee, W. C., A. E. Emerson, O. Park, T. Park & K. P. Schmidt 1949. *Principles of animal ecology.* Philadelphia: Saunders.

Allen, E. B. & R. T. T. Forman 1976. Plant species removal and old-field community structure and stability. *Ecology* **57**, 1233–43.

Anderson, A. B. 1981. White-sand vegetation of Brazilian Amazonia. *Biotropica* **13**, 199–210.

Anderson, D. J. 1967. Cyclical succession in *Dryas* communities. *Journal of Ecology* **55**, 629–35.

Anderson, D. J. 1982. Environmentally adaptive traits in arid zone plants. In *Evolution of the flora and fauna of arid Australia*, W. R. Barker & P. J. Greenslade (eds), 133–9. Frewville: Peacock Publications.

Anderson, M. C. 1966. Ecological groupings of plants. *Nature* **212**, 54–6.

Angevine, M. W. & B. F. Chabot 1979. Seed germination syndromes in higher plants. In *Topics in plant population biology*, O. T. Solbrig, S. Jain, G. B. Johnson & P. H. Raven (eds), 188–231. New York: Columbia University Press.

Antonovics, J., A. D. Bradshaw & R. G. Turner 1971. Heavy metal tolerance in plants. *Advances in Ecological Research* **7**, 1–85.

Armesto, J. J., J. D. Mitchell & C. Villagran 1986. A comparison of spatial patterns of trees in some tropical and temperate forests. *Biotropica* **18**, 1–11.

Ashton, D H. & R. N. Webb 1977. The ecology of granite outcrops at Wilson's Promontory, Victoria. *Australian Journal of Ecology* **2**, 269–96.

Ashton, D. H. & E. J. Willis 1982. Antagonisms in the regeneration of *Eucalyptus regnans* in the mature forest. In *The plant community as a working mechanism*, E. I. Newman (ed.), 113–28. British Ecological Society Special Publication No. 1. Oxford: Blackwell.

Ashton, P. S. 1969. Speciation among tropical forest trees: some deductions in the light of recent evidence. In *Speciation in tropical environments*, R. H. Lowe-McConnell (ed.), 155–96. London: Linnean Society, Academic Press.

Ashton, P. S. 1977. A contribution of rainforest research to evolutionary theory. *Annals of the Missouri Botanical Garden* **64**, 694–705.

Aston, B. C. 1916. The vegetation of the Tarawera Mountain, New Zealand. *Transactions of the New Zealand Institute* **48**, 304–14.

Atkinson, I. A. E. 1970. Successional trends in the coastal and lowland forest of Mauna Loa and Kilauea volcanoes, Hawaii. *Pacific Science* **24**, 387–400.

Auclair, A. N. & G. Cottam 1971. Dynamics of black cherry (*Prunus serotina*) in Southern Wisconsin oak forests. *Ecological Monographs* **41**, 153–77.

Aughenbaugh, J. E. 1935. Replacement of chestnut in Pennsylvania. *Bulletin of the Pennsylvania Department of Waters* **54**, 1–38.

Augspurger, C. 1983. Offspring recruitment around tropical trees: changes in cohort distance with time. *Oikos* **40**, 189–96.

Augspurger, C. K. 1984a. Light requirements of neotropical tree seedlings: a comparative study of growth and survival. *Journal of Ecology* **72**, 777–95.

Augspurger, C. K. 1984b. Seedling survival of tropical tree species: interactions of dispersal distance, light gaps and pathogens. *Ecology* **65**, 1705–12.

Austin, M. P. 1977. Use of ordination and other multivariate descriptive methods to study succession. *Vegetatio* **35**, 165–75.

Austin, M. P. 1980. An exploratory analysis of grassland dynamics: an example of lawn succession. In *Succession*, E. Van der Maarel (ed.), Symposium on Advances in Vegetation Sciences, Nijmegen, The Netherlands, May 1979, 89–94. The Hague: Junk.

Austin, M. P. 1981. Permanent quadrats: an interface for theory and practice. In *Vegetation dynamics in grasslands, heathlands and Mediterranean ligneous formations*, P. Poissonet, F. Romane, M. P. Austin & E. van der Maarel (eds), 1–10. The Hague: Junk.

Austin, M. P., P. Greig-Smith & P. S. Ashton 1972. The application of quantitative methods to vegetation survey III. A reexamination of rain forest data from Brunei. *Journal of Ecology* **60**, 305–34.

Aweto, A. O. 1981. Secondary succession and soil fertility restoration in south-western Nigeria II. Soil fertility restoration. *Journal of Ecology* **69**, 609–14.

Backer, C. A. 1909. De flora van het eiland Krakatau. *Jaarverslag van den Topographischen Dienst in Nederlands-Indië, Batavia (Java)* 1908.

Backer, C. A. 1929. *The problem of Krakatoa as seen by a botanist*. Van Ingen, Java: Backer.

Bacone, J., F. A. Bazzaz & W. R. Boggess 1976. Correlated photosynthetic responses and habitat factors of two successional tree species. *Oecologia* **23**, 63–74.

Bagnold, R. A. 1971. *Physics of blown sand & desert dunes*. London: Chapman & Hall.

Bakelaar, R. G. & E. G. Odum 1978. Community and population level responses to fertilization in an old-field ecosystem. *Ecology* **59**, 660–5.

Baker, H. G., K. S. Bawa, G. W. Frankie & P. A. Opler 1983. Reproductive biology of plants in tropical forests. In *Tropical rain forest ecosystems*, F. B. Golley (ed.), 183–215. Ecosystems of the World 14A. Amsterdam: Elsevier.

Baker, W. L. 1972. *Eastern forest insects*. United States Department of Agriculture.

Ballard, J. & G. M. Woodwell 1977. The flux of water and nutrients to the ground water under a field to forest sere. In *The belowground ecosystem*, J. K. Marshall (ed.), 185–8. Range Science Department Science Series No. 26. Fort Collins: Colorado State University.

Barber, K. E. 1981. *Peat stratigraphy and climatic change: a palaeoecological test of the theory of cyclic peat bog regeneration*. Rotterdam: Balkema.

Barbour, M. G. 1973. Desert dogma reexamined: root/shoot productivity and plant spacing. *American Midland Naturalist* **89**, 41–57.

Barbour, M. G., J. H. Burk & W. D. Pitts 1980. *Terrestrial plant ecology*. Menlo Park, CA: Benjamin/Cummings.

Barclay-Estrup, P. & C. H. Gimingham 1969. The description and interpretation of cyclical changes in a heath community I. Vegetation change in relation to the *Calluna* cycle. *Journal of Ecology* **57**, 737–58.

Bard, G. E. 1952. Secondary succession on the Piedmont of New Jersey. *Ecological Monographs* **22**, 195–215.

Barden, L. S. 1979. Tree replacement in small canopy gaps of *Tsuga canadensis* forest in the Southern Appalachians, Tennessee. *Oecologia* **44**, 141–2.

Barden, L. S. 1980. Tree replacement in a cove hardwood forest of the Southern Appalachians. *Oikos* **35**, 16–9.

Barker, W. R. & P. J. Greenslade (eds) 1982. *Evolution of the flora and fauna of arid Australia*. Frewville: Peacock Publications.

Barrow, M. D., A. B. Costin & P. Lake 1968. Cyclical changes in an Australian fjaeldmark community. *Journal of Ecology* **56**, 89–96.

Bartholomew, B. 1970. Bare zone between California shrub and grassland communities: the toll of animals. *Science* **170**, 1210–21.

Bartholomew, B. 1971. Role of animals in suppression of herbs by shrubs. *Science* **173**, 462–3.

Baskin, J. M. & C. C. Baskin 1985. The annual dormancy cycle in buried weed seeds: a continuum, *BioScience* **35**, 492–8.

Bazzaz, F. A. 1968. Succession on abandoned fields in the Shawnee Hills, southern Illinois. *Ecology* **49**, 924–36.

Bazzaz, F. A. 1974. Ecophysiology of *Ambrosia artemisiifolia*: a successional dominant. *Ecology* **55**, 112–9.

Bazzaz, F. A. 1979. The physiological ecology of plant succession. *Annual Review of Ecology and Systematics* **10**, 351–71.

Bazzaz, F. A. 1987. Experimental studies on the evolution of niche in successional plant populations. In *Colonization, succession and stability*, A. J. Gray, M. J. Crawley & P. J. Edwards (eds), 245–72. 26th Symposium, British Ecological Society. Oxford: Blackwell.

Bazzaz, F. A. & S. T. A. Pickett 1980. The physiological ecology of tropical succession: a comparative review. *Annual Review of Ecology and Systematics* **11**, 287–310.

Beals, E. W. 1969. Vegetational change along altitudinal gradients. *Science* **165**, 981–5.

Beatley, J. C. 1974. Phenological events and their environmental triggers in Mojave Desert ecosystems. *Ecology* **55**, 856–63.

Beatley, J. C. 1980. Fluctuations and stability in climax shrub and woodland vegetation of the Mojave, Great Basin and Transition Deserts of Southern Nevada. *Israel Journal of Botany* **28**, 149–68.

Beeftink, W. G. 1979. The structure of salt marsh communities in relation to environmental disturbances. In *Ecological processes in coastal environments*, 77–93. R. L. Jefferies & A. J. Davy (eds). 19th Symposium of the British Ecological Society. Oxford: Blackwell.

Beeftink, W. G. (ed.) 1980. *Vegetation dynamics*. Proceedings of the Second Symposium of the Working Group on Succession Research on Permanent Plots. Verseke, 1–3 October 1975. The Hague: Junk.

Benninghof, W. S. 1952. Interaction of vegetation and soil frost phenomena. *Arctic* **5**, 34–44.

Bieleski, R. L. 1973. Phosphate pools, phosphate transport and phosphate availability. *Annual Review of Plant Physiology* **24**, 225–52.

Billings, W. D. 1938. The structure and development of old field shortleaf pine stands and certain associated physical properties of the soil. *Ecological Monographs* **8**, 437–99.

Billings, W. D. 1941. Quantitative correlations between vegetational changes and soil development. *Ecology* **22**, 448–56.

Billings, W. D. 1974a. Environment: concept and reality. In *Vegetation and environment*, B. R. Strain & W. D. Billings (eds), 9–38. Handbook of Vegetation Science Part 6. The Hague: Junk.

Billings, W. D. 1974b. Arctic and alpine vegetation: plant adaptations to cold summer climates. In *Arctic and alpine environments*, J. D. Ives & R. G. Barry (eds), 403–43. London: Methuen.

Billings, W. D. & H. A. Mooney 1959. An apparent frost hummock-sorted polygon cycle in the alpine tundra of Wyoming. *Ecology* **40**, 16–20.

Bird, E. C. F. & M. M. Barson 1982. Stability of mangrove systems. In *Mangrove ecosystems in Australia*, B. F. Clough (ed.), Canberra: Australian Institute of Marine Science/Australian National University Press.

Birkeland, P. W. & E. E. Larson 1978. *Putnam's geology*, 3rd edn. New York: Oxford University Press.

Birks, H. H. & H. J. B. Birks 1980. *Quaternary palaeoecology*. London: Edward Arnold.

Birks, H. J. B. 1980a. The present flora and vegetation of the moraines of the Klutlan Glacier, Yukon Territory, Canada: a study in plant succession. *Quaternary Research* **14**, 60–86.

Birks, H. J. B. 1980b. Modern pollen assemblages and vegetational history of the moraines of the Klutlan Glacier and its surroundings, Yukon Territory, Canada. *Quaternary Research* **14**, 101–29.

Biswell, H. H. 1974. Effects of fire on chaparral. In *Fire and ecosystems*, T. T. Kozlowski & C. E. Ahlgren (eds), 321–65. New York: Academic Press.

Black, C. A. 1968. *Soil–plant relationships*. New York: Wiley.

Black, M. & J. Edelman 1970. *Plant growth*. London: Heinemann.

Blais, J. R. 1968. Budworm attacks: regional variation in susceptibility, based on outbreak histories. *Forestry Chronicle* **44** (3), 17–23.

Bledsoe, L. J., G. M. van Dyne 1971. A compartment model simulation of secondary succession. In *Systems analysis and simulation in ecology*. B. C. Patten (ed.), 479–511. New York: Academic Press.

Bliss, L. C. 1962. Adaptations of Arctic and alpine plants to environmental conditions. *Arctic* **15**, 117–44.

Bliss, L. C. 1985. Alpine. In *Physiological ecology of North American plant communities*, B. F. Chabot & H. A. Mooney (eds), 41–65. New York: Chapman & Hall.

Bliss, L. C., J. Svoboda & D. I. Bliss 1984. Polar deserts, their plant cover and plant production in the Canadian High Arctic. *Holarctic Ecology* **7**, 305–24.

Boardman, N. K. 1977. Comparative photosynthesis of sun and shade plants. *Annual Review of Plant Physiology* **28**, 355–77.

Boerner, R. E. 1982. Fire and nutrient cycling in temperate ecosystems. *BioScience* **32**, 187–92.

Bold, H. C. 1967. *Morphology of plants*, 2nd edn. New York: Harper & Row.

Bold, H. C. 1977. *The plant kingdom*, 4th edn. Englewood Cliffs, NJ: Prentice–Hall.

Bond, G. 1951. Fixation of nitrogen associated with the root nodules of *Myrica gale*, with special reference to its pH relations and ecological significance. *Annals of Botany* **15**, 447–59.

492

Booth, W. E. 1941. Algae as pioneers in plant succession and their importance in erosion control. *Ecology* **22**, 38–46.

Boring, L. R., C. D. Monk & W. T. Swank 1981. Early regeneration of a clear-cut Southern Appalachian forest. *Ecology* **62**, 1244–53.

Bormann, F. H. 1953. Factors determining the role of loblolly pine and sweetgum in early old-field succession in the Piedmont of North Carolina. *Ecological Monographs* **23**, 339–58.

Bormann, F. H. 1956. Ecological implications of changes in the photosynthetic response of *Pinus taeda* seedlings during ontogeny. *Ecology* **37**, 70–5.

Bormann, F. H. 1966. The structure, function and ecological significance of root grafts in *Pinus strobus* and its ecological implications. *Ecology* **40**, 678–91.

Bormann, F. H. & M. F. Buell 1964. Old age stands of hemlock–northern hardwood forest in Central Vermont. *Bulletin of the Torrey Botanical Club* **91**, 451–65.

Bormann, F. H. & G. E. Likens 1979a. *Pattern and process in a forested ecosystem.* New York: Springer.

Bormann, F. H. & G. E. Likens 1979b. Catastrophic disturbance and the steady state in northern hardwood forests. *American Scientist* **67**, 660–9.

Bormann, F. H., T. G. Siccama, G. E. Likens & R. H. Whittaker 1970. The Hubbard Brook ecosystem study: composition and dynamics of the tree stratum. *Ecological Monographs* **40**, 377–88.

Botkin, D. B., J. F. Janak & J. R. Wallis 1972. Some ecological consequences of a computer model of forest growth. *Journal of Ecology* **60**, 849–72.

Boughey, A. S. 1963. Interactions between animals, vegetation and fire in Southern Rhodesia. *Ohio Journal of Science* **63**, 193–209.

Boughey, A. S., P. E. Munro, J. Meikeljohn, R. M. Strang & M. J. Swift 1964. Antiobiotic reactions between African savanna species. *Nature* **203**, 1302–3.

Bourdeau, P. 1954. Oak seedling ecology determining segregation of species in Piedmont oak-hickory forests. *Ecological Monographs* **24**, 297–320.

Bourdeau, P. F. & M. L. Laverick 1958. Tolerance and photosynthetic adaptability to light intensity in white pine, red pine, hemlock and *Ailanthus* seedlings. *Forest Science* **4**, 196–207.

Bourliere, F. 1983. Animal species diversity in tropical forests. In *Tropical rain forest ecosystems*, F. B. Golley (ed.), 77–91. Ecosystems of the World 14A. Amsterdam: Elsevier.

Bowen, G. D. 1980. Coping with low nutrients. In *The biology of Australian native plants*, J. S. Pate & A. J. McComb (eds). 33–64. Perth: University of Western Australia Press.

Braun, E. L. 1950. *Deciduous forests of eastern North America.* Philadelphia: Blakiston.

Bray, J. R. 1956. Gap phase replacement in a maple–basswood forest. *Ecology* **37**, 598–600.

Bray, J. R. 1958. Notes toward an ecological theory. *Ecology* **39**, 770–6.

Bray, J. R. & J. T. Curtis 1957. An ordination of the upland forest communities of southern Wisconsin. *Ecological Monographs* **27**, 325–49.

Brereton, A. J. 1971. The structure of the species populations in the initial stages of salt-marsh succession. *Journal of Ecology* **59**, 321–38.

Britton, M. E. 1966. *Vegetation of the Arctic tundra.* Corvallis: Oregon State University Press.

Brokaw, N. V. 1985. Tree falls, regrowth and community structure in tropical forests. In *The ecology of natural disturbance and patch dynamics*, S. T. A. Pickett & A. S. White (eds), 54–77. New York: Academic Press.

Brokaw, N. V. 1987. Gap-phase regeneration of three pioneer tree species in a tropical forest. *Journal of Ecology* **75**, 9–19.

Buckman, H. O. & N. C. Brady 1969. *The nature and properties of soils*, 7th edn. New York: Macmillan.

Buell, M. F. & H. F. Buell 1941. Surface level fluctuation in Cedar Creek Bog, Minnesota. *Ecology* **22**, 317–21.

Buell, M. F. & H. F. Buell 1975. Moat bogs in the Itasca Park area, Minnesota. *Bulletin of the Torrey Botanical Club* **102**, 6–9.

Buell, M. F., H. F. Buell & J. A. Small 1954. Fire in the history of Mettler's Woods. *Bulletin of the Torrey Botanical Club* **81**, 253–5.

Buell, M. F., H. F. Buell & W. A. Reiners 1968. Radial mat growth in Cedar Creek Bog, Minnesota. *Ecology* **49**, 1198–9.

Buell, M. F., H. F. Buell, J. A. Small & T. G. Siccama 1971. Invasion of trees in secondary succession in the New Jersey Piedmont. *Bulletin of the Torrey Botanical Club* **98**, 67–74.

Bunting, B. T. 1967. *The geography of the soil*, 2nd edn. London: Hutchinson.

Buol, S. W., F. D. Hole & R. J. McCracken 1973. *Soil genesis and classification.* Ames: Iowa State University Press.

Burdon, J. J. 1982. Intra-specific diversity in a natural population of *Trifolium repens*. *Journal of Ecology* **68**, 717–36.

Burges, A. & D. P. Drover 1953. The rate of podzol development in sands of the Woy Woy District, New South Wales. *Australian Journal of Botany* **1**, 83–94.

493

Burgess, P. F. 1969. Ecological factors in hill and mountain forests of the states of Malaya. *Malayan Nature Journal* **22**, 136–43.

Burke, W. 1974. Regeneration of podocarps on Mt Tarawera, Rotorua. *New Zealand Journal of Botany* **12**, 219–26.

Burkholder, P. R. 1952. Cooperation and conflict among primitive organisms. *American Scientist* **40**, 601–31.

Burrows, C. J. 1973. The ecological niches of *Leptospermum scoparium* and *L. ericoides* (Angiospermae: Myrtaceae). *Mauri Ora* **1**, 5–12.

Burrows, C. J. 1974. Plant macrofossils from Late-Devensian deposits at Nant Ffrancon, Caernarvonshire. *New Phytologist* **73**, 1003–33.

Burrows, C. J. 1977. Riverbed vegetation. In *Cass: history and science in the Cass District, Canterbury, New Zealand*, C. J. Burrows (ed.), 215–25. Christchurch: Botany Department, University of Canterbury.

Burrows, C. J. & V. L. Burrows 1976. *Procedures for the study of snow avalanche chronology using growth layers of woody plants*. INSTAAR Univ. Colorado, Boulder, Occas. Paper No. 23.

Burrows, C. J. & A. T. Dobson 1972. Mires of the Manapouri–Te Anau lowlands. *Proceedings of the New Zealand Ecological Society* **19**, 75–99.

Burrows, C. J. & D. E. Greenland 1979. An analysis of the evidence for climatic change in New Zealand in the last thousand years: evidence from diverse natural phenomena and from instrumental records. *Journal of the Royal Society of New Zealand* **9**, 321–73.

Burrows, C. J., D. R. McQueen, A. E. Esler & P. Wardle 1979. New Zealand heathlands. In *Heathlands and related shrublands*, R. L. Specht (ed.), 339–64. Ecosystems of the World 9A. Amsterdam: Elsevier.

Bykov, B. A. 1974. Fluctuations in the semi desert and desert vegetation of the Turanian Plain. In *Vegetation dynamics*, R. Knapp (ed.), 243–52. Handbook of Vegetation Science. Part 8. The Hague: Junk.

Cain, S. A. 1935. Studies on virgin hardwood forest III. Warrens Woods, a beech–maple climax forest in Berrien County, Michigan. *Ecology* **16**, 500–13.

Caldwell, M. M., R. W. White, R. T. Moore & L. B. Camp 1977. Carbon balance, productivity and water use of cold-winter desert shrub communities dominated by C_3 and C_4 species. *Oecologia* **29**, 275–300.

Caldwell, M. M., J. H. Richards, J. H. Manwaring & D. M. Eissenstat 1987. Rapid shifts in phosphate acquisition show direct competition between neighbouring plants. *Nature* **327**, 615–6.

Campbell, D. H. 1909. The new flora of Krakatoa. *American Naturalist* **43**, 449–60.

Campbell, E. O., J. C. Heine & W. A. Pullar 1973. Identification of plant fragments and pollen from peat deposits in Rangitaiki Plains and Maketu Basins. *New Zealand Journal of Botany* **11**, 317–30.

Casparie, W. A. 1972. Bog development in Southeastern Drenthe (The Netherlands). *Vegetatio* **25**, 1–272.

Castri, F. di, D. W. Goodall & R. L. Specht (eds) 1981. *Mediterranean-type shrublands*. Ecosystems of the World II. Amsterdam: Elsevier.

Cates, R. G. & G. H. Orians 1975. Successional status and the palatability of plants to generalized herbivores. *Ecology* **56**, 410–8.

Chandler, R. F. 1942. The time required for a podsol profile formation as evidenced by the Mendenhall glacial deposits near Juneau, Alaska. *Proceedings of the Soil Science Society of America* **7**, 454–9.

Chapin, F. S. 1980. Mineral nutrition of wild plants. *Annual Review of Ecology and Systematics* **11**, 233–60.

Chapman, A. G. 1942. Forests of the Illinoian till plain of South-eastern Indiana. *Ecology* **23**, 189–98.

Chapman, R. F. 1976. *A biology of locusts*. London: Edward Arnold.

Chapman, V. J. 1966. *Salt marshes and salt deserts of the world*, 2nd edn. London: Leonard Hill.

Chapman, V. J. (ed.). 1977. *Wet coastal ecosystems*. Ecosystems of the World 1. Amsterdam: Elsevier.

Chesson, P. L. & R. R. Warner 1981. Environmental variability promotes coexistence in lottery competitive systems. *American Naturalist* **117**, 923–43.

Chew, R. M. 1974. Consumers as regulators of ecosystems: an alternative to energetics. *Ohio Journal of Science* **74**, 359–70.

Chippindall, L. C. & A. O. Crook 1976–1978. *240 Grasses of southern Africa*. Salisbury: M. O. Collins.

Christensen, N. L. 1977. Change in structure, pattern and diversity associated with climax forest maturation in Piedmont, North Carolina. *American Midland Naturalist* **97**, 176–88.

Christensen, N. L. & C. H. Muller 1975a. Effects of fire on factors controlling plant growth in *Adenostoma* chaparral. *Ecological Monographs* **45**, 29–55.

Christensen, N. L. & C. H. Muller 1975b. Relative importance of factors controlling germination and seedling survival in *Adenostoma* chaparral. *American Midland Naturalist* **93**, 71–8.

Christensen, N. L. & R. K. Peet 1981. Secondary forest succession on the North Carolina Piedmont. In *Forest succession: concepts and application*, D. C. West, H. H. Shugart & D. B. Botkin (eds), 230–45. New York: Springer.

Churchill, E. D. & H. C. Hanson 1958. The concept of climax in Arctic and alpine vegetation. *Botanical Review* **24**, 127–91.

494

Clark, D. A. & D. B. Clark 1984. Spacing dynamics of a tropical rain forest tree: evaluation of the Janzen–Connell model. *American Naturalist* **124**, 769–88.

Clarkson, B. 1980. *Vegetation of Mt Tarawera*. M.Sc. Thesis, University of Waikato, New Zealand.

Clausen, J. C. 1962. *Stages in the evolution of plant species*. New York: Hafner.

Clausen, J. C., D. D. Keck & W. M. Hiesey 1948. *Experimental studies on the nature of species III*. Carnegie Inst., Washington, Publ. No. 581.

Clayton, L. 1965. Karst topography on stagnant glaciers. *Journal of Glaciology* **5**, 107–12.

Clements, F. E. 1904. The development and structure of vegetation. *Botanical Survey of Nebraska* **3**, 1–175.

Clements, F. E. 1905. *Research methods in ecology*. Lincoln, NB: University Publishing Co.

Clements, F. E. 1910. Life history of lodgepole burn forests. *United Stated Department of Agriculture Forest Service Bulletin* **79**, 1–56.

Clements, F. E. 1916. *Plant succession: an analysis of the development of vegetation*. Carnegie Inst., Washington, Publ. No. 242.

Clements, F. E. 1928. *Plant succession and indicators*. New York: Wilson.

Clements, F. E. 1936. Nature and structure of the climax. *Journal of Ecology* **24**, 252–84.

Cline, A. C. & S. H. Spurr 1942. The virgin upland forest of central New England. *Harvard Forest Bulletin* **21**, 1–58.

Clymo, R. S. 1963. Ion exchange in *Sphagnum* and its relation to bog ecology. *Annals of Botany* (n.s.) **27**, 309–24.

Clymo, R. S. 1965. Experiments on breakdown of *Sphagnum* in two bogs. *Journal of Ecology* **53**, 747–58.

Cockayne, L. 1928. *The vegetation of New Zealand*, 2nd edn. Leipzig: Engelmann.

Coile, T. S. 1940. Soil changes associated with loblolly pine succession on abandoned agricultural land of the Piedmont Plateau. *Duke University School of Forestry Bulletin* No. 5, 85–98.

Cole, D. W. & M. Rapp 1981. Elemental cycling in forest ecosystems. In *Dynamic properties of forest ecosystems*, D. E. Reichle (ed.), 341–409. I.B.P. 23. London: Cambridge University Press.

Coley, P. D. 1983. Herbivory and defensive characteristics of tree species in a lowland tropical forest. *Ecological Monographs* **53**, 209–33.

Colinvaux, P. A. 1973. *Introduction to ecology*. New York: Wiley.

Collier, B., G. W. Cox, A. W. Johnson & P. C. Miller 1973. *Dynamic ecology* Englewood Cliffs, NJ: Prentice-Hall.

Collins, S. 1961. Benefits to understorey from canopy defoliation by gypsy moth larvae. *Ecology* **42**, 836–8.

Collins, S. L., J. L. Vankat & J. V. Perino 1979. Potential tree species dynamics in the arbor–vitae association of Cedar Bog, a west-central Ohio fen. *Bulletin of the Torrey Botanical Club* **106**, 290–8.

Colwell, R. K. & D. J. Futuyma 1971. On the measurement of niche breadth and overlap. *Ecology* **52**, 567–76.

Connell, J. H. 1978. Diversity in tropical rain forests and coral reefs. *Science* **199**, 1302–10.

Connell, J. H. 1979. Tropical rain forests and coral reefs as open, non-equilibrium systems. In *Population dynamics*, R. M. Anderson, B. D. Turner & L. R. Taylor (eds), 141–64. 20th Symposium British Ecological Society. Oxford: Blackwell.

Connell, J. H. & E. Orias 1964. The ecological regulation of species diversity. *American Naturalist* **98**, 399–414.

Connell, J. H. & R. O. Slatyer 1977. Mechanisms of succession in natural communities and their role in community stability and organization. *American Naturalist* **111**, 1119–44.

Conway, V. M. 1949. The bogs of central Minnesota. *Ecological Monographs* **19**, 173–206.

Conway, V. M. 1954. Stratigraphy and pollen analysis of southern Pennine blanket peats. *Journal of Ecology* **42**, 117–47.

Cook, R. E. 1980. The biology of seeds in the soil. In *Demography and evolution in plant populations*, O. T. Solbrig (ed.), 107–29. Oxford: Blackwell.

Cooper, A. W. 1981. Above-ground biomass accumulation and net primary production during the first 70 years of succession in *Populus grandidentata* stands on poor sites in northern Lower Michigan. In *Forest succession: concepts and application*, D. C. West, H. H. Shugart & D. B. Botkin (eds), 339–60. New York: Springer.

Cooper, K. M. 1982. Mycorrhizal fungi – their role in plant establishment. *International Plant Propagators Society Combined Proceedings for 1982*.

Cooper, R. & E. D. Rudolph 1953. The role of lichens in soil formation and plant succession. *Ecology* **34**, 805–7.

Cooper, W. S. 1913. The climax forest of Isle Royale, Lake Superior, and its development. *Botanical Gazette* **55**, 1–44, 115–40, 189–235.

Cooper, W. S. 1916. Plant succession in the Mt Robson region, British Columbia, *Plant World* **19**, 211–38.

Cooper, W. S. 1923a, b, c. The recent ecological history of Glacier Bay, Alaska I, II, III. *Ecology* **4**, 93–128, 223–46, 355–65.

Cooper, W. S. 1926. The fundamentals of vegetational change. *Ecology* **7**, 391–413.

Cooper, W. S. 1931. A third expedition to Glacier Bay, Alaska. *Ecology* **12**, 61–95.

Cooper, W. S. 1937. The problem of Glacier Bay, Alaska, a study of glacier variations. *Geographical Review* **27**, 37–62.

Cooper, W. S. 1939. A fourth expedition to Glacier Bay, Alaska. *Ecology* **20**, 130–55.

Cooper, W. S. 1942. An isolated colony of plants on a glacier-clad mountain. *Bulletin of the Torrey Botanical Club* **69**, 429–33.

Corré, J. J. 1979. L'équilibre des biocénoses végétales salées en Basse Camargue. In *Ecological processes in coastal environments*, R. L. Jefferies & A. J. Drury (eds), 65–76. 19th Symposium of the British Ecological Society. Oxford: Blackwell.

Coupland, R. T. 1958. The effects of fluctuations in weather upon the grasslands of the Great Plains. *Botanical Review* **24**, 273–317.

Coupland, R. T. 1974. Fluctuations in North American grassland vegetation. In *Vegetation dynamics*, R. Knapp (ed.), 233–42. Handbook of Vegetation Science, Part 8. The Hague: Junk.

Cowles, H. C. 1899a, b, c, d. The ecological relations of the vegetation on the sand dunes of Lake Michigan. Parts 1, 2, 3, 4. *Botanical Gazette* **27**, 95–117, 167–202, 281–308, 361–91.

Cowles, H. C. 1901. The physiographic ecology of Chicago and vicinity. *Botanical Gazette* **31**, 73–108, 145–82.

Cowles, H. C. 1910. The fundamental causes of succession among plant associations. *Report of the British Association for the Advancement of Science* **1909**, 668–70.

Cowles, H. C. 1911. The causes of vegetative cycles. *Botanical Gazette* **51**, 161–83.

Cox, T. L., W. F. Harris, B. S. Ausmus & N. T. Edwards 1977. The role of roots in biogeochemical cycles in eastern deciduous forests. In *The belowground ecosystem: a synthesis of plant-associated processes*, J. K. Marshall (ed.), 321–6. Range Science Department, Science Series No. 26. Fort Collins: Colorado State University.

Crawley, M. J. 1983. *Herbivory*. Oxford: Blackwell Scientific.

Crawley, M. J. 1985. Reduction of oak fecundity by low-density herbivore populations. *Nature* **314**, 163–4.

Crawley, M. J. 1986. The structure of plant communities. In *Plant ecology*, M. J. Crawley (ed.), 1–50. Oxford: Blackwell.

Crocker, R. L. & B. A. Dickson 1957. Soil development on the recessional moraines of the Herbert and Mendenhall glaciers, South-East Alaska. *Journal of Ecology* **45**, 169–85.

Crocker, R. L. & J. Major 1955. Soil development in relation to vegetation and surface age, Glacier Bay, Alaska. *Journal of Ecology* **43**, 427–48.

Cromack, K. 1981. Below-ground processes in forest succession. In *Forest succession: concepts and application*, D. C. West, H. H. Shugart & D. B. Botkin (eds), 361–73. New York: Springer.

Crouch, G. L. 1976. Wild animal damage to forests in the United States and Canada. *Proceedings of XVI IUFRO World Congress, Oslo, Norway* 468–78.

Crow, T. R. 1980. A rainforest chronicle – a 30 year record of change in structure and composition at El Verde, Puerto-Rico. *Biotropica* **12**, 42–55.

Crutchfield, J. P., J. D. Farmer, N. H. Packard & R. S. Shaw 1986. Chaos. *Scientific American* **255** (December), 38–49.

Cumming, D. M. 1982. The influence of large herbivores on savanna structure in Africa. In *Ecology of tropical savannas*, B. J. Huntley & B. H. Walker (eds), 217–45. Heidelberg: Springer.

Curtis, J. T. 1959. *The vegetation of Wisconsin*. Madison: University of Wisconsin Press.

Curtis, J. T. & G. Cottam 1950. Antibiotic and autotoxic effects in prairie sunflower. *Bulletin of the Torrey Botanical Club* **77**, 187–91.

Dachnowski, A. 1912. The successions of vegetation in Ohio lakes and peat deposits. *Plant World* **15**, 25.

Dahl, E. 1956. *Rondane: mountain vegetation in south Norway and its relation to the environment*. Oslo: Aschehoug.

Dainty, J. 1979. The ionic and water relations of plants which adjust to a fluctuating saline environment. In *Ecological processes in coastal environments*, R. L. Jefferies & A. J. Davy (eds), 201–9. 19th Symposium of the British Ecological Society. Oxford: Blackwell.

Daly, G. T. 1966. Nitrogen fixation by nodulated *Alnus rugosa*. *Canadian Journal of Botany* **44**, 1607–21.

Damman, A. W. H. 1971. Effect of vegetation changes on the fertility of a Newfoundland forest site. *Ecological Monographs* **41**, 253–70.

Danielson, A. 1970. Pollen analytical late Quaternary studies in the Ra district of Ostfold, South-east Norway. *Arbok for Universitet i Bergen: Matematisk Naturvitenskapelig Serie* **1969**, No. 14.

Dansereau, P. 1957. *Biogeography: an ecological perspective*. New York: Ronald Press.

Dansereau, P. 1958. A universal system for recording vegetation. In *Contributions de l'institut botanique de l'Université de Montréal*, **72**, 1–58.

Daubenmire, R. F. 1936. The 'Big Woods' of Minnesota: its structure and relation to climate, fire and soils. *Ecological Monograph* **6**, 235–67.

Daubenmire, R. 1968a. Ecology of fire in grasslands. *Advances in Ecological Research* **5**, 209–66.

Daubenmire, R. 1968b. *Plant communities*. New York: Harper & Row.

Davis, J. H. 1940. *The ecology and geologic role of mangroves in Florida*. Carnegie Inst. Washington, Publ. No. 517, 303–412.

Davis, M. B. 1981. Quaternary history and the stability of forest communities. In *Forest succession: concepts and applications*, D. C. West, H. H. Shugart & D. B. Botkin (eds), 132–53. New York: Springer.

Davis, R. M. & J. E. Cantlon 1969. Effect of size of area open to colonization on species composition in early old-field succession. *Bulletin of the Torrey Botanical Club* **96**, 660–73.

Day, G. M. 1953. The Indian as an ecological factor in the northeastern forest. *Ecology* **34**, 329–46.

De Angelis, D. L. 1980. Energy flow, nutrient cycling and ecosystem resilience. *Ecology* **61**, 764–71.

Decker, H. F. 1966. Plants. In *Soil development and ecological succession in a deglaciated area of Muir Inlet: South-east Alaska*, A. Mirsky (ed.), 73–95. Ohio State University Institute of Polar Studies, Report No. 20, Columbus, Ohio.

Degn, H., A. V. Holden & L. F. Olsen 1987. *Chaos in biological systems*. New York: Plenum Press.

Denaeyer-de Smet, S. 1971. Tenuers en éléments biogènes des tapis végétaux dans les forêts caducifoliées d'Europe. In *Productivity of forest ecosystems*, P. Duvigneaud (ed.), 515–25. Paris: UNESCO.

Denslow, J. S. 1980a. Gap partitioning among tropical rain-forest trees. *Biotropica* (Supplement) **12**, 47–55.

Denslow, J. S. 1980b. Patterns of plant species diversity under different plant disturbance regimes. *Oecologia* **46**, 18–21.

Denslow, J. S. 1987. Tropical rainforest gaps and tree species diversity. *Annual Review of Ecology and Systematics* **18**, 431–51.

De Vos, A. 1969. Ecological conditions affecting the production of wild herbivorous mammals on grasslands. *Advances in Ecological Research* **6**, 137–84.

De Vries, D. M. 1953. Objective combinations of species. *Acta Botanica Neerlandica* **1**, 497–9.

Diamond, J. M. 1975. Assembly of species communities. in *Ecology and evolution of communities*, M. L. Cody & J. M. Diamond (eds), 342–444. Cambridge, MA: Belknap Press.

Diamond, J. M. 1986. Carnivore dominance and herbivore coexistence in Africa. *Nature* **320**, 112.

Diamond, J. M. & T. J. Case 1986. *Community ecology*. New York: Harper & Row.

Dickinson, C. H. & G. J. Pugh 1974. *Biology of plant litter decomposition*. London: Academic Press.

Dietrich, R. V. & B. J. Skinner 1979. *Rocks and rock minerals*. New York: Wiley.

Diller, O. D. 1935. The relation of temperature and precipitation to the growth of beech in northern Indiana. *Ecology* **16**, 72–81.

Dix, R. L. 1957. Sugar maple in forest succession at Washington, D.C. *Ecology* **38**, 663–5.

Dix, R. L. 1964. A history of biotic and climatic changes within the North American grassland. in *Grazing in terrestrial and marine environments*, D. J. Crisp (ed.), 71–89. 4th Symposium of the British Ecological Society. Oxford: Blackwell.

Dobzhansky, T. 1950. Evolution in the tropics. *American Scientist* **38**, 209–21.

Docters van Leeuwen, W. M. 1936. Krakatau, 1883 to 1933. *Annales du Jardin Botanique de Buitenzorg* **46–47**, 1–507.

Doty, M. S. 1967. Contrasts between the pioneer populating process on land and shore. *Bulletin of the Southern California Academy of Science* **66**, 175–94.

Druce, A. P. & C. Ogle 1978. *Vascular plants of Mt Tarawera*. Photocopied handwritten list.

Drury, W. H. 1956. Bog flats and physiographic processes in the upper Kuskokwim River region, Alaska. *Contributions of the Gray Herbarium, Harvard* **178**, 1–239.

Drury, W. H. & I. C. Nisbet 1973. Succession. *Journal of the Arnold Arboretum* **54**, 331–68.

Dunbar, M. J. 1972. *Ecological development in polar regions. A study in evolution*. Englewood Cliff, NJ: Prentice-Hall.

Du Rietz, G. E. 1919. Review of Clement's 'Plant Succession'. *Svensk Botaniska Tidskrift* **13**, 117–4.

Duvigneaud, P. (ed.) 1971a. *Productivity of forest ecosystems*. Paris: UNESCO.

Duvigneaud, P. 1971b. Concepts sur la productivité primaire des écosystèmes forestiers. In *Productivity of forest ecosystems*, P. Duvigneaud (ed.), Paris: UNESCO.

Duvigneaud, P. & S. Denaeyer-de Smet 1970. Biological cycling of minerals in temperate deciduous forests. In *Analysis of temperate forest ecosystems*, D. E. Reichle (ed.), 199–225. Ecological Studies I. Berlin: Springer.

Duvigneaud, P. & S. Denaeyer-de Smet 1971. Cycle des éléments biogènes dans les écosystèmes forestiers d'Europe. In *productivity of forest ecosystems*, P. Duvigneaud (ed.), 527–42. Paris: UNESCO.

Duvigneaud, P. & P. Kestemont 1977. *Productivité biologique en Belgique*. SCOPE Travaux de la Section Belge du Programme Biologique International.

Duvigneaud, P. P. Kestemont & P. Ambroes 1971. Productivité primaire des forêts tempérées d'essences feuillues caducifoliées en Europe occidentale. In *Productivity of forest ecosystems* P. Duvigneaud (ed.), 259–70. Paris: UNESCO.

Edwards, C. A., D. E. Reichle & D. A. Crossley 1970. The role of soil invertebrates in turnover of organic matter and nutrients. In *Analysis of temperate forest ecosystems*, D. E. Reichle (ed.), 147–72. Berlin: Springer.

Eggler, W. A. 1971. Quantitative studies of vegetation on sixteen young lava flows on the island of Hawaii. *Tropical Ecology* **12**, 66–100.

Egler, F. E. 1954. Vegetation science concepts. I. Initial floristic composition – a factor in old-field vegetation development. *Vegetatio* **4**, 412–7.

Egner, H. & E. Eriksson 1955. Current data on the chemical composition of air and precipitation. *Tellus* **7**, 134–9.

Ehleringer, J. 1985. Annuals and perennials of warm deserts. In *Physiological ecology of North American plant communities*, B. F. Chabot & H. A. Mooney (eds), 162–80. New York: Chapman & Hall.

Ehrlich, P. R. & P. H. Raven 1969. Differentiation of populations. *Science* **165**, 1228–32.

Einarsson, E. 1967. The colonization of Surtsey, the new volcanic island, by vascular plants. *Aquilo* (*Botany*) **6**, 172–82.

Ellenberg, H. 1971. Integrated experimental ecology. *Ecological Studies* **2**, 1–214.

Ellenberg, H. 1974. Zeigerwerte der gefässpflanzen Mitteleuropas. *Scripta Geobotanica* **9**, 7–122.

Ellenberg, H. 1978. *Die Vegetation Mitteleuropas mit den Alpen*, 2nd edn. Stuttgart: Ulmer.

Ellis, J. E. (ed.) 1977. *Grassland ecosystems of North America*. Stroudsburg, PA: Dowden, Hutchinson & Ross.

Ellison, L. 1960. Influence of grazing on plant succession of rangelands. *Botanical Review* **26**, 1–78.

Emanuel, W. R., H. H. Shugart & D. C. West 1978. Spectral analysis and forest dynamics: the effects of perturbations on long-term dynamics. In *Time series and ecological processes*, H. H. Shugart (ed.), 195–210. Proceedings of the Society of Industrial and Applied Mathematics Conference, Alta, Utah, 27 June–1 July 1978.

Enright, N. J. 1978. The interrelationships between plant species distribution and properties of soils undergoing podsolization in a coastal area of South West Australia. *Australian Journal of Ecology* **3**, 389–401.

Ernst, A. 1908. *The new flora of the volcanic islands of Krakatau* (transl. A. C. Seward). Cambridge: Cambridge University Press.

Erwin, T. L. 1983. Beetles and other insects of tropical forest canopies at Manaus, Brazil, sampled by insecticidal fogging. In *Tropical rain forest: ecology and management*, S. L. Sutton, T. C. Whitmore & A. C. Chadwick (eds), 59–75. Oxford: Blackwell.

Esau, K. 1965. *Plant anatomy*, 2nd edn. New York: Wiley.

Esler, A. 1970. Manawatu sand dune vegetation. *Proceedings of the New Zealand Ecological Society* **17**, 41–6.

Etherington, J. R. 1982. *Environment and plant ecology*, 2nd edn. New York: Wiley.

Etherington, J. R. 1983. *Wetland ecology*. Studies in Biology No. 154. London: Edward Arnold.

Evenari, M., I. Noy-Meir & D. W. Goodall 1986. *Hot deserts and arid shrublands*. Ecosystems of the World 12B. Amsterdam: Elsevier.

Faegri, K. & J. Iversen 1975. *Textbook of pollen analysis*. Copenhagen: Munksgaard.

Faegri, K. & L. Van der Pijl 1971. *The principles of pollination ecology* New York: Pergamon.

Falinski, J. B. 1986. *Vegetation dynamics in temperate lowland primeval forest.* (*Ecological studies in Bialowieza forest, Poland*). Dordrecht: Junk.

Federer, C. A. 1976. Differing diffusive resistance and leaf developments may cause differing transpiration among hardwoods in the spring. *Forest Science* **22**, 359–64.

Feeny, P. 1970. Seasonal change in oak leaf tannins and nutrients as a cause of spring feeding by winter moth caterpillars. *Ecology* **51**, 565–81.

Fenner, M. 1985. *Seed ecology*. Outline Studies in Ecology. London: Chapman & Hall.

Ferrell, W. K. 1953. Effect of environmental conditions on survival and growth of forest tree seedlings under field conditions in the Piedmont region of North Carolina. *Ecology* **34**, 667–88.

Field, W. O. 1947. Glacier recession in Muir Inlet, Glacier Bay, Alaska. *Geographical Review* **37**, 369–99.

Field, W. O. (ed.) 1975. *Mountain glaciers of the Northern Hemisphere*. Hanover, NH: United States Army Corps of Engineers, Cold Regions Research and Engineering Laboratory.

Finegan, B. 1984. Forest succession. *Nature* **312**, 109–14.

Fitter, A. H. & R. K. Hay 1981. *Environmental physiology of plants*. London: Academic Press.

Flenley, J. R. 1979. *The equatorial rain forest: a geological history*. London: Butterworth.

Flenley, J. R. 1980. *The Krakatoa centenary expedition. Interim report*. Dept Geography, Univ. Hull, Misc. Ser. No. 24.

Flenley, J. R. & K. Richards (eds) 1982. *The Krakatoa centenary expedition. Final report*. Dept Geography, Univ. Hull, Misc. Series No. 25.

Forcier, L. K. 1975. Reproductive strategies and the co-occurrence of climax tree species. *Science* **189**, 808–9.

Forest Service, United States Department of Agriculture 1948. *Woody plant seed manual*, Washington, DC: US Government Printing Office.

498

Foster, R. B. 1977. *Tachigalia versicolor* is a suicidal neotropical tree. *Nature* **268**, 624–6.

Foster, R. B. 1980. Heterogeneity and disturbance in tropical vegetation. In *Conservation biology: an evolutionary-ecological perspective*, M. E. Soule & B. A. Wilcox (eds), 75–92. Sunderland, MA: Sinauer.

Foster, R. B. & N. Brokaw 1982. General character of the vegetation. In *The ecology of a tropical forest: seasonal rhythms and long-term changes*, E. G. Leigh, A. S. Rand & D. S. Windsor (eds), 151–72. London: Oxford University Press.

Foster, R. B., B. J. Arce & T. S. Wachter 1986. Dispersal and the sequential plant communities in an Amazonian Peru floodplain. In *Frugivores and seed dispersal*, 357–70. A. Estrada & T. H. Fleming (eds), Dordrecht: Junk.

Fowells, H. A. 1965. *Silvics of forest trees of the United States*. United States Dept Agric. Handbook No. 271, Washington, DC.

Fowler, N. 1986. The role of competition in plant communities in arid and semiarid regions. *Annual Review of Ecology and Systematics* **17**, 89–110.

Fox, F. J. 1977. Alternation and coexistence of tree species. *American Naturalist* **111**, 69–89.

Francis, P. 1976. *Volcanoes*. London: Penguin.

Franklin, J. F. & M. A. Hemstrom 1981. Aspects of succession in the coniferous forests of the Pacific Northwest. In *Forest succession: concepts and application*, D. C. West, H. H. Shugart & D. B. Botkin (eds), 212–29. New York: Springer.

Freas, K. E. & P. R. Kemp 1983. Some relationships between environmental reliability and seed dormancy in desert annual plants. *Journal of Ecology* **71**, 211–7.

Friedmann, E. I. & M. Galun 1974. Desert algae, lichens and fungi. In *Desert biology* Vol. 2, G. W. Brown (ed.), 165–212. New York: Academic Press.

Fridriksson, S. 1975. *Surtsey, evolution of life on a volcanic island*. London: Butterworth.

Fridriksson, S. 1982. Life develops on Surtsey. *Endeavour* (n.s.) **6**, 100–7.

Fritts, H. C. 1976. *Tree rings and climate*. London: Academic Press.

Froment, A., M. Tanghe, P. Duvigneaud, A. Galoux, S. Denaeyer-de Smet, G. Schnock, J. Grulois, F. Mommaerts-Billet & J. P. Vanseveren 1971. Le chenaie melangée calcicole de Virelles-Blaimont, en Haute Belgique. In *Productivity of forest ecosystems*, P. Duvigneaud (ed.), 635–65. Paris: UNESCO.

Fuller, G. D. 1914. Evaporation and soil moisture in relation to the succession of plant associations. *Botanical Gazette* **58**, 193–234.

Fuller, W. H. 1974. Desert soils. In *Desert biology*, Vol 2, G. W. Brown (ed.), New York: Academic Press.

Gadgil, R. L. & P. D. Gadgil 1975. Suppression of litter decomposition by mycorrhizal roots of *Pinus radiata*. *New Zealand Journal of Forestry Science* **5**, 33–41.

Galoux, A. 1971. Flux et transferts d'energie au niveau des écosystemes forestiers. In *Productivity of forest ecosystems*, P. Duvigneaud (ed.), 21–40. Paris: UNESCO.

Gams, H. 1918. Principienfragen der Vegetationsforschung. *Vierteljahrsschrift für Naturforschende Gesellschaft Zurich* **63**, 293–493.

Gant, R. E. & E. E. Clebsch 1975. The allelopathic influences of *Sassafras albidum* in old-field succession in Tennessee. *Ecology* **56**, 604–15.

Garbutt, K. & A. R. Zangerl 1983. Application of genotype–environment interaction analysis to niche quantification. *Ecology* **64**, 1292–6.

Garwood, N. C. 1986. Constraints on the timing of seed germination in a tropical forest. In *Frugivores and seed dispersal*, A. Estrada & T. H. Fleming (eds), 247–56. The Hague: Junk.

Gates, D. H., L. A. Stoddart & C. W. Cook 1956. Soil as a factor influencing plant distribution in salt-deserts of Utah. *Ecological Monographs* **26**, 155–75.

Gause, G. F. 1934. *The struggle for existence*. Baltimore: Williams & Wilkins.

Gentry, A. H. 1982. Patterns of neotropical species diversity. *Evolutionary Biology* **15**, 1–84.

Gentry, A. H. 1988. Tree species of upper Amazonian forests. *Proceedings of the National Academy of Science, USA* **85**, 156.

George, M. F., S. G. Hong & M. J. Burke 1977. Cold hardiness and deep supercooling of hardwoods: its occurrence in provenance collections of red oak, yellow birch, black walnut and black cherry. *Ecology* **58**, 674–80.

Gibson, C. W. D., T. C. Guilford, C. Hambler & P. H. Sterling 1983. Transition matrix models and succession after release from grazing on Aldabra atoll. *Vegetatio* **52**, 151–9.

Gilbert, L. E. & P. H. Raven (eds) 1975. *Coevolution of plants and animals*. Austin, TX: University of Texas Press.

Gill, A. M. 1975. Fire and the Australian flora: a review. *Australian Forestry* **38**, 4–25.

Gill, A. M., R. H. Groves & I. R. Noble 1981. *Fire and the Australian biota*. Canberra: Australian Academy of Science.

Gill, D. E. 1986. Individual plants as genetic mosaics: ecological organisms versus evolutionary individuals. In *Plant ecology*, M. J. Crawley (ed.), 321–44. Oxford: Blackwell.

Giller, P. S. 1984. *Community structure and the niche*. Outline Studies in Ecology. London: Chapman & Hall.

Gimingham, C. H., S. B. Chapman & N. R. Webb 1979. European heathlands. In *Heathlands and related shrublands*, R. L. Specht (ed.), 365–414. Ecosystems of the World 9A. Amsterdam: Elsevier.

Gleason, H. A. 1917. The structure and development of the plant association. *Bulletin of the Torrey Botanical Club* **43**, 463–81.

Gleason, H. A. 1926. The individualistic concept of the plant association. *Bulletin of the Torrey Botanical Club* **53**, 7–26.

Gleason, H. A. 1927. Further views on the succession concept. *Ecology* **8**, 299–327.

Gleason, H. A. 1939. The individualistic concept of the plant association. *American Midland Naturalist* **21**, 92–110.

Gleason, H. A. 1968. *The new Britton and Brown illustrated flora of the northeastern United States and adjacent Canada*, Vols 1, 2 & 3. New York: Hafner.

Glenn-Lewin, D. C. 1980. The individualistic nature of plant community development. *Vegetatio* **43**, 141–6.

Gliessman, S. R. & C. H. Muller 1978. The allelopathic mechanisms of dominance in bracken (*Pteridium aquilinum*) in southern California. *Journal of Chemical Ecology* **4**, 337–62.

Godwin, H., D. Walker & E. H. Willis 1957. Radiocarbon dating and post-glacial vegetation history: Scaleby Moss. *Proceedings of the Royal Society* **147**, 352–66.

Goldthwait, R. P. 1963. *Dating the little ice age in Glacier Bay, Alaska*. Rep. Int. Geol. Congr, 21 session, Norden, 1960, part 27, 37–46.

Golley, F. B. 1960. Energy dynamics of a food chain of an old-field community. *Ecological Monographs* **30**, 187–206.

Golley, F. B. 1961. Energy values of ecological materials. *Ecology* **42**, 581–4.

Golley, F. B. 1977. *Ecological succession*. Benchmark Papers in Ecology. Stroudsburg, PA: Dowden, Hutchinson & Ross.

Golley, F. B. 1983. Nutrient cycling and nutrient conservation. In *Tropical rain forest ecosystems*, F. B. Golley (ed.), 137–56. Ecosystems of the World 14A. Amsterdam: Elsevier.

Gomez-Pompa, A. & C. Vazquez-Yanes 1981. Successional studies of a rain forest in Mexico. In *Forest succession: concepts and applications*, D. C. West, H. H. Shugart & D. B. Botkin (eds), 246–66. New York: Springer.

Good, E. E. 1966. Mammals. In *Soil development and ecological succession in a deglaciated area of Muir Inlet, Southwest Alaska*, A. Mirsky (ed.), 147–58. Ohio St. Univ. Inst. Polar Studies, Rep. No. 20, Columbus, Ohio.

Good, N. F. 1968. A study of natural replacement of chestnut in six stands in the highlands of New Jersey. *Bulletin of the Torrey Botanical Club* **95**, 240–53.

Goodall, D. W. 1963. The continuum and the individualistic association. *Vegetatio* **11**, 297–316.

Gore, A. J. 1961. Factors limiting plant growth on high-level blanket peat. I. Calcium and phosphate. *Journal of Ecology* **49**, 399–402.

Gorham, E. 1961. Factors influencing supply of major ions to inland waters, with special reference to the atmosphere. *Geological Society of America Bulletin* **72**, 795–840.

Gorham, E., P. M. Vitousek & W. A. Reiners 1979. The regulation of chemical budgets over the course of terrestrial ecosystem succession. *Annual Review of Ecology and Systematics* **10**, 53–84.

Grace, J. 1983. *Plant–Atmosphere relationships*. Outline Studies in Ecology. London: Chapman & Hall.

Grace, J., E. D. Ford & P. G. Jarvis (eds) 1981. *Plants and their atmospheric environment*. British Ecological Society 21st Symposium, Edinburgh 1979. Oxford: Blackwell.

Graebner, P. 1895. Studien über die norddeutsche Heide. *Engler Botanisches Jahrbuch* **20**, 500.

Graebner, P. 1901. Die Heide Norddeutschlands. In *Die Vegetation Der Erde* Bd 5, A. Engler & O. Drude (eds), Leipzig: Engelmann.

Graham, B. F. & F. H. Bormann 1966. Natural root grafts. *Botanical Review* **32**, 255–92.

Gray, A. J., R. G. Parsell & R. Scott 1979. The genetic structure of plant populations in relation to the development of salt marshes. In *Ecological processes in coastal environments*, R. L. Jefferies & A. J. Davy (eds), 43–64. 19th Symposium of the British Ecological Society. Oxford: Blackwell.

Gray, A. J., M. J. Crawley & P. J. Edwards 1987. *Colonization, succession and stability*. 26th Symposium of the British Ecological Society. Oxford: Blackwell.

Green, D. S. 1983. The efficacy of dispersal in relation to safe site density. *Oecologia* **56**, 356–8.

Greenfield, L. G. 1981. An enzyme sequence to digest the soil organic matter in mull and mor soils. *Mauri Ora* **9**, 3–13.

Greenland, D. J. & P. H. Nye 1959. Increases in the carbon and nitrogen contents of tropical soils under natural fallows. *Journal of Soil Science* **10**, 284–91.

Griggs, R. F. 1934. The problem of Arctic vegetation. *Washington Academy of Science Journal* **24**, 153–75.

REFERENCES

Grime, J. P. 1966. Shade avoidance and tolerance in flowering plants. In *Light as an ecological factor*, R. Bainbridge, G. C. Evans & O. Rackham (eds), 281–301. British Ecological Society Symposium No. 6. Oxford: Blackwell.

Grime, J. P. 1977. Evidence for the existence of three primary strategies in plants and its relevance to ecological and evolutionary theory. *American Naturalist* **111**, 1169–94.

Grime, J. P. 1979. *Plant strategies and vegetation processes*. New York: Wiley.

Grime, J. P. & B. C. Jarvis 1975. Shade avoidance and shade tolerance in flowering plants. Effects of light on the germination of species of contrasted ecology. In *Light as an ecological factor*, Vol. II, R. Bainbridge, G. C. Evans & O. Rackham (eds), 525–32. British Ecological Society Symposium No. 16. Oxford: Blackwell.

Grime, J. P. & D. W. Jeffrey 1965. Seedling establishment in vertical gradients of sunlight. *Journal of Ecology* **53**, 621–42.

Grime, J. P., J. M. Mackey, S. H. Hillier & D. J. Read 1987. Floristic diversity in a model system using experimental microcosms. *Nature* **328**, 420–3.

Groves, R. H. 1981a. Nutrient cycling in heathlands. In *Heathlands and related shrublands: analytical studies*, R. L. Specht (ed.), 151–64. Ecosystems of the World 9B. Amsterdam: Elsevier.

Groves, R. H. 1981b. Heathland soils and their fertility status. In *Heathlands and related shrublands*, R. L. Specht (ed.), 143–50. Ecosystems of the World 9B. Amsterdam: Elsevier.

Grubb, P. J. 1976. A theoretical background to the conservation of ecologically distinct groups of annuals and biennials in the chalk grassland ecosystem. *Biological Conservation* **10**, 53–76.

Grubb, P. J. 1977. The maintenance of species-richness in plant communities: the importance of the regeneration niche. *Biological Reviews* **52**, 107–45.

Grubb, P. J. 1985. Plant populations and vegetation in relation to habitat, disturbance and competition: problems of generalization. In *The population structure of vegetation*, J. White (ed.), 595–622. Dordrecht: Junk.

Grubb, P. J., D. Kelly & J. Mitchley 1982. The control of relative abundance in communities of herbaceous plants. In *The plant community as a working mechanism*, E. I. Newman (ed.), 79–97. British Ecological Society Special Publication No. 1. Oxford: Blackwell.

Gutierrez, L. T. & W. R. Fey 1980. *Ecosystem succession*. Cambridge, MA: MIT Press.

Hadley, N. F. 1973. Desert species and adaptation. *American Scientist* **60**, 338–47.

Hale, M. E. 1967. *The biology of lichens*. London: Edward Arnold.

Halligan, J. P. 1973. Bare areas associated with shrub stands in grassland: the case of *Artemisia californica*. *BioScience* **23**, 429–32.

Hanes, T. L. 1981. California chaparral. In *Mediterranean-type shrublands*, F. di Castri, D. W. Goodall & R. L. Specht (eds), 139–74. Ecosystems of the World II. The Hague: Elsevier.

Hanson, H. C. 1953. Vegetation types in northwestern Alaska and comparisons with communities in other Arctic regions. *Ecology* **34**, 111–40.

Harborne, J. B. 1982. *Introduction to ecological biochemistry*. London: Academic Press.

Harper, J. L. 1961. Approaches to the study of plant competition. In *Mechanisms in biological competition*, F. L. Milthorpe (ed.), 1–39. Symposium of the Society for Experimental Biology 15. Cambridge: CUP.

Harper, J. L. 1977. *Population biology of plants*. London: Academic Press.

Harper, J. L. 1978. The demography of plants with clonal growth. In *Structure and functioning of plant populations*, A. H. J. Freysen & J. W. Woldendorp (eds), 27–48. Amsterdam: North–Holland.

Harper, J. L., P. H. Lovell & K. G. Moore 1970. The shapes and sizes of seeds. *Annual Review of Ecology and Systematics* **1**, 327–56.

Harper, J. L., J. T. Williams & G. R. Sagar 1965. The behaviour of seeds in the soil. I. The heterogeneity of soil surfaces and its role in determining the establishment of plants from seed. *Journal of Ecology* **53**, 273–86.

Harris, W. F., R. S. Kinerson & N. T. Edwards 1977. Comparisons of below-ground biomass of natural deciduous forests and loblolly pine plantations. In *The below-ground ecosystem: a synthesis of plant-associated processes*, J. K. Marshall (ed.), 29–38. Range Science Department, Science Series No. 26. Fort Collins: Colorado State University.

Harrison, A. F. 1979. Phosphorus cycles of forest and upland grassland ecosystems and some effects of land management practices. In *Phosphorus in the environment: its chemistry and biochemistry*, 175–99. CIBA Foundation Symposium 57 (n.s.). Amsterdam: Excerpta Medica.

Harte, J. & D. Levy 1975. On the vulnerability of ecosystems disturbed by man. In *Unifying concepts in ecology*, W. H. Van Dobben & R. H. Lowe-McConnell (eds), 208–23. The Hague: Junk.

Hartshorn, G. S. 1978. Treefalls and tropical forest dynamics. In *Tropical trees as living systems*, P. B. Tomlinson & M. H. Zimmerman (eds), 617–38. Cambridge: Cambridge University Press.

Hartshorn, G. S. 1980. Neotropical forest dynamics. In *Tropical succession* J. Ewel (ed.), *Biotropica* (Supplement) **12**, 23–30.

Harwood, W. E. & W. D. Jackson 1975. Atmospheric losses of four plant nutrients during a forest fire. *Australian Forestry* **38**, 92–9.

Hasselo, H. N. & J. T. Swarbrick 1960. The eruption of the Cameroons Mountain in 1950. *West Africa Scientific Association* **6**, 96–101.

Heath, G. W., M. K. Arnold & C. A. Edwards 1966. Studies in leaf litter breakdown. I. Breakdown rates of leaves of different species. *Pedobiologia* **6**, 1–12.

Heilman, P. E. 1966. Change in distribution and availability of nitrogen with forest succession on north slopes in interior Alaska. *Ecology* **47**, 825–31.

Heilman, P. E. 1968. Relationship of availability of phosphorus and cations to forest succession and bog formation in interior Alaska. *Ecology* **49**, 331–6.

Heinselman, M. L. 1963. Forest sites, bog processes and peatland types in the glacial Lake Agassiz region, Minnesota. *Ecological Monographs* **33**, 327–74.

Heinselman, M. L. 1970. Landscape evolution, peatland types and the environment in the Lake Agassiz Peatlands Natural Area, Minnesota. *Ecological Monographs* **40**, 235–61.

Heinselman, M. L. 1975. Boreal peatlands in relation to environment. In *Coupling of land and water systems*, A. D. Hasler (ed.), 93–103. Ecological Studies 10. Heidelberg: Springer.

Heinselman, M. L. 1981a. Fire intensity and frequency as factors in the distribution and structure of northern ecosystems. In *Fire regimes and ecosystem properties*, 7–57. Proceedings of Conference, US Dept Agric. Forest Service Tech. Rep. WO-26.

Heinselman, M. L. 1981b. Fire and succession in the conifer forests of northern North America. In *Forest succession: concepts and application*, D. C. West, H. H. Shugart & D. B. Botkin (eds), 374–405. New York: Springer.

Held, M. E. 1983. Pattern of beech regeneration in the east-central United States. *Bulletin of the Torrey Botanical Club* **110**, 55–62.

Held, M. E. & J. E. Winstead 1975. Basal area and climax status in mesic forest systems. *Annals of Botany* **39**, 1147–8.

Hermann, R. K. 1977. Growth and production of tree roots: a review. In *The below-ground ecosystem: a synthesis of plant-associated processes*, J. K. Marshall (ed.), 7–28. Range Science Department, Science Series No. 26. Fort Collins: Colorado State University.

Hibbert, F. A. & V. R. Switsur 1976. Radiocarbon dating of Flandrian pollen zones in Wales and Northern England. *New Phytologist* **77**, 793–807.

Hickman, J. C. 1977. Energy allocation and niche differentiation in four co-existing annual species of *Polygonum* in western North America. *Journal of Ecology* **65**, 317–26.

Hickman, J. C. 1979. The basic biology of plant numbers. In *Topics in plant population biology*, O. J. Solbrig, S. Jain, G. B. Johnson & P. H. Raven (eds), 232–63. New York: Columbia University Press.

Hicks, D. J. & B. F. Chabot 1985. Deciduous forest. In *Physiological ecology of North American plant communities*, B. F. Chabot & H. A. Mooney (eds), 257–77. New York: Chapman & Hall.

Hill, M. O. 1979a, b. DECORANA; TWINSPAN. Cornell Ecology Program. Ithaca, New York: Cornell University.

Hillel, D. & N. Tadmor 1962. Water regimes and vegetation in the Central Negev Highlands of Israel. *Ecology* **43**, 33–41.

Holdridge, L. R. 1967. *Life zone ecology*. San Jose, Costa Rica: Tropical Science Center.

Holdridge, L. R. *et al.* 1971. Forest environments in tropical life zones. Oxford: Pergamon Press.

Holmgren, P., P. G. Jarvis & M. S. Jarvis 1965. Resistances to carbon dioxide and water vapour transfer in leaves of different plant species. *Physiologia Plantarum* **18**, 557–73.

Holt, B. R. 1972. Effect of arrival time on recruitment, mortality and reproduction in successional plant populations. *Ecology* **53**, 668–73.

Hopkins, D. M. & R. S. Sigafoos 1951. Frost action and vegetation patterns on Seward Peninsula, Alaska. *United States Geological Survey Bulletin* **974C**, 51–101.

Horn, H. S. 1971. *The adaptive geometry of trees*. Monographs in Population Biology No. 3. Princeton, NJ: Princeton University Press.

Horn, H. S. 1974. The ecology of secondary succession. *Annual Review of Ecology and Systematics* **5**, 25–37.

Horn, H. S. 1975a. Forest succession. *Scientific American* (May), 91–8.

Horn, H. 1975b. Markovian properties of forest succession. In *Ecology and evolution of communities*, M. L. Cody & J. M. Diamond (eds), 196–211. London: Belknap Press.

Horn, H. S. 1976. Succession. In *Theoretical ecology: principles and applications*, R. M. May (ed.), 187–204. London: Blackwell.

Horn, H. S. 1981. Some causes of variety in patterns of secondary succession. In *Forest succession: concepts and application*, D. C. West, H. H. Shugart & D. B. Botkin (eds), 24–35. New York: Springer.

Horton, J. S. & L. L. Kraebel 1955. Development of vegetation after fire in the chamise chaparral of southern California. *Ecology* **36**, 244–62.

Hough, A. F. 1936. A climax forest community on East Tionesta Creek in northwestern Pennsylvania. *Ecology* 17, 1–28.

Hough, A. F. 1965. A twenty-year record of understorey vegetational change in a virgin Pennsylvania forest. *Ecology* 46, 370–3.

Howe, H. F. 1983. Annual variation of a neotropical seed-dispersal system. In *Tropical rain forest: ecology and management*, S. L. Sutton, T. L. Whitmore & A. C. Chadwick (eds), 211–28. Br. Ecol. Soc., Spec. Publ. No. 2. Oxford: Blackwell.

Howe, H. F. & D. J. Smallwood 1982. Ecology of seed dispersal. *Annual Review of Ecology and Systematics* 13, 201–18.

Hubbell, S. P. 1980. Seed predation and the coexistence of tree species in tropical forests. *Oikos* 35, 21.

Hubbell, S. P. & R. B. Foster 1983. Diversity of canopy trees in a neotropical forest and implications for conservation. In *Tropical rain forest: ecology and management*, S. L. Sutton, T. C. Whitmore & A. C. Chadwick (eds), 25–42. Br. Ecol. Soc., Spec. Publ. No. 2. Oxford: Blackwell.

Hubbell, S. P. & R. B. Foster 1986a. Canopy gaps and the dynamics of a neotropical forest. In *Plant ecology*, M. J. Crawley (ed.), 77–96. Oxford: Blackwell.

Hubbell, S. P. & R. B. Foster 1986b. Biology, chance and history and the structure of tropical rain forest tree communities. In *Community ecology*, J. Diamond & T. J. Case (eds), 314–29. New York: Harper & Row.

Huber, O. 1978. Light compensation point of vascular plants of a tropical cloud forest and an ecological interpretation. *Photosynthetica* 12, 382–90.

Hult, R. 1885. Blekinges vegetation: ett bidrag till växtformationernas utveckling historie. *Meddelanden Societas pro Fauna et Flora, Fennica* 12, 161–251.

Hult, R. 1887. Die alpinen Pflanzenformationen des nördlichtsen Finnlans. *Meddelanden Societas pro Fauna et Flora, Fennica* 14, 153–228.

Huston, M. 1979. A general hypothesis of species diversity. *American Naturalist* 113, 81–101.

Huston, M. 1980. Soil nutrients and tree species richness in Costa Rican forests. *Journal of Biogeography* 7, 147–57.

Huston, M. & T. Smith 1987. Plant succession: life history and competition. *American Naturalist* 130, 168–98.

Hutchinson, G. E. 1958. Concluding remarks. *Cold Spring Harbor Symposium in Quantitative Biology* 22, 415–27.

Hutchinson, G. E. 1965. The niche: an abstractly inhabited hypervolume. In *The ecological theatre and the evolutionary play*, 26–78. New Haven, CT: Yale University Press.

Hutchinson, G. E. 1978. *An introduction to population ecology*. New Haven, CT: Yale University Press.

Iverson, L. R. & M. K. Wali 1982. Buried viable seeds and their relation to revegetation after surface mining. *Journal of Range Management* 35, 648–52.

Jacks, G. V. 1965. The role of organisms in the early stages of soil formation. In *Experimental pedology*, E. G. Hallsworth & D. V. Crawford (eds), 219–26. Proceedings of the 11th Easter School in Agricultural Science, University of Nottingham 1964. London: Butterworth.

Jackson, T. A. 1969. *The role of pioneer lichens in the chemical weathering of recent volcanic rocks on the island of Hawaii*. Ph.D. Thesis, University of Missouri, Columbia.

Jackson, W. A. & R. J. Volk 1970. Photorespiration. *Annual Review of Plant Physiology* 21, 385–432.

Jacobson, G. L. & H. J. B. Birks 1980. Soil development on recent end moraines of the Klutlan Glacier, Yukon Territory, Canada. *Quaternary Research* 14, 87–100.

Janos, D. P. 1980. Vesicular–arbuscular mycorrhizae affect lowland tropical rainforest plant growth. *Ecology* 61, 151–62.

Janos, D. P. 1983. Tropical mycorrhizas, nutrient cycles and plant growth. In *Tropical rain forest ecology and management*, S. L. Sutton, T. C. Whitmore & A. C. Chadwick (eds), 327–45. British Ecological Society Special Publication No. 2. Oxford: Blackwell.

Janzen, D. H. 1970. Herbivores and the number of tree species in tropical forests. *American Naturalist* 104, 501–28.

Janzen, D. H. 1975. *Ecology of plants in the Tropics*. London: Edward Arnold.

Janzen, D. H. 1983. Food webs: who eats what, why, how and with what effects in a tropical forest? In *Tropical rain forest ecosystems*, F. B. Golley (ed.), 167–82. Ecosystems of the World 14A. Amsterdam: Elsevier.

Jefferies, R. L. & A. J. Davy 1979. *Ecological processes in coastal environments*. 19th Symposium of the British Ecological Society, Norwich, 1977. Oxford: Blackwell.

Jeffers, J. N. R. 1978. *An introduction to systems analysis: with ecological applications*. London: Edward Arnold.

Jenik, J. 1986. Forest succession: theoretical concepts. In *Forest dynamics research in western and central Europe*, J. Fanta (ed.), 7–16. IUFRO – Subject Group S1–01–00, Ecosystems. Wageningen: Pudoc.

Jenny, H. 1941. *Factors of soil formation*. New York: McGraw–Hill.

Jenny, H. 1958. Role of the plant factor in the pedogenic functions. *Ecology* 39, 5–16.

Jenny, H. 1961. Derivation of state factor equations of soil and ecosystems. *Proceedings of the Soil Science Society of America* 25, 385–8.

Jenny, H. 1980. *Soil genesis with ecological perspectives*. Ecology Studies 37. New York: Springer.

Jordan, C. F. 1981. *Tropical ecology*. Benchmark Papers in Ecology 10. Stroudsberg PA: Hutchinson & Ross.

Jordan, C. F. & R. Herrera 1981. Tropical rain forests: are nutrients really critical? *American Naturalist* **117**, 167–80.

Jordan, C. F. & J. R. Kline 1972. Mineral cycling: some basic concepts and their application in a tropical rain forest. *Annual Review of Ecology and Systematics* **3**, 33–50.

Julander, O. 1945. Drought resistance in range and pasture grasses. *Plant Physiology* **20**, 573–99.

Kaminsky, R. 1981. The microbial origin of the allelopathic potential of *Adenostoma fasciculatum*. *Ecological Monographs* **51**, 365–82.

Keeley, J. E. & P. H. Zedler 1978. Reproduction of chaparral shrubs after fire: a comparison of sprouting and seeding strategies. *American Midland Naturalist* **99**, 142–61.

Keever, C. 1950. Causes of succession on old fields of the Piedmont, North Carolina. *Ecological Monographs* **20**, 229–50.

Keever, C. 1953. Present composition of some stands of the former oak–chestnut forest in the southern Blue Ridge Mountains. *Ecology* **34**, 44–54.

Keever, C. 1979. Mechanisms of plant succession in old fields of Lancaster County, Pennsylvania. *Bulletin of the Torrey Botanical Club* **106**, 299–308.

Kershaw, K. A. 1963. Pattern in vegetation and its causality. *Ecology* **44**, 377–88.

Kershaw, K. A. 1973. *Quantitative and dynamic plant ecology*, 2nd edn. London: Edward Arnold.

King, T. J. & S. R. Woodell 1984. Are regular patterns in desert shrubs artefacts of sampling? *Journal of Ecology* **72**, 295–8.

Kira, T. & T. Shidei 1967. Primary production and turnover of organic matter in different forest ecosystems of the Western Pacific. *Japanese Journal of Ecology* **17**, 70–87.

Kivilaan, A. & R. S. Bandurski 1981. The one hundred year period for Dr Beal's seed viability experiment. *American Journal of Botany* **68**, 1290–2.

Knapp, R. (ed.) 1974a. *Vegetation dynamics*. Handbook of Vegetation Science Part 8. The Hague: Junk.

Knapp, R. 1974b. Mutual influence between plants, allelopathy, competition and vegetation changes. In *Vegetation dynamics*, R. Knapp (ed.), 111–22. Handbook of Vegetation Science Part 8. The Hague: Junk.

Knight, D. H. 1975. A phytosociological analysis of species-rich tropical forest on Barro Colorado Island, Panama. *Ecological Monographs* **45**, 259–84.

Korchagin, A. A. & V. G. Karpov 1974. Fluctuations in coniferous taiga communities. In *Vegetation dynamics*, 225–32. Handbook of Vegetation Science Part 8. The Hague: Junk.

Korstian, C. F. & T. S. Coile 1938. Plant competition in forest stands. *Duke University School of Forestry Bulletin* **3**, 1–125.

Korstian, C. F. & P. W. Stickel 1927. The natural replacement of blight-killed chestnut in the hardwood forests of the Northeast. *Journal of Agricultural Research* **34**, 631–48.

Kozlowski, T. T. 1949. Light and water in relation to growth and competition of Piedmont forest tree species. *Ecological Monographs* **19**, 207–31.

Kozlowski, T. T. & C. E. Ahlgren (eds) 1974. *Fire and ecosystems*. New York: Academic Press.

Kramer, P. J. & J. P. Decker 1944. Relation between light intensity and rate of photosynthesis of loblolly pine and certain hardwoods. *Plant Physiology* **19**, 350–8.

Kramer, P. J., H. J. Oosting & C. F. Korstian 1952. Survival of pine and hardwood seedlings in forest and open. *Ecology* **33**, 427–30.

Kratz, T. K. & C. B. De Witt 1986. Internal factors controlling peatland-lake ecosystem development. *Ecology* **67**, 100–7.

Kuijt, J. 1969. *The biology of parasitic flowering plants*. Berkeley, CA: University of California Press.

Lambrechtsen, N. C. 1972. *What grass is that?* New Zealand Department of Scientific and Industrial Research. Information Series No. 87. Wellington: Government Printer.

Lamont, B. B. 1981. Specialized roots of non-symbiotic origin in heathlands. In *Heathlands and related shrublands: analytical studies*, R. L. Specht (ed.), 183–96. Ecosystems of the World 9B. Amsterdam: Elsevier.

Landwehr, J. & J. J. Barkman 1966. *Atlas van de Nederlandse Bladmossen*. Amsterdam: Koninklijke Nederlandse Natuurhistorische Vereniging.

Lange, O. L., P. S. Nobel, C. B. Osmond & H. Ziegler 1981. *Physiological plant ecology*. Vol. I: *Responses to the plant environment*. Berlin: Springer.

Langford, A. N. & M. F. Buell 1969. Integration and stability in the plant association. *Advances in Ecological Research* **6**, 83–135.

Larcher, W. 1980. *Physiological plant ecology*, 2nd edn. Berlin: Springer.

504

Lawlor, L. R. 1980. Structure and stability in natural and randomly constructed competitive communities. *American Naturalist* **116**, 394–400.

Lawrence, D. B. 1950. Glacier fluctuation for six centuries in south-east Alaska and its relation to solar activity. *Geographical Review* **40**, 191–223.

Lawrence, D. B. 1951. Recent glacier history of Glacier Bay, Alaska and development of vegetation and glacier terrain with special reference to the importance of alder in the succession. *American Philosophical Society Yearbook* **1950**, 175–6.

Lawrence, D. B. 1953. *Development of vegetation and soil in south-east Alaska with special reference to the accumulation of nitrogen.* Final Rep. ONR Project NR160–83.

Lawrence, D. B. 1958. Glaciers and vegetation in south-east Alaska. *American Scientist* **46**, 89–122.

Lawrence, D. B. 1979. Primary versus secondary succession at Glacier Bay National Monument, Southeastern Alaska. In *Proceedings of the first conference on scientific research in the National Parks, New Orleans, November 1976,* R. M. Linn (ed.), US Dept Interior, Natn. Park Service Trans. Proc. Ser. No. 5.

Lawrence, D. B. & L. Hulbert 1960. Growth stimulation of adjacent plants by lupin and alder on recent glacier deposits in South east Alaska. *Bulletin of the Ecological Society of America* **31**, 58.

Lawrence, D. B., R. E. Schoenicke, A. Quispel & G. Bond 1967. The role of *Dryas drummondii* in vegetation development following ice recession at Glacier Bay, Alaska, with special reference to its nitrogen fixation by root nodules. *Journal of Ecology* **55**, 793–813.

Lee, W. G. & A. E. Hewitt 1982. Soil changes associated with development of vegetation on an ultramafic scree, northwest Otago, New Zealand. *Journal of the Royal Society of New Zealand* **12**, 229–42.

Leigh, E. G. 1982. Why are there so many kinds of tropical trees? In *The ecology of a tropical rain forest: seasonal rhythms and long-term changes,* E. G. Leigh, A. S. Rand & D. M. Windsor (eds), 63–7. London: Oxford University Press.

Leighton, M. & D. R. Leighton 1983. Vertebrate responses to fruiting seasonality within a Bornean rain forest. In *Tropical rain forest: ecology and management,* S. L. Sutton, T. C. Whitmore & A. C. Chadwick (eds), 181–96. British Ecological Society, Special Publication No. 2. Oxford: Blackwell.

Leisman, G. A. 1953. The rate of organic matter accumulation on the sedge mat zones of bogs in the Itasca State Park region of Minnesota. *Ecology* **34**, 81–101.

Leisman, G. A. 1957. Further data on the rate of organic matter accumulation in bogs, using depth of accumulation around *Larix* seedlings. *Ecology* **38**, 361.

Leopold, A. C. 1961. Senescence in plant development. *Science* **134**, 1727–32.

Leopold, A. C. 1964. *Plant growth and development.* New York: McGraw-Hill.

Leopold, A. C. 1980. Aging and senescence in plant development. In *Senescence in plants,* K. V. Thimann (ed.), 1–12. Boca Raton, FL: CRC Press.

Levandowsky, M. & B. S. White 1977. Randomness, time scales and the evolution of biological communities. *Evolutionary Biology* **10**, 69–161.

Levitt, J. 1978. An overview of freezing injury and survival and its interrelationships to other stresses. In *Plant cold hardiness and freezing stress,* P. H. Li & A. Sakai (eds), New York: Academic Press.

Lewontin, R. C. 1969. The meaning of stability. *Brookhaven Symposia in Biology* **22**, 13–24.

Lindeman, R. L. 1941. The developmental history of Cedar Creek Bog, Minnesota. *American Midland Naturalist* **25**, 101–12.

Lindeman, R. L. 1942. The trophic–dynamic aspect of ecology. *Ecology* **23**, 399–418.

Livingston, R. B. & M. L. Allessio 1968. Buried viable seed in successional field and forest stands, Harvard Forest, Massachusetts. *Bulletin of the Torrey Botanical Club* **95**, 58–69.

Lloyd, F. E. 1942. *The carnivorous plants.* New York: Ronald Press.

Loach, K. 1967. Shade tolerance in tree seedlings I. Leaf photosynthesis and respiration in plants raised under artificial shade. *New Phytologist* **66**, 607–21.

Loewe, F. 1966. Climate. In *Soil development and ecological succession in a deglaciated area of Muir Inlet, Southwest Alaska,* A. Mirsky (ed.), 19–28. Ohio State University Institute of Polar Studies Report No. 20, Columbus, Ohio.

Longman, K. A. & J. Jenik 1974. *Tropical rainforest and its environment.* London: Longman.

Lorimer, C. G. 1977. The presettlement forest and natural disturbance cycle of northeastern Maine. *Ecology* **38**, 139–48.

Loucks, O. L. 1962. Ordinating forest communities by means of environmental scalars and phytosociological indices. *Ecological Monographs* **32**, 137–66.

Loucks, O. L. 1970. Evolution of diversity, efficiency and community stability. *American Zoologist* **10**, 17–25.

Loucks, O. L., A. R. Ek, W. C. Johnson & R. A. Monserud 1981. Growth, aging and succession. In *Dynamic properties of forest ecosystems,* D. Reichle (ed.), 37–86. I.B.P.23. Cambridge: Cambridge University Press.

Ludlum, D. M. 1963. *Early American hurricanes 1492–1870.* Boston, MA: American Meteorological Society.

Lugo, A. E. & S.C. Snedaker 1974. The ecology of mangroves. *Annual Review of Ecology and Systematics* **5**, 39–64.

Lutz, H. J. 1928. Trends and silvicultural significance of upland forest succession in southern New England. *Yale University School of Forestry Bulletin* **22**, 1–68.

Lutz, H. J. 1930a. Observations on the invasion of newly-formed glacial moraines by trees. *Ecology* **11**, 562–7.

Lutz, H. J. 1930b. The vegetation of Heart's Content, a virgin forest in northwestern Pennsylvania. *Ecology* **11**, 1–29.

Lyons, J. M. 1973. Chilling injury in plants. *Annual Review of Plant Physiology* **24**, 445–66.

Mabberley, D. J. 1983. *Tropical rain forest ecology*. Glasgow: Blackie.

Mabry, T. J., J. H. Hunziker & D.R. Difeo (eds) 1977. *Creosote bush*. Stroudsburg, PA: Dowden, Hutchinson & Ross.

McAndrews, J. H. 1967. Pollen analysis and vegetational history of the Itasca region, Minnesota. In *Quaternary palaeoecology*, E.J. Cushing & H. E. Wright (eds). New Haven, CT: Yale University Press.

McArthur, R. H. & E. O. Wilson 1967. *The theory of island biogeography*. Princeton, NJ: Princeton University Press.

McCaleb, S. B. & W. D. Lee 1956. Soils of North Carolina I. *Soil Science* **82**, 419–31.

McCleary, J. A. 1968. The biology of desert plants. In *Desert biology*, Vol. 1, G. W. Brown (ed.), 141–94. New York: Academic Press.

McCormick, J. 1968. Succession. *Via* **1**, 1–16.

McCormick, J. F. & R. B. Platt 1980. Recovery of an Appalachian forest following the chestnut blight or 'Catherine Keever – you were right'. *American Midland Naturalist* **104**, 264–73.

MacDonnell, N. 1978. The prevalence of chaos. *Nature* **271**, 305–6.

McDonnell, M. J. & E. W. Stiles 1983. The structural complexity of old-field vegetation and the recruitment of bird-dispersed plant species. *Oecologia* **56**, 109–16.

McIntosh, R. P. 1970. Community, competition and adaptation. *Quarterly Review of Biology* **45**, 259–80.

McIntosh, R. P. 1980a. The relationship between succession and the recovery process in ecosystems. In *The recovery process in damaged ecosystems* J. Cairns (ed.), 11–62. Ann Arbor, MI: Ann Arbor Scientific Publications.

McIntosh, R. 1980b. The background and some current problems of theoretical ecology. *Synthèse* **43**, 195–255.

McIntosh, R. P. 1981. Succession and ecological theory. In *Forest succession: concepts and application*, D. C. West, H. H. Shugart & D. B. Botkin (eds), 10–23. New York: Springer.

McIntosh, R. P. 1983. Pioneer support for ecology. *Bioscience* **33**, 107–12.

Mackinnon, J. A. 1972. *The behaviour and ecology of the orang utan (Pongo pymaeus)*. D.Phil. Thesis, Oxford University.

McMillan, C. 1959. The role of ecotypic variation in the distribution of the central grassland of North America. *Ecological Monographs* **29**, 285–308.

McMillan, C. 1969. Ecotypes and ecosystem function. *Bioscience* **19**, 131–4.

McMillen, G. C. & J. H. McClennon 1979. Leaf angle: an adaptive feature of sun and shade leaves. *Botanical Gazette* **140**, 437–42.

McNaughton, S. J. 1978. Serengeti ungulates: feeding selectivity influences the effectiveness of plant defence guilds. *Science* **99**, 806–7.

McNaughton, S. J. 1979. Grassland–herbivore dynamics. In *Serengeti, dynamics of an ecosystem*, A. R. E. Sinclair & M. Norton-Griffiths (eds), 46–81. Chicago: University of Chicago Press.

McNaughton, S. J. & L. L. Wolf 1970. Dominance and the niche in ecological systems. *Science* **167**, 131–8.

McNaughton, S. J. & L. L. Wolf 1979. *General ecology*. New York: Holt, Rhinehart & Winston.

McPherson, J. K. & G. L. Thompson 1972. Competitive and allelopathic suppression of understorey by Oklahoma oak forests. *Bulletin of the Torrey Botanical Club* **99**, 293–300.

Maissurow, D. K. 1941. The role of fire in the perpetuation of virgin forests of northern Wisconsin. *Journal of Forestry* **39**, 201–7.

Major, J. 1951. A functional factorial approach to plant ecology. *Ecology* **32**, 392–412.

Major, J. 1974. Differences in duration of successional seres. In *Vegetation dynamics*, R. Knapp (ed.), 155–9. Handbook of Vegetation Science Part 8. The Hague: Junk.

Makepeace, W., A. T. Dobson & D. Scott 1985. Interference pnenomena due to mouse-ear and king-devil hawk-weed. *New Zealand Journal of Botany* **23**, 79–90.

Malajczuk, N. & B. B. Lamont 1981. Specialized roots of symbiotic origin in heathlands. In *Heathlands and related shrublands: analytical studies*, R.L. Specht (ed.), 165–82. Ecosystems of the World 9B. Amsterdam: Elsevier.

Malmer, N. 1975. Development of bog mires. In *Coupling of land and water systems*, A. D. Hasler (ed.), 85–91. Ecological Studies 10. Heidelberg: Springer.

Margalef, R. 1963. Successions of populations. *Advancing Frontiers of Plant Science* **2**, 137–188.

Margalef, R. 1968. *Perspectives in ecological theory*. Chicago: University of Chicago Press.

Margaris, N. S. 1981. Adaptive strategies in plants dominating Mediterranean-type ecosystems. In *Mediterranean-type shrublands*, F. di Castri, D.W. Goodall & R.L. Specht (eds), 309–16. Ecosystems of the World 11. Amsterdam: Elsevier.

Marks, P. L. 1974. The role of pin cherry (*Prunus pensylvanica*) in the maintenance of stability in northern hardwood ecosystems. *Ecological Monographs* **44**, 73–88.

Marks, P. L. 1975. On the relation between extension growth and successional status of deciduous trees of the northeastern United States. *Bulletin of the Torrey Botanical Club* **102**, 172–7.

Marks, P. L. & F.H. Bormann 1972. Revegetation following forest cutting: mechanisms for return to steady-state nutrient cycling. *Science* **176**, 914–5.

Marquis, D.A. 1974. *The impact of deer browsing on Allegheny hardwood regeneration*. US Dept Agric. Forest Serv. Res. Paper NE-308. Upper Darby, PA: Northeastern Forest Experiment Station.

Marquis, D. A. 1975a. *The Allegheny hardwood forests of Pennsylvania*. Upper Darby: PA: US Dept Agric. Northeastern Forest Exp. Stn, Tech. Rep. NE–15.

Marquis, D. A. 1975b. Seed storage and germination under northern hardwood forests. *Canadian Journal of Forestry Research* **5**, 478–84.

Martin, J. L. 1955. Observations on the origin and early development of a plant community following a forest fire. *Forestry Chronicle* **31**, 154–61.

Mason, C. F. 1976. *Decomposition*. Studies in Biology No. 74. London: Edward Arnold.

Matthews, J. A. 1979a. A study of the variability of some successional and climax plant assemblage-types, using multiple discriminant analysis. *Journal of Ecology* **67**, 255–71.

Matthews, J. A. 1979b. The vegetation of the Storbreen gletschervorfeld, Jotunheimen, Norway II. Approaches involving ordination and general conclusions. *Journal of Biogeography* **6**, 133–67.

Matthews, J. A. 1979c. A study of the variability of some successional and climax plant assemblage types, using multiple discriminant analysis. *Journal of Ecology* **67**, 255–71.

Mattson, W. J. & N. D. Addy 1975. Phytophagous insects as regulators of forest primary production. *Science* **190**, 515–22.

May, R. M. 1975. *Stability and complexity in model ecosystems*, 2nd edn. Princeton, NJ: Princeton University Press.

May, R. M. 1981. *Theoretical ecology* 2nd edn. Oxford: Blackwell.

Mayer-Kress, G. (ed.) 1986. *Dimensions and entropies in chaotic systems*. Berlin: Springer.

Medina, E., H. A. Mooney & C. Vazquez-Yanes (eds) 1984. *Physiological ecology of plants of the wet tropics*. The Hague: Junk.

Mellinger, M. V. & S. J. McNaughton 1975. Structure and function of successional vascular plant communities in Central New York. *Ecological Monographs* **45**, 161–82.

Miles, J. 1979. *Vegetation dynamics*. Outline Studies in Ecology. London: Chapman & Hall.

Miles, J. 1981. Problems in heathland and grassland dynamics. In *Vegetation dynamics in grasslands, heathlands and Mediterranean ligneous formations*, P. Poissonet, F. Romane, M. A. Austin, E. Van der Maarel & W. Schmidt (eds), 61–74. The Hague: Junk.

Mirsky, A. (ed.) 1966. *Soil development and ecological succession in a deglaciated area of Muir Inlet, Southwest Alaska*. Rep. No. 20 Ohio State University Institute of Polar Studies, Columbus, Ohio.

Mokma, D. L., M. L. Jackson & J. K. Syers 1973. Mineralogy of a chronosequence of soils from greywacke and mica-schist alluvium. *New Zealand Journal of Science* **16**, 769–97.

Moleski, F. L. 1976. *Condensed tannins in soil: inputs and effects on microbial populations*. Ph.D. Dissertation, University of Oklahoma, Norman. Dissertation Abstract 76–15816.

Mooney, H. A. 1975. Plant physiological ecology – a synthetic view. In *Physiological adaptation to the environment*, F. J. Vernberg (ed.), 19–36. New York: Intext.

Mooney, H. A. & S. L. Gulman 1982. Constraints on leaf structure and function in reference to herbivory. *Bioscience* **32**, 198–206.

Mooney, H. A. & P. C. Miller 1985. Chaparral. In *Physiological ecology of North American plant communities*, B. F. Chabot & H. A. Mooney (eds), 213–31. New York: Chapman & Hall.

Moore, A. W. 1966. Non-symbiotic nitrogen-fixation in soil and soil–plant systems. *Soils and Fertilizers* **29**, 113–29.

Moore, P. D. & D. J. Bellamy 1974. *Peatlands*. London: Elek Science.

Mosse, B. 1978. Mycorrhiza and plant growth. In *Structure and function of plant populations*, A. J. H. Freyson & J. W. Woldendorp (eds), 269–98. Amsterdam: North–Holland.

Mott, D. G. 1976. The consequences of applying no control to epidemic spruce budworm in eastern spruce–fir. In *Proceedings of the symposium on the spruce budworm, Alexandria, Virginia*, 67–72. Washington, DC: US Dept Agric. Forest Service Misc. Pubn No. 1327.

Mueller, I. M. 1941. An experimental study of rhizomes of certain prairie plants. *Ecological Monographs* **11**, 165–88.

Muller, C. H. 1940. Plant succession in the *Larrea–Flourensia* climax. *Ecology* **21**, 206–12.

Muller, C. H. 1952. Plant succession in arctic heath and tundra in northern Scandinavia. *Bulletin of the Torrey Botanical Club* **79**, 296–309.

Muller, C. H., R. B. Hanawalt & J. K. McPherson 1968. Allelopathic control of herb growth in the fire cycle of California chaparral. *Bulletin of the Torrey Botanical Club* **95**, 25–31.

Muller, W. H. & C.H. Muller 1956. Association patterns involving desert plants that contain toxic products. *American Journal of Botany* **43**, 354–61.

Mutch, R.W. 1970. Wildland fires and ecosystems – an hypothesis. *Ecology* **51**, 1046–51.

Nadkarni, N. 1981. Canopy roots: convergent evolution in rainforest nutrient cycles. *Science* **214**, 1023–4.

Nanson, G.C. & H. F. Beach 1977. Forest succession and sedimentation on a meandering river floodplain, northeast British Columbia. *Canadian Journal of Biogeography* **4**, 229–502.

Naveh, Z. 1974. Effects of fire in the Mediterranean region. In *Fire and ecosystems*, T. T. Kozlowski & C.E. Ahlgren (eds), 401–34. New York: Academic Press.

Nelson, T.C. 1955. Chestnut replacement in the southern highlands. *Ecology* **36**, 352–3.

Newman, E.I. 1982. Niche separation and species diversity in terrestrial vegetation. In *The plant community as a working mechanism*, E.I. Newman (ed.), 61–77. British Ecological Society Special Publication No. 1. Oxford: Blackwell.

Newman, E.I. & R. D. Rovira 1975. Allelopathy among some British grassland species. *Journal of Ecology* **63**, 727–37.

Nicholls, J. L. 1959. The volcanic eruptions of Mt Tarawera and Lake Rotomahana and effects on surrounding forests. *New Zealand Journal of Forestry* **8**, 133–42.

Nichols, G.E. 1917. The interpretation and application of certain terms and concepts in the ecological classification of plant communities. *The Plant World* **20**, 305–319, 341–53.

Nichols, G. E. 1923. A working basis for the ecological classification of plant communities. *Ecology* **4**, 11–23, 154–80.

Niering, W. A. & F. E. Egler 1955. A shrub community of *Viburnum lentago*, stable for twenty-five years. *Ecology* **36**, 356–60.

Niering, W. A., R H. Goodwin & S. Taylor 1970. Prescribed burning in southern New England: introduction to long-range studies. *Ecology* **51**, 267–86.

Nix, H. A. 1978. Determinants of environmental tolerance limits in plants. In *Biology and quaternary environments*, D. Walker & J. C. Guppy (eds), 195–206. Canberra: Australian Academy of Science.

Noble, I. R. & R. O. Slatyer 1977. The effects of disturbance on plant succession. *Proceedings of the Ecological Society of Australia* **10**, 135–45.

Noble, I. R. & R. O. Slatyer 1980. The use of vital attributes to predict successional changes in plant communities subject to recurrent disturbance. *Vegetatio* **43**, 5–21.

Noble, J. C. 1981. The significance of fire in the biology and evolutionary ecology of mallee *Eucalyptus* populations. In *Fire and the Australian biota*, A. M. Gill, R. H. Groves & I. R. Noble (eds), 153–9. Canberra: Australian Academy of Science.

Noble, J. C. 1982. The significance of fire in the biology and evolutionary ecology of mallee *Eucalyptus* populations. In *Evolution of the flora and fauna of arid Australia*, W. R. Barker & P. J. Greenslade (eds), 153–9. Frewville: Peacock.

Noble, M. & C. D. Sandgren 1976. A floristic survey of Muir Point, Glacier Bay National Monument, Alaska. *Bulletin of the Torrey Botanical Club* **103**, 132–6.

Nooden, L. D. 1980. Senescence in the whole plant. In *Senescence in plants*, K. V. Thimann (ed.), 219–58. Boca Raton, FL: CRC Press.

Northeastern Soil Research Committee 1954. *The changing fertility of New England soils*. US Dept Agric., Agric. Info. Bull. 113.

Norton, D. & C. J. Burrows 1979. Adventive plants at Temple Basin ski-field, Arthurs Pass National Park. *Mauri Ora* **7**, 103–15.

Nye, P. H. 1961. Organic matter and nutrient cycles under moist tropical forest. *Plant and Soil* **13**, 333–46.

Nye, P. H. & D. J. Greenland 1960. *Soils under shifting cultivation*. Commonwealth Bureau Soils, Harpenden, Tech. Commun. No. 51.

Odum, E. P. 1960. Organic production and turnover in old-field succession. *Ecology* **41**, 34–49.

Odum, E. P. 1963. *Ecology*. Modern Biology Series. New York: Holt, Rinehart & Winston.

Odum, E. P. 1969. The strategy of ecosystem development. *Science* **164**, 262–70.

Odum, E. P. 1971. *Fundamentals of ecology*, 3rd edn. Philadelphia: Saunders.

Oliver, C. D. 1980. Forest development in North America following major disturbances. *Forest Ecology and Management* **3**, 153–68.

Olmsted, N. W. & J. D. Curtis 1947. Seeds of the forest floor. *Ecology* **28**, 49–52.

Olson, J. S. 1958. Rates of succession and soil changes on southern Lake Michigan sand dunes. *Botanical Gazette* **119**, 125–70.

Olson, J. S. 1963. Energy storage and the balance of producers and decomposers in ecological systems. *Ecology* **44**, 322–31.

Olson, A. R. 1965. *Natural changes in some Connecticut woodlands during 30 years*. Connecticut Agric. Exp. Stn, New Haven, Bull. No. 669.

O'Neill, R. V. 1975. Modeling in the eastern deciduous forest biome. In *Systems analysis and simulation in ecology*, Vol. 3, B. C. Patten (ed.), 49–72. New York: Academic Press.

O'Neill, R. V. 1976a. Ecosystem persistence and heterotrophic regulation. *Ecology* **57**, 1244–53.

O'Neill, R. V. 1976b. Paradigms of ecosystem analysis. In *Ecological theory and ecosystem models*, S. A. Levin (ed.), 16–9. Indianapolis, IN: Institute of Ecology.

O'Neill, R. V., D. I. De Angelis, J. B. Waide & T. F. Allen 1986. *A hierarchical concept of the ecosystem*. Princeton, NJ: Princeton University Press.

O'Neill, R. V., W. F. Harris, B. S. Ausmus & D. E. Reichle 1975. A theoretical basis for ecosystem analysis with particular reference to element cycling. In *Mineral cycling in southeastern ecosystems*, F. G. Howell, J. B. Gentry & M. H. Smith (eds), 28–40. Energy Research and Development Administration Symposium, Washington, DC.

Oomes, M. J. & W. T. Elberse 1976. Germination of six grassland herbs in microsites with different water contents. *Journal of Ecology* **64**, 745–55.

Oosting, H. J. 1942. An ecological analysis of the plant communities of Piedmont, North Carolina. *American Midland Naturalist* **28**, 1–126.

Oosting, H. J. 1956. *The study of plant communities: an introduction to plant ecology*, 2nd edn. San Francisco: Freeman.

Oosting, H. J. & M. E. Humphreys 1940. Buried viable seed in a successional series of old fields and forest soils. *Bulletin of the Torrey Botanical Club* **67**, 253–73.

Oosting, H. J. & P. J. Kramer 1946. Water and light in relation to pine reproduction. *Ecology* **27**, 47–53.

Orians, G. H. 1975. Diversity, stability and maturity in natural ecosystems. In *Unifying concepts in ecology*, W. H. Van Dobben & R. H. Lowe-McConnell (eds), 139–50. The Hague: Junk.

Orloci, L. 1978. *Multivariate analysis in vegetation research*, 2nd edn. The Hague: Junk.

Osvald, H. 1930. Sveriges mossetyper. *Svenska Geografiska Arbok* **1930**, 117–40.

Ovington, J. D. 1959. The circulation of minerals in plantations of *Pinus sylvestris*. *Annals of Botany* (n.s.) **23**, 229–39.

Owen, D. F. 1980. How plants may benefit from the animals that eat them. *Oikos* **35**, 230–35.

Owen, D. F. & R. G. Wiegert 1982. Grasses and grazers: is there a mutualism? *Oikos* **38**, 258.

Ozanne, P. G. & R. L. Specht 1981. Mineral nutrition of heathlands; phosphorus toxicity. In *Heathlands and related shrublands: analytical studies*, R. L. Specht (ed.), 209–14. Ecosystems of the World 9B. Amsterdam: Elsevier.

Paijmans, K. (ed.) 1976. *New Guinea vegetation*. Canberra: CSIRO/Australian National University Press.

Park, G. N. 1970. Concepts in vegetation–soil system dynamics 1. Stability, climax, maturity and steady-state. *Tuatara* **18**, 132–44.

Parker, G. R. & W. T. Swank 1982. Tree species response to clear-cutting a southern Appalachian watershed. *American Midland Naturalist* **108**, 304–10.

Parrish, J. A. & F. A. Bazzaz 1976. Underground niche separation in successional plants. *Ecology* **57**, 1281–8.

Parrish, J. A. & F. A. Bazzaz 1979. Difference in pollination niche relationships in early and late successional plant communities. *Ecology* **60**, 597–610.

Parrish, J. A. D. & F. A. Bazzaz 1982a. Response of plants from three successional communities on a nutrient gradient. *Journal of Ecology* **70**, 233–48.

Parrish, J. A. D. & F. A. Bazzaz 1982b. Competitive interactions in plant communities of different successional ages. *Ecology* **63**, 314–20.

Parrish, J. A. & F. A. Bazzaz 1982c. Niche responses of early and late successional tree seedlings on three resource gradients. *Bulletin of the Torrey Botanical Club* **109**, 451–6.

Peet, R. K. 1981. Changes in biomass and production during secondary forest succession. In *Forest succession: concepts and application*, D. C. West, H. H. Shugart & D. B. Botkin (eds), 324–38. New York: Springer.

Peet, R. K. & N. L. Christensen 1980a. Succession: a population process. *Vegetatio* **43**, 131–40.

Peet, R. K. & N. L. Christensen 1980b. Hardwood forest vegetation of the North Carolina Piedmont. *Veroffentlickung Geobotanisches Institut Rübel, Zurich* **69**, 14–39.

Persson, S. 1980. Succession in a South Swedish deciduous wood: a numerical approach. In *Succession. Symposium on advances in vegetation science Nijmegen, The Netherlands, May 1979*, E. van der Maarel (ed.), 103–22. The Hague: Junk.

509

Peterson, D. L. & F. A. Bazzaz 1978. Life cycle characteristics of *Aster pilosus* in early successional habitats. *Ecology* **59**, 1005–13.

Pewe, T. L. 1966. *Permafrost and its effects on life of the North.* Corvallis: Oregon State University Press.

Phillips, D. L. & W. T. Swank 1982. Successional responses to the death of chestnut at the Coweeta Hydrologic Laboratory. *Abstract, Ecological Society of America Meeting, Pennsylvania State University 1982.*

Phillips, J. 1934, 1935. Succession, development, the climax and the complex organism: an analysis of concepts. *Journal of Ecology* **22**, 554–71; **23**, 210–46, 488–508.

Phillips, J. 1974. Effects of fire in forest and savanna ecosystems of sub-Saharan Africa. In *Fire and ecosystems*, T. T. Kozlowski & C. E. Ahlgren (eds), 435–81. New York: Academic Press.

Phares, R. 1971. Growth of red oak (*Quercus rubra*) seedlings in relation to light and nutrients. *Ecology* **52**, 669–72.

Pianka, E. R. 1970. On r- and K-selection. *American Naturalist* **104**, 592–7.

Pianka, E. R. 1974. *Evolutionary ecology.* New York: Harper & Row.

Pianka, E. R. 1981. Competition and niche theory. In *Theoretical ecology*, 2nd edn, R. M. May (ed.), 114–41. Oxford: Blackwell.

Pickett, S. T. 1983. Differential adaptation of tropical species to canopy gaps and its role in community dynamics. *Tropical Ecology* **24**, 68–84.

Pickett, S. T. & F. A. Bazzaz 1978. Organization of an assemblage of early successional species on a soil moisture gradient. *Ecology* **59**, 1248–55.

Pickett, S. T., S. L. Collins & J. J. Armesto 1987. A hierarchical consideration of causes and mechanisms of succession. *Vegetatio* **69**, 109–14.

Pielou, E. C. 1975. *Ecological diversity.* New York: Wiley.

Pielou, E. C. 1981. The usefulness of ecological models: a stock-taking. *Quarterly Review of Biology* **56**, 17–31.

Pike, L. H. 1978. The importance of epiphytic lichens in mineral cycling. *Bryologist* **81**, 247–57.

Pimm, S. L. 1982. *Food webs.* London: Chapman & Hall.

Pinder, J. E. 1975. Effects of species removal in an old-field plant community. *Ecology* **56**, 747–51.

Platt, W. J. 1975. The colonization and formation of equilibrium plant species associations on badger disturbances in a tall-grass prairie. *Ecological Monographs* **45**, 285–305.

Platt, W. J. & I. M. Weis 1977. An experimental study of competition among fugitive prairie plants. *Ecology* **66**, 708–20.

Poissonet, P., F. Romane, M. A. Austin, E. Van der Maarel & W. Schmidt 1981. *Vegetation dynamics in grasslands, heathlands and Mediterranean ligneous formations.* The Hague: Junk.

Polunin, N. 1955. Aspects of Arctic botany. *American Scientist* **43**, 307–22.

Pomeroy, L. R. 1970. The strategy of mineral cycling. *Annual Review of Ecology and Systematics* **1**, 171–90.

Poore, M. E. 1955a, b, c. The use of phytosociological methods in ecological investigations, parts I, II, III. *Journal of Ecology* **43**, 226–44, 245–69, 606–51.

Poore, M. E. D. 1956. The use of phytosociological methods in ecological investigations, IV. *Journal of Ecology* **44**, 28–50.

Poore, M. E. 1964. Integration in the plant community. *Journal of Ecology* **52**, 213–26.

Poore, M. E. D. 1968. Studies in Malaysian rain forest 1. The forest on Triassic sediments in Jengka Forest Reserve. *Journal of Ecology* **56**, 143–96.

Postgate, J. 1980. Foreword. In *Recent advances in biological nitrogen fixation*, N. S. Subba Rao (ed.), i–xiv. London: Edward Arnold.

Potzger, J. E. & R. C. Friesner 1934. Some comparisons between virgin forest and adjacent areas of secondary succession. *Butler University Botanical Studies* **3**, 85–98.

Pound, R. & F. E. Clements 1898. *The phytogeography of Nebraska 1. General survey.* Lincoln NB: North.

Prance, G. T. (ed.), 1982. *Biological diversification in the Tropics.* New York: Columbia University Press.

Precht, H., J. Christophersen, H. Hensel & W. Larcher (eds) 1973. *Temperature and life*, 2nd edn. New York: Springer.

Priester, D. S. & M. T. Pennington 1978. *Inhibitory effects of broomsedge extracts on the growth of young loblolly pine seedlings.* US Dept Agric. Forest Service Res. Paper SE–182.

Putwain, P. D. & J. L. Harper 1970. Studies in the dynamics of plant populations III. The influence of associated species on populations of *Rumex acetosa* and *R. acetosella* in grassland. *Journal of Ecology* **58**, 251–64.

Putz, F. E. 1983. Treefall pits and mounds, buried seeds and the importance of soil disturbance to pioneer trees on Barro Colorado, Panama. *Ecology* **64**, 1069–74.

Putz, F. E. & K. Milton 1982. Tree mortality rates on Barro Colorado Island. In *The ecology of a Tropical forest: seasonal rhythms and long-term changes*, E. G. Leigh, A. S. Rand & D. M. Windsor (eds), 95–100. Washington, DC: Smithsonian Institute Press.

Quarterman, E. 1957. Early plant succession on abandoned cropland in the central basin of Tennessee. *Ecology* **38**, 300–9.

Quezel, P. 1976. Les chênes sclerophylles en région méditerranéenne. *Options Méditerranéens* **35**, 25–39.

Rabotnov, T. A. 1974. Differences between fluctuations and succession. In *Vegetation dynamics*, R. Knapp (ed.), 19–24. Handbook of Vegetation Science Part 8. The Hague: Junk.

Rabotnov, T. A. 1985. Dynamics of plant coenotic populations. In *The population structure of vegetation*, J. White (ed.), 121–42. Handbook of Vegetation Science Part 3. Dordrecht: Junk.

Ramensky, L. G. 1924. The basic lawfulness in the structure of the vegetation cover. (in Russian.) *Vestnik Obytnogo Dela Voronezh* 37–73. (Also an abstract, in German: Die Grundgesetz mässigkeiten im Aufbau der vegetations-decke. *Botanisches Zentralblatt* (N.F.) **7**, 453–5.)

Ralston, C. W. & A. B. Prince 1965. Accumulation of dry matter and nutrients by pine and hardwood forests in the lower Piedmont of North Carolina. In *Forest–soil relationships in North America*, C. T. Youngberg (ed.), 77–94. Corvallis: Oregon State University Press.

Ranwell, D. S. 1972. *Ecology of salt marshes and sand dunes*. London: Chapman & Hall.

Raunkaier, C. 1934. *The life forms of plants and statistical plant geography*. Oxford: Clarendon Press.

Raup, H. M. 1951. Vegetation and cryoplanation. *Ohio Journal of Science* **51**, 105–16.

Raup, H. M. 1957. Vegetational adjustment to the instability of the site. *Proceedings and Papers of the 6th Technical Meeting of the International Union for Conservation of Nature and Nature Reserves, Edinburgh*, 36–48.

Raup, H. M. 1964. Some problems in ecological theory and their relation to conservation. *Journal of Ecology* **52** (Supplement), 19–28.

Raup, H. M. 1971. The vegetational relations of weathering, frost action and patterned ground processes. *Meddedel-ser om Grönland* **194**, 1–92.

Raup, H. M. 1975. Species versatility in shore habitats. *Journal of the Arnold Arboretum* **56**, 126–63.

Raup, H. M. 1981. Physical disturbance in the life of plants. In *Biotic crises in ecological and evolutionary time*, M. H. Nitecki (ed.), 39–52. New York: Academic Press.

Raven, P. H., R.F. Evert & H. Curtis 1985. *Biology of plants*, 3rd edn. New York: Worth.

Reardon, P. O. C.L. Leinweber & L. B. Merrill 1972. The effect of bovine saliva on grasses. *Journal of Animal Science* **34**, 897–8.

Reed, F. C. P. 1977. Plant species number, biomass accumulation and productivity of a differentially fertilized Michigan old-field. *Oecologia* **30**, 43–53.

Regehr, D.L. & F.A. Bazzaz 1979. On the population dynamics of *Erigeron canadensis*, a successional winter annual. *Journal of Ecology* **67**, 923–33.

Reichle, D. E. 1971. Energy and nutrient metabolism of soil and litter invertebrates. In *Productivity of Forest Ecosystems*, P. Duvigneaud (ed.). Proceedings of the Brussels Symposium. Paris: UNESCO.

Reichle, D. E., J. F. Franklin & D.W. Goodall (eds) 1975a. *Productivity of world ecosystems*. Washington: National Academy of Sciences.

Reichle, D. F., R. V. O'Neill & W. F. Harris 1975b. Principles of energy and material exchange in ecosystems. In *Unifying concepts in ecology*, W. H. Van Dobben & R.H. Lowe-McConnell (eds), 27–43. The Hague: Junk.

Reid, C. P. P. & F. W. Woods 1969. Translocation of C14 labelled compounds in mycorrhizae and its implications in interplant nutrient cycling. *Ecology* **50**, 179–87.

Reimold, R. J. 1977. Mangals and salt marshes of Eastern United States. In *Wet coastal ecosystems*, V. J. Chapman (ed.), 157–66. Ecosystems of the World 1. Amsterdam: Elsevier.

Reiners, W. A., I. A. Worley & D. B. Lawrence 1971. Plant diversity in a chronosequence at Glacier Bay, Alaska. *Ecology* **52**, 55–69.

Rhoades, D. F. 1977. Integrated antiherbivore, antidesiccant and ultraviolet screening properties of creosote bush resin. *Biochemical Systematics and Ecology* **5**, 281–90.

Rhoades, D. F. 1979. Evolution of plant chemical defense against herbivores. In *Herbivores: their interaction with secondary plant metabolites*, G. A. Rosenthal & D. H. Janzen (eds), 3–54. New York: Academic Press.

Rice, E. L. 1972. Allelopathic effects of *Andropogon virginicus* and its persistence in old fields. *American Journal of Botany* **59**, 752–5.

Rice, E. L. 1974. *Allelopathy*. New York: Academic Press.

Rice, E. L. 1979. Allelopathy – an update. *Botanical Review* **45**, 15–109.

Rice, E. L. & R. L. Parenti 1967. Inhibition of nitrogen-fixing and nitrifying bacteria by seed plants. V. Inhibitors produced by *Bromus japonicus. Southwestern Naturalist* **12**, 97–103.

Richards, N. A. & C. E. Farnsworth 1971. Effects of cutting level on regeneration of northern hardwoods protected from deer. *Journal of Forestry* **69**, 230–3.

Richards, P. W. 1964. *The tropical rainforest. An ecological study*, 2nd reprint. Cambridge: Cambridge University Press.

Richards, P. W. 1969. Speciation in the tropical rain forest and the concept of the niche. In *Speciation in Tropical Environments*, R. H. Lowe-McConnell (ed.), 149–54. London: Linnaean Society/Academic Press.

Richardson, J. A. 1958. The effect of temperature on the growth of plants on pit heaps. *Journal of Ecology* **46**, 537–46.

Ricklefs, R. E. 1973. *Ecology*, Newton, MA: Chiron Press.

Ridley, H. N. 1930. *The dispersal of plants throughout the world*. London: Reeve.

Rigg, G. B. 1940. The development of *Sphagnum* bogs in North America. *Botanical Review* **6**, 666–93.

Rigg, G. B. 1951. The development of *Sphagnum* bogs in North America. *Botanical Review* **17**, 109–31.

Risser, P. G., E. C. Birney, H. D. Blocker, S. W. May, W. J. Parton & J. A. Wiens 1981. *The true prairie ecosystem*. US IBP Synthesis Series 16. Stroudsburg, PA: Hutchinson & Ross.

Ritchie, I. M. 1968. *A study of forest ecosystem development*. M.Agric.Sc. Thesis, Lincoln College, New Zealand.

Roberts, H. A. 1981. Seed banks in soils. *Advances in Applied Biology* **6**, 1–55.

Roberts, H. A. & P. M. Feast 1973. Emergence and longevity of seeds of annual weeds in cultivated and undisturbed soil. *Journal of Applied Ecology* **10**, 133–43.

Root, R. B. 1967. The niche exploitation pattern of the blue-gray gnatcatcher. *Ecological Monographs* **37**, 317–48.

Root, R. B. 1973. Organization of the plant–arthropod association in simple and diverse habitats: the fauna of collards (*Brassica oleracea*) *Ecological Monographs* **43**, 95–124.

Rosenthal, G. A. & D. H. Janzen 1979. *Herbivores: their interaction with secondary plant metabolites*. New York: Academic Press.

Royama, T. 1984. Population dynamics of the spruce budworm, *Choristoneura fumiferana*. *Ecological Monographs* **54**, 429–62.

Runge, E. C. A. 1973. Soil development sequences and energy models. *Soil Science* **115**, 183–93.

Runkle, J. R. 1982. Patterns of disturbance in some old-growth mesic forests of eastern North America. *Ecology* **63**, 1533–46.

Sagar, G. R. & J. L. Harper 1961. Controlled interference with natural populations of *Plantago lanceolata, P. major* and *P. media*. *Weed Research* **1**, 163–76.

Salisbury, E. J. 1925. Note on the edaphic succession in some dune soils with special reference to the time factor. *Journal of Ecology* **13**, 322–8.

Salisbury, E. J. 1942. *The reproductive capacity of plants*. London: Bell & Sons.

Salisbury, E. J. 1952. *Downs and dunes: their plant life and its environment*. London: Bell & Sons.

Salisbury, E. 1961. *Weeds and aliens*. London: Collins.

Salo, J., R. Kalliola, I. Hakkinen, Y. Makinen, P. Niemela, M. Puhaka & P. Coley. 1986. River dynamics and the diversity of Amazon lowland forest. *Nature* **322**, 254–8.

Salt, G. W. 1979. A comment on the use of the term emergent properties. *American Naturalist* **113**, 145–8.

Sarukhan, J. 1974. Studies on plant demography. *Ranunculus repens, R. bulbosus* and *R. acris*. II. Reproductive strategies and seed population dynamics. *Journal of Ecology* **62**, 151–77.

Schaffer, W. M. & E. G. Leigh 1976. The prospective role of mathematical theory in plant ecology. *Systematic Biology* **1**, 233–45.

Schnock, G. 1967a. Course annuel de la temperature de l'habitat (sol et atmosphère) et période de végétation (1965). *Bulletin Institut Royale des Sciences naturelles de Belgique: Biologie* **43** (35), 1–15.

Schnock, G. 1967b. Thermisme comparé de l'habitat dans la forêt et la prairie permanente. *Bulletin Institut Royale des Sciences naturelles de Belgique: Biologie* **43** (36), 1–17.

Schnock, G. 1971. Le bilan de l'eau dans l'écosystème forêt. Application à une chênaie mélangée de haute Belgique. In *Productivity of forest ecosystems*, P. Duvigneaud (ed.), Paris: UNESCO.

Schoenike, R. E. 1957. Influence of mountain avens (*Dryas drummondii*) on growth of young cotton woods (*Populus trichocarpa*) at Glacier Bay, Alaska. *Proceedings of the Minnesota Academy of Science* **25**, 55–8.

Schopmeyr, C. S. 1974. *Seeds of woody plants in the United States*. US Dept. Agric. Handbook No. 450, Washington, DC.

Scott, D. 1974. Description of relationships between plants and environment. In *Vegetation and environment*, B. R. Strain & W. D. Billings (eds), 49–72. Handbook of Vegetation Science Part 6. The Hague: Junk.

Scott, D. 1961. Influence of tussock grasses on zonation of accompanying smaller species. *New Zealand Journal of Science* **4**, 116–22.

Seitz, R. 1986. Siberian fire as 'nuclear winter' guide. *Nature* **323**, 116–7.

Selleck, G. W. 1960. The climax concept. *Botanical Review* **26**, 534–45.

Shafi, M. I. & G. A. Yarranton 1977. Vegetational heterogeneity during a secondary (post fire) succession. *Canadian Journal of Botany* **51**, 73–90.

Shanks, R. E. & J. S. Olson 1961. First-year breakdown of leaf litter in southern Appalachian forests. *Science* **134**, 194–5.

Sharitz, R. R. & J. F. McCormick 1973. Population dynamics of two competing annual plant species. *Ecology* **54**, 723–39.

Sharpe, D. M., K. Cromack, W. C. Johnson & B. S. Ausmus 1980. A regional approach to litter dynamics in southern Appalachian forests. *Canadian Journal of Forestry Research* **10**, 395–404.

Sheldon, J. C. 1974. The behaviour of seeds in soil III. The influence of seed morphology and the behaviour of seedlings on the establishment of plants from surface lying seeds. *Journal of Ecology* **62**, 47–66.

Shelford, V. E. & S. Olson 1935. Sere, climax and influent animals with special reference to the transcontinental coniferous forest of North America. *Ecology* **16**, 375–401.

Shimwell, D. W. 1971. *Description and classification of vegetation*. London: Sidgwick & Jackson.

Shreve, F. 1914. *A montane rain forest*. Carnegie Inst., Washington, Pub. No. 199, 1–110.

Shreve, F. 1915. *The vegetation of a desert mountain range, as conditioned by climatic factors*. Carnegie Inst., Washington, Pub. No. 217, 1–112.

Shreve, F. 1917. The establishment of desert perennials. *Journal of Ecology* **5**, 210–6.

Shreve, F. 1942. The desert vegetation of North America. *Botanical Review* **8**, 195–246.

Shreve, F. & A. L. Hinckley 1937. Thirty years of change in desert vegetation. *Ecology* **18**, 463–78.

Shugart, H. H. 1984. *A theory of forest dynamics*. New York: Springer.

Shugart, H. H. & J. M. Hett 1973. Succession: similarities of species turnover rates. *Science* **180**, 1379–81.

Shugart, H. H. & I. R. Noble 1981. A computer model of succession and fire response of the high-altitude *Eucalyptus* forest of the Brindabella Range, Australian Capital Territory. *Australian Journal of Ecology* **6**, 149–64.

Shugart, H. H. & D. C. West 1977. Development of an Appalachian deciduous forest succession model and its application to assessment of the chestnut blight. *Journal of Environmental Management* **5**, 161–79.

Shukla, R. P. & P. S. Ramakrishnan 1984. Leaf dynamics of tropical trees related to successional status. *New Phytologist* **97**, 697–706.

Siccama, T. G., F. H. Bormann & G. E. Likens 1970. The Hubbard Brook ecosystem study: productivity, nutrients and phytosociology of the herbaceous layer. *Ecological Monographs* **40**, 389–402.

Sigafoos, R. S. 1952. Frost action as a primary physical factor in tundra plant communities. *Ecology* **33**, 480–7.

Sigafoos, R. S. & E. L. Hendricks 1969. The time interval between stabilization of alpine glacial deposits and establishment of tree seedlings. *US Geological Survey Professional Paper* **650–1B**, B89–B93.

Silvertown, J. W. 1980a. Leaf-canopy induced seed dormancy in a grassland flora. *New Phytologist* **85**, 109–18.

Silvertown, J. W. 1980b. The evolutionary ecology of mast seeding in trees. *Biological Journal of the Linnaean Society* **14**, 235–50.

Silvertown, J. W. 1982. *Introduction to plant population ecology*. London: Longman.

Silvertown, J. 1985. Fertile arguments in the desert. *Ecology* **316**, 298.

Silvester, W. B. 1969. *Nitrogen fixation by root nodules of Coriaria*. Ph.D. Thesis, Lincoln College, New Zealand.

Silvester, W. B. 1977. Dinitrogen fixation by plant associations excluding legumes. In *A treatise of dinitrogen fixation*, R. W. Hardy & A. H. Gibson (eds), 141–257. Section IV: *Agronomy and Ecology*. New York: Wiley.

Singh, R. N. 1961. *Role of blue-green algae in nitrogen economy of Indian agriculture*. New Delhi: Indian Council of Agricultural Research.

Sjörs, H. 1950. On the relation between vegetation and electrolytes in north Swedish mire water. *Oikos* **2**, 241–58.

Sjörs, H. 1961. Surface patterns in Boreal peatland. *Endeavour* **20**, 217–24.

Sjörs, H. 1980. An arrangement of changes along gradients, with examples from succession in boreal peatland. *Vegetatio* **43**, 1–4.

Skutch, A. F. 1929. Early stages of plant succession following forest fire. *Ecology* **10**, 177–90.

Slatyer, R. O. 1967. *Plant–water relationships*. London: Academic Press.

Smathers, G. A. & D. Mueller-Dombois 1974. *Invasion and recovery of vegetation after a volcanic eruption in Hawaii*. Hawaiian Natn. Park Service Sci. Monog. Ser. No. 5.

Smeck, N. E. 1973. Phosphorus: an indicator of pedogenetic weathering processes. *Soil Science* **115**, 199–206.

Smith, A. P. 1979. The paradox of autotoxicity in plants. *Evolutionary Theory* **4**, 173–80.

Smith, H. 1981. Light quality as an ecological factor. In *Plants and their atmospheric environment*, J. Grace, E. D. Ford & P. G. Jarvis (eds), 93–110. British Ecological Society 21st Symposium, Edinburgh, 1979. Oxford: Blackwell.

Smith, J. M. B. 1978. Dispersal and establishment of plants. in *Biology and quaternary environments*, D. Walker & J. C. Guppy (eds), 207–23. Canberra: Australian Academy of Science.

Sohlberg, E. H. & L. C. Bliss 1984. Microscale pattern of vascular plant distribution in two high arctic plant communities. *Canadian Journal of Botany* **62**, 2033–42.

Soil Survey staff 1975. *Soil taxonomy*. US Dept Agric. Handbook No. 436. Washington, DC.

Sparkes, C. H. & M. F. Buell 1955. Microclimatological features of an old field and an oak-hickory forest in New Jersey. *Ecology* **36**, 363–4.

513

Specht, R. L. (ed.) 1979. *Heathland and related shrublands*. Ecosystems of the World 9A. Amsterdam: Elsevier.

Specht, R. L. (ed.) 1981. *Heathlands and related shrublands*. Ecosystems of the World 9B. Amsterdam: Elsevier.

Sprague, R. & D. B. Lawrence 1960. The fungi on deglaciated Alaskan terrain of known age III. *Washington State University (Pullman) Research Studies* **28**, 1–20.

Sprugel, D. G. 1976. Dynamic structure of wave-generated *Abies balsamea* forest in the northeastern United States. *Journal of Ecology* **64**, 889–912.

Spurr, S. H. 1956. Natural restocking of forests following the 1938 hurricane in central New England. *Ecology* **37**, 443–51.

Spurr, S. H. & B. V. Barnes 1980. *Forest ecology*, 3rd edn. New York: Wiley.

Stapanian, M. A. 1986. Seed dispersal by birds and squirrels in the deciduous forests of the United States. In *Frugivores and seed dispersal*, A. Estrada & T. H. Fleming (eds), 225–36. The Hague: Junk.

Stark, N. 1978. Man, tropical forests and the biological life of a soil. *Biotropica* **10**, 1–10.

Stark, N. & C. F. Jordan 1978. Nutrient retention by the root mat of an Amazonian rain forest. *Ecology* **59**, 434–7.

Stearns, F. W. 1949. Ninety years change in a northern hardwood forest in Wisconsin. *Ecology* **30**, 350–8.

Stebbins, G. L. 1950. *Variation and evolution in plants*. New York: Columbia University Press.

Stebbins, G. L. 1981. Coevolution of grasses and herbivores. *Annals of the Missouri Botanical Garden* **68**, 75–86.

Steers, J. A. 1977. Physiography. In *Wet coastal ecosystems*, V. J. Chapman (ed.), 31–60. Ecosystems of the World 1. Amsterdam: Elsevier.

Stephens, E. P. 1956. The uprooting of trees: a forest process. *Soil Science Society of America Proceedings* **20**, 113–6.

Stephens, G. R. 1981. *Defoliation and mortality in Connecticut forests*. Connecticut Agric. Exp. Stn, New Haven, Bull. No. 796.

Stephens, G. R. & P. E. Waggoner 1970. *The forests anticipated from 40 years of natural transitions in mixed hardwoods*. Connecticut Agric. Exp. Stn, New Haven, Bull. No. 707.

Stephens, G. R. & P. E. Waggoner 1980. *A half-century of natural transitions in mixed hardwood forests*. Connecticut Agric. Exp. Stn, New Haven, Bull. No. 783.

Stevens, P. R. 1963. *A chronosequence of soils and vegetation near the Franz Josef glacier*. M.Agr.Sc. Thesis. Lincoln College, New Zealand.

Stevens, P. R. 1968. *A chronosequence of soils near the Franz Josef glacier*. Ph.D. Thesis, Lincoln College, New Zealand.

Stevens, P. R. & T. W. Walker 1970. The chronosequence concept and soil formation. *The Quarterly Review of Biology* **45**, 333–50.

Stewart, D. R. & J. Stewart 1971. Comparative food preferences of five East African ungulates at different seasons. In *The scientific management of animal and plant communities for conservation*, E. Duffey & A. S. Watt (eds), 351–66. 11th Symposium of the British Ecological Society. Oxford: Blackwell.

Stewart, G. R., F. Larcher, I. Ahmad & J. A. Lee 1979. Nitrogen metabolism and salt-tolerance in higher plant halophytes. In *Ecological processes in coastal environments*, R. L. Jefferies & A. J. Drury (eds), 211–28. 19th Symposium of the British Ecological Society. Oxford: Blackwell.

Stewart, W. D. P. 1967. Nitrogen-fixing plants. *Science* **158**, 1426–32.

Stone, E. C., R. F. Grah & P. J. Zinke 1972. Preservation of the primeval redwoods in the Redwood National Park, Part 1. *American Forests* **78**, 50–5.

Strong, D. R. 1977. Epiphyte loads, tree falls and perennial forest disruption: a mechanism for maintaining higher tree species richness in the tropics without animals. *Journal of Biogeography* **4**, 215–8.

Subba Rao, N. S. (ed.) 1980. *Recent advances in biological nitrogen fixation*. London: Edward Arnold.

Sutherland, W. J. & A. R. Watkinson 1986. Do plants evolve differently? *Nature* **320**, 305.

Sutton, S. L. 1983. The spatial distribution of flying insects in tropical rain forests. In *Tropical rain forest: ecology and management*, S. L. Sutton, T. C. Whitmore & A. C. Chadwick (eds), 77–91. Oxford: Blackwell.

Swan, F. R. 1970. Post-fire response of four plant communities in South-Central New York State. *Ecology* **51**, 1074–82.

Sweeney, J. R. 1956. Responses of vegetation to fire. *University of California Publications in Botany* **28**, 143–250.

Sweeney, J. R. 1968. Ecology of some 'fire type' vegetation in northern California. In *Proceedings of Tall Timbers fire ecology conference, Hoberg, California, November 9–10, 1967*, 111–25. Tallahassee, FL: Tall Timbers Research Station.

Swieringa, S. & R. E. Wilson 1972. Phenodynamic analyses of two first-year old fields. *American Journal of Botany* **59**, 367–72.

Swift, M. J., O. W. Heal & J. M. Anderson 1979. *Decomposition in terrestrial ecosystems*. Studies in Ecology 5. Oxford: Blackwell.

Sydes, C. & J. P. Grime 1981. Effects of tree litter on herbaceous vegetation in deciduous woodland I. *Journal of Ecology* **69**, 237–62.

Syers, J. K. & T. W. Walker 1969a, b. Phosphorus transformations in a chronosequence of soils developed on wind blown sand in New Zealand, I & II. *Journal of Soil Science* **20**, 57–64, 219–24.

Szeicz, G. 1974. Solar radiation in plant canopies. *Journal of Applied Ecology* **11**, 1117–56.

Tamm, C. O. 1954. Some observations on the nutrient turn-over in a bog community dominated by *Eriophorum vaginatum*. *Oikos* **5**, 189–94.

Tamm, C. O. 1975. Plant nutrients as limiting factors in ecosystem dynamics. In *Productivity of world ecosystems*, D. E. Reichle, J. F. Franklin & D. W. Goodall (eds), 123–32. Washington, DC: National Academy of Sciences.

Tansley, A. G. 1916. The development of vegetation. A review of Clements' 'Plant Succession' 1916. *Journal of Ecology* **4**, 198–204.

Tansley, A. G. 1920. The classification of vegetation and the concept of development. *Journal of Ecology* **8**, 118–49.

Tansley, A. G. 1929. Succession: the concept and its values. *Proceedings of the 4th international congress of plant sciences, Ithaca, 1926*, 677–86.

Tansley, A. G. 1935. The use and abuse of vegetational concepts and terms. *Ecology* **16**, 284–307.

Tansley, A. G. 1939. *The British islands and their vegetation*. London: Cambridge University Press.

Tarr, R. S. 1909. The Yakutat Bay Region, Alaska. *US Geological Survey Professional Papers* **64**, 1–183.

Taylor, B. W. 1957. Plant succession on recent volcanoes in Papua. *Journal of Ecology* **45**, 233–43.

Taylor, N. H. & I. J. Pohlen 1962. *Soil survey method*. New Zealand Department of Scientific and Industrial Research, Soil Bureau Bulletin No. 25. Wellington: Government Printer.

Taylor, R. J. & R. W. Pearcy 1976. Seasonal patterns of CO_2 exchange characteristics of understorey plants from a deciduous forest. *Canadian Journal of Botany* **54**, 1094–103.

Terborgh, J. 1986. Community aspects of frugivory in tropical forests. In *Frugivores and seed dispersal*, A. Estrada & T. H. Fleming (eds), 371–84. Dordrecht: Junk.

Thom, B. G. 1967. Mangrove ecology and deltaic geomorphology: Tabasco, Mexico. *Journal of Ecology* **55**, 301–43.

Thompson, C. H. 1980. *Podzols on coastal sand dunes at Cooloola, Queensland*. Australian Soc. Soil Sci. Natn. Conf. Paper No. 117, Sydney, 1980.

Thompson, C. H. 1981. Podzol chronosequences on coastal dunes of eastern Australia. *Nature* **291**, 59–61.

Thompson, D. Q. & R. H. Smith 1971. The forest primeval in the Northeast – a great myth? *Proc. annual Tall Timbers fire ecol. conf.* No. 10, 255–66.

Thompson, J. N. 1982. *Interaction and evolution*. New York: Wiley.

Tierson, W. C., E. F. Patric & D. F. Behrend 1966. Influence of white-tailed deer on the logged northern hardwood forest. *Journal of Forestry* **64**, 801–5.

Tilman, D. 1978. Cherries, ants and tent caterpillars: timing of nectar production in relation to susceptibility of caterpillars to ant predation. *Ecology* **59**, 686–92.

Tilman, D. 1982. *Resource competition and community structure*. Monographs in Population Biology No. 17. Princeton, NJ: Princeton University Press.

Timmins, S. M. 1981. *The vegetation of Mount Tarawera, Rotorua, New Zealand. A study using Landsat digital data*. M.Sc. Thesis, University of Waikato, New Zealand

Ting, I. P. & M. Gibbs (eds) 1982. *Crassulacean acid metabolism*. Rockville, MD: American Society of Plant Physiologists.

Tisdale, E. W., M. A. Fosberg & C. F. Poulton 1966. Vegetation and soil development on a recently glaciated area near Mount Robson, British Columbia. *Ecology* **47**, 517–23.

Tolstead, W. L. 1941. Plant communities and secondary succession in south-central South Dakota. *Ecology* **22**, 322–8.

Trappe, J. M. & R. D. Fogel 1977. Ecosystematic functions of mycorrhizae. In *The below-ground ecosystem: a synthesis of plant-associated processes*, J. K. Marshall (ed.), 205–14. Range Science Department, Science Series No. 26. Fort Collins, CO: Colorado State University.

Trautman, M. B. 1966. Birds. In *Soil development and ecological succession in a deglaciated area of Muir Inlet, Southwest Alaska*, A. Mirsky (ed.), 121–46. Ohio St. Univ. Inst. Polar Studies. Rep. No. 20. Columbus, Ohio.

Treub, M. 1888. Notice sur la nouvelle flore de Krakatau. *Annales du Jardin Botanique de Buitenzorg* **7**, 213–23.

Troll, C. 1978. *Geoecological relations between the Southern Temperate Zone and Tropical mountains*. Wiesbaden: F. Steiner.

Turkington, R. A. & J. L. Harper 1979. The growth, distribution and neighbour relationships of *Trifolium repens* in a permanent pasture 4. Fine-scale biotic differentiation. *Canadian Journal of Botany* **57**, 245–54.

Turkington, R. A., P. B. Cavers & L. W. Aarssen 1977. Neighbour relationships in grass–legume communities. 1. Interspecific contacts in four grassland communities near London, Ontario. *Canadian Journal of Botany* **55**, 2701–11.

Turner, E. P. 1928. A brief account of the reestablishment of vegetation on Tarawera mountain since the eruption of 1886. *Transactions of the New Zealand Institute* **59**, 60–6.

Ugolini, F. C. 1966. Soils. In *Soil development and ecological succession in a deglaciated area of Muir Inlet, Southwest Alaska*, A. Mirsky (ed.), 29–58. Ohio St. Univ. Inst. Polar Studies. Rep. No. 20. Columbus, Ohio.

Ugolini, F. C. 1968. Soil development and alder invasion in a recently deglaciated area of Glacier Bay, Alaska. In *Biology of alder*, J. M. Trappe, J. F. Franklin, R. F. Tarrant & G. M. Hansen (eds), 115–40. Portland OR: US Forest Service Pacific Northwest Forest and Range Experiment Station.

Ugolini, F. C. & D. H. Mann 1979. Biopedological origin of peatlands in southeast Alaska. *Nature* **281**, 366–8.

Urban, D. L., R. V. O'Neill & H. H. Shugart 1987. Landscape ecology. *BioScience* **37**, 119–27.

Usher, M. B. 1981. Modelling ecological succession, with particular reference to Markovian models. In *Vegetation dynamics in grasslands, heathlands and Mediterranean ligneous formations*, P. Poissonet, F. Romane, M. P. Austin & E. van der Maarel (eds), 11–18. The Hague: Junk.

Vaartaja, O. 1952. Forest humus quality and light conditions as factors influencing damping-off. *Phytopathology* **42**, 501–6.

Vaartaja, O. 1959. Evidence of photoperiodic ecotypes in trees. *Ecological Monographs* **29**, 91–111.

Valentine, J. W. 1968. The evolution of ecological units above the population level. *Journal of Paleontology* **42**, 253–67.

Van Borssum Waalkes, J. 1960. Botanical observations on the Krakatau Islands in 1951 and 1952. *Annales Bogoriensis* **4**, 5–64.

Van Cleve, K. & L. A. Viereck 1972. Distribution of selected chemical elements in even-aged alder (*Alnus*) ecosystems near Fairbanks, Alaska. *Arctic and Alpine Research* **4**, 239–55.

Van Cleve, K. & L. A. Viereck 1981. Forest succession in relation to nutrient cycling in the Boreal forest of Alaska. In *Forest succession: concepts and application*, D. C. West, H. H. Shugart & D. B. Botkin (eds), 185–211. New York: Springer.

Van Cleve, K., L. A. Viereck & R. L. Schleutner 1970. Accumulation of nitrogen in alder ecosystems developed on the Tanana River flood plain near Fairbanks, Alaska. *Arctic and Alpine Research* **3**, 101–14.

Van den Bergh, J. P. & W. G. Braakhekke 1978. Coexistence of plant species by niche differentiation. In *Structure and functioning of plant populations*, A. H. J. Freysen & J. W. Woldendorp (eds), 125–38. Amsterdam: North–Holland.

Van der Maarel, E. (ed.) 1980. *Succession*. Symposium on advances in vegetation sciences, Nijmegen, May 1979. The Hague: Junk.

Van der Pijl, L. 1972. *Principles of dispersal in higher plants*. Berlin: Springer.

Van Hulst, R. 1979a. On the dynamics of vegetation: succession in model communities. *Vegetatio* **39**, 85–96.

Van Hulst, R. 1979b. On the dynamics of vegetation: Markov chains as models of succession. *Vegetatio* **40**, 3–14.

Vankat, J. L., W. H. Blackwell & W. E. Hopkins 1975. The dynamics of Hueston Woods and a review of the successional status of the southern beech–maple forest. *Castanea* **40**, 290–308.

Van Steenis, C. G. G. J. 1969. Plant speciation in Malesia, with special reference to the theory of non-adaptive saltatory evolution. *Biological Journal of the Linnaean Society* **1**, 97–133.

Van Valen, L. M. 1980. Evolution as a zero-sum game for energy. *Evolutionary Theory* **4**, 289–300.

Vasek, F. C. 1980a. Creosote bush: long-lived clones in the Mojave Desert. *American Journal of Botany* **67**, 246–55.

Vasek, F. C. 1980b. Early successional stages in Mojave Desert scrub vegetation. *Israel Journal of Botany* **28**, 133–48.

Vazquez-Yanes, C. & H. Smith 1982. Phytochrome control of seed germination in the tropical rain forest pioneer trees *Cecropia obtusifolia* and *Piper auritum* and its ecological significance. *New Phytologist* **92**, 477–85.

Veblen, T. T. 1979. Structure and dynamics of *Nothofagus* forests near timberline in south-central Chile. *Ecology* **60**, 937–45.

Veblen, T. T., D. H. Ashton, F. M. Schlegel & A. T. Veblen 1977. Plant succession in a timberline depressed by vulcanism in south-central Chile. *Journal of Biogeography* **4**, 275–94.

Veblen, T. T., F. M. Schlegel & R. Escobar 1980. Structure and dynamics of old-growth *Nothofagus* forests in the Valdivian Andes, Chile. *Journal of Ecology* **68**, 1–31.

Venable, D. L. & L. Lawlor 1980. Delayed germination and dispersal in desert annuals: escape in space and time. *Oecologia* **46**, 272–82.

Viereck, L. A. 1966. Plant succession and soil development on gravel outwash of the Muldrow glacier, Alaska. *Ecological Monographs* **36**, 181–99.

Viereck, L. A. 1970. Forest succession and soil development adjacent to the Chena River in interior Alaska. *Arctic and Alpine Research* **2**, 1–26.

Vitousek, P. M. & W. A. Reiners 1975. Ecosystem succession and nutrient retention: a hypothesis. *BioScience* **25**, 376–81.

Vogel, S. 1968. 'Sun leaves' and 'shade leaves': differences in convective heat dissipation. *Ecology* **49**, 1203–4.

516

Vogl, R. J. 1974. Effects of fires on grasslands. In *Fire and ecosystems*, T. T. Kozlowski & C. E. Ahlgren (eds), 139–94. New York: Academic Press.

Vogl, R. 1980. The ecological factors that produce perturbation-dependent ecosystems. In *The recovery process in damaged ecosystems*, J. Cairns (ed.), 63–94. Ann Arbor, MI: Ann Arbor Scientific Publications.

Wali, M. (ed.) 1975. *Prairie: a multiple view*. Grand Forks: University of North Dakota Press.

Walker, B. H. 1981. Is succession a viable concept in African savanna ecosystems? In *Plant succession: concepts and application*, D. C. West, H. H. Shugart & D. B. Botkin (eds), 431–48. New York: Springer.

Walker, D. 1970. Direction and rate in some British post-glacial hydroseres. In *Studies in the vegetational history of the British Isles*, D. Walker & R. G. West (eds), 117–39. London: Cambridge University Press.

Walker, D. 1982. The development of resilience in burned vegetation. In *The plant community as a working mechanism*, E. I. Newman (ed.), 27–43. Br. Ecol. Soc. Spec. Publ. No. 1. Oxford: Blackwell.

Walker, E. P. 1964. *Mammals of the world*. Baltimore: Johns Hopkins Press.

Walker, J., C. H. Thompson, I. F. Fergus & B. R. Tunstall 1980. Plant succession and soil development in coastal sand dunes of subtropical Eastern Australia. In *Forest succession: concepts and applications*, H. H. Shugart, D. Botkin & D. West (eds), 107–31. New York: Springer.

Walker, T. W. & J. K. Syers 1976. The fate of phosphorus during pedogenesis. *Geoderma* **15**, 1–19.

Wallace, L. L. & E. L. Dunn 1980. Comparative photosynthesis of three gap phase successional tree species. *Oecologia* **45**, 331–40.

Walter, H. & E. Stadelmann 1974. A new approach to the water relations of desert plants. In *Desert biology*, G. W. Brown (ed.), Vol. 2, 213–310. New York: Academic Press.

Wardle, J. A. 1984. *The New Zealand beeches: ecology, utilization and management*. Christchurch: New Zealand Forest Service.

Wardle, P. 1973. Variations of the glaciers of Westland National Park and the Hooker Range, New Zealand. *New Zealand Journal of Botany* **11**, 349–88.

Wardle, P. 1977. Plant communities of Westland National Park (New Zealand) and neighbouring lowland and coastal areas. *New Zealand Journal of Botany* **15**, 323–98.

Wardle, P. 1980. Plant succession in Westland National Park and its vicinity. *New Zealand Journal of Botany* **18**, 221–32.

Warington, K. 1936. The effect of constant and fluctuating temperatures on the germination of the weed seeds in arable soil. *Journal of Ecology* **24**, 104–185.

Warming, E. 1891. De psammophile vormationer i Danmark. *Videnskabelige Meddelelser Fra den Naturhistorisk Forening i Kjobenhavn* **1891**, 153.

Watson, E. V. 1968. *British mosses and liverworts*. Cambridge: Cambridge University Press.

Watt, A. S. 1925. On the ecology of British beechwoods with special reference to their regeneration II. The development and structure of beech communities on the South Downs. *Journal of Ecology* **13**, 27–73.

Watt, A. S. 1947. Pattern and process in the plant community. *Journal of Ecology* **35**, 1–22.

Watt, A. S. 1955. Bracken versus heather, a study of plant sociology. *Journal of Ecology* **43**, 490–506.

Watt, A. S. 1960. Population changes in acidophilous grass heath in Breckland 1936–1957. *Journal of Ecology* **48**, 605–29.

Watt, A. S. 1962. The effect of excluding rabbits from grassland A (*Xerobrometum*) in Breckland, 1936–60. *Journal of Ecology* **50**, 181–98.

Watt, K. E. F. 1975. Critique and comparison of biome ecosystem modeling. In *Systems analysis and simulation in ecology*, B. C. Pattern (ed.), Vol. III, 139–55. New York: Academic Press.

Weaver, J. E. 1942. Role of seedlings in recovery of midwestern ranges from drought. *Ecology* **23**, 275–94.

Weaver, J. E. 1950. Stabilization of midwestern grassland. *Ecological Monographs* **20**, 251–70.

Weaver, J. E. 1958. Summary and interpretation of underground development in natural grassland communities. *Ecological Monographs* **28**, 55–78.

Weaver, J. E. 1968. *Prairie plants and their environment: a fifty-year study in the Midwest*. Lincoln, NB: University of Nebraska Press.

Weaver, J. E. & F. W. Albertson 1940. Deterioration of Midwestern ranges. *Ecology* **21**, 216–36.

Weaver, J. E. & F. W. Albertson 1956. *Grasslands of the Great Plains: their nature and use*. Lincoln, NB: Johsen.

Weaver, J. E. & F. E. Clements 1929. *Plant ecology*. New York: McGraw-Hill.

Weaver, J. E. & F. E. Clements 1938. *Plant ecology*, 2nd edn. New York: McGraw-Hill.

Webb, L. J., J. G. Tracey & K. P. Haydock 1967. A factor toxic to seedlings of the same species associated with living roots of the nongregarious subtropical rainforest tree *Grevillea robusta*. *Journal of Applied Ecology* **4**, 13–25.

Weber, C. A. 1910. Was lehrt die die Aufbau den Moore Nordwestdeutschlands über den Wechsel des Klimas in postglazialer Zeit? *Zeitschrift für Deutschen Geologischen Gesellschaft* **62**, 143–62.

Webster, J. B., J. B. Waide & B. D. Patten 1974. Nutrient recycling and the stability of ecosystems. in *Mineral cycling in Southeastern ecosystems*, F. G. Howell, J. B. Gentry & M. H. Smith (eds), 1–27. Springfield, VA: Energy Research and Development Administration Symposium Series.

Wein, R. S. & J. M. Moore 1977. Fire history and rotations in the New Brunswick Acadia forest. *Canadian Journal of Forest Research* **7**, 285–94.

Wells, P. V. 1976. A climax index for broadleaf forest: an *n*-dimensional ecomorphological model of succession. In *Central hardwood forest conference*, J. S. Fralish, G. J. Weaver & R. C. Schlesinger (eds), 131–76. Carbondale, IL: Department of Forestry, Southern Illinois University.

Went, F. W. 1948. Ecology of desert plants I. Observations on germination in the Joshua Tree National Monument, California. *Ecology* **29**, 242–53.

Went, F. W. 1949. Ecology of desert plants II. The effect of rain and temperature on germination and growth. *Ecology* **30**, 1–13.

Werner, P. A. & W. J. Platt 1976. Ecological relationships of co-occurring goldenrods (Solidago: Compositae). *American Naturalist* **110**, 959–71.

Wernstedt, F. L. 1972. *World climatic data*. Lemont, PA: Climatic Data Press.

Wesson, G. & P. F. Wareing 1969. The role of light in germination of naturally-occurring populations of buried weed seeds. *Journal of Experimental Botany* **20**, 402–13.

West, D. C., H. H. Shugart & D. B. Botkin. 1981. *Forest succession: concepts and applications*. New York: Springer.

West, N. E. (ed.) 1983. *Temperate deserts of the world*. Ecosystems of the World 5. Amsterdam: Elsevier.

West, R. G. 1971. *Studying the past by pollen analysis*. Oxford Biology Readers No. 10. London: Oxford University Press.

Wetzel, R. G. 1975. *Limnology*. Philadelphia: Saunders.

Whalley, R. D. B. & A. A. Davidson 1968. Physiological aspects of drought dormancy in grasses. *Proceedings of the Ecological Society of Australia* **3**, 17–9.

Whatley, J. M. & F. R. Whatley 1980. *Light and plant life*. Studies in Biology No. 124. London: Edward Arnold.

White, F. 1976. The underground forests of Africa: a preliminary review. *Gardens Bulletin* **39**, 57–71.

White, J. 1979. The plant as a metapopulation. *Annual Review of Ecology and Systematics* **10**, 109–45.

White, J. (ed.) 1985. *The population structure of vegetation*. Handbook of Vegetation Science Part 3. Dordrecht: Junk.

White, P. S. 1979. Pattern, process and natural disturbance in vegetation. *Botanical Review* **45**, 229–99.

Whiteside, M. C., J. P. Bradbury & S. J. Tarapchack 1980. Limnology of the Klutlan moraines, Yukon Territory, Canada. *Quaternary Research* **14**, 130–48.

Whitmore, T. C. 1975. *Tropical rain forests of the Far East*. Oxford: Oxford University Press.

Whitmore, T. C. 1982. On pattern and process in forest. In *The plant community as a working mechanism*, E. I. Newman (ed.), 45–59. British Ecological Society Special Publication No. 1. Oxford: Blackwell.

Whitmore, T. C. 1985. Forest succession. *Nature* **315**, 692.

Whitney, G. G. 1976. The bifurcation ratio as an indicator of adaptive strategy in woody plant species. *Bulletin of the Torrey Botanical Club* **103**, 67–72.

Whitney, G. G. 1984. Fifty years of change in the arboreal vegetation of Heart's Content, an old-growth hemlock–white pine–northern hardwood stand. *Ecology* **65**, 403–8.

Whittaker, R. H. 1951. A criticism of the plant association and climatic climax concepts. *Northwest Science* **25**, 17–31.

Whittaker, R. H. 1953. A consideration of climax theory: the climax as a population and pattern. *Ecological Monographs* **23**, 41–78.

Whittaker, R. H. 1956. Vegetation of the Great Smoky Mountains. *Ecological Monographs* **26**, 1–80.

Whittaker, R. H. 1962. Classification of natural communities. *Botanical Review* **28**, 1–239.

Whittaker, R. H. 1965. Dominance and diversity in land plant communities. *Science* **147**, 250–60.

Whittaker, R. H. 1967. Gradient analysis of vegetation. *Biological Reviews* **41**, 207–64.

Whittaker, R. H. 1969a. New concepts of kingdoms of organisms. *Science* **163**, 150–60.

Whittaker, R. H. 1969b. Evolution of diversity in plant communities. In *Diversity and stability in ecological systems*, G. M. Woodwell & H. H. Smith (eds), 178–95. Brookhaven Symposium in Biology No. 22. New York: Upton.

Whittaker, R. H. 1972. Evolution and measurement of species diversity. *Taxon* **21**, 213–51.

Whittaker, R. H. 1973. *Ordination and classification of communities*. Handbook of Vegetation Science, Part 5. The Hague: Junk.

Whittaker, R. H. 1974. Climax concepts and recognition. In *Vegetation dynamics* R. Knapp (ed.), 137–54. Handbook of Vegetation Science Part 8. The Hague: Junk.

Whittaker, R. H. 1975a. Functional aspects of succession in deciduous forests. In *Sukzessionforschung*, W. Schmidt (ed.), 377–96. Berichte der Internationalen Symposien der Internationalen Vereinigung für Vegetationskunde Herausgegeben von Reinhold Tuxen. Vaduz: J. Cramer.

Whittaker, R. H. 1975b. *Communities and ecosystems*, 2nd edn. New York: Macmillan.

REFERENCES

Whittaker, R. H. 1977. Evolution of species diversity in land communities. *Evolutionary Biology* **10**, 1–67.

Whittaker, R. H. & S. A. Levin 1975. *Niche theory and application*. Benchmark Papers in Ecology 3. New York: Wiley.

Whittaker, R. H. & G. E. Likens 1975. The biosphere and man. In *Primary productivity of the biosphere*, H. Lieth & R. H. Whittaker (eds), 305–28. Berlin: Springer.

Whittaker, R. H. & W. A. Niering 1965. Vegetation of the Santa Catalina Mountains, Arizona. I. A gradient analysis of the south slope. *Ecology* **46**, 429–52.

Whittaker, R. H. & G. M. Woodwell 1968. Dimension and production relations of trees and shrubs in the Brookhaven forest, New York. *Journal of Ecology* **56**, 1–25.

Whittaker, R. J., M. B. Bush & K. Richards. 1989. Plant recolonization and vegetation succession on the Krakatau Islands, Indonesia. *Ecological Monographs* **59**, 59–123.

Wieland, N. K. & F. A. Bazzaz 1975. Physiological ecology of three codominant succession annuals. *Ecology* **56**, 681–8.

Williams, W. T., J. M. Lambert & G. N. Lance 1966. Multivariate methods in plant ecology V. Similarity analyses and information analysis. *Journal of Ecology* **54**, 427–45.

Williamson, G. B. 1975. Pattern and seral composition in an old-growth beech–maple forest. *Ecology* **56**, 727–31.

Wilm, H. G. 1936. The relation of successional development to the silviculture of forest burn communities in southern New York. *Ecology* **17**, 283–91.

Wilson, E. O. 1969. The species equilibrium. *Brookhaven Symposia in Biology* **22**, 38–47.

Wilson, E. O. 1981. The central problems of sociobiology. In *Theoretical ecology*, 2nd edn. R. M. May (ed.), 205–17. Oxford: Blackwell.

Wilson, R. E. 1970. The role of allelopathy in old-field succession on grassland areas of central Oklahoma. *Proceedings of the Symposium on Prairie Restoration* **1968**, 24–5.

Winterringer, G. S. & A. G. Vestal 1956. Rock ledge vegetation in southern Illinois. *Ecological Monographs* **26**, 105–30.

Woldendorp, J. W. 1978. The rhizosphere as part of the plant–soil system. In *Structure and functioning of plant populations*, A. H. J. Freyson & J. W. Woldendorp (eds), 237–68. Amsterdam: North–Holland.

Woods, D. B. & N. C. Turner 1971. Stomatal response to changing light by four tree species of varying shade tolerance. *New Phytologist* **70**, 77–84.

Woods, F. W. 1970. Interspecific transfer of organic materials by root systems of woody plants. *Journal of Applied Ecology* **7**, 481–6.

Woods, F. W. & R. E. Shanks 1959. Natural replacement of chestnut by other species in the Great Smoky Mountains National Park. *Ecology* **40**, 349–61.

Woods, K. D. 1979. Reciprocal replacement and the maintenance of codominance in a beech–maple forest. *Oikos* **33**, 31–9.

Woods, K. D. & R. H. Whittaker 1981. Canopy–understorey interaction and the internal dynamics of mature hardwood and hemlock–hardwood forests. In *Forest succession: concepts and application*, D. C. West, H. H. Shugart & D. B. Botkin (eds), 305–23. New York: Springer.

Woodwell, G. M. 1963. The ecological effects of radiation. *Scientific American* **208**, 40–9.

Woolhouse, H. W. 1981. Soil acidity, aluminium toxicity and related problems in the nutrient environment of heathlands. In *Heathlands and related shrublands: analytical studies*, R. L. Specht (ed.), 215–24. Ecosystems of the World 9B. Amsterdam: Elsevier.

Worley, I. A. 1973. The 'black crust' phenomenon in Upper Glacier Bay, Alaska. *Northwest Science* **47**, 20–9.

Wratten, S. D., P. Goddard & P. J. Edwards 1981. British trees and insects: the role of palatability. *American Naturalist* **118**, 916–9.

Wuenscher, J. E. 1969. Niche specification and competition modelling. *Journal of Theoretical Biology* **25**, 436–43.

Wuenscher, J. E. 1974. The ecological niche and vegetation dynamics. in *Vegetation and environment*, B. R. Strain & W. D. Billings (eds), 39–45. Handbook of Vegetation Science Part 6. The Hague: Junk.

Wuenscher, J. E. & T. T. Kozlowski 1970. Carbon dioxide transfer resistance as a factor in shade-tolerance of tree seedlings. *Canadian Journal of Botany* **48**, 453–6.

Wuenscher, J. E. & T. T. Kozlowski 1971a. The response of transpiration resistance to leaf temperature as a desiccation resistance mechanism in tree seedlings. *Physiologia Plantarum* **24**, 254–9.

Wuenscher, J. E. & T. T. Kozlowski 1971b. Relationship of gas-exchange resistance to tree seedling ecology. *Ecology* **52**, 1016–23.

Yarranton, G. A. & R. G. Morrison 1974. Spatial dynamics of a primary succession: nucleation. *Journal of Ecology* **62**, 417–28.

Zach, L. W. 1950. A northern climax, forest or muskeg? *Ecology* **31**, 304–6.

Zadoks, J. C. 1978. The biotic environment of plants. In *Structure and functioning of plant populations*, A. H. J. Freysen & J. W. Woldendorp (eds), 299–315. Amsterdam: North–Holland.

Zedler, J. & P. Zedler 1969. Association of species and their relationship to microtopography within old fields. *Ecology* **50**, 432–42.

Zedler, P. H. 1981. Vegetation change in chaparral and desert communities in San Diego County, California. In *Forest succession: concepts and application*, D. C. West, H. H. Shugart & D.B. Botkin (eds), 406–30. New York: Springer.

Zinke, P. J. 1962. The pattern of influence of individual forest trees on soil properties. *Ecology* **43**, 130–3.

ADDITIONAL REFERENCES, IN ALPHABETICAL ORDER

Baker, F.S. 1949. A revised tolerance table. *Journal of Forestry* **47**, 179–81.

Beeftink, W.G. 1977. The coastal salt marshes of Western and Northern Europe: An ecological and phytosociological approach. In *Wet Coastal Ecosystems*, V. J. Chapman (ed.), 109–55. Ecosystems of the World 1. Amsterdam: Elsevier.

Dodson, C. H., & A. H. Gentry 1978. Flora of the Rio Palenque Science Center, Los Rios Province, Ecuador. *Selbyana* **4**, 1–625.

Duvigneaud, P. 1967. La productivité primaire des écosystèmes terrestres. In *Problèmes de Productivité Biologique*, M. Lamotte, F. Bourlière (eds), Paris: Masson.

Epstein, E. 1973. Roots. *Scientific American* **228**, 48–61.

Eriksson, E. 1960. The yearly circulation of chloride and sulphur in nature: meteorological, geochemical and pedological implications II. *Tellus* **12**, 63–109.

Evenari, M. 1986. The desert environment. In *Hot Deserts and Arid Shrublands*, M. Evenari, I. Noy-Meir, D. W. Goodall (eds), 1–22. Ecosystems of the World 12A. Amsterdam: Elsevier.

Fridriksson, S. 1970. The colonization of vascular plants on Surtsey in 1968. *Surtsey Research Progress Report V*, 10–14. Surtsey Research Society, Reykjavik.

Fridriksson, S. 1982. Vascular plants on Surtsey 1977–80. *Surtsey Research Report IX*, 46–58. Reykjavik: Surtsey Research Society.

Fridriksson, S. 1987. Plant colonization of a volcanic island, Surtsey, Iceland. *Arctic and Alpine Research* **19**, 425–31.

Gleason, H. A. 1968. *The new Britton and Brown illustrated flora of the Northeastern United States and adjacent Canada.* Vols 1, 2, 3. New York: Hafner Publishing Co.

Goldthwait, R. P. 1966. Glacial history. In *Soil development and ecological succession in a deglaciated area of Muir Inlet, Southeast Alaska.* A. Mirsky (ed.), 1–18. Report No. 20. Institute of Polar Studies, Ohio State University, Columbus, Ohio.

Hubbell, S. P. 1979. Tree dispersion, abundance and diversity in a tropical dry forest. *Science* **203**, 1299–1309.

Imms, A. D. 1957. A *General textbook of entomology.* (9th Edition, revised by O. W. Richards, R.G. Davies). London: Methuen.

Janzen, D.H. 1969. Seed eaters versus seed size, number, toxicity and dispersal. *Evolution* **23**, 1–27.

Matthews, J. A. 1978. Plant colonization patterns on a gletschervorfeld, southern Norway: A mesoscale geographical approach to vegetation change and phytometric dating. *Boreas* **7**, 155–78.

Miles, J. 1987. Vegetation succession: past and present perceptions. In *Colonization, succession and stability*, A. J. Gray, M. J. Crawley, P. J. Edwards, 1–29. 26th Symposium of the British Ecological Society. Oxford: Blackwell.

Muller, C.H. & R. del Moral 1966. Soil toxicity induced by terpenes from *Salvia leucophylla. Bulletin of the Torrey Botanical Club* **93**, 130–7.

Norton, D.A. & D. Kelly 1988. Mast seeding over 33 years by *Dacrydium cupressinum* (rimu) (Podocarpaceae) in New Zealand: the importance of economies of scale. *Functional Ecology* **2**, 399–408.

Packham, J. R. & D.J. Harding 1982. *Ecology of woodland processes.* London: Edward Arnold.

Pickett, S. T. A. 1976. Succession: An evolutionary interpretation. *American Naturalist* **110**, 107–19.

Poole, A. L. & N. M. Adams 1963. *Trees and shrubs of New Zealand.* Wellington: Government Printer.

Rabotnov, T. A. 1966. Peculiarities of the structure of polydominant meadow communities. *Vegetatio* **13**, 109–16.

Spurr, S. H. 1956. Forest associations in the Harvard Forest. *Ecological Monographs* **26**, 245–63.

Spurr, H. 1964. *Forest ecology.* New York: Ronald Press.

Walter, H. 1971. *Ecology of tropical and subtropical vegetation.* Edinburgh: Oliver and Boyd.

REFERENCES

Whittaker, R. H. 1960. Vegetation of the Siskyou Mountains, Oregon and California. *Ecological Monographs* **30**, 279–38.

Wilson, D. S. 1980. *The natural selection of populations and communities.* Menlo Park, CA: Benjamin/Cummings.

Glossary

aa Volcanic lava with rough, subangular to blocky surface texture.

abiotic Having no connection with living organisms or life processes.

abscission Physiological process of regular loss by plants of certain organs (e.g. leaves, short shoots).

acclimation Process of gradual acquisition by plants of *hardiness* (q.v.) for severe conditions. (≈ *hardening* q.v.).

achene A single-seeded, dry, indehiscent fruit e.g. of *Ranunculus*, many members of Asteraceae, or the grains of grasses, Poaceae.

acid An inorganic or organic compound which forms hydrogen ions when dissociated (e.g. in solution).

actinobacterium Member of a group of soil-living moneran (q.v.) microorganisms, formerly known as actino-mycetes.

adolescent A juvenile plant older than a seedling, but not yet mature (saplings and poles of tree species).

adsorption Relatively loose binding of ions to soil colloids (or other sites) (verb *adsorb*).

adult An individual plant, capable of seed production (or sexual reproduction if not a seed plant). (cf. *adolescent*, *juvenile* q.v., *seedling*).

adventitious bud A dormant shoot bud on stem or root, which may become activated if the plant is damaged, or habitat conditions change.

aerenchyma Loose, spongy *parenchyma* tissue (q.v.) in aquatic plants. It permits air to diffuse through stems and roots.

A-horizon The uppermost soil layer where biological activity and *leaching* (q.v.) are intense. Organic matter accumulates in this horizon.

alfisol Forest or meadow soil in cool, moist climate, not strongly leached (q.v.).

Algae A major group of mainly aquatic *eucaryotic* (q.v.) organisms classed by some as plants and by others in a kingdom in their own right.

allelopathy (Preferred term *phytotoxicity* (q.v.)). (adjective *allelopathic*).

allogenic An influence on plants caused by a factor quite independent of the plants. (cf. *autogenic* q.v.).

allopatric speciation Evolutionary divergence of species when populations of an ancestral species are physically isolated.

allopolyploid Condition in species which have originated by hybridization, with doubling of chromosome sets, which ensures the presence of homologous chromosome pairs.

amensalism A relationship between two species of organism in which one is harmed but the other is unaffected.

amphibole A member of a group of ferromagnesian silicate minerals.

andesite A greyish volcanic rock rich in plagioclase (feldspar), biotite (mica), hornblende (pyroxene) and augite (amphibole).

Angiosperm Seed plant with flowers, where the ovules are enveloped within tissues which form an ovary.

Animalia Kingdom of eucaryote, multicellular organisms which ingest food and digest it internally.

anion A negatively charged ion. (cf. *cation* q.v.).

annual A plant species which germinates, completes its life processes and dies within a year.

apatite One of a group of calcium phosphate minerals.

apomixis Reproduction of plants by seeds which have formed without any sexual process (i.e. without meiosis and syngamy). (adjective *apomictic*).

aquifer A water-bearing, underground rock layer.

aridisol Soil of desert or sub-arid regions.

argillite A sedimentary rock consisting of silt or clay-sized cemented grains.

ascospore Spore produced in the sporangium (ascus) of a member of the Acsomycete group of fungi.

association A vegetation classification unit. American: characterized by *physiognomic dominant* species (q.v.); European: characterized by species peculiar to the community.

autecology The study of the environmental relationships of single organisms.

autogenic Any influence on plants which results from plant processes. (cf. *allogenic* q.v.)

autoinhibition Limitations placed on plants by their own activities (e.g. *phytotoxic* effects (q.v.), shading, soil modification).

autosuccession (preferred term *direct replacement* q.v.).

autotoxicity Phytotoxic autoinhibitory effects.

autotrophism Ability of green plants and some microbes to synthesize their own food through one of the photosynthetic processes, or (bacteria), chemosynthetic processes (adjective *autotrophic*).

awn Short or long bristle extending from one of the bracts investing a single grass floret; often involved with seed dispersal and the positioning of the seed for germination.

bacterium Members of a group of universally-distributed moneran microorganisms. (plural bacteria).

bar A unit of pressure or suction measurement (= 0.987 atmosphere; $= 10^5$ Pascal).

barrens Landscape where rock debris surfaces, sand etc. dominate and plants, if present, are very sparsely distributed.

basalt A dark volcanic rock composed of calcium-plagioclase (feldspar), pyroxene and olivine.

base Compound or metal which, when in solution in water, forms hydroxyl ions. Bases include the *cations* K^+, Mg^{++}, Ca^{++}.

benefaction Facultative relationship between a pair of organisms where one benefits and the other is unaffected (cf. *mutual benefaction* q.v.).

B-horizon The soil layer beneath the A-horizon. It is the place of strongest weathering and *illuviation* (q.v.).

biennial A plant species which germinates and grows vegetatively in its first year, then flowers and dies in its second year.

bio-energetics The study of the entry of energy to biological systems and its transformations and ultimate fate there.

biome A large-scale *ecosystem* (q.v.) comprising vegetation and associated animals e.g. Arctic tundra; subtropical savannah; tropical rain-forest (cf. *coenose* q.v.).

biomass Weight of living material (expressed as dry matter) of an individual organism, population, community or region.

biota The organisms (plants, animals, microbes) in any region.

blanket peatland An area where a layer of peat overlies most of the terrain irrespective of slope.

bog A *mire* (q.v.) with underlying acid peat.

B.P. Before present, i.e. before 1950 A.D., the base date for the radiocarbon dating method.

breccia A rock-type consisting of angular clasts bound together, with or without a relatively fine matrix.

Bryophyta A eucaryotic plant group with relatively conspicuous leafy or *thalloid gametophyte* generation (q.v.) and smaller 'parasitic' *sporophyte generation* (q.v.); (mosses, liverworts, hornworts).

bulbil A small vegetative offshoot from some plant species, which may become detached and take root, thus initiating a new plant.

CAM Crassulacean acid metabolism. A photosynthetic system found in many succulent plants. The plants open stomata in the dark and take in CO_2, fix it and metabolize it further in daylight the next day.

C3 The Calvin photosynthetic system found in most plants in which the initial product is a three-carbon compound. *Photorespiration* (q.v.) occurs in bright light and high temperatures.

C4 The Hatch-Slack photosynthetic system, found in tropical and subtropical grasses and some other plants, in which the initial product is a four-carbon compound and productivity is not limited by photorespiration.

calciphile A plant species requiring base-rich soil conditions and often found on calcareous substrates.

calorie A unit of energy measurement. 1 gram-calorie is the amount of energy which raises 1 cubic centimetre of pure water through a temperature of 1°C.

calorific value The energy content of any organic substance in gram-calories.

cambium *Meristematic* region (q.v.) which gives rise to new tissues in plant stems and roots.

canopy The uppermost layer of leafy stems in vegetation.

carnivore An animal (or plant) which eats animals.

carr Shrubby or low forest cover in wetlands.

carrying capacity The ability of a given area to support a particular *biomass* (q.v.) or population-size of organisms in *steady state* (q.v.).

cation A positively-charged ion. (cf. *anion* q.v.).

cation-exchange capacity The amount of simple ions (Na^+, K^+, Mg^{++}, Ca^{++}, NH_4^+, H^+) which a soil yields when leached with a buffered salt solution (e.g. ammonium acetate).

chaparral Shrubby vegetation comprising many species of relatively small-leaved shrubs which burn readily, occurring in California, U.S.A.

chernozem A Russian term for a grassland soil with a characteristic deep, black A horizon, enriched by decayed plant roots.

C-horizon The soil layer beneath the B-horizon, consisting of unweathered parent material.

chlorophyll Green pigment in plants vitally involved in the process of photosynthesis.

chlorosis Yellowing of the foliage of plants. A sign of nutrient deficiency or disease.

523

chronosequence A sequence consisting of different-aged areas of landforms, soils and vegetation where environmental conditions have been relatively uniform as the systems developed. Surfaces and vegetation of different age are taken to represent a developmental sequence.

chronoseries The various developmental phases in a chronosequence, as expressed on the ground at one time.

clast A term referring to particles of rock of any size; from clay to boulders.

clay Very small clasts (< 0.002 mm diam.) important in determining the physical and chemical properties of soils.

clay mineral A secondary mineral with properties such as ion-exchange, adhesiveness and swelling when wet.

climax vegetation (Preferred term *mature vegetation* (q.v.)).

clone Population of organisms derived sexually from one original parent.

closed vegetation Cover of vegetation where plant canopies touch, i.e. there is no open ground space apparent.

C/N ratio The ratio of carbon to nitrogen in soils. A ratio of > 12 indicates that the soil will not supply nitrogen readily to plant roots.

co-dominant One of two or more *dominant* (q.v.) plant species in a stand.

coenocline A plant community gradient corresponding to a spatial environmental factor gradient.

coenose A major ecological unit (cf. *ecosystem*, *biome*) comprising plants, animals, microbes and the inorganic components of their environment.

coevolution Evolution of pairs of organisms in a kind of 'arms race', driven by their reciprocal interactions.

cohort A population of organisms of any species which originated at the same time.

colloid A very fine particle (clay or humus) with ion exchange properties and other properties important in determining soil character. (adjective *colloidal*).

colluvium Rock debris or soil which has moved downhill under the impetus of gravitational force.

colonist A plant species which can establish in open, unvegetated sites.

colonization The process of plant establishment on open unvegetated sites.

commensalism A relationship between two organisms where one is benefited and the other is neither harmed nor benefited. The relationship is obligate for the species which benefits.

compensation point The light intensity at which photosynthetic energy gains by plants are balanced by respiratory losses.

competition Contest for resources or direct interference between neighbour plant species, affecting their performance.

conidium An asexual fungal sport (= conidiospore). (plural *conidia*).

community An abstract grouping of stands consisting of similar plant species assemblages.

constancy A lack of change in some parameter of a system (numbers or proportions of species, taxonomic composition, etc.).

continuum A system (e.g. vegetation) which exhibits continuous spatial (or temporal) intergradation.

cover A projection of the foliage and branch systems of plants onto the ground, usually expressed as a percentage of the total area.

cuticle The outermost, non-cellular, layer of a leaf surface.

cutin The material of cuticle. (verb *cutinize*).

cyanobacterium Member of a group of moneran microorganisms, widely distributed in soil and water, formerly known as blue-green algae.

cycle A pattern of vegetation change where species replacements occur in simple or complex sequence, eventually returning to the original species composition (= cyclic replacement).

determinism Conceptual model in which relationships are fixed and events occur according to a set, predictable programme. (adjective *deterministic*). (cf. *stochastic* q.v.).

detritivore Animal which obtains its nutriment and energy by consuming detritus.

detritus Dead plant and animal remains.

dicotyledon Member of a group of Angiosperm plants which originate from seedlings that have two seed-leaves.

diploid Cells, tissues or organisms having paired sets of chromosomes (cf. *haploid* q.v.).

direct gradient analysis Method employing extensive sampling on natural gradients to analyse distributions of plant species with respect to the distribution patterns of the main environmental variables.

direct replacement A dynamic vegetation pattern where a species (or species-assemblage) immediately replaces itself after disturbance.

direct succession (Preferred term *direct replacement* q.v.)

disturbance Environmental factor, almost always *allogenic* (q.v.) which causes damage to plants or their substrate.

diversity The relative richness of numbers of species in different communities. Measured by numbers of indices, of which the simplest is numbers of species per unit area.

dolomite A sedimentary rock consisting of magnesium and calcium carbonate.

524

dominant A plant species which is the most prominent in a stand, contributes the most biomass, makes most environmental demands and influences various subordinate plants profoundly.

dormancy The capacity of organisms, or specific organs such as buds or seeds, to go into a state of rest during unfavourable periods. Specific treatments (often climatic) induce and overcome dormancy. (adjective *dormant*).

dynamic equilibrium A state of relative stability in a biological system. Short-term changes may occur, while the overall species-composition remains constant in the long-term (cf. **steady-state, mature** q.v.).

ecocline A spatial environmental factor gradient and a plant community gradient, considered together.

ecological amplitude For a species, the range of ecological tolerance, in relation to all environmental gradients which influence it ($\simeq$ **fundamental niche** q.v.).

ecological guild A group of species, the individuals of which exploit the same kinds of resources in a similar way.

ecology The discipline which encompasses the study of organisms in relation to their environment and to one another.

ecomorphology Features of the form of organisms which can be expressly linked to their relations with environment. (adjective *ecomorphologic*).

ecophysiology The bio-physico-chemical functioning of organisms with particular reference to interactions with variations in their environment.

ecosystem A group of organisms which live together, closely interacting with their external environment and with one another. Ecosystems may be of any size from tiny e.g. those around a root hair, or in a flower, to vast (called *coenoses*, or *biomes* q.v.).

ecotone A zone of transition, usually narrow, between two distinct plant communities.

electromagnetic spectrum The range of wave-motions found in the Universe: long-wave (radio) ($1\,cm - 2000\,m$) $\leftarrow$ visible light ($0.4 - 0.8\ \mu m$) $\rightarrow$ short wave (gamma rays) ($< 6 \times 10^{-6} \mu m$).

eluviation (preferred term *leaching* q.v.).

emergent Trees in forest, the crowns of which project above the main canopy level.

emergent properties Features of a biological system which amount to more than the sum of the effects created by its individual constituent species.

endogenous factor (preferred term *autogenic* q.v.).

enhancement A general term for situations where the presence of one species brings about benefits to another (cf. *commensalism, mutualism, benefaction*).

entisol Recent or undeveloped soil.

epigeal In germinating seeds, the situation where the cotyledons rise above the level of the soil surface (cf. *hypogeal* q.v.).

epiphyte A plant species which grows non-parasitically on stems or branches of another species.

environmental pressure Factor which is damaging or stressful to plants.

eucaryote An organism, the cell or cells of which have nuclei (cf. *procaryote* q.v.).

eustatic Referring to movements of sea-level independent of relative movements in the height of land-masses.

eutrophic Water which is relatively rich in nutrient cations, phosphorus, etc.

evapotranspiration Combined evaporation and transpiration from areas of land covered by vegetation.

exchangeable bases The cations Na^+, K^+, Mg^{++}, Ca^{++}, NH_4^+.

exogenous factor (preferred term *allogenic* q.v.).

facilitation model A description of one of the mechanisms bringing about vegetation change ($=$ autogenic change). Species occupying a site cause changed conditions facilitating the establishment of other species which replace the residents.

feldspar A silicate mineral group containing forms rich in potassium (orthoclase) and others rich in sodium and calcium (plagioclase).

fen A moderately to highly nutrient-rich wetland.

fern Member of a group of vascular plants with small, inconspicuous gametophytes and larger, conspicuous sporophytes which produce spores in groups (sori) on their leaves.

field capacity The soil water content (held in capillaries and on the surfaces of soil aggregates) two days after rain (*water potential* (q.v.) of about -0.15 bar).

field layer Herbaceous plant cover on the ground in forest.

fjord A glacial valley which has been invaded by the sea.

flora The complement of plant species in a region.

floret A small flower, part of an inflorescence in Angiosperm families such as the Asteraceae and Poaceae (singly or in groups in *spikelets* in the latter).

fluctuation A pattern of vegetation change where species replacements occur in response to more or less irregular shifts in environmental conditions ($=$ fluctuating replacement).

forb A dicotyledonous herb.

forest   Vegetation in which the dominant cover plants are trees.

formation   A large unit of vegetation classification characterized by particular plant forms (grassland, forest, etc.).

fossil   Remains or traces of organisms preserved in the rocks (includes pollen grains in peat bogs, gases, etc.).

frass   Insect faeces.

fruit   The mature ovary of an Angiosperm, containing one or more seeds.

fumarole   A vent in a volcanic area which emits steam or other gases.

fundamental niche   The 'space' occupied by a species in the multi-dimensional environmental gradient complex affecting it, irrespective of competitor plants ($\simeq$ *ecological amplitude* q.v., cf. *realized niche* q.v.).

fungus   Member of a group of *saprophytic* (q.v.) or *parasitic* (q.v.) eucaryote microbe organisms with threadlike spreading vegetative form (mycelium, consisting of *hyphae* q.v.) and more specialized *sporangiophores* (q.v.) (plural *fungi*).

gametophyte   In plant groups with alternation of generations, the part of the life history which carries *gametes* (q.v.).

gamete   Male (sperm or equivalents) and female (ova) cells, with haploid chromosome complement. Male and female gametes fuse to form zygotes, thus restoring the diploid chromosome complement.

genotype   The inherent genetic complement of any individual organism. (adjective *genotypic*).

gibbsite   A clay mineral.

gley   A Russian term for a waterlogged soil with characteristic grey, blueish, whiteish or greenish B horizon colours, often rust-mottled. (adjective *gleyed*).

gneiss   A course-grained metamorphic rock with irregular bands of light quartz and feldspar alternating with dark amphiboles and mica.

goethite   A clay mineral.

granite   A plutonic rock containing crystals of quartz, feldspar, mica and some darker minerals.

grassland   Vegetation type where the dominant cover plants are grasses.

grazing   In the strict sense, the consumption of grass by animals. Often the term is extended to apply to the feeding of animals (including aquatic species) on other plant parts, by any means, including sap-sucking, wood-boring. Seed-eating, fruit-eating and nectar-taking are usually excluded.

gross annual primary production   The production (biomass gained per unit area) of an area of vegetation, not taking into account respiratory losses.

group selection   The selection of populations or groups of populations with distinctive gene pools and phenotypes, rather than selection of individuals within a population.

Gymnosperm   A seed plant in which ovules are borne more or less exposed on leaf-like scales (often in cones).

halloysite   A clay mineral.

halophyte   Plant species tolerant of high concentrations of sodium chloride.

haploid   Cells, tissues or organisms having single chromosome sets (cf. *diploid* q.v.).

hardening   Acquisition by plants of tolerance to climatic or other extremes as conditions gradually become more severe ($\simeq$ *acclimation*).

hardiness   General term for relative condition of tolerance by plants of severe conditions.

haustorium   A specialized organ by which a parasitic plant attaches to its host and through which the parasite draws nutriment.

heathland   A land system with infertile soils and often shrubby vegetation in which heaths (families Ericaceae and Epacridaceae) are usually important.

hemiparasite   A parasitic plant which is photosynthetic to some extent.

herb   A non-woody Angiosperm. (Most ferns, lycopods and Bryophytes may also be regarded as herbs).

herbivore   An animal which feeds on plants.

heterotroph   An organism which obtains its sustenance by feeding on autotrophs (alive or dead), or other heterotrophs (which in turn have fed on autotrophs).

histosol   Organic soil in a water-logged site.

holism   Philosophic viewpoint which considers structure and function of biological systems as wholes when attempting to explain biological phenomena. (cf. *reductionism* q.v.).

homeostasis   Self-regulating stability of a biological system.

humus   Amorphous, dark organic material in soil A-horizons, formed from much-decayed plant litter and animal detritus.

hyaline cells   Hollow cells in the leaves of *Sphagnum* moss species. They enable the plants to take up large amounts of water.

hypha   Thread-like branching fungal tissue which extends through whatever substrate the fungus is living on. (plural *hyphae*).

hypogeal   Condition when the cotyledons of germinating seeds remain below the soil surface. (cf. *epigeal* q.v.).

526

ignimbrite Pale-coloured massive *rhyolitic rock* (q.v.) formed from consolidated, hot, volcanic ash flows.

illite A clay mineral.

illuviation Process of accumulation of materials (iron and aluminium compounds, finely-divided organic matter etc.) which have passed down through a soil in solution, or in suspension.

inbreeding Self-fertilization, or crossing of individuals derived from the same parent, resulting in increased homozygosity (i.e. increase in numbers of identical alleles on both homologous chromosomes). (cf. *outbreeding* q.v.).

inceptisol Young soil on a new surface.

indirect gradient analysis (Preferred term *ordination* q.v.).

individualistic concept The view that plant communities are fortuitous floristic assemblages determined by the availability of species in adjacent sites; the species occur together because their environmental requirements overlap.

inertia The ability of a biological system to resist or recover from external disturbances.

inhibition model A model explaining one of the patterns of vegetation change. Resident species on a site resist invasion and no change occurs until they die, thus allowing others to enter the vacant spaces.

ion A positive or negatively charged atom or group of atoms.

ion-exchange complex In soils the places (clay colloids, humus colloids, colloidal films on coarser clasts) where ion exchange with plant root hairs or soil solution takes place.

isostatic rebound Rise in land after release from ice (or water) loading.

juvenile A young plant, not yet sexually reproducing (either a *seedling*, or an *adolescent* q.v.).

kaolinite A clay mineral.

kinetic model The view that vegetation is normally in a continuous state of flux, arising from disturbance, and that stable 'equilibrium' vegetation is a rarely expressed and relatively unusual condition.

K-selection Selection experienced when populations are near carrying capacity (q.v.). It is manifest in plant species which are slow-growing, long-lived and strongly competitive, specializing in maintenance and producing relatively small numbers of often poorly-dispersed propagules. (cf. *r-selection* q.v.).

lag deposit A residue of coarse clasts left after finer materials have been eroded away.

lapilli Small lava fragments (about 5–10 mm diameter) ejected during a volcanic eruption.

laterite A reddish, iron-rich soil, developed in tropical and sub-tropical regions.

leaching The process of washing of soluble or suspended materials downward through soil profiles by percolating water.

leaf area index The amount of leaf area of vegetation per unit area of ground.

legume A member of one of the families of Angiosperms Papilionaceae, Mimosaceae or Caesalpiniaceae.

lichen A mutualistic association between a fungus and either a green alga, or a cyanobacterium. Lichens form distinctive *thalli* (q.v.) Although they are not strictly plants they can be considered to be part of the vegetation.

life style The way that a plant species behaves in its habitat.

light saturation Light intensities above which plants cannot increase photosynthetic rates.

lignotuber An often large, swollen, woody structure formed at and below ground level by some shrubs and trees that are subject to frequent burning. After a fire, new stems sprout from dormant buds in the lignotuber.

limestone A sedimentary rock mainly consisting of calcium carbonate (often derived from the shells of small marine animals).

limonite A clay mineral.

littoral Pertaining to sea-coast sites.

liverwort One of a group of *Bryophyte* plants (q.v.) having gametophytes with thalloid or leafy form.

loess Wind-blown silt.

lognormal distribution A model describing the relative importance relationships of species in some relatively species-rich communities. In these there are many species of intermediate importance (measured e.g. as biomass), and few very important or very rare species.

lycopod A group of vascular plants with relatively inconspicuous gametophytes and more conspicuous sporophytes. The sporangia are usually gathered together into strobili (cones).

macroclimate The large-scale climate above the level of the vegetation.

macrofossil A fossil large enough for general structure to be visible with the naked eye.

macrophyte The large plants of aquatic sites.

macroscopic Having structure visible with the naked eye.

Markov chain A statistical series in which future conditions derive from the probabilistic behaviour of the conditions prevailing at present.

master factor An overriding environmental factor which influences all or most other factors.

mast year phenomenon Periodic seed production, with the seed years being separated by series of non-seeding years.

maturation   The process of development of mature features. In vegetation the accomplishment of relative *steady state* (q.v.)

mature vegetation   Vegetation where the dominant species replace themselves over two or more generations.

meristem   Regions in plants which give rise to new cells (root and shoot apices, cambia, nodes and, in monocotyledons, at the bases of leaves). (adjective *meristematic*).

mesa   A flat-topped plateau which remains when surrounding terrain has been eroded away.

mesic   An implication that microclimatic conditions are relatively mild, and soils well supplied with water (cf. mesophytic; cf. *xeric* q.v.).

mesophyll resistance   In leaves, the internal resistance to diffusion of CO_2 and other gases.

mesophyte   A plant requiring mild, well-watered conditions. There is often an implication of requirements for high soil fertility also. (adjective *mesophytic*). (cf. *xerophyte* q.v.).

microbe   One of the *microscopic* (q.v.) organisms. Especially applied to *Monerans* (q.v.) and *Fungi* (q.v.) but sometimes extended to include microscopic Algae. (adjective *microbial*).

microclimate   The climate below the level of the vegetation canopy, or within the spread of individual plants, or in smaller spaces.

micropyle   Small pore in seeds through which water and gases enter.

microenvironment   Small-scale environment immediately surrounding particular organisms.

microfossil   *Fossil* (q.v.), the structure of which can only be seen with a lens or microscope.

microscopic   Structure visible only with a lens or microscope.

microsite   A small-scale site such as that perceived, e.g. by a germinating seed.

mineralization   The decomposition of organic detritus to its ultimate elemental constituents.

mire   A vegetated wetland, often with underlying peat.

modal vegetation   The most usual vegetation cover in a region.

moder   A forest soil exhibiting characteristics intermediate between the extremes of *mull* (q.v.) and *mor* (q.v.).

mollisol   Soil with deep, fertile organic A horizon found in semi-arid grassland localities (= *chernozem* (q.v.), chestnut soil).

Monera   The groups of procaryotic (i.e. without cell nuclei) organisms, Bacteria, Actinobacteria and Cyanobacteria.

monocarpic   Flowering once, then dying.

monoclimax   The idea that all vegetation in a region will ultimately converge on a single formation type, and that climate is the overall controlling factor determining that vegetation type (cf. *polyclimax* q.v.).

monocotyledon   Angiosperm plant group, the seedlings of which have only one seed leaf.

monolayer   Vegetation cover with a dense canopy comprising a single stratum of leafy branches.

monopodial   A stem in which growth continues from year to year by an apical *meristem* (q.v.) on the same axis (cf. *sympodial* q.v.)

monotypic   Vegetation stands consisting of only one plant species (or one clearly dominant species and only small subordinates such as Bryophytes).

montmorillinite   A clay mineral.

mor   A forest soil characterized by a thick layer of acid, slowly decomposing litter, above the mineral soil.

moraine   A ridge of bouldery debris formed around the margin of a glacier.

moss   One of a group of *Bryophyte* plants (q.v.) having leafy gametophytes.

mull   A forest soil characterized by base-rich, fast-decomposing litter, the humus from which becomes closely incorporated into the mineral soil.

multilayer   Vegetation cover where the plants form an open canopy with several strata of leafy branches.

muskeg   Bog vegetation in northern North America.

mutual benefaction   Facultative relationship between pairs of species where each benefits from the other's presence (cf. *benefaction* q.v.).

mutualism   Obligate relationship between pairs of species where each benefits from the other's presence. (adjective *mutualistic*).

mycelium   Interwoven network of fungal *hyphae* (q.v.).

mycorrhiza   A mutualistic relationship of certain Fungi with plant roots. (adjective *mycorrhizal*).

nastic movement   Adjustment of positions of leaves or flowers of some plant species in relation to the sun's position.

natural vegetation   Vegetation not managed or disturbed by humans.

n–dimensional niche   The 'position' which a species occupies within the complex of environmental gradients which affect it.

necrosis   Death of parts of a plant through nutrient deficiency or disease.

nett annual primary production   The amount of dry matter produced per plant, or per unit area, after subtraction of respiratory losses.

névé   The upper basin of a glacier where snow accumulates and is converted into ice.

niche concept   The idea that species each have an individual response to the environmental conditions occurring in their habitats. (cf. *fundamental niche* q.v., *realized niche* q.v.).

niche preemption   A model describing the relative importance relationships of species in some relatively species-poor communities. In these a dominant occupies most of the niche space and the next most important species most of the remaining niche space and so on, in a pattern similar to a geometric series.

niche proportions   The 'shape' of the niche of any species.

nitrification   The biological formation of nitrate through a series of steps beginning with the fixation of elemental nitrogen.

nitrogen fixation   Conversion, by moneran microorganisms, of inert elemental nitrogen into nitrogen compounds.

nitrophile   A plant species which lives on nitrate-rich soils, such as in bird roost areas.

non-vascular   Referring to the condition in Bryophytes and Algae, which lack a true conducting system.

nucleation   Regular vegetation patterns which develop around individual colonist plants.

nuée ardente   Incandescent ash cloud erupted from a volcano.

'old fields'   Fields which were once cultivated, then abandoned.

oligotrophic   Water which is low in nutrients and usually acidic.

olivine   A ferromagnesian mineral.

open vegetation   Vegetation pattern where individual plants or patches of plants form a mosaic with unvegetated ground.

ordination   The systematic ordering of vegetation samples on a number of abstract geometric axes by comparison of their species composition (and possibly other parameters).

organic   Material which is living or has once been alive.

outbreeding   The maintenance of heterozygosity through cross-fertilization (achieved in plants by dioecy, protandry, protogyny and, in some species, by incompatibility of a plant's pollen with its own stigma). (= outcrossing). (cf. *inbreeding* q.v.).

outwash   Alluvium deposited by a river issuing from a glacier.

oxisol   Strongly-weathered, iron sesquioxide-rich soil type in tropical and subtropical regions. ($\approx$ *laterite* q.v.).

pahoehoe   Basaltic volcanic lava which sets with a relatively smooth surface.

paludification   The process of conversion of dry land into a mire through progressive waterlogging and peat accumulation.

parasite   An organism with an intimate and usually specific, obligate relationship with another, host, organism. The parasite feeds on its host, the host receives no benefit.

parenchyma   Relatively unspecialized plant tissue, with thin-walled, more or less isodiametric cells.

patterned ground   In polar and alpine regions the small-scale landforms (such as polygons and stone stripes) which form in response to frequent freeze and thaw.

pathogen   A virus, bacterium or fungus which attacks plants.

peat   A more or less compressed and anaerobic accumulation of dead plant remains.

permafrost   In the high Arctic the permanently frozen ground that is apparent below a thin upper soil layer which thaws out in summer.

permanent wilting   The state when a soil is unable to supply further water to plants and they wilt and die (soil *water potential* (q.v.) of about -15 bar).

persistence   The survival time of a system or some component of it.

pH   A measure of hydrogen ion concentration ($\log_{10}[H^{+}]$. Extreme acidity 0 – Neutrality 7 – Extreme alkalinity 14. Each unit increase represents a tenfold decrease in hydrogen-ion concentration.

phenology   The study of seasonal changes in plant activity and processes. (adjective *phenological*).

phenotype   The outwardly expressed form of an organism. (adjective *phenotypic*).

phloem   Tissue in vascular plants by which carbohydrate and other photosynthetic products are conducted from the leaves to the rest of the plant.

photoperiod   Daylength. Commonly one of the cues for initiation or termination of certain plant physiological processes (flowering, bud growth, hardening).

photorespiration   Respiratory dissipation of the products of photosynthesis in *C3* plants (q.v.) when light intensities and temperatures are too high.

phototropism   Bending of growing shoots towards a light source through elongation of cells on the shaded side of the stem.

physiognomic dominant   The species which appears to be the most prominent in a plant community by virtue of its size and abundance. Not necessarily the true ecological *dominant* (q.v.)

physiognomy   The general appearance of vegetation arising from its stature, texture, colour etc.

phytochrome   A pigmented protein in plants which senses and responds to differences in red/far red light ratios and is associated with control of a range of plant processes (e.g. flowering, diurnal rhythms, seed germination, leaf expansion).

phytotoxicity   The production of chemical compounds by plants which can inhibit other plants (or themselves). It is difficult to demonstrate, unequivocally, that this actually occurs in nature (= allelopathy).

pioneer   An early colonist plant species on new sites.

Plantae   The plant kingdom, incorporating eucaryotic organisms characterized by the occurrence of chlorophylls and the photosynthetic process.

plant succession   Directional sequence of plant species populations on a site. (Preferred term *sequential replacement* q.v.).

pneumatophore   'Breathing root' of some mangrove species. A peg-like organ which projects vertically from the water-logged substrate.

podocarp   Member of a group of tropical and Southern Hemisphere gymnosperms, the Podocarpaceae.

podsol   A Russian term for a soil with a distinct layer of acid litter and humus, resting on mineral soil. The uppermost mineral soil horizon is strongly leached and often white in colour. (Verb *podsolize*).

pole   A young tree greater than 3 m tall.

polycarpic   Referring to a perennial plant species which flowers more than once. (cf. *monocarpic* q.v.).

polyclimax   The concept of control of stable vegetation types by a variety of factors in addition to climate.

polygon   In polar or alpine regions, small-scale landforms consisting or garlands of stones around a central, finer-textured zone and resulting from freeze-thaw processes.

polymorphism   Genetically-controlled pattern taking the form of distinct proportions of variants of a particular character (e.g. flower colour, seed size).

population   A group of individuals of a species living together, all of which have the potential to interbreed. Extended populations include individuals which contribute seeds or pollen to a site.

primary succession   *Succession* (q.v.) occurring on new land surfaces.

procaryote   An organism which has no cell nuclei.

process   A set of causally connected events.

productivity   Gaining of biomass by plants through the photosynthetic process. (or by heterotrophic nutrition of *parasites*, *saprophytes* q.v.).

propagule   Any unit by which plants multiply themselves (seeds, *bulbils* (q.v.) and other vegetative offset organs such as *rhizomes* (q.v.), *stolons* (q.v.) (etc.).

proteoid root   Specialized root found in members of family Proteaceae thought to permit enhanced nutrient uptake in infertile soils.

Protista   The kingdom of relatively undifferentiated unicellular (or sometimes colonial), procaryote organisms from which the other kingdoms of higher, multicellular organisms have diverged.

proximate causes   The causal factors most closely related to particular natural processes (in this case vegetation change).

Pteridophyta   The group of vascular plants which includes ferns, lycopods, horsetails and psilophytes.

pumice   Rhyolitic volcanic glass which hardened as gases passed through it, leaving it porous.

pyroxene   One of a group of complex silicate minerals.

quartzite   A sedimentary or metamorphic rock composed of cemented quartz grains.

radicle   Root of a seedling.

raised bog   Bog where peat accumulation has raised the surface above the local water table.

ramet   An individual derived by vegetative reproduction. A member of a *clone* (q.v.).

realized niche   The *niche* (q.v.) of species as it is constrained by the competitive influences of neighbour species. (cf. *fundamental niche* q.v.).

recurrence horizon   A distinct stratum in the sediments of a peat bog, resulting from a period of drying and oxidation of the surface peat, followed by further peat accumulation on return to wet conditions.

reductionism   The philosophic viewpoint that all phenomena (in this case biological phenomena) are explicable only in terms of the functioning of the component parts (species, organs, cells). (cf. *holism* q.v.).

reedswamp   Relatively fertile mire, characterized by the dominance of tall, emergent monocotyledon plants such as *Typha* or *Phragmites*.

regeneration   Replacement of old individual plants by younger members of the same species. Also, the normal growth of individuals through youth to maturity (and eventually old age).

resilience   The speed with which a system returns to its former state after a disturbance.

resource-use competition   Competition between species for the resources needed for growth (light, nutrients, water, gases).

Rhizobium   A genus of bacteria noted for their ability to form mutualistic relationships with roots of *legumes* (q.v.) (in the form of root nodules containing masses of the bacteria) and to fix atmospheric nitrogen.

rhizoid Root-like organ of Bryophytes which functions in the same way as a root in taking up water and nutrients.

rhizome Specialized underground stem of plants which extends laterally, giving rise to new above-ground shoots.

rhyolite A light-coloured, fine-grained volcanic rock rich in quartz and feldspars.

r-selection Selection experienced when populations are colonizing open, uninhabited sites. It is manifest in species which are fast-growing, short-lived and weakly competitive, specializing in reproductive increase and producing large numbers of well-dispersed propagules (cf. **K-selection** q.v.).

sandstone A sedimentary rock consisting of sand-sized grains cemented together.

sapling A young tree, up to about 3 m high.

saprophage An animal which feeds on dead remains of plants or animals.

saprophyte A specialized plant, often with little or no **chlorophyll** (q.v.), which obtains its nutriment from dead remains of plants through the agency of a mycorrhizal fungus.

savannah Grassland vegetation with scattered low trees or shrubs.

schist A foliated metamorphic rock, often rich in mica and with quartz bands.

sclerophyll Plant with thick, leathery, fibrous leaves.

scoria Fragments of rough volcanic lava ejected explosively from a volcano. From 10 mm to 25 cm diameter.

scrub Vegetation dominated by shrubs.

secondary succession Sequential vegetation change following the disturbance of established vegetation.

seed bank Populations of dormant seeds in the soil, or retained in dry fruit on parent plants.

seed plant Plant from one of the two groups **Gymnosperms** (q.v.) or **Angiosperms** (q.v.).

semi-natural vegetation Vegetation which is modified to some degree by human activities (e.g. by grazing of stock, harvesting of a particular wild plant crop, periodic human-lit fire) but in which natural processes otherwise prevail.

senescence The occurrence in plant species of programmed decline into old-age and death. Evident in **monocarpic** species (q.v.) and probably in some other perennial **polycarpic** species (q.v.).

sequential replacement A pattern of vegetation change where there is a chain of species-by-species replacements on an area of ground. (≈ *plant succession*).

seral (preferred term **transitional** q.v.).

serotinous cone Cone in some species of pines and other Gymnosperms where the cone scales are held together by gum and open, releasing seeds, only after a fire (which melts the gum).

sesquioxide Oxide of one of the trivalent cations (e.g. iron or aluminium).

silica Silicon dioxide SiO_2.

soil horizon Distinct layer evident in a **soil profile** (q.v.).

soil profile The sequence from the soil surface to below the bottom of the weathered zone.

solifluction Slow flow of wet soil on slopes in cold regions, usually most evident during the snow thaw or ground thaw period.

solum The soil (implying the complete soil profile).

somatic mutation Mutation of non-gametic (i.e. vegetative) cells in plants giving rise, e.g., to new forms of tissues in leaves or stems.

spodosol Leached, acid forest soil, in cool to cold, moist climate.

sporangium The spore-containing organ of Fungi, Bryophyta and some vascular plants. (plural *sporangia*).

sporangiophore In Fungi, a specialized hypha bearing one or more sporangia.

spore A small, usually tough-walled resistant organ, formed by Fungi and other microbes and by Bryophytes and Pteridophytes. Spores have functions in dispersal, outlasting unfavourable periods and giving rise to new generations of organisms.

sporophyte generation In Bryophytes and Pteridophytes the **diploid** (q.v.) plants which give rise to spores.

stability A condition of relative lack of change in a biological system (cf. constancy, persistence, inertia and resilience; also **steady state**, **dynamic equilibrium**, **mature vegetation** q.v.).

stand A patch of vegetation of relatively homogenous species composition.

steady-state Referring to a system in relative equilibrium (i.e. one that does not change for a period of time).

stochastic A variation or process involving random, unpredictable elements, where the outcome must be estimated in terms of probabilities (cf. **deterministic** q.v.).

stolon Specialized overground stem of plants which extends laterally, putting down roots at intervals and giving rise to new shoot systems.

stomatal resistance The resistance of stomata (dependent on their size and number) to the passage of gases.

stratigraphy The discipline which uses stratified sediments, and their included fossils, to determine earth history.

stress Physiological constraint placed on plants by extremes of environment such as high or low temperature, drought, lack of nutrients or high salinity.

style of vegetation change One of the patterns of vegetation change recognized by temporal shifts in population composition.

subcanopy   Individual plants or strata of plants which are beneath the canopy layer. (cf. *understorey* q.v.).

subordinate species   Plant species which are inconspicuous and/or not abundant in comparison to the relative abundance and conspicuousness of *dominant* species (q.v.).

summer annual   An annual plant species which germinates in spring and completes its life processes before the winter.

sunfleck   A well-lit patch at ground level in forest, beneath a small canopy gap. Sunflecks shift position relatively rapidly throughout the day.

suppressed juvenile   Young tree which remains stunted beneath a forest canopy (possibly for many years) until a gap occurs, when it begins to grow rapidly.

swamp   A wetland characterized by fertile water and relatively tall, emergent vegetation. ($\simeq$ fen).

symbiont   One of a pair of species with obligate interrelationships (positive or negative).

symbiosis   The relationship between symbionts. (adjective *symbiotic*).

sympodial   A stem system in which the apex dies or ends in an inflorescence and new leader growth occurs from axillary buds. (cf. *monopodial* q.v.).

synecology   The branch of ecology which studies communities.

systems analysis   The description of complex systems (e.g. biological systems) and the processes occurring in them by means of multi-dimensional computer simulations.

taiga   Coniferous forest in northern North America and Eurasia.

talus   Sheets of fine to coarse *colluvial* rock debris (q.v.) on slopes (= scree).

taxocene   A community in which numbers of species from the same higher taxonomic group (genus, family) occur together and exploit the same kinds of resources.

taxon   A unit of classification at any level (e.g. species, genus, family etc).

tectonic   Relating to major earth movements.

tephra   Volcanic ash deposited by air-fall.

thallus   Relatively simple plant form, not differentiated into stems and leaves, found in Algae, some Bryophyta and in lichens. (plural *thalli*).

thalloid   Like a thallus.

thermoperiod   The period of time that temperature remains above a particular threshold level.

till   The debris deposited by a glacier.

tiller   The branch of a grass (and some other monocotyledon plants).

'tolerance'   Classification of tree species according to their relative ability to regenerate on a gradient from open, well-lit sites to closed, deeply shaded sites.

tolerance model   A description of one of the mechanisms involved with vegetation change. Species which occupy a site have little or no effect on recruitment of other species, which grow to maturity irrespective of their presence. Long-lived species will become more prominent as short-lived species die.

toposequence   A distinct spatial pattern present on a topographic gradient.

transitional   Referring to species in a vegetation sequence: those species which occur after the pioneer phase, but which last for only one generation and are then replaced by other species.

transition probability   The probability that a plant species will be replaced by its own kind, or by another species (judged by the identity of adjacent juveniles). Used to predict future composition of vegetation.

transpiration stream   The up-flowing column of water maintained in the vascular system of plants by the negative pressure gradient from root hairs to leaves.

trophic level   The classification of organisms according to their mode of feeding: primary or autotrophic; secondary or heterotrophic (herbivorous); tertiary or heterotrophic (carnivorous); saprotrophic (detritivorous).

turbation (of soil)   Disturbance, usually by a regularly occurring factor such as freeze and thaw.

turion   A small vegetative resting organ formed by some aquatic plant species.

tuff   Volcanic ash, often occurring as cemented layers within sheets of coarser volcanic debris.

tundra   Treeless landscape in sub-polar and alpine regions with grass, sedge and low shrub vegetation.

turgor   The hydrostatic pressure which keeps individual cells and tissues in plants 'inflated'.

understorey   In forest, all plants or strata of plants which are lower than the canopy layer.

ungulate   A hoofed mammal.

udult   Red-yellow soil (ultisol) in which deeper layers are dry for 3 months in most years.

ustult   Red-yellow soil (ultisol) in sites where dry periods are of short duration.

vascular plant   A plant with a specialized conducting system. (cf. *xylem*, *phloem* q.v.).

vegetation   The plant cover of an area.

vegetation change   Any dynamic vegetation pattern where dominant populations of one or more species on a site are being replaced by new populations of the same or different species.

vegetation continuum   The concept that vegetation composition varies continuously (in space or time), in relation to continuity in the environmental gradients which influence species distribution.

vegetative proliferation   Production by plants of vegetative offshoots. Vegetative spread is also implied.

vermiculite   A clay mineral.

vertisol   Mixed or inverted soils.

vine   A plant with long, flexible stems which climbs on and is supported by other plants.

virus   A submicroscopic (visible with the electron microscope) obligate parasite consisting of nucleic acid and protein.

vitalism   The belief that living things are imbued with a 'life force', or vital principle.

vivipary   Germination of seeds on the parent plant. They may or may not fall off and take root.

volume-weight sample   A soil sample taken as a definite volume, from which the density of the sample can be determined.

water potential   The conceptual system describing the negative pressure gradient which maintains the *transpiration stream* (q.v.) in plants. Water potential of pure water is 0 **bar** (q.v.); of soil at *field capacity* (q.v.) ~ -0.15 bar; of root hairs up to -5 bar; of leaves ~ -10 bar; of moist air ~ -100 bar; of dry air ~ -1000 bar.

wetland   Any vegetated area where the main environmental determinant is water.

wind throw   The felling of trees by strong wind.

winter annual   An annual plant which germinates in autumn, overwinters as a small plant with a few basal leaves, then completes its life processes in the following spring and early summer.

xeric   Very dry (referring to habitat conditions) (cf. *mesic* q.v.).

xerophyte   A plant specialized for life in very dry sites. (cf. *mesophyte* q.v.).

xylem   Conducting system in *vascular* plants (q.v.) which carries water and dissolved nutrients from the roots to the rest of the plant.

Index

547

Processes of Vegetation Change presents a wide-ranging, well-illustrated account of vegetation dynamics, a vital subject for plant ecologists and land managers. The main thrust of the text is the development of novel concepts and theories which relate to the full range of change phenomena that occur in natural vegetation and which emphasize the important and distinctive roles of individual plant species.

Early chapters concentrate on the essential background information relevant to change processes, with examples drawn from a wide range of localities. Particular attention is paid to the nature of reciprocal interactions between plants and their habitats and other organisms. Events that occur during the colonization of new sites and the consequent modifications of habitat conditions are carefully outlined. Long-term sequences of change are explored using chrono-sequences and fossil data from wetland sites. Cyclical replacement sequences, direct replacements and fluctuations are considered in the context of short and long-term changes, with due attention given to the conceptual importance of appropriate scales of time and space. These ideas are developed into a coherent expression of the complexity of the interactive processes operating in the plant/environment system. Current theories of vegetation change are threaded through the text, which leads up to a final section presenting a new theoretical framework based on proximate causes, which affords a fresh appreciation of vegetation change mechanisms.

The text contains many original illustrations and makes use of worldwide example data from the tropics to sub-polar regions and from sea shores, through deserts, grasslands and forests to alpine regions, but with particular emphasis on North American locations. The wealth of detail and clear presentation make this book an essential reading item for students of plant ecology, forestry and land systems management.

Colin J. Burrows is currently working at the Department of Plant and Microbial Sciences, University of Canterbury, Christchurch, New Zealand.

Franz Josef Glacier, Westland, New Zealand. The glacier has receded from the foreground area and shrunk downward in recent decades, exposing bedrock and rock debris which are being rapidly colonized by mosses, lichens, herbs and shrubs.

Photograph by C. J. Burrows
Cover design by Juan Hayward

ISBN 0 04580012 X HB
ISBN 0 04580013 8 PB

ISBN 0-04-580012-X

9 780045 800124